The Solar Dynamics Observatory

Phillip Chamberlin • William Dean Pesnell •
Barbara Thompson
Editors

The Solar Dynamics Observatory

Previously published in *Solar Physics* Volume 275,
Issues 1–2, 2012

Editors
Phillip Chamberlin
NASA Goddard Space Flight Center
Greenbelt, MD
USA

Barbara Thompson
NASA Goddard Space Flight Center
Greenbelt, MD
USA

William Dean Pesnell
NASA Goddard Space Flight Center
Greenbelt, MD
USA

ISBN 978-1-4614-3672-0
Springer New York Dordrecht Heidelberg London

Library of Congress Control Number: 2012933455

Cover illustration: The cover image shows the Solar Dynamics Observatory in the cleanroom at Goddard Space Flight Center undergoing final tests and preparation just prior to shipping to Florida for launch. Credit: NASA/B. Lambert

Printed on acid-free paper

Springer is part of Springer Science+Business Media (www.springer.com)

Contents

Preface
R.R. Fisher **1**

INTRODUCTION

The *Solar Dynamics Observatory* (SDO)
W.D. Pesnell · B.J. Thompson · P.C. Chamberlin **3**

AIA

The *Atmospheric Imaging Assembly* (AIA) on the *Solar Dynamics Observatory* (SDO)
J.R. Lemen · A.M. Title · D.J. Akin · P.F. Boerner · C. Chou · J.F. Drake · D.W. Duncan · C.G. Edwards · F.M. Friedlaender · G.F. Heyman · N.E. Hurlburt · N.L. Katz · G.D. Kushner · M. Levay · R.W. Lindgren · D.P. Mathur · E.L. McFeaters · S. Mitchell · R.A. Rehse · C.J. Schrijver · L.A. Springer · R.A. Stern · T.D. Tarbell · J.-P. Wuelser · C.J. Wolfson · C. Yanari · J.A. Bookbinder · P.N. Cheimets · D. Caldwell · E.E. Deluca · R. Gates · L. Golub · S. Park · W.A. Podgorski · R.I. Bush · P.H. Scherrer · M.A. Gummin · P. Smith · G. Auker · P. Jerram · P. Pool · R. Soufli · D.L. Windt · S. Beardsley · M. Clapp · J. Lang · N. Waltham **17**

Initial Calibration of the *Atmospheric Imaging Assembly* (AIA) on the *Solar Dynamics Observatory* (SDO)
P. Boerner · C. Edwards · J. Lemen · A. Rausch · C. Schrijver · R. Shine · L. Shing · R. Stern · T. Tarbell · A. Title · C.J. Wolfson · R. Soufli · E. Spiller · E. Gullikson · D. McKenzie · D. Windt · L. Golub · W. Podgorski · P. Testa · M. Weber **41**

Heliophysics Event Knowledgebase for the *Solar Dynamics Observatory* (SDO) and Beyond
N. Hurlburt · M. Cheung · C. Schrijver · L. Chang · S. Freeland · S. Green · C. Heck · A. Jaffey · A. Kobashi · D. Schiff · J. Serafin · R. Seguin · G. Slater · A. Somani · R. Timmons **67**

Computer Vision for the *Solar Dynamics Observatory* (SDO)
P.C.H. Martens · G.D.R. Attrill · A.R. Davey · A. Engell · S. Farid · P.C. Grigis · J. Kasper · K. Korreck · S.H. Saar · A. Savcheva · Y. Su · P. Testa · M. Wills-Davey · P.N. Bernasconi · N.-E. Raouafi · V.A. Delouille · J.F. Hochedez · J.W. Cirtain · C.E. DeForest · R.A. Angryk · I. De Moortel · T. Wiegelmann · M.K. Georgoulis · R.T.J. McAteer · R.P. Timmons **79**

EVE

***Extreme Ultraviolet Variability Experiment* (EVE) on the *Solar Dynamics Observatory* (SDO): Overview of Science Objectives, Instrument Design, Data Products, and Model Developments**
T.N. Woods · F.G. Eparvier · R. Hock · A.R. Jones · D. Woodraska · D. Judge · L. Didkovsky · J. Lean · J. Mariska · H. Warren · D. McMullin · P. Chamberlin · G. Berthiaume · S. Bailey · T. Fuller-Rowell · J. Sojka · W.K. Tobiska · R. Viereck **115**

***Extreme Ultraviolet Variability Experiment* (EVE) *Multiple EUV Grating Spectrographs* (MEGS): Radiometric Calibrations and Results**
R.A. Hock · P.C. Chamberlin · T.N. Woods · D. Crotser · F.G. Eparvier · D.L. Woodraska · E.C. Woods **145**

***EUV SpectroPhotometer* (ESP) in *Extreme Ultraviolet Variability Experiment* (EVE): Algorithms and Calibrations**
L. Didkovsky · D. Judge · S. Wieman · T. Woods · A. Jones **179**

HMI

The *Helioseismic and Magnetic Imager* (HMI) Investigation for the *Solar Dynamics Observatory* (SDO)
P.H. Scherrer · J. Schou · R.I. Bush · A.G. Kosovichev · R.S. Bogart · J.T. Hoeksema · Y. Liu · T.L. Duvall Jr. · J. Zhao · A.M. Title · C.J. Schrijver · T.D. Tarbell · S. Tomczyk **207**

Design and Ground Calibration of the *Helioseismic and Magnetic Imager* (HMI) Instrument on the *Solar Dynamics Observatory* (SDO)
J. Schou · P.H. Scherrer · R.I. Bush · R. Wachter · S. Couvidat · M.C. Rabello-Soares · R.S. Bogart · J.T. Hoeksema · Y. Liu · T.L. Duvall Jr. · D.J. Akin · B.A. Allard · J.W. Miles · R. Rairden · R.A. Shine · T.D. Tarbell · A.M. Title · C.J. Wolfson · D.F. Elmore · A.A. Norton · S. Tomczyk **229**

Image Quality of the *Helioseismic and Magnetic Imager* (HMI) Onboard the *Solar Dynamics Observatory* (SDO)
R. Wachter · J. Schou · M.C. Rabello-Soares · J.W. Miles · T.L. Duvall Jr. · R.I. Bush **261**

Wavelength Dependence of the *Helioseismic and Magnetic Imager* (HMI) Instrument onboard the *Solar Dynamics Observatory* (SDO)
S. Couvidat · J. Schou · R.A. Shine · R.I. Bush · J.W. Miles · P.H. Scherrer · R.L. Rairden **285**

Polarization Calibration of the *Helioseismic and Magnetic Imager* (HMI) onboard the *Solar Dynamics Observatory* (SDO)
J. Schou · J.M. Borrero · A.A. Norton · S. Tomczyk · D. Elmore · G.L. Card **327**

Implementation and Comparison of Acoustic Travel-Time Measurement Procedures for the *Solar Dynamics Observatory/Helioseismic and Magnetic Imager* Time – Distance Helioseismology Pipeline
S. Couvidat · J. Zhao · A.C. Birch · A.G. Kosovichev · T.L. Duvall Jr. · K. Parchevsky · P.H. Scherrer **357**

Time–Distance Helioseismology Data-Analysis Pipeline for *Helioseismic and Magnetic Imager* Onboard *Solar Dynamics Observatory* (SDO/HMI) and Its Initial Results
J. Zhao · S. Couvidat · R.S. Bogart · K.V. Parchevsky · A.C. Birch · T.L. Duvall Jr. · J.G. Beck · A.G. Kosovichev · P.H. Scherrer **375**

E/PO

The *Solar Dynamics Observatory* (SDO) Education and Outreach (E/PO) Program: Changing Perceptions One Program at a Time
E. Drobnes · A. Littleton · W.D. Pesnell · K. Beck · S. Buhr · R. Durscher · S. Hill · M. McCaffrey · D.E. McKenzie · D. Myers · D. Scherrer · M. Wawro · A. Wolt **391**

Solar Phys (2012) 275:1–2
DOI 10.1007/s11207-011-9914-3

Preface

R.R. Fisher

Received: 17 November 2011 / Accepted: 17 November 2011 / Published online: 6 January 2012

My initial exposure to early data from the *Solar Dynamics Observatory* (SDO) took place late one morning in the Spring of 2010 and eventually ended in the early evening. As night fell, I found myself feeling excited, a little relieved, and hugely impressed by the scientific and technical contributions of those who developed the SDO systems. I was also confounded by what I had seen.

The data were so rich! The temporal and spatial content of the imagery was overwhelming. It took until the next morning for me to realize that the new data were, in fact, revolutionary. In half a day, my understanding of the dynamic Sun had been swept away by the SDO systems. The mission's quest for new observing capability was wildly successful! I could only speculate on the SDO's future impact on scientific research.

In its first year SDO has given us unprecedented views of filament eruptions, measured the total energy emitted by flares, and started watching for active regions before they erupt through the solar surface. Each is a vital piece in our understanding of solar activity and of how that activity creates space weather. Filament eruptions are the source of coronal mass ejections, while solar flares increase satellite drag and disrupt communications. Prediction of solar activity is a goal of SDO and knowing where they will erupt is a major step toward short-term prediction.

SDO is the first space-based mission of NASA's Living With a Star program. It carries three instruments dedicated to improving our understanding of the production, evolution, and destruction of the solar magnetic field. The SDO science investigation teams are committed to rapidly providing their data to scientists, space-weather organizations, and the public.

SDO provides full-disk images of the Sun in many wavelengths, starting in the soft X-ray and moving though the UV into the visible. Magnetic-field maps and Doppler-velocity maps

The Solar Dynamics Observatory
Guest Editors: W. Dean Pesnell, Phillip C. Chamberlin, and Barbara J. Thompson

R.R. Fisher (✉)
NASA/Goddard Space Flight Center, Greenbelt, MD, USA
e-mail: richard.r.fisher@nasa.gov

are now obtained at cadences never before achieved. The EUV spectral irradiance is measured at an unprecedented rate allowing the estimation of the total energy radiated by even small, C-class flares. This irradiance is also used to calibrate the EUV images, so that the radiance from coronal loops can be measured and converted into temperature maps. By revealing global changes in the solar atmosphere over spatial scales ranging from the resolution limit to fractions of a solar diameter, SDO is writing a new chapter in our understanding of the Sun and of how space weather is created by solar activity.

This volume on the *Solar Dynamics Observatory* (SDO) mission continues a tradition of *Solar Physics* to dedicate topical issues to major space science missions. I encourage you to read the articles, use SDO data in your research, and simply enjoy looking at the Sun in the splendor that SDO is revealing.

Solar Phys (2012) 275:3–15
DOI 10.1007/s11207-011-9841-3

The *Solar Dynamics Observatory* (SDO)

W. Dean Pesnell · B.J. Thompson · P.C. Chamberlin

Received: 16 June 2011 / Accepted: 7 August 2011 / Published online: 18 October 2011

Abstract The *Solar Dynamics Observatory* (SDO) was launched on 11 February 2010 at 15:23 UT from Kennedy Space Center aboard an Atlas V 401 (AV-021) launch vehicle. A series of apogee-motor firings lifted SDO from an initial geosynchronous transfer orbit into a circular geosynchronous orbit inclined by 28° about the longitude of the SDO-dedicated ground station in New Mexico. SDO began returning science data on 1 May 2010. SDO is the first space-weather mission in NASA's *Living With a Star* (LWS) Program. SDO's main goal is to understand, driving toward a predictive capability, those solar variations that influence life on Earth and humanity's technological systems. The SDO science investigations will determine how the Sun's magnetic field is generated and structured, how this stored magnetic energy is released into the heliosphere and geospace as the solar wind, energetic particles, and variations in the solar irradiance. Insights gained from SDO investigations will also lead to an increased understanding of the role that solar variability plays in changes in Earth's atmospheric chemistry and climate. The SDO mission includes three scientific investigations (the *Atmospheric Imaging Assembly* (AIA), *Extreme Ultraviolet Variability Experiment* (EVE), and *Helioseismic and Magnetic Imager* (HMI)), a spacecraft bus, and a dedicated ground station to handle the telemetry. The Goddard Space Flight Center built and will operate the spacecraft during its planned five-year mission life; this includes: commanding the spacecraft, receiving the science data, and forwarding that data to the science teams. The science investigations teams at Stanford University, Lockheed Martin Solar Astrophysics Laboratory (LMSAL), and University of Colorado Laboratory for Atmospheric and Space Physics (LASP) will process, analyze, distribute, and archive the science data. We will describe the building of SDO and the science that it will provide to NASA.

The Solar Dynamics Observatory
Guest Editors: W. Dean Pesnell, Phillip C. Chamberlin, and Barbara J. Thompson

W.D. Pesnell (✉) · B.J. Thompson · P.C. Chamberlin
NASA Goddard Space Flight Center, Greenbelt, MD, USA
e-mail: william.d.pesnell@nasa.gov

B.J. Thompson
e-mail: barbara.j.thompson@nasa.gov

P.C. Chamberlin
e-mail: phillip.c.chamberlin@nasa.gov

Keywords SDO · Solar cycle · Helioseismology · Coronal · Space weather

1. Preface: *Living with a Star* and the *Solar Dynamics Observatory*

The goal of NASA's *Living With a Star* (LWS) Program is to provide the scientific understanding needed to address those aspects of heliophysics science that may affect life and society, where heliophysics is the study of the Sun and how it influences and drives changes in all objects within its reach. The ultimate goal is to develop the ability to predict conditions at Earth and in the interplanetary medium due to the Sun's ever-changing output, a subject called space weather.

LWS is designed to give us a better understanding of the causes of space weather, whose effects can disable satellites, cause power-grid failures, and disrupt global positioning system and other communications signals. The LWS program includes coordinated strategic missions, missions of opportunity, a targeted research and technology program, a space-environment testbed flight opportunity, and partnerships with other agencies and nations. Each LWS mission is designed to answer specific science questions needed to understand the interconnected systems that impact us.

Radiation Belt Storm Probes (RBSP), the second LWS mission, will be launched in 2012 to study the acceleration to high energies of electrons in the Earth's radiation belts. *Balloon Array for Radiation Belt Relativistic Electron Losses* (BARREL) is a multiple-balloon investigation of the Earth's radiation belts that will fly from Antarctic stations during the southern Summers of 2012 and 2013. Two solar missions, *Solar Probe Plus* and *Solar Orbiter* (a joint mission with ESA), are planned for launch later this decade. They will study the coronal heating and solar-wind acceleration problems from platforms flying close to the Sun and out of the Ecliptic.

LWS is a part of the *International Living With a Star* (ILWS) Program, an organization bringing together researchers and space-weather operators to further develop our ability to understand and predict space weather.

2. Introduction

The *Solar Dynamics Observatory* (SDO) is designed to provide the data and scientific understanding necessary to predict solar activity, from anticipating whether flares and CMEs will occur the next day to the level of solar activity in future solar cycles. Monitoring the topology of magnetic fields as they are formed within the Sun itself through the outer atmosphere, or corona, by high-resolution images may provide the precursors for predicting flares and CMEs. A key insight that SDO hopes to provide is the topological configuration that is needed to drive reconnection and reorganization in these built-up and stressed magnetic fields – the energy release and source that drives the solar eruptive events.

Longer timescales require a knowledge of how magnetic field is transported, amplified, and destroyed inside the Sun and ultimately ejected from the interior – the solar dynamo. Between these limiting timescales another goal of SDO is to predict where and when active regions will emerge, how those magnetic fields erupt and decay, and how the host of other phenomena related to the solar magnetic field come and go.

Solar Cycle 24 started in December 2008 and is predicted to rise to a below-average peak in 2013. This affords an excellent laboratory for SDO. All previous space-based missions with significant duration sampled Solar Cycles 20 – 23, which had above-average levels of activity. The latter half of the 20th century is estimated to have the highest levels of solar activity in the past 10 000 years (Solanki *et al.*, 2004). Solar Cycle 23 was well-measured by the still-operating SOHO and *Hinode* missions, with STEREO starting at the beginning of the extended solar-minimum period.

Two instruments on SDO concentrate on the energy radiated in the extreme ultraviolet. Solar Extreme Ultraviolet (EUV: 1 – 122 nm) photons are the dominant cooling radiation of the solar corona and are also the dominant source of heating for Earth's upper atmosphere. When the Sun is active, EUV emissions can rise and fall by factors of hundreds (and X-rays by factors of thousands) in just a matter of seconds. These surges heat the Earth's upper atmosphere, inflating it and increasing the drag on satellites. EUV photons also can break the bonds of atmospheric atoms and molecules, creating a layer of ions that alters and sometimes severely disturbs radio communications and global positioning system navigation.

SDO will also measure variations inside the Sun, providing the information needed to understand the internal motions that generate the magnetic field, and then the magnetic field as it emerges through the solar surface. The combination of the various measurements allows SDO to study the lifecycle of the solar magnetic field.

To achieve its goals SDO will transmit approximately 150 000 high-resolution full-Sun images and 9000 EUV spectra, or 1.5 terabyte of science data, to the ground each day during the prime mission. This will be converted into images, Dopplergrams, magnetograms, and spectra. Over the five-year prime mission, SDO will return roughly 3 – 4 petabytes of raw data.

The rapid cadence and improved spatial resolution of the SDO instruments is already providing new science results. Images from AIA quickly showed how the global magnetic field of the Sun must relax after even a modest flare and filament (Schrijver and Title, 2011). The low level of flaring during the first year of SDO science operations allowed the discovery that the energy released in many flares is poorly estimated by the GOES X-ray radiometers (Woods *et al.*, 2011b). Energy is radiated at longer wavelengths and for a longer time – a significant finding for people modeling the response of the terrestrial atmosphere to flares.

3. History

The *Solar Dynamics Observatory* is a flagship mission that represents a substantial enhancement of capabilities demonstrated by earlier NASA missions and other science projects. Those earlier measurements revealed new details about the solar interior and the mechanisms driving the solar dynamo. Unlike the line-of-sight (LOS) magnetographs flown on earlier missions, SDO carries a full-disk imaging vector magnetograph with a data cadence sufficient to see changes in the magnetic-field strength and direction that could be related to flares or CMEs. Another needed capability was rapid-cadence images of the solar corona with wavelength coverage to span temperatures from the chromosphere (10 000 K) to flares (2×10^7 K). Full-disk images with spatial resolution of $\approx 1''$ are needed to both resolve known spatial scales and to sample the variations across the disk. The solar images from *Yohkoh*, SOHO, and *Skylab* were used by solar physicists to understand solar activity and by space-weather forecasters to monitor solar activity. Additional bandpasses to better estimate the plasma temperature and higher cadence were both required to understand the response of the corona to the injected magnetic field.

One direct connection between solar activity and the Earth is through the emissions at extreme ultraviolet wavelengths (1 – 122 nm). The utility of these measurements was demonstrated by the SEE instrument on the TIMED satellite (Woods *et al.*, 2005). An instrument with higher duty cycle and cadence would measure the total-energy content in flares while also providing the data to calibrate the coronal images and measuring the radiant energy output of the Sun that creates the ionosphere.

Enhancing the helioseismic and LOS magnetic-field capabilities demonstrated by the MDI instrument on SOHO and the ground-based GONG network was part of a proposal to NASA entitled SONAR (or Hale). This mission had advanced sufficiently far toward realization that it was moved to the LWS program and revised to become SDO. The initial organization of the LWS program by the LWS Science Architecture Team occurred in parallel with the development of SDO.

A Science Definition Team for SDO was formed in November 2000 with David Hathaway (NASA/MSFC) as the chair and Barbara Thompson (NASA/GSFC) as the Study Scientist. The team met several times until July 2001, when the SDT Report (Hathaway *et al.*, 2001) that codified all of the measurement and science requirements was released. This report formed the basis for the NASA Announcement of Opportunity AO 02-OSS-01 soliciting science investigations, which was released 18 January 2002.

The HMI and EVE Science Investigation Teams (SITs) were selected on 19 August 2002 while the AIA SIT was selected on 7 November 2003. The observatory was built and completed testing by September 2008, when delays in the Atlas V launcher line caused it to be left in storage until June 2009. At that time SDO was moved to Florida for launch preparations and launched on 11 February 2010. Science operations began on 1 May 2010 and will continue for the baseline lifetime of five years. There are no consumables on the spacecraft that will limit the useful life of the mission. It is estimated that over 100 years of propellant is left in the tanks.

SDO has several features unique to a NASA science mission:

i) A sustained high rate of science-data production and the movement of those data within the spacecraft. This requires reliable electronics that are robust against radiation-induced effects while maintaining timing tolerances better than 1 ns.
ii) The continuous downlink of science data, using a highly automated data pipeline to process the data for immediate use by the space-weather community as well as SDO science investigations.
iii) The extremely accurate pointing and stability required of the spacecraft to allow registration of successive images.
iv) The long (five years) mission life with high-duty cycle instruments.

To date all of these features but the fourth are working and only the long lifetime has yet to be proven.

Because SDO was designed to measure high-resolution data for a five-year mission, the science data resolve the eruptive events, such as flares and coronal mass ejections, while building a well-characterized database for long-term studies. By combining SDO data with those of other missions, such as SOHO, TRACE, *Hinode*, and TIMED, we will be able to construct a picture of the Sun over an entire 22-year magnetic cycle. It is these long-term research projects that should solve the puzzle of the solar cycle and the space weather it creates at Earth.

SDO was built by a team made up of the Goddard Space Flight Center, Stanford University, in Palo Alto, California, the Lockheed Martin Solar Astrophysics Laboratory (LMSAL), also in Palo Alto, California, and the Laboratory for Atmospheric and Space Physics (LASP), University of Colorado in Boulder, Colorado. Goddard was responsible for designing and building the spacecraft and many of the components of the spacecraft. The other team members are the Science Investigation Teams, who are responsible for designing and

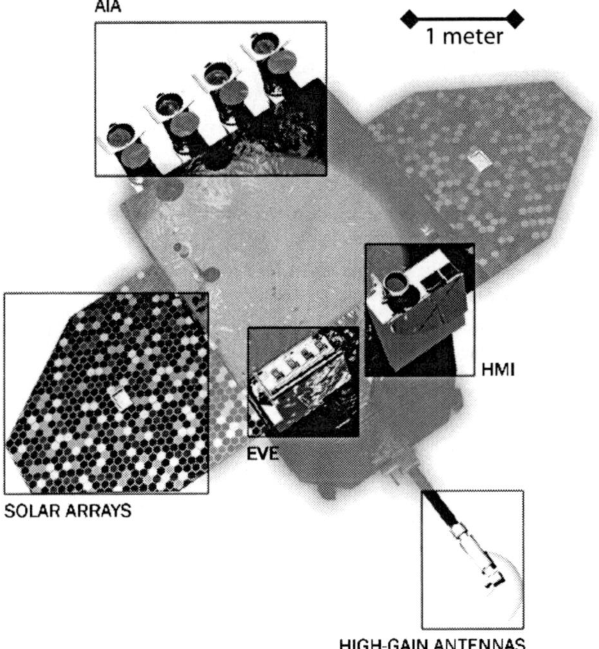

Figure 1 The SDO spacecraft with the instruments, high-gain antennas, and solar arrays highlighted. The thrusters and main engine are located on the opposite face of the satellite.

building their instrument, for running the instruments after launch, and performing science investigations related to the data produced by the instruments.

4. The Spacecraft Summary

The SDO spacecraft is shown in Figure 1. A summary of relevant information is:

Satellite: Three-axis stabilized and fully redundant spacecraft.

Duration: Five-year science mission and sufficient propellant to operate for an additional five years.

Mass: Total mass at launch was 3000 kg: instruments 300 kg, spacecraft 1300 kg, and fuel 1400 kg.

Dimensions: The overall length of the spacecraft along the Sun-pointing axis is 4.7 m, and each side is 2.2 m. The span of the spacecraft is 6.1 m along the extended solar panels and 6.0 m along the deployed high-gain antennas.

Solar Array: The 6.6 m^2 of solar panels produce 1500 W of power with an estimated efficiency of 16% after five years on orbit. The "homeplate" shape prevents the solar panel from blocking the high-gain antennas.

High-gain Antennas: The high-gain antennas rotate once each orbit to track the ground station.

Maximum Downlink Rate: The spacecraft has a continuous, high-rate science-data downlink rate of 150 Mbps at a Ka-Band frequency of about 26 GHz; includes 130 Mbps of data and 20 Mbps of encoding overhead.

Mission Operations Center: Located at Goddard Space Flight Center in Maryland.

SDO Ground Station: Located at White Sands Complex in New Mexico.

Table 1 The SDO Level 1 science requirements.

Scientific Questions
What mechanisms drive the quasi-periodic 11-year cycle of solar activity?
How is active region magnetic flux synthesized, concentrated, and dispersed across the solar surface?
How does magnetic reconnection on small scales reorganize the large-scale field topology and current systems, and how significant is it in heating the corona and accelerating the solar wind?
Where do the observed variations in the Sun's extreme ultraviolet spectral irradiance arise, and how do they relate to the magnetic-activity cycles?
What magnetic-field configurations lead to the coronal mass ejections, filament eruptions, and flares that produce energetic particles and radiation?
Can the structure and dynamics of the solar wind near Earth be determined from the magnetic-field configuration and atmospheric structure near the solar surface?
When will activity occur, and is it possible to make accurate and reliable forecasts of space weather and climate?

Table 2 The instrument science objectives.

Gauge the energy from solar activity that is input into the Earth's atmosphere and near-Earth space
Probe the dynamics of the near-surface shear layer of the Sun to observe the local strong-flux regions before they reach the photosphere
Probe the internal processes that govern the solar cycle
Observe the initiation and progression of dynamic processes in the chromosphere and corona to understand their connection to the Sun's magnetic field.

Table 3 SDO measurement objectives.

Make measurements over a significant portion of a solar cycle to capture the solar variations that exist on all time scales from seconds (solar eruptive events) to months (active region evolution, solar rotation) to years (solar cycle)
Measure the extreme ultraviolet spectral irradiance of the Sun at a rapid cadence
Measure the Doppler shifts due to oscillation velocities over the entire visible disk
Make high-resolution measurements of the longitudinal and vector magnetic field over the entire visible disk
Make images of the chromosphere and inner corona at several temperatures at a rapid cadence.

5. Science Goals

The scientific goals of the SDO project, which were first posed in the SDO Science Definition Team report (Hathaway *et al.*, 2001), are to improve our understanding of the seven science questions listed in Table 1. The instruments must be designed to meet the instrument science objectives listed in Table 2 using the measurements listed in Table 3. All of the questions are related to developing an understanding of the solar magnetic field. Three of the science questions emphasize predicting solar activity.

To satisfy these science goals, the Science Investigation Teams described next have created specific research topics and goals.

Table 4 The SDO science investigations.

Investigation	Abbrev.	Returned Data
Heliospheric and Magnetic Imager	HMI	Full-disk Dopplergrams Full-disk LOS magnetograms Full-disk vector magneograms
Atmospheric Imaging Assembly	AIA	Rapid-cadence, full-disk EUV solar images
Extreme Ultraviolet Variability Experiment	EVE	Rapid-cadence EUV spectral irradiance

6. Science Investigation Teams

The science payload of SDO comprises three instrument suites built by the Science Investigation Teams (SIT). The SITs are summarized in Table 4. The instrument suites are described in detail in other articles in this Topical Issue, so we provide only a brief summary of each.

6.1. *Atmospheric Imaging Assembly*

The *Atmospheric Imaging Assembly* (AIA) is an array of four telescopes that observes the surface and atmosphere of the Sun. The AIA Principal Investigator is Alan Title of LMSAL. AIA was built to obtain full-disk images of the solar atmosphere with a field of view of at least $40'$ and a two-pixel resolution of $1.2''$. Filters on the telescopes cover ten different wavelength bands that include seven extreme ultraviolet, two ultraviolet, and one visible-light band to reveal key aspects of solar activity. The wavelength bands were chosen to yield diagnostics over the range of 6000 K to 3×10^6 K. AIA uses multilayer coatings combined with foil filters to isolate the desired spectral bandpasses for each telescope. AIA was built by LMSAL, Palo Alto, California. A detailed description of AIA can be found in Lemen *et al.* (2011).

6.2. *Extreme Ultraviolet Variability Experiment*

The *Extreme ultraviolet Variability Experiment* (EVE) measures fluctuations in the Sun's ultraviolet output. Tom Woods of LASP is the Principal Investigator. EVE will measure the solar spectral irradiance in the most variable and unpredictable part of the solar spectrum from 0.1 to 105 nm as well as 121.6 nm. EVE has three parts: MEGS, ESP, and SAM. MEGS combines grating spectrometers with two CCDs to create images of the spectral irradiance in various wavelength ranges. These individual sections are then combined to give the spectrum between 6.5 to 105 nm at 0.1 nm spectral resolution. A silicon photodiode strategically placed inside the MEGS-B channel measures the irradiance of Lyman-α at 121.6 nm, the single brightest line in the EUV. ESP is a series of radiometers placed behind a transmission grating that measures the irradiance in several wavelength bands (0.1 – 5.9, 17.2 – 20.6, 23.1 – 27.6, 28.0 – 31.6, and 34.0 – 38.1 nm). It is similar to the SEM instrument on SOHO. SAM is a pinhole camera used with the MEGS-A CCD to measure individual X-ray photons in the wavelength range 0.1 – 7 nm.

This design will allow solar scientists to continuously monitor EUV emissions at a ten-second cadence. EVE was built by LASP at the University of Colorado. A detailed description of the EVE instrument and science investigation can be found in Woods *et al.* (2011a).

6.3. *Helioseismic and Magnetic Imager*

The *Helioseismic and Magnetic Imager* (HMI) will map magnetic and velocity fields at the surface of the Sun. A key goal of this experiment is to decipher the physics of the solar magnetic dynamo. The Principal Investigator for HMI is Phil Scherrer of Stanford University. HMI was built by LMSAL, Palo Alto, California. A detailed description of the HMI science investigation is given in Schou *et al.* (2011).

HMI measures the Doppler shift of the Fe I 617.3 nm spectral line to determine the photospheric surface velocity over the Sun's entire visible disk. HMI creates a full-disk photospheric velocity measurement (or Dopplergram) every 45 seconds with a two-pixel resolution of $1''$, a noise level $\approx 25\ \mathrm{m\,s^{-1}}$, a data recovery of 95% of the Dopplergrams, and a data completeness of 99% for each Dopplergram. These Dopplergrams are then used to study the inside of the Sun.

HMI uses the Zeeman effect of the same spectral line to measure the Stokes parameters required to create full-disk longitudinal magnetic-field measurements (LOS magnetogram) and full-disk vector photospheric magnetic-field maps (vector magnetogram). The LOS magnetograms have a cadence of 45 seconds with a two-pixel resolution of $1''$ resolution, a noise level of 17 G, and a dynamic range of ± 3 kG. The vector magnetograms have a 12-minute cadence with a polarization accuracy of no less than 0.3%. HMI thus provides the first rapid-cadence measurements of the strength and direction of the solar magnetic field over the visible disk of the Sun.

7. Science Data Capture Requirement

An important consideration during the post-launch science phase, or Phase E, is the science data-capture requirement, driven primarily by the needs of the helioseismic studies. Long, uninterrupted periods of continuous observation provide the best data for the helioseismology algorithms that are used to study the internal structure of the convection zone. The development of predictive models for short-term phenomena such as flares and CMEs also relies on a low-latency, continuous dataset. SDO is designed to return science data for at least 22 individual 72-day periods over the five-year mission. For one 72-day period to be considered complete, the ground system must capture 95% of all possible observation time. The remaining 5% is covered in the SDO Science Data Capture Budget, which accounts for science-data outages and interruptions expected during the mission. These outages and interruptions include:

- Planned operations, such as thruster-based, station-keeping maneuvers, and high-gain antenna (HGA) handovers. SDO uses thrusters to maintain the longitude and inclination of its orbit and to control the momentum in the reaction wheels. HGA handovers occur near eclipses when a single antenna cannot maintain contact with SDOGS while keeping the observatory oriented along the solar-rotation axis (solar North).
- Instrument calibration, roll, and off-point maneuvers. Each instrument has a set of calibration maneuvers that are performed on a regular basis. Some, such as daily flatfields and internal calibrations, affect only a single instrument. Others, such as roll maneuvers, affect all of the instruments.
- Eclipses of the Sun by the Earth and transits of the Moon across the Sun. For several weeks around the Equinoxes, SDO will experience eclipses when the Earth passes between the satellite and the Sun. This causes a loss of 44 hours each year. Lunar transits

occur at least once per year and a maximum of four times a year during the SDO mission. Transits of Mercury (on 9 May 2016 and 11 November 2019) and Venus (on 6 June 2012) across the solar disks are counted toward good science data as they are used for calibration purposes (100 hours per year).

- Random interruptions caused by energetic charged particles (Single Event Upsets and Solar Energetic Electrons: 34 hours per year).
- Unplanned interruptions at SDOGS include signal attenuation due to rain, interference of direct solar radio emissions when SDO is directly in line with the Sun and the ground station, and equipment problems (112.5 hours per year).

Whenever a maneuver moves the instruments off of the Sun, it is necessary to include time for thermal and alignment recovery.

The Flight Operations Team (FOT) and instrument operations teams are jointly responsible for tracking science data losses. Planned events causing data loss will be scheduled in order to satisfy the requirements for data capture and completeness. For data to be considered complete, the ground system must forward sufficient downlinked HMI data to the JSOC to allow HMI to construct 95% of anticipated Dopplergrams with 99% of the pixels containing science data. The AIA and EVE instruments have less stringent data-capture requirements of 90%. In the event that more than 5% (HMI) or 10% (AIA and EVE) of the data are lost before a 72-day period is completed, the PIs may jointly elect to start a new 72-day period during which the data-capture requirement will be enforced. After one year of science operations, we have completed five 72-day observation periods without an interruption. As of 1 April 2011 the ground system is forwarding 99.97% of the transmitted Instrument Multiplexing Protocol Data Units (IM_PDUs) to the Science Operation Centers (SOCs) and HMI has recovered 97% of the Dopplergrams and magnetograms. As anticipated, the eclipse seasons suffer from the greatest loss of data.

7.1. Ground System and Mission Operations

SDO uses a dedicated ground station with dual antennas to meet science-data downlink completeness requirements. The ground system can receive a science-data downlink rate of 130 Mbps for 24 hours per day and seven days per week to achieve full mission success. Once on the ground, the science data are assembled into files and transmitted to the instrument SOCs. A local, 30-day science data cache is available at the ground station. The ground station also provides S-band frequency command, telemetry and tracking functions in support of SDO mission operations. The SDO Ground System is described by Tann, Pages, and Silva (2005).

Because the instruments view the full disk of the Sun, there is no need to have an elaborate commanding process. SDO is pointed at the center of the Sun except for calibration and spacecraft maneuvers. Thus the data are quite uniform, simplifying the data access, *i.e.* no campaigns. The high-duty cycle allows external users to obtain concurrent observations by "observing the database" of SDO rather than coordinating special observing sequences.

8. SDO Orbit and Mission Phases

The rapid cadence and continuous coverage required for SDO science observations led to placing the satellite into an inclined geosynchronous orbit. This allows for a continuous, high data rate, contact with a single, dedicated ground station. When viewed from the ground station, the orbit resembles an elongated figure-eight (or annalemma) with a width of $(i/4)\sin i = 3.3°$ as it orbits the Earth once per day. Relying on a single site reduces the complexity of the ground system by removing the need to combine data from multiple, widely spaced antennas scattered around the globe.

Nearly continuous observations of the Sun can be obtained from other orbits, such as a Sun-synchronous, low-Earth orbit (LEO). If SDO had been placed into a LEO, it would have been necessary to store large volumes of scientific data onboard until a downlink opportunity was available, and multiple sites around the globe would be needed to downlink the data. However, no space-qualified data recorder with the capability to handle this large data volume exists. The large data rate of SDO, the unavailability of a data recorder, and the ability to continuously stream data from the spacecraft if a geosynchronous orbit were selected led to the selection of the inclined geosynchronous orbit.

Disadvantages of the inclined geosynchronous orbit include higher launch costs (relative to LEO) and two Earth-shadow (eclipse) seasons each year. During these two – three week long eclipse periods, SDO will experience a daily interruption of solar observations, and these interruptions have been included in SDO's data-capture budget (see Section 7). There will also be three lunar transits per year from this orbit.

This inclined geosynchronous orbit is located on the outer edges of Earth's radiation belt, where the radiation dose can be quite high. Additional shielding was added to reduce the effects of exposure to this ionizing radiation.

The combined Atlas V/Centaur launch vehicle placed SDO into a geosynchronous transfer orbit with an apogee altitude of 35 350 km and a perigee altitude of 2500 km. A series of nine apogee-motor burns were used to raise perigee altitude to 35 350 km, and then three thruster motor burns were used to raise both apogee and perigee altitudes to their final values of 35 800 km (Figure 2). Excessive fuel-slosh during the early main engine burns delayed the reaching of the final orbit until 16 March 2010. Once SDO was on-orbit the main engine was disabled, leaving only the thrusters to perform maneuvers.

Two types of on-orbit maneuver are used. One maneuver changes the spacecraft velocity, known as a delta-V maneuver, and is used to keep the spacecraft within its assigned orbit. The other maneuver, called a delta-H (where "H" is the symbol for angular momentum), is used to dissipate the momentum stored during regular operations. At the end of its mission, SDO will be moved to a disposal orbit outside of the geosynchronous belt and all energy sources depleted.

9. SDO Data

Science data from SDO are transmitted via Ka-band to the dedicated ground station in New Mexico. From there the data stream is converted into telemetry files and forwarded to the SOCs. Each instrument is assigned a fraction of the telemetry stream. Because continuous contact is maintained with the spacecraft, the science-data return is maximized.

Data from the SDO mission consist of images, full-disk maps, and spectra. Each SDO science investigation team maintains a data archive of its science data products for the life of the SDO mission, and have made these data publicly available. Given the volume of data produced by SDO, a variety of methods have been developed to obtain data at both reduced and full resolutions. The data latency for near-realtime data has been 15 minutes during the mission operations phase.

Full-resolution science data from HMI and AIA are served by the Joint-SOC (JSOC) at Stanford University (http://jsoc.stanford.edu/ajax/lookdata.html). The data are provided as FITS formatted files (a description of the FITS format can be found at Pence *et al.*,

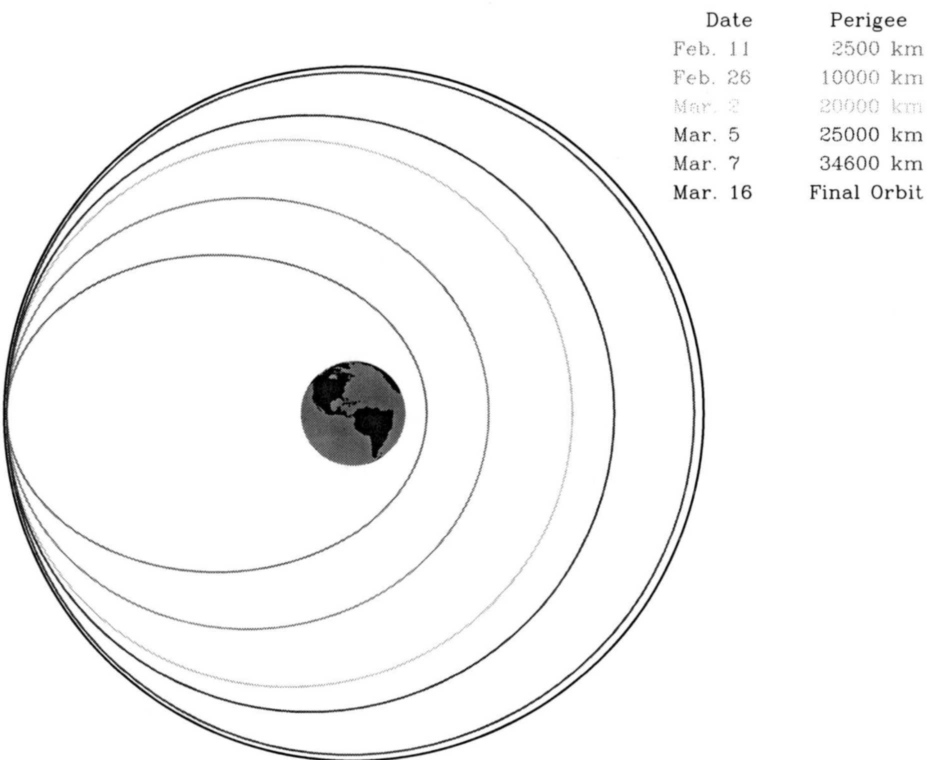

Figure 2 The raising of the SDO orbit from GTO to inclined geosynchronous. The initial GTO, two intermediate orbits, and the estimated final orbit are shown to scale. Apogee Motor Firings (AMFs) all took place at the apogee of the orbit (the left side of the diagram) and were designed to lift the perigee up to a geosynchronous altitude. A few trim maneuvers then moved SDO into its final orbit. Perigee values are altitudes (height above the Earth).

2010). Readers for files in the FITS formats for most languages can be found at http://heasarc.gsfc.nasa.gov/docs/software/fitsio/fitsio.html. The HMI full-disk maps, such as Dopplergrams and magnetograms, along with supporting images are served at Level 1.5. AIA data can be obtained as Level 1.5 images or as Level 1.0 data that must be promoted to Level 1.5 by user software. Descriptions of the AIA data-product levels can be found in Lemen *et al.* (2011). Routines for this purpose are available in the IDL language through the SolarSoft package (Freeland and Handy, 1998) available at http://www.lmsal.com/solarsoft/. A synoptic sequence of all AIA bandpasses in 1K × 1K FITS files with three-minute cadence is also provided for long-term analysis.

Full-resolution science data from EVE are served by the EVE SOC at the University of Colorado (http://lasp.colorado.edu/eve/data/data_access.htm). These data are available in compressed FITS files. The EVE team also produces daily averaged data at three resolutions. The EVE team provides a FITS reader routine in the IDL language that can be used to read their data files.

9.1. Rules of the Road for Data Use

The SDO science investigators agree to abide by the "Rules of the Road" developed for the Sun–Earth Connection and its successor, the Heliophysics Science Division. These rules are available at http://lwsde.gsfc.nasa.gov/Rules_Revised20030327.html and repeated here:

i) The Principal Investigators (PI) shall, in a timely manner, make available to the science-data user community (Users) data and access methods to reach the scientifically useful data and provide analysis tools equivalent to the level that the PI uses.
ii) The Principal Investigators (PI) shall make available appropriate data products to the public that assist the PI's Education and Public Outreach responsibilities.
iii) The PI shall ensure that all scientifically important data are archived to ensure long-term accessibility of the data and their correct and independent usability.
iv) The PI shall notify Users of updates to processing software and calibrations via metadata and other appropriate documentation.
v) Users should consult with the PI to ensure that the Users are accessing the most recent available versions of the data and analysis routines.
vi) Browse products are not intended for science analysis or publication and should not be used for those purposes without consent of the PI.
vii) Users should acknowledge the sources of data used in all publications, presentations, and reports.
viii) Users should transmit to the PI a copy of each manuscript that uses the PI's data upon submission of that manuscript for consideration of publication. On publication the citation should be transmitted to the PI and any other providers of data.
ix) Users are encouraged to make tools of general utility and/or value-added data products widely available to the community. Users are encouraged to notify the PI of such utilities or products. The User should also clearly label the product as being different from the original PI-produced data product.
x) The editors and referees of scientific journals should avail themselves of the expertise of the PI while a data set is still unfamiliar to the community, and when it is uncertain whether authors have employed the most up-to-date data and calibrations.

9.2. Browse and Public Access Data

SDO data are made available to the public in common graphics formats at several websites. For example, the mission website at http://sdo.gsfc.nasa.gov/ serves images in the JPEG format at a roughly 15-minute cadence and several spatial resolutions. Movies covering the previous 48 hours are available in both mpeg format and as self-updating kiosk movies. The instrument teams provide access to similar graphics formats at more rapid cadences. Helioviewer (http://helioviewer.org/) and jHelioviewer (http://jhelioviewer.org/) allow one to browse the AIA images, making and sharing movies from an archive of images in the JPEG2000 format.

Press releases, graphics of wide interest, and other announcements are available from NASA Headquarters at http://www.nasa.gov/sdo.

9.3. Final Data Archive

At the end of the SDO mission, each SIT will deliver the data obtained during the prime mission to a data archive for permanent storage.

10. Summary

The *Solar Dynamics Observatory* is setting the standard for new rapid-cadence science data that enables the study of the energetics of solar events in the corona and the internal structure and surface magnetic of the Sun. SDO data are being combined with that from other satellites in NASA Heliophysics Great Observatory, such as STEREO, SOHO, *Wind*, TIMED, and ACE, to develop an understanding of how solar activity is generated and how it can affect the Earth.

Many new, significant discoveries have already occurred, with more to follow as this mission progresses. In particular, researchers hope to learn how storms get started near the solar surface and how they propagate upward through the solar atmosphere toward Earth and elsewhere in the solar system. Scientists will use SDO data to help them understand how the Sun's changing magnetic fields are created and how they evolve to release the energy that heats the corona and creates the eruptions that are the storms of solar activity and the seeds of space weather.

Acknowledgements This work was supported by NASA's *Solar Dynamics Observatory* (SDO).

References

Freeland, S.L., Handy, B.N.: 1998, *Solar Phys.* **182**, 497.

Hathaway, D., Antiochos, S., Bogdan, T., Davila, J., Dere, K., Fleck, B., *et al.*: 2001, Solar Dyanmics Observatory: Report of the Science Definition Team. Technical Report NP-2001-12-410-GSFC, NASA. http://sdo.gsfc.nasa.gov/assets/docs/sdo_sdt_report.pdf.

Lemen, J.R., Title, A.M., Akin, D.J., Boerner, P.F., Chou, C., Drake, J.F., Duncan, D.W., Edwards, C.G., Friedlaender, F.M., Heyman, G.F., Hurlburt, N.E., Katz, N.L., Kushner, G.D., Levay, M., Lindgren, R.W., Mathur, D.P., McFeaters, E.L., Mitchell, S., Rehse, R.A., Schrijver, C.J., Springer, L.A., Stern, R.A., Tarbell, T.D., Wuelser, J.-P., Wolfson, C.J., Yanari, C.: 2011, *Solar Phys.* doi:10.1007/s11207-011-9776-8.

Pence, W.D., Chiappetti, L., Page, C.G., Shaw, R.A., Stobie, E.: 2010, *Astron. Astrophys.* **524**, A42. doi:10.1051/0004-6361/201015362.

Schou, J., Scherrer, P.H., Bush, R.I., Wachter, R., Couvidat, S., Rabello-Soares, M.C., Bogart, R.S., Hoeksema, J.T., Liu, Y., Duvall Jr., T.L., Akin, D.J., Allard, B.A., Miles, J.W., Rairden, R., Shine, R.A., Tarbell, T.D., Title, A.M., Wolfson, C.J., Elmore, D.F., Norton, A.A., Tomczyk, S.: 2011, *Solar Phys.* doi:10.1007/s11207-011-9842-2.

Schrijver, C.J., Title, A.M.: 2011, *J. Geophys. Res.* **116**, A04108. doi:10.1029/2010JA016224.

Solanki, S.K., Usoskin, I.G., Kromer, B., Schüssler, M., Beer, J.: 2004, *Nature* **431**, 1084. doi:10.1038/nature02995.

Tann, H.K., Pages, R.J., Silva, C.J.: 2005, In: *Space Systems Engineering Conference* **GT-SSEC.C.5**. http://hdl.handle.net/1853/8029.

Woods, T.N., Eparvier, F.G., Bailey, S.M., Chamberlin, P.C., Lean, J., Rottman, G.J., Solomon, S.C., Tobiska, W.K., Woodraska, D.L.: 2005, *J. Geophys. Res.* **110**, A01312. doi:10.1029/2004JA010765.

Woods, T.N., Eparvier, F.G., Hock, R., Jones, A.R., Woodraska, D., Judge, D., Didkovsky, L., Lean, J., Mariska, J., Warren, H., McMullin, D., Chamberlin, P., Berthiaume, G., Bailey, S., Fuller-Rowell, T., Sojka, J., Tobiska, W.K., Viereck, R.: 2011a, *Solar Phys.* doi:10.1007/s11207-009-9487-6.

Woods, T.N., Hock, R., Eparvier, F., Jones, A.R., Chamberlin, P.C., Klimchuk, J.A., Didkovsky, L., Judge, D., Mariska, J., Warren, H., Schrijver, C.J., Webb, D.F., Bailey, S., Tobiska, W.K.: 2011b, *Astrophys. J.* **739**, 59. doi:10.1088/0004-637X/739/2/59.

Solar Phys (2012) 275:17–40
DOI 10.1007/s11207-011-9776-8

THE SOLAR DYNAMICS OBSERVATORY

The *Atmospheric Imaging Assembly* (AIA) on the *Solar Dynamics Observatory* (SDO)

James R. Lemen · Alan M. Title · David J. Akin · Paul F. Boerner · Catherine Chou · Jerry F. Drake · Dexter W. Duncan · Christopher G. Edwards · Frank M. Friedlaender · Gary F. Heyman · Neal E. Hurlburt · Noah L. Katz · Gary D. Kushner · Michael Levay · Russell W. Lindgren · Dnyanesh P. Mathur · Edward L. McFeaters · Sarah Mitchell · Roger A. Rehse · Carolus J. Schrijver · Larry A. Springer · Robert A. Stern · Theodore D. Tarbell · Jean-Pierre Wuelser · C. Jacob Wolfson · Carl Yanari · Jay A. Bookbinder · Peter N. Cheimets · David Caldwell · Edward E. Deluca · Richard Gates · Leon Golub · Sang Park · William A. Podgorski · Rock I. Bush · Philip H. Scherrer · Mark A. Gummin · Peter Smith · Gary Auker · Paul Jerram · Peter Pool · Regina Soufli · David L. Windt · Sarah Beardsley · Matthew Clapp · James Lang · Nicholas Waltham

Received: 6 December 2010 / Accepted: 11 April 2011 / Published online: 2 June 2011

Abstract The *Atmospheric Imaging Assembly* (AIA) provides multiple simultaneous high-resolution full-disk images of the corona and transition region up to 0.5 $R_\odot$ above the solar limb with 1.5-arcsec spatial resolution and 12-second temporal resolution. The AIA consists of four telescopes that employ normal-incidence, multilayer-coated optics to provide

The Solar Dynamics Observatory
Guest Editors: W. Dean Pesnell, Phillip C. Chamberlin, and Barbara J. Thompson

J.R. Lemen (✉) · A.M. Title · D.J. Akin · P.F. Boerner · C. Chou · J.F. Drake · D.W. Duncan · C.G. Edwards · F.M. Friedlaender · G.F. Heyman · N.E. Hurlburt · N.L. Katz · G.D. Kushner · M. Levay · R.W. Lindgren · D.P. Mathur · E.L. McFeaters · S. Mitchell · R.A. Rehse · C.J. Schrijver · L.A. Springer · R.A. Stern · T.D. Tarbell · J.-P. Wuelser · C.J. Wolfson · C. Yanari
Solar and Astrophysics Laboratory, Lockheed Martin Advanced Technology Center, Bldg. 252, Org. ADBS, 3251 Hanover St., Palo Alto, CA 94304, USA
e-mail: lemen@lmsal.com

Present address:
L.A. Springer
Department of Physics, Montana State University-Bozeman, P.O. Box 173840, Bozeman, MT 59717, USA

J.A. Bookbinder · P.N. Cheimets · D. Caldwell · E.E. Deluca · R. Gates · L. Golub · S. Park · W.A. Podgorski
Smithsonian Astrophysical Observatory, 60 Garden Street, Cambridge, MA 02138, USA

R.I. Bush · P.H. Scherrer
W. W. Hansen Experimental Physics Laboratory, Center for Space Science and Astrophysics, Stanford University, Stanford, CA 94305, USA

M.A. Gummin
Alias Aerospace, Inc., 1731 Saint Andrews Court, St. Helena, CA 94574, USA

narrow-band imaging of seven extreme ultraviolet (EUV) band passes centered on specific lines: Fe XVIII (94 Å), Fe VIII, XXI (131 Å), Fe IX (171 Å), Fe XII, XXIV (193 Å), Fe XIV (211 Å), He II (304 Å), and Fe XVI (335 Å). One telescope observes C IV (near 1600 Å) and the nearby continuum (1700 Å) and has a filter that observes in the visible to enable coalignment with images from other telescopes. The temperature diagnostics of the EUV emissions cover the range from 6×10^4 K to 2×10^7 K. The AIA was launched as a part of NASA's *Solar Dynamics Observatory* (SDO) mission on 11 February 2010. AIA will advance our understanding of the mechanisms of solar variability and of how the Sun's energy is stored and released into the heliosphere and geospace.

Keywords Solar corona · Solar instrumentation · Solar imaging · Extreme ultraviolet

1. Introduction

The primary goal of the *Solar Dynamics Observatory* is to understand the physics of solar variations that influence life and society. It achieves that goal by targeted basic research focused on determining how and why the Sun varies, and on improving our understanding of how the Sun drives global change and space weather. As one of the SDO instruments designed to meet this goal, the *Atmospheric Imaging Assembly* (AIA) focuses on the evolution of the magnetic environment in the Sun's atmosphere, and its interaction with embedded and surrounding plasma. Figure 1 shows the four AIA telescopes mounted to the SDO spacecraft, and Figure 2 shows schematically the layout of the telescopes with respect to their wavelength band passes.

The images of the corona taken by *Yohkoh*, the *Extreme ultraviolet Imaging Telescope* (EIT) on the *Solar and Heliospheric Observatory* (SOHO), the *Sun Earth Connection Coronal and Heliospheric Investigation* (SECCHI) on the *Solar Terrestrial Relations Observatory* (STEREO), and the *Transition Region and Coronal Explorer* (TRACE) have shown that all coronal structures evolve in density, temperature, and position on time scales as short as seconds. Conditions within coronal-loop volumes and open magnetic structures appear to depend primarily on "local" conditions, *i.e.* on conditions determined by the path of the field line, or loop, from end to end in the photosphere. Combined with the marked temporal evolution of the atmospheres contained within coronal loops or open-field regions on relatively short time scales, EUV images are dissimilar for different band passes. Apart from

P. Smith
Harvard University-Smithsonian Astrophysical Observatory, 60 Garden Street, Cambridge, MA 02138, USA

G. Auker · P. Jerram · P. Pool
e2v technologies, 106 Waterhouse Lane, Chelmsford, Essex CM1 2QU, UK

R. Soufli
Lawrence Livermore National Laboratory, Livermore, CA 94550, USA

D.L. Windt
Reflective X-ray Optics LLC, New York, NY 10027, USA

S. Beardsley · M. Clapp · J. Lang · N. Waltham
Rutherford Appleton Laboratory, Harwell Business Innovation Campus, Didcot, Oxon, OX11 0QX, UK

Figure 1 The AIA telescopes with their guide telescopes as seen mounted on the SDO spacecraft's instrument module during integration in the clean room at NASA's Goddard Space Flight Center.

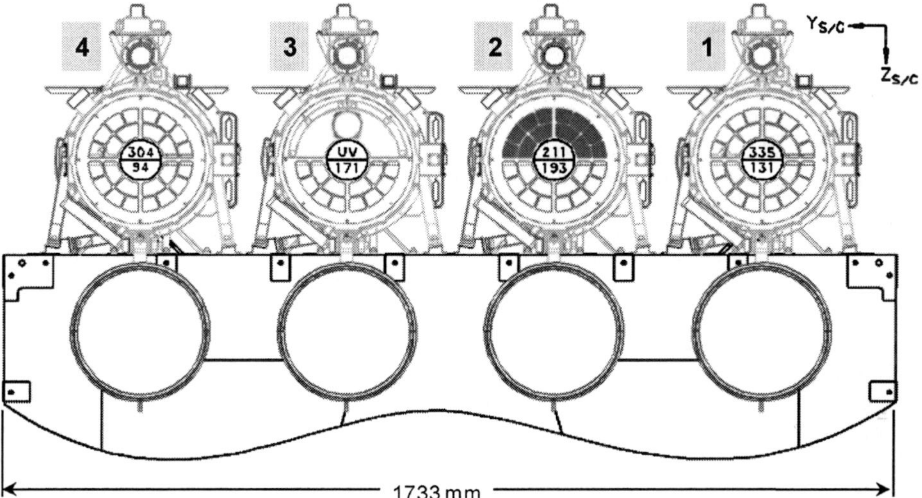

Figure 2 The layout of the wavelength channels or band passes in each of the four AIA telescopes. Telescope 2 has an aperture blade to select between wavelength channels. The other telescopes rely on filters in filter wheels to select between channels. The top half of telescope number 3 has a MgF_2 window with a coating centered at 1600 Å.

thermal evolution, there is an abundance of waves, flows, and impulsive phenomena that occur on short time scales of minutes or less.

The coronal magnetic field itself evolves on time scales that range from seconds to years. Slowest is the evolution of the largest-scale field. On shorter time scales, electrical currents that are induced in, and carried with, the magnetic field may build up over weeks or months, while release of those stresses may take only a fraction of a minute. *Yohkoh* and SOHO have shown, moreover, that there are no purely local field topologies: the short- and long-term evolution of the corona is affected by both nearby and distant magnetic sources.

AIA was designed to study these characteristics of the Sun's dynamic magnetic field and the coronal response to it. AIA provides the following essential capabilities:

i) a view of the entire corona at the best feasible resolutions compatible with SDO's constraints, providing coverage of the full thermal range of the corona;

ii) a high signal-to-noise ratio for two- to three-second exposures that reaches 100 in quiescent conditions for the low-temperature coronal-imaging channels and during flaring in the higher-temperature channels, with a dynamic range of up to 10 000; and
iii) essentially uninterrupted viewing for months at a time at a temporal resolution of approximately 10 to 12 seconds, and sometimes faster to study energetic transient phenomena.

These capabilities are met by a design that includes the following.

i) Four 20-cm, dual-channel, normal-incidence telescopes, which observe a 41 arcmin field of view in ten EUV and UV channels, with 0.6-arcsec pixels and 4096 × 4096 CCDs. Four of the EUV wavelength bands open new perspectives on the solar corona, having never been imaged or imaged only during brief rocket flights. The set of six EUV channels that observe ionized iron allow the construction of relatively narrow-band temperature maps of the solar corona from below 1 MK to above 20 MK.
ii) Detectors with a full well of at least 150 000 electrons, with typically 18 e^- $photon^{-1}$ (for 193 Å), and a readout noise of <25 e^-, and data compression that is nearly lossless.
iii) A standard baseline observing program running most of the time, while observing continuously from SDO's geo-synchronous orbit. AIA has the capability to adjust its observing program to changing solar conditions in order to implement observing programs that are optimized to meet the requirements of specific scientific objectives. This allows, for example, a two-second cadence in a reduced field of view using four wavelengths for flare studies.

With these capabilities, AIA enables us to observe the changing topology of the magnetic field even as the coronal plasma is changing in temperature. In its standard operating mode, the AIA images the entire Sun ten times faster per wavelength than TRACE at twice the number of coronal channels, with a 16-fold increase in the number of pixels per image. Compared with the full-disk SOHO/EIT (Delaboudinière *et al.*, 1995), AIA runs at 700× the usual EIT image frequency, with 16 times more pixels per image, in typically eight instead of four channels. Compared to EIT and TRACE, the AIA thus represents an increase in the information rate for coronal observations by a factor of 400 to 22 000. AIA also advances on the *Hinode*/*X-Ray Telescope* (XRT) (Golub *et al.*, 2007) and SECCHI/*Extreme Ultra-Violet Imager* (EUVI) (Howard *et al.*, 2008) telescope designs, each which contained 2k × 2k pixel CCDs with one and 1.6-arcsec pixel sizes, respectively. And SDO's *Helioseismic and Magnetic Imager* (HMI) provides, among other things, high spatial resolution line-of-sight magnetograms at a cadence suitable for the evolution of the photosphere that is an order of magnitude faster than SOHO/MDI (Schou *et al.*, 2011).

The basic AIA observables are full-Sun intensities at a range of wavelengths. In combination with the higher-level and metadata products these comprise the data archive which is freely accessible to the research community; the archive and its interfaces (including those for data requests and archive queries) are described by Hurlburt *et al.* (2011).

2. Science Overview

The AIA investigation covers a broad range of science objectives that focus on five core research themes that both advance solar and heliospheric physics in general and provide advanced warning of coronal and inner-heliospheric disturbances of interest to the Living With a Star program, *i.e.*, global change, space weather, human exploration of space, and technological infrastructure in space and on Earth.

i) Energy input, storage, and release: the 3D dynamic coronal structure, including reconnection and the effects of coronal currents.
ii) Coronal heating and irradiance: origins of the thermal structure and coronal emission, to understand the basic properties of the solar coronal plasma and field, and the spatially-resolved input to solar spectral irradiance, as observed by, *e.g.*, SDO's *Extreme ultraviolet Variability Explorer* (EVE).
iii) Transients: sources of radiation and energetic particles.
iv) Connections to geospace: material and magnetic field output of the Sun.
v) Coronal seismology: a diagnostic to access sub-resolution coronal physics.

These five themes guide the AIA science investigation, and their fundamental observational needs have shaped the design of the instrument.

2.1. Energy Input, Storage, and Release: The Dynamic Coronal Structure

All coronal activity relies on energy dissipation within the plasma contained in the Sun's magnetic field. One of the primary objectives of the NASA's Living with a Star program, and of the AIA science investigation in particular, is to understand how energy is brought into the coronal field, how it is stored there, and how it is released. One key part of that objective is to understand the geometry of the magnetic field, and its evolution subject to external forcing and internal energy dissipation.

Plasma motions in and below the solar photosphere force the embedded magnetic field to move with the flows. These motions apply stress to the outer-atmospheric field on a broad range of spatial and temporal scales. This stress propagates throughout the outer atmosphere in the form of a variety of wave types and frequencies, as well as a broad spectrum of slower changes associated with evolving induced currents. Key questions to be answered are: How are these perturbations imposed upon the corona? How do they propagate? How do the electric currents close either within the atmosphere or through the photosphere?

TRACE observations reveal that the stresses that are applied to the coronal field by the photospheric motions lead to a magnetic field that is only moderately tangled on scales of the granular motions up to the largest scales, despite theoretical expectations that field-line braids should persist for long times in a coronal environment where reconnection was expected to proceed only slowly. Instead of a highly tangled field, however, TRACE images show that small-scale braids, driven by convective motions from the arcsecond scale of the granulation upward, are only seen around filaments and prominences. Even then, the images suggest that loop twists rarely exceed about half a turn. This implies rapid dissipation of the induced currents, which is mimicked by some recent numerical simulations. The fidelity of these simulations does not approach the real corona, however, with anticipated (but maybe not realized) magnetic Reynolds numbers for the bulk of the coronal volume many orders of magnitude larger than in the simulated volume. In contrast to the apparently efficient smoothing out of the small-scale complexity, many filament configurations erupt, requiring substantial amounts of stored energy. Why do small-scale twists dissipate readily, whereas large-scale stresses persist for long times in filament and sigmoid-field configurations which can eventually lead to instabilities?

2.2. Coronal Heating and Irradiance: Thermal Structure and Emission

Solar radiation at UV, EUV, and SXR wavelengths plays a significant role in the determining the physical properties of the Earth's upper atmosphere. Variations in solar radiation at these wavelengths drive changes in the density and ionization of the Earth's thermosphere and ionosphere, impacting the performance of ground-based communications systems and spacecraft in low-Earth orbit. Long-term observations of the solar irradiance with the required accuracy have proven difficult, however, and much of our knowledge of the solar irradiance and its variability remains uncertain.

Observations over the past several decades have clearly established that solar variability at UV, EUV, and SXR wavelengths is tied to the variability of the Sun's surface magnetic fields. Past studies of the Sun and Sun-like stars has revealed that the primary determinant of the radiation leaving the corona is the total magnetic flux threading the photosphere. At present, however, we do not possess a detailed understanding of how magnetic energy is released within the solar corona. Thus we cannot use our knowledge of the Sun's magnetic fields to model and predict changes in the solar irradiance. Such calculations are an invaluable aid to interpreting and extending direct measurements of the solar irradiance. One of the primary objectives of the SDO mission is to develop a physical understanding of solar irradiance and its variability, specifically at (E)UV wavelengths, with close collaboration between the science teams of AIA and of SDO's *Extreme ultraviolet Variability Experiment* (EVE: Woods *et al.*, 2011). AIA images will show the locations and structural/thermal changes associated with EUV irradiance changes measured with high precision by EVE.

2.3. Transients: Sources of Radiation and Energetic Particles

Most of the time, the coronal field appears to evolve smoothly. At times, however, massive explosions or eruptions disturb the coronal field. Many are not strong enough to rupture the encapsulating coronal magnetic field, and only the energetic radiation escapes toward Earth. Some, in contrast, are associated with opening magnetic fields into the heliosphere; these coronal eruptions appear to be the counterparts of coronal mass ejections. What triggers the eruptions or explosions? What determines whether the field confines the eruption to the corona or allows coupling into the heliosphere?

Flares, CMEs, filament destabilizations, sprays, and other transients are sources of radiation and energetic particles, and therefore prime drivers of violent solar weather. Transients result from the sudden conversion of magnetic energy into bulk, thermal, and non-thermal energy as the magnetic field reconnects. Theories of 3D reconnection are still in early stages of development, and advances in its understanding are hampered by observational limitations.

AIA is designed to make the next leap forward in understanding transient initiation and evolution. Its high-cadence, full-disk, multi-temperature observations are revealing the reorganizing field in the initial phases of flares and of filament eruptions, as well as the later evolution as the field relaxes into its new state. Particularly promising is the inclusion of four EUV pass bands that observe the evolution of the coronal plasma for the first time at near-arcsecond resolution for temperatures between 3 to 20 MK (94, 131, 211, and 335 Å).

2.4. Connections to Geospace: Material and Magnetic-Field Output of the Sun

The solar wind, the embedded magnetic field, and eruptive perturbations in the form of CMEs drive the variations in the space surrounding the Earth and other planets. The dynamic connections between Sun and geospace are a cornerstone of the ILWS program. To understand how the Sun's variability affects life and society, we must understand how the products of this variability are transported into and through the heliosphere, and how they interact with the Earth's magnetic field and atmosphere. The AIA investigation is expected to make substantial quantitative advances in many areas relevant to this problem. At the foundation of this expectation lies the improved understanding of the global coronal field and its extension into the heliosphere. In Sections 2.2 and 2.3 we discussed this for the pathways for escape of energetic particles into the heliosphere, the irradiance that affects the ionosphere and below, and the triggering of flares and CMEs. Here we focus on the magnetized solar wind and its perturbations.

The solar wind flows out radially into the heliosphere dragging along the magnetic-field lines that are forced "open" within the first few radii from the Sun. The successes of simple concepts such as the potential-field source-surface (PFSS) model suggest that more realistic and detailed models should explain why this is. What determines which field lines will open? Why does the PFSS work so well? The latter is particularly interesting because the PFSS model does not incorporate field dynamics. How is field opened, and how is it closed again as the connections evolve? This is somehow connected to the properties of the solar wind, which we know depend on the field geometry and strength. But although that dependence is empirically constrained, we still need to learn what determines the physical properties of the solar wind.

The eruptive coronal mass ejections perturb the background solar wind. How do they evolve through the coronal field? How do they couple into, and propagate through, the heliosphere on their way to the Earth and the other planets? HMI and AIA data will enable modelers to addresses these questions by advancing beyond PFSS models to full MHD models of the coronal and inner heliosphere.

2.5. Coronal Seismology: A Diagnostic to Study Coronal Waves and Oscillations

Recent observations from SOHO and TRACE show a variety of oscillation modes in the transition region and corona. These observations opened up the promising new field of coronal seismology. By studying the properties, excitation, propagation and decay of these oscillations and waves, we can reveal fundamental physical properties of the solar transition region and corona, such as the magnetic field, density, temperature, and viscosity.

Examples of seismic responses of the corona have been found in large-scale coronal Moreton and EIT waves, in polar plumes, sunspot fields, upper transition-region moss, and what appear to be ordinary loops. Perhaps most striking are the transverse oscillations seen by TRACE. Longitudinal oscillations and waves have also been observed in the upper transition-region moss and in coronal loops, both in cooler loops with TRACE and in hot (10^7 K) loops with the *Solar Ultraviolet Measurements of Emitted Radiation* (SUMER) spectrometer.

There are still many unresolved issues about these waves and oscillations. How do they get excited? Why does only a fraction of the observed flares lead to clear oscillations? Are longitudinal waves in coronal loops with five-minute periods related to p modes? How do any of these waves propagate? Which of the many possible theoretical mechanisms can explain the unexpectedly rapid decay of some of these waves? These are some of the questions that we need to answer before we can seismically probe coronal physics.

The AIA design enables us to develop the necessary improved understanding of the properties and other unresolved issues of the observed waves. The high cadence of AIA will extend the parameter space to higher frequencies (by a factor of two – four compared with TRACE). The broad, simultaneous temperature coverage positions us to study waves in parts of active regions that have not yet been seen. Guided by significant advances in the theory of these waves (from 3D MHD simulations), and complemented with spectroscopic measurements of densities and line-of-sight velocities (*e.g.* by the *EUV Imaging Spectrometer* on *Hinode*), and reliable magnetic-field extrapolations (*e.g.* from HMI and AIA), AIA will fully exploit the potential of coronal seismology.

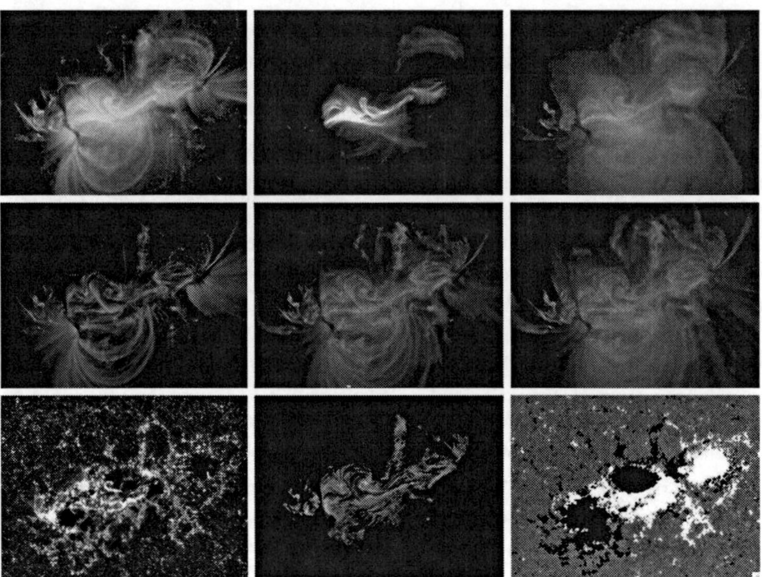

Figure 3 Images of an active region observed on 15 February 2011 at 01:45 UT with AIA. The top row panels are observations from the 131-, 94-, and 335-Å channels (from left to right). The middle row panels are from the 171-, 193-, and 211-Å channels, and the bottom row panels are from 1600 and 304 Å (left and middle, respectively). The bottom right corner panel is an HMI line-of-sight magnetogram showing the same field of view. The AIA images have been processed to Level 1.5 and each contains 480 × 360 pixels, which corresponds to 288 × 216 arcsecs.

3. Instrument Overview

The AIA instrument consists of four generalized Cassegrain telescopes that are optimized to observed narrow band passes in the EUV in order to observe solar emissions from the transition region and corona. An active region observed on 15 February 2011 is shown in Figure 3 for the seven EUV channels, the 1600-Å channel, and an HMI line-of-sight magnetogram. This mosaic of images illustrates how the various instrument-response functions sample the temperature-dependent structure of the solar corona. Table 1 lists the primary ions for each band pass and their characteristic emission temperatures, along with the types of solar features that may be observed.

Each AIA f/20 telescope has a 20-cm primary mirror and an active secondary mirror. Key parameters of the telescopes are given in Table 2. The telescope design is fully baffled to prevent charged particles from reaching the CCD, and the aperture pupil is located by a mask that is mounted in front of the primary mirror. Each telescope field of view is approximately 41 arcmin circular diameter. By design, the CCD corners do not receive solar emission from the optics (being shaded by the filter-wheel mechanism). These dark regions provide useful means to monitor detector noise levels and to check for the presence of energetic particles. The telescope mirrors have multilayer coatings that are optimized for the selected EUV wavelength of interest. Three of the telescopes, numbers 1, 2, and 4, have two different EUV band passes. Telescope number 3's mirror has a 171-Å band pass on one half and the other half has a broad-band UV coating. At the focal plane are back-thinned CCD sensors with 4096 × 4096 pixels; each 12-μm pixel corresponds to 0.6 arcsec. Entrance filters at the telescope aperture block visible and IR radiation. Filters in a filter-wheel mechanism located

Table 1 The primary ions observed by AIA. Many are species of iron covering more than a decade in coronal temperatures.

Channel	Primary ion(s)	Region of atmosphere	Char. log(T)
4500 Å	continuum	photosphere	3.7
1700 Å	continuum	temperature minimum, photosphere	3.7
304 Å	He II	chromosphere, transition region	4.7
1600 Å	C IV + cont.	transition region, upper photosphere	5.0
171 Å	Fe IX	quiet corona, upper transition region	5.8
193 Å	Fe XII, XXIV	corona and hot flare plasma	6.2, 7.3
211 Å	Fe XIV	active-region corona	6.3
335 Å	Fe XVI	active-region corona	6.4
94 Å	Fe XVIII	flaring corona	6.8
131 Å	Fe VIII, XXI	transition region, flaring corona	5.6, 7.0

Table 2 The AIA instrument characteristics.

Mirrors	Multilayer-coated Zerodur
Primary diameter	20 cm
Effective focal length	4.125 m
Field of view	41 × 41 arcmin (along detector axes)
	46 × 46 arcmin (along detector diagonal)
Pixel size/Resolution	0.6 arcsec (12 μm)/1.5 arcsec
CCD detector	4096 × 4096, thinned, back-illuminated
Detector full well	150 000 electrons
Effective pointing stability (with image stabilization system)	0.12 arcsec RMS
Cadence(Full-frame readout)	
All telescopes	8 wavelengths in 10 to 12 seconds
Typical exposure times	0.5 to 3 seconds
Flight computer	BAe RAD 6000
Mass	
Four telescopes	112 kg
AIA electronics box	26 kg
Main wiring harness	17 kg
Instrument power	160 W
Science Telemetry	
Interface to spacecraft	67 Mbps
Ground capture	≈2 Tbytes (uncompressed) per day

in front of the focal plane are used to select the wavelength channel of interest in three of the telescopes and the fourth telescope contains a selector mechanism to choose the wavelength. A mechanical shutter is used to regulate the exposure time. The active secondary is pointed in response to signals from a dedicated guide telescope (GT), which is mounted on the side of the telescope tube. The design of the AIA telescope employs many features that were successful for TRACE (Handy *et al.*, 1999) but in comparison, each AIA telescope has 16 times more pixels than TRACE and covers the whole solar disk. AIA has sufficient

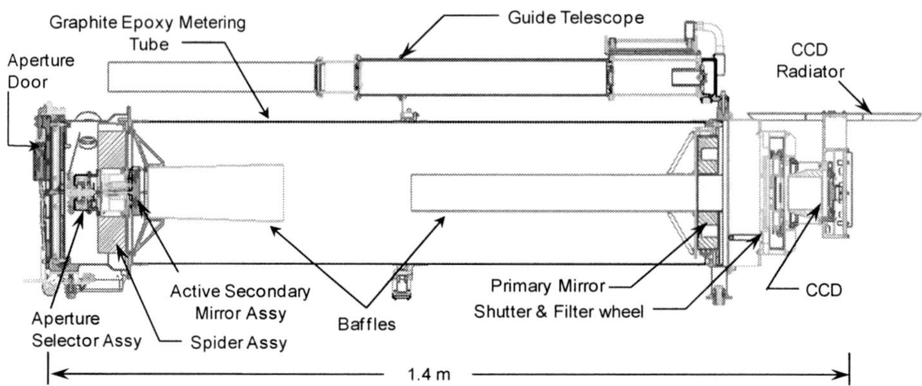

Figure 4 A cross sectional view of AIA telescope number 2 and its guide telescope. The aperture door protects the entrance filters during launch operations. Each of the four AIA telescopes has its own guide telescope, which provides a signal for the active secondary to stabilize the image on the CCD.

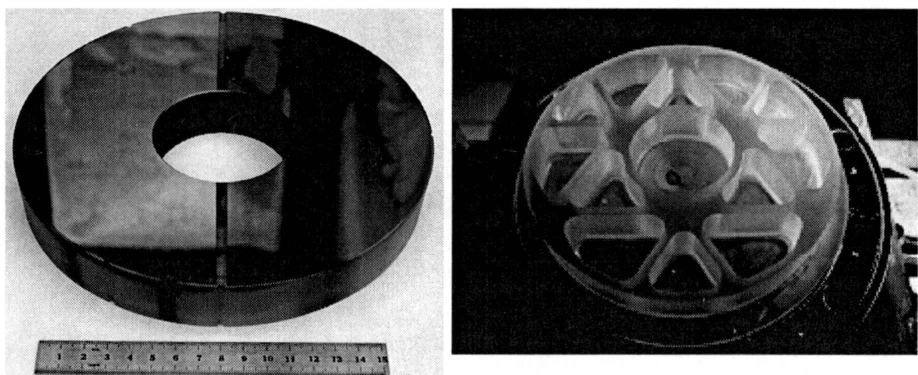

Figure 5 An AIA primary mirror (left) coated with two EUV coatings and a view of the backside of the primary mirror showing the light-weighting. The outer diameter of the primary mirror is 200 mm and the inner diameter is 65 mm. The mirror is held by a low-stress mount that has flexures bonded to the outer rim of the substrate.

sensitivity to obtain a cadence that surpasses TRACE by more than a factor of ten. AIA routinely gathers as much data in three days as the TRACE mission did during its 12-year mission. A cross sectional view of AIA is provided in Figure 4. Further details about the telescope design, optics mounts, and filters are given by Cheimets *et al.* (2009).

3.1. Mirrors and Multilayer Coatings

The AIA telescope mirrors (Figure 5) are fabricated on Zerodur™ substrates, which provide a low coefficient of thermal expansion. They are figured and polished to a micro roughness of <5 Å rms in the spatial frequency range from 10^{-3} to 5×10^{-2} nm^{-1} as measured by atomic-force microscopy at Lawrence Livermore National Laboratory (LLNL, Soufli *et al.*, 2007). Three of the primary flight substrates (94/304, 171/UV, and 131/335 Å) and three of the secondary flight substrates (171/UV, 193/211, and 131/335 Å) were fabricated by the Sagem Corporation. Sagem super-polished the substrates to a sphere, and then an ion-beam

Table 3 Multilayer and filter properties. The metal filters are supported on a 82% transmitting nickel mesh.

Telescope	Central Wavelength (Å)	Mirror Coating	Entrance Filter	Focal-plane Filter	Redundant Focal-plane Filter
1	131.0	Mo/Si	2000 Å Zr	2000 Å Zr	3000 Å Z + 4000 Å polyimide
1	335.4	SiC/Si	1500 Å Al	1500 Å Al	2500 Å Al
2	193.5	Mo/Si	1500 Å Al	1500 Å Al	2500 Å Al
2	211.3	Mo/Si	1500 Å Al	1500 Å Al	2500 Å Al
3	171.1	Mo/Si	1500 Å Al	1500 Å Al	2500 Å Al
3	1600	Al/MgF_2	MgF_2 Window	MgF_2	–
3	1700	Al/MgF_2	MgF_2 Window	Fused silica	–
3	4500	Al/MgF_2	MgF_2 Window	Fused silica	–
4	93.9	Mo/Y	2000 Å Zr	2000 Å Zr	3000 Å Zr
4	303.8	SiC/Si	1500 Å Al	1500 Å Al	2500 Å Al

etch was used to achieve the desired asphere, while preserving the required super-polished surface. L-3 Communications-Tinsley Laboratories produced the flight primary (193/211 Å) and secondary (94/304 Å) substrates using a classic, computer-controlled grind and super-polishing approach.

Different multilayer coatings were applied to each half of each telescope optic to achieve the desired central wavelength (see Table 3). Reflective X-ray Optics applied the 94-, 131-, and 335-Å flight coatings and LLNL applied the 171-, 193-, 211-Å flight coatings (Soufli *et al.*, 2005). LLNL led the reflectance calibration and mapping of all AIA EUV flight mirrors, using the reflectometer at beam line 6.3.2 of the Advanced Light Source synchrotron at Lawrence Berkeley National Laboratory (Underwood *et al.*, 1997; Gullikson, Mrowka, and Kaufmann, 2001).

The UV channel in AIA telescope 3 is based on the TRACE design and makes use of a MgF_2 window with a coating centered at 1600 Å. The mirrors were coated by the Acton Research Corporation (now a division of Princeton Instruments) with aluminum with an MgF_2 protective overcoat. Three filters in the telescope number 3 filter wheel, also manufactured by Acton, select the band passes of interest: 1600 Å on MgF_2, which includes the C IV lines, 1700 Å on fused silica, which suppresses the C IV lines, and a $\approx$500-Å FWHM band pass filter centered at 4500 Å.

The mechanical mounts of the mirrors were designed to maintain arc-second alignment during launch and to ensure that thermally induced stresses during on-orbit operations do not introduce aberrations that would degrade the imaging performance (Podgorski *et al.*, 2009). The primary-mirror mount contains titanium flexures that are bonded to the mirror in three locations on the side of the substrate. The flexures are attached to a titanium ring that is the interface to the aft end of the metering tube. The mount design can withstand temperature changes of up to $\pm 12^\circ$C without compromising the imaging performance. The secondary mirror is clamped in three locations and is pressed by a spring onto three PZT actuators that provide imaging stabilization.

3.2. Filters

The AIA design makes use of metal entrance filters at the aperture of each telescope and in a filter wheel located in front of each focal plane. The filters suppress unwanted UV, visible, and IR radiation for the EUV channels. Two filter materials are used for the filters, aluminum and zirconium, whose transmissions are illustrated in Figure 2 of Boerner *et al.* (2011). Aluminum is used for 171 Å and longer wavelengths and zirconium is used for the two shorter wavelengths. Because of the filter-transmission properties, the selection of aluminum or zirconium in the filter wheel is sufficient to select between the shorter or longer band pass in AIA telescope number 1, which contains 131 Å and 335 Å, and telescope number 4, which contains 94 Å and 304 Å. The redundant, thicker zirconium filter in telescope number 1 also includes a 4000-Å layer of polyimide to provide the option of additional attenuation at 131 Å during very bright flares. The filters and their respective thicknesses are given in Table 3. Each focal-plane filter contains redundant thicker filters in case the thin filters develop pinholes. The filters are supported on a nickel mesh with a 70 line-per-inch spacing. The mesh of the front filter results in a diffraction pattern that is visible in high-contrast scenes such as solar flares (Lin, Nightingale, and Tarbell, 2001). The mesh in the focal-plane filters results in a faint shadow pattern in the focal plane, which is removed by flat fielding. All metal filters were manufactured by Luxel Corporation. The focal-plane filter transmissions were measured in Lockheed Martin Solar and Astrophysics Laboratory's (LMSAL) calibration facility (Boerner *et al.*, 2011). Entrance-filter transmissions are computed using Luxel Corporation-provided filter-material thicknesses.

3.3. CCD Detector and Camera System

The AIA and HMI instruments use CCDs that have the same architecture. This enables the camera-electronics boxes to be interchangeable between the two instruments. The CCDs are custom designed e2v technologies CCD203-82 devices, the AIA ones are back-thinned and back-illuminated while the HMI detectors are front-illuminated. The CCDs are operated non-inverted to obtain good full-well capacity (>150 ke$^-$ pixel^{-1}). The pixel size is 12 μm and each detector has 4096×4096 pixels (Figure 6). There are four readout amplifiers on each device and during normal operations, all four 2048×2048 quadrants are read out simultaneously. The quantum efficiencies of the AIA CCDs have been characterized in the laboratory at LMSAL (Boerner *et al.*, 2011). Laboratory exposure-dose tests conducted at 304 Å revealed modest reduction in quantum efficiency with high dosage. This slight reduction is partially reversible by warming the device, consistent with the presence of trapped charge in the oxide layer.

The camera electronics (Figure 7) were developed at the Rutherford Appleton Laboratory (Waltham *et al.*, 2011). The design is based on previous flight designs, most recently for the cameras of the STEREO/SECCHI instrument (Howard *et al.*, 2008). There are four AIA camera-electronics boxes, one for each telescope, and each camera has four interfaces to read out simultaneously the signals from the four CCD amplifiers. Each quadrant is read at a rate of 2 Mpixels s^{-1} with less than 25 electrons of read noise. The camera implements correlated double sampling and analog-to-digital functions in an RAL-designed, radiation-hardened ASIC. Each camera communicates with the AIA electronics box by an IEEE1355/SpaceWire link, which enables data transmission rates of up to 100 Mbits s^{-1}. The camera has internal dedicated DC–DC power converters to convert the SDO spacecraft 28 V primary power to the required voltages. The camera has programmable control of the

Figure 6 A front-illuminated CCD203-82 device mounted in its nickel-plated Invar 36 package. The AIA flight-version CCD is back-thinned and back-illuminated.

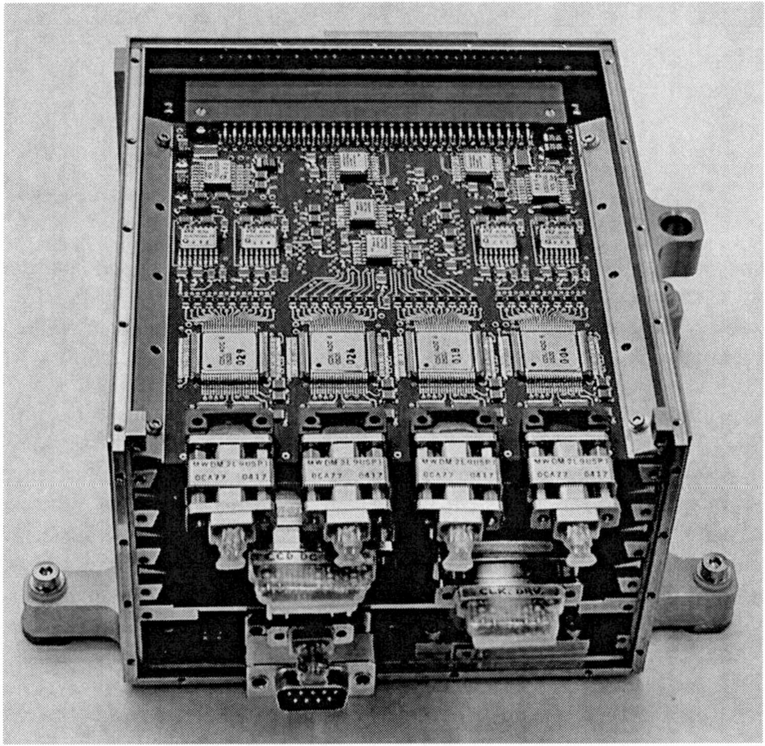

Figure 7 The camera-electronics box (CEB) shown is used in both the AIA and HMI instruments. AIA contains four cameras and HMI two. The camera can operate at up to 2 Mpixels s^{-1} and is interfaced to the AIA electronics box via a SpaceWire link.

waveform generator to provide different operating modes – such as continuous readout, full-frame or windowed, clearing, summing, and exposure – and it provides on-orbit adjustment of gain and pedestal offset.

Table 4 Key Parameters of the AIA Guide Telescope and Image Stabilization System

Wavelength and band pass width	570 nm, 50 nm band
Focal length	1665 mm
Linear range of GT	>±95 arcsecs
Acquisition range	1570 arcsec
Noise equivalent angle	<0.12 arcsec
ISS mirror range	±46 arcsec (in AIA telescope focal plane)
Low frequency jitter reduction factor	50
ISS bandwidth (cross-over)	30 Hz

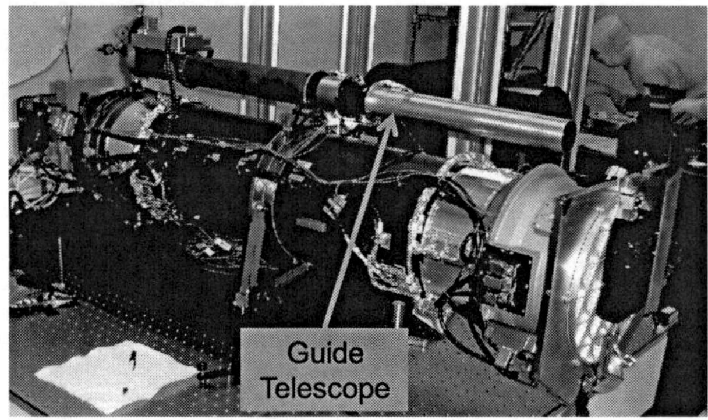

Figure 8 The Guide Telescope shown mounted on the AIA number 2 telescope.

Decontamination heaters are attached to each CCD through a switch controlled by the spacecraft electronics. The CCD heaters were activated shortly after launch for a month-long instrument bakeout. They will be used periodically during the SDO mission to clear the buildup of contaminants as necessary and to eliminate trapped charge in the oxide layers.

3.4. Guide Telescope and Image-Stabilization System

Each EUV telescope has its own guide telescope (GT) which provides an error signal to its image stabilization system (ISS). The GT is based on the TRACE GT and STEREO/SECCHI designs and consists of an achromatic refractor with a band-pass entrance filter that is centered at 570 nm and a Barlow-lens assembly. Four redundant photodiodes positioned behind an occulter measure the position of the solar limb. Unlike TRACE, the AIA/GT has no prisms to off-point the Sun because the AIA instrument has a full-disk field of view. Other differences compared to TRACE include increased shielding for the preamplifier and diodes, and radiation-hardened glass for the entrance window. The key parameters for the GT are shown in Table 4. The GT mounted to telescope number 2 is shown in Figure 8.

The signals for the diodes are amplified and provided to the limb-tracker electronics in the main AIA electronics box (AEB) which in turn controls the three piezoelectric transducers (PZTs) that activate the secondary mirror. Spacecraft jitter in pitch and yaw is reduced to less than 0.24 arcsecs (1 σ) in each axis and is effective for frequencies less than ≈10 Hz. Each PZT can tilt the secondary mirror by ±46 arcsecs in the telescope focal plane (*i.e.*, in the image). Like TRACE, the ISS has gain adjustments that may be calibrated on orbit from data taken while slewing the observatory across the solar limbs. Calibration observations are planned to be accomplished monthly.

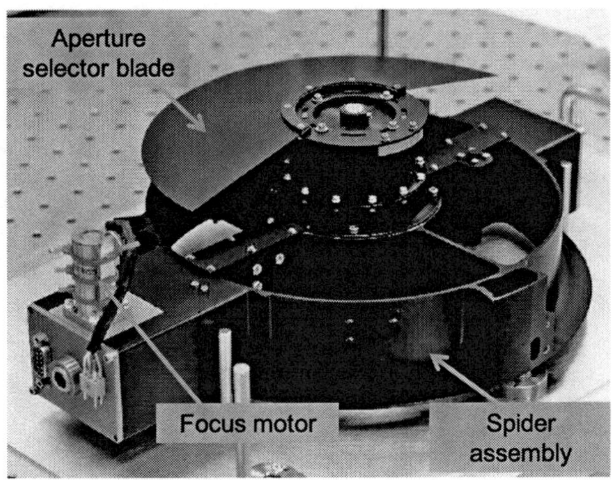

Figure 9 The aperture selector mounted on the telescope number 2 secondary mirror spider assembly. The focus motor can be seen on the left.

The AIA flight software samples the GT signals every 4 ms with a 12-bit ADC and this signal is provided to the SDO spacecraft attitude control system (ACS) software to provide a fine Sun-sensor signal. Any one of the GTs may be selected as the sensor input for the spacecraft ACS. If all four diodes are illuminated, then the GT is in its linear range and the error signals are valid. If fewer than four diodes are illuminated, then Sun-presence flags are set for the appropriate diodes and these may be used by the ACS to acquire the Sun.

3.5. Mechanisms

AIA contains a total of 17 mechanisms consisting of five distinct types. Each telescope contains a front door, a focus mechanism, a filter wheel, and a shutter mechanism. Telescope number 2 also contains an aperture selector.

The front door is designed to protect the entrance filters during launch from debris and from acoustic loads. The AIA design shares heritage with the TRACE and *Hinode*/XRT designs. The door is latched shut with a high-output paraffin actuator which has redundant heater circuits. A Viton O-ring ensures a good seal when closed. The door hinge is spring-loaded and the door is slightly bowed to promote the 180-degree opening when on orbit. Although the paraffin actuators may be used multiple times, there is no means to close the door once on orbit, so it is considered a "one-shot" operation. All four of the AIA doors deployed successfully at the end of the bake-out period, approximately one month after launch.

AIA telescope number 2 has an aperture selector which enables only one half of the aperture to be exposed at one time. The aperture selector blade is 211 mm in diameter and is required to select between 193 Å and 211 Å (Figure 9). In the other telescopes, the selection of the filter in the filter wheel can be used to select the desired channel, but since both of these band passes require aluminum filters, the aperture selector was necessary. The Kollmorgen RB-01501 motor can move from one position to the other in less than a second. This approach is the same as was employed for TRACE and STEREO/SECCHI. This mechanism and all the other mechanisms (apart from the door) were life tested in a vacuum for over 60 million operations, approximately double the number expected during the five-year prime SDO mission.

Each telescope contains a focus mechanism that adjusts the position of the secondary mirror relative to the primary along the optical axis by up to ±800 µm in 2.2-µm steps. As

Figure 10 The AIA filter-wheel mechanism. The wheel has five openings containing various filters for each telescope as described in Table 3.

Figure 11 The AIA shutter mechanism consisting of its blade and shutter motor. The narrow slot is used for 5-ms exposures and the larger opening is used for 80-ms and longer exposures.

the optical design has a magnification of three, this corresponds to an approximately 22-μm change in focus per step. The design is based on the TRACE focus mechanism and employs a stepper motor manufactured by CDA InterCorp to drive a screw that pushes a level arm attached to the secondary mirror cell. The focus can be adjusted as often as every exposure if necessary.

The filter wheel is based on the *Triana/Earth Polycromatic Imaging Camera* (EPIC) and *Hinode*/XRT designs and consists of a thin, brushless DC motor manufactured by H. Magnetics (Figure 10). To the motor is added an optical encoder, and the mechanism is operated as a stepper motor with 324 steps per revolution. A one-position move requires about one second. For telescope numbers 1, 2, and 4 the filter wheel contains four metal (focal-plane) filters and the fifth position is left open for calibration purposes. Telescope number 3 contains two metal filters and three glass filters (for the UV channels).

The shutter mechanism consists of a 159-mm diameter thin blade mounted to a Kollmorgen RB-00704-G02 motor (Figure 11). The housing and optical encoder are a LMSAL design and the bearings are from Timken MPB. This design has flight heritage on several missions including SOHO/MDI, TRACE, STEREO/SECCHI, and *Hinode/Focal Plane Package* (FPP) and XRT. The blade rotates to open the shutter, exposing the CCD. The blade has a narrow opening that provides a 5-ms exposure when rotated over the CCD and a wide opening that provides an 80-ms exposure. Longer exposures are obtained by stopping the wide opening in front of the CCD for a specified length of time that is controlled by the flight software. Exposure times longer than 5 ms and shorter than 80 ms are acquired by sweeping the shutter's narrow opening over the CCD multiple times.

3.6. Electronics and Software

The AIA electronics shares much of its design with HMI and has selective redundancy. The AIA electronics box (AEB) provides power and housekeeping (MIL-STD-1553) interfaces to the spacecraft in addition to the science-data interface by way of two redundant IEEE 1355 SpaceWire ports. The AEB contains redundant DC–DC power systems and redundant RAD6000 CPU cards, the latter being procured from BAE Systems. A block diagram of the electronics system is shown in Figure 12.

The flight software was developed using the real-time multi-tasking VxWorks environment. The software is responsible for receiving commands from the spacecraft and controls the flow of housekeeping data and science data to the spacecraft. It also is responsible for the interfaces to the mechanisms, the heaters, the guide telescope, and the cameras.

The image sequencer controls the scheduling of images in the four telescopes and is configured by tables that can be uploaded from the ground. Under typically planned operations, the basic time step is set to 12 seconds and each telescope is able to acquire two images within each 12-second window, and so eight full-images are acquired every 12 seconds. Because the four cameras share two interfaces to the spacecraft, there are various constraints on how long an exposure can be in a particular telescope. Under nominal operations exposures up to 2.9 seconds are possible. The sequencer design is flexible to support the AIA science observing objectives and is very robust. It is inherently deterministic, provides accurate temporal scheduling, and was the basis for the HMI sequencer design. The flight software has the TRACE automatic exposure-control algorithm implemented to enable the control of exposure times in case of flares.

3.7. Instrument Calibration

The AIA calibration is described by Boerner *et al.* (2011). The instrument effective area has been estimated from component-level calibration measurements. Responses to solar emissions are computed assuming the CHIANTI solar spectral model (Dere *et al.*, 1997, 2009). The response functions for the six EUV band passes that are dominated by iron emission lines are shown in Figure 13. One of the channels in telescope 4 is centered on the emission from He II at 303.8 Å. This band also contains emission from nearby Si XI at 303.3 Å, which may contribute up to 20% of the total intensity detected in this channel when observing active regions (Thompson and Brekke, 2000). AIA calibration data and computed response functions may be accessed on-line (see Boerner *et al.*, 2011) to facilitate detailed or specific analyses.

4. Instrument Operations

The AIA is designed to operate in a regular, synoptic fashion, and it is anticipated that the observing program will be changed infrequently. The initial baseline observing program acquires a full-frame EUV image and one UV or visible-light image every 12 seconds. The basic cadence can be adjusted and may be shortened to 10 seconds if warranted by active solar conditions. The AIA camera electronics support the read out of small regions of interest, which reduces the observing cadence to 2 seconds for selected wavelength channels in order to achieve special campaign observing objectives.

At the baseline 12-second cadence, the data acquisition exceeds the AIA telemetry allocation by a factor of 2.2, thus necessitating the use of on-board compression. AIA has two

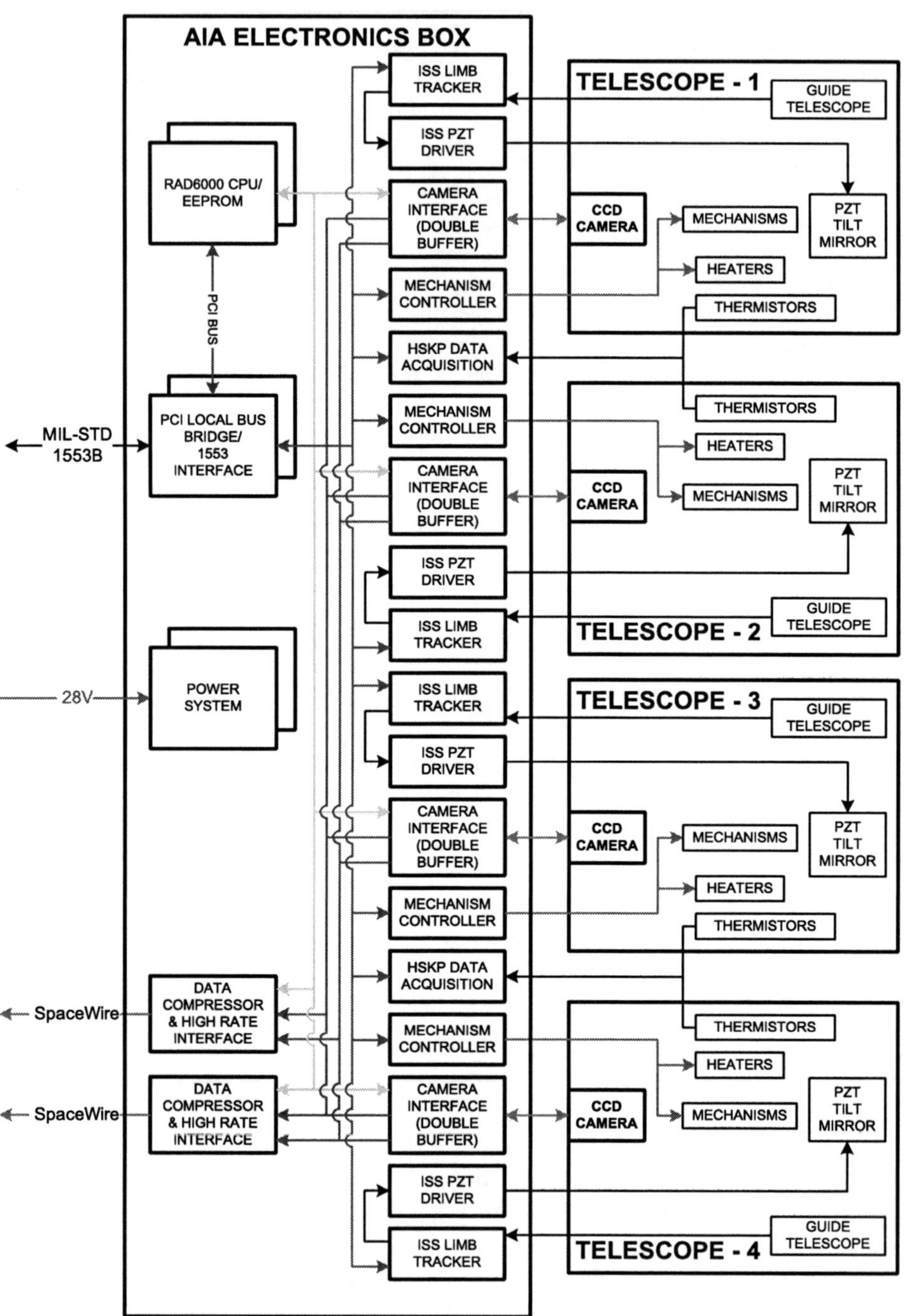

Figure 12 The AIA electronics block diagram. The AIA electronics box interfaces include the spacecraft (to the left) and the four AIA telescopes (to the right).

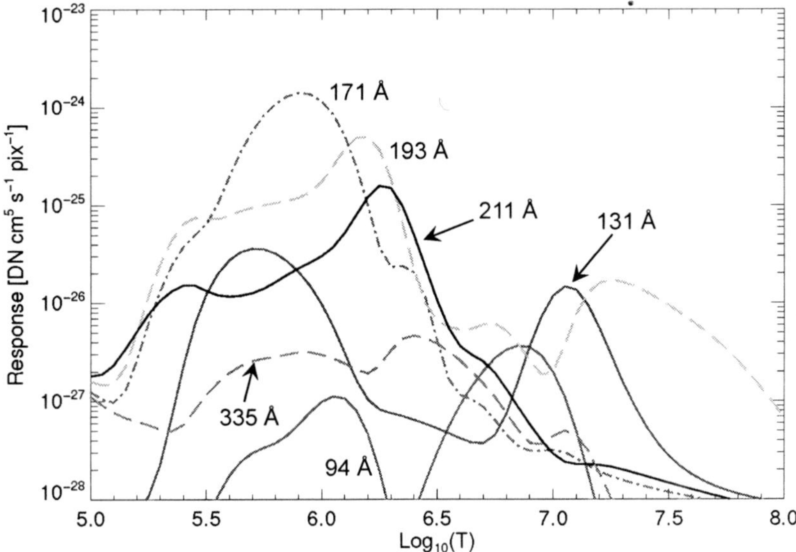

Figure 13 Temperature response functions for the six EUV channels that are dominated by iron emission lines calculated from the effective-area functions and assuming the CHIANTI model for the solar emissivity. This figure corresponds to Figure 11 of Boerner *et al.* (2011) and Figure 8 of Boerner *et al.* may be consulted for the corresponding effective-area functions.

data compression/high-speed interface cards that perform data compression in hardware and then transmit the compressed data to the spacecraft interface. Two compression algorithms are implemented in field-programmable-gate arrays whose parameters can be fine-tuned depending on the wavelength channel. The primary algorithm is lossless Rice compression, which typically achieves a factor of two reduction in data volume for EUV images. A non-lossless algorithm relies on a look-up table that is configurable from the ground. The baseline table is a scaled square-root function. More details are provided by Boerner *et al.* (2011). For the baseline 12-second observing mode, the longest exposure that can be acquired is limited to 2.9 seconds as 3.1 seconds is required for camera readout and other system overhead, and then this repeated in the next 6 seconds for the other companion telescope channel.

An automatic exposure control is implemented similar to that which is used on TRACE. A count-rate histogram is acquired in the hardware on the data compression/high-rate interface board prior to compression. The flight software retrieves the histogram and quickly analyzes whether the next exposure time should be made larger or smaller. Shorter exposure times during impulsive flares helps to reduce saturation caused by the acquired signal exceeding detector full well.

Science operations for both AIA and HMI are controlled from the Instrument Operations Center at LMSAL. Tables are prepared that configure the observing sequence and other instrument parameters, and are transmitted on a secure Internet connection to the SDO mission operations center at GSFC before being transmitted to the spacecraft. Routine operations typically require only a few command loads during the work week. Special operations to establish calibration parameters for telescope alignment, focus, and ISS are scheduled to occur periodically.

5. Data Processing

The SDO mission does not record science data on orbit, but rather transmits telemetry continuously to two SDO ground stations located near White Sands, New Mexico. From there, the AIA and HMI data are transferred by dedicated broad-band circuits to the Stanford University campus where the Level 0 data are permanently archived in the Joint SDO Operations Center (JSOC) science-data processing (SDP) facility. Telemetry data typically arrive at the JSOC-SDP minutes after they are acquired on orbit. If a problem in ground transmission occurs, the retransmission is expected to occur within one to two days. A local storage at the SDO ground system in New Mexico has 30 days of storage as a contingency for short-term problems with the network or at the JSOC-SDP. AIA images are assembled from the telemetry packet data and decompressed, and then stored using lossless Rice compression. Instrument and spacecraft housekeeping data are received in various telemetry packets and these are decoded and archived in the JSOC-SDP for use in later data processing. These data as archived in the SDP are considered the permanent HMI and AIA data archive and are archived onto 800 Gbyte tapes with a duplicate stored at LMSAL.

The data recorded at the JSOC-SDP are labeled as Level 0. Parallel data pipeline computing systems process the HMI and AIA Level 0 data to produce Level 1 and higher data products. The processing steps to produce Level 1 and Level 1.5 AIA data are described below. Level 1.5 includes a correction for plate scale, roll, and telescope coalignment. The data record management system (DRMS) provides a sophisticated and efficient system to retrieve and distribute AIA data and associated metadata to co-investigators and to the solar community.

5.1. Level 1 Data Processing

Level 0 images are stored in the data archive as 16-bit integer 4096×4096 arrays, even if only a portion of the CCD is read by the camera. The data are Rice compressed in order to reduce the ground storage requirements. Processing Level 0 data to Level 1 involves several steps. If the camera readout format includes "over-scan" rows and columns, these are removed. The over-scan rows and columns are not real, but contain the charge remaining after 2048 parallel transfers (for one quarter of the CCD) or 2048 serial transfers, and thus provide a means by which CCD charge transfer efficiency may be monitored on orbit. It is planned to only occasionally include over-scan rows and columns in AIA operations for CCD engineering purposes.

Step 2 is the removal of a dark image to account for the digital offset of the camera, CCD read noise, and dark current. The on-orbit temperature of the CCDs is typically less than $-65°\mathrm{C}$, and thus dark current is negligible for images with exposures of a few seconds or less.

Step 3 applies a flat-field correction to the AIA images. A flat-field image is constructed for each telescope channel to account for detector non-uniformities, gain differences between CCD quadrants, vignetting, and shadowing caused by the focal-plane filter nickel support mesh. The detector non-uniformities are determined from multiple images acquired on orbit with different solar pointing coordinates, so that the resulting flat-field image represents only the instrumental effects. These are routinely updated because the AIA back-illuminated CCD silicon surface is not completely passivated and the CCD charge-collection efficiency drops as charge produced by observed solar emissions is trapped in the device oxide layer. Ground testing has demonstrated that significant restoration of sensitivity is possible by warming the CCD, and this is planned for on-orbit operations. Thus, new flat-field images will always need to be constructed following such CCD warming operations, and the Level 1 processing keeps track of the appropriate flat-field data to use, depending on the epoch of the acquired data that is being processed.

Step 4 makes corrections to individual pixels according to two different algorithms. The first algorithm corrects permanently "bad" pixels that do not respond correctly to light. Each CCD has a small number of such pixels, much less than 0.1%, and their values are replaced with interpolated values from neighboring pixels. As the SDO mission progresses, we expect that there will be additional permanently bad pixels. Thus, the processing pipeline will maintain a time-dependent log of bad pixels for each CCD so that older data can be reprocessed correctly using the appropriate bad pixel list for the epoch of the acquired data. The second algorithm seeks to remove the "spikes" that appear as the result of the interaction of energetic particles that either deposit energy directly in the CCD or interact with the instrument structure and give rise to high-energy photons that are detected in the CCD. The spikes typically look like isolated features that can be as small as one pixel or may in some cases leave a streak of deposited energy in many adjacent pixels. The AIA "despike" algorithm was adopted from one used by the TRACE program. Spikes are detected by comparing the value of each pixel with that of neighboring pixels and flagging occurrences where the differences are not statistically consistent with what is expected given the telescope point-response function. Once the "spikes" are found, they are replaced with the median value of neighboring pixels. The addresses of these pixels and their original values are stored separately so that the algorithm can be easily reversed. For example, for the study of very faint features or for very compact flares it may be desirable to omit the despiking algorithm, as the procedure may introduce undesirable artifacts in the final images. The approach adopted for the pipeline processing makes it possible to "undo" the despiking step.

In Step 5 the AIA images are flipped so as to put solar North at the top of the array. The metadata are updated to add the required information about the image and a small amount of spacecraft housekeeping data is included, the most important being the roll angle. The attitude information is consistent with the world coordinate system (Thompson, 2006). The camera gain (e^{-} DN^{-1}) and the effective area (Boerner *et al.*, 2011) are included with the metadata.

Much of the information used to process AIA data to Level 1 depends on calibration data or algorithms that are time dependent. These include the camera gain (e^{-} DN^{-1}), effective area, pointing information, bad-pixel list, flat field, and average band pass effective area. Information is included with the metadata to identify the algorithm or calibration data used to produce the Level 1 data products. As newer calibration data becomes available, it is possible to reprocess the Level 1 data using the most recently available calibrations.

Level 1 data are typically exported from the JSOC as FITS files with the data represented as 32-bit floating values. If it is desired to reverse the despiking procedure it is also necessary to request the list of corresponding despike pixels addresses and their original data values.

The Level 1 processing produces calibrated AIA data in all respects except for the deconvolution of the instrument point spread function, which is made up of the contributions from the mirrors and the CCD, and from the diffraction caused by the entrance filters. As described in Section 3.1, the entrance filters are either aluminum or zirconium layers supported by a 70 line-per-inch nickel square-pattern mesh. EUV light passing through these filters is diffracted by the mesh. The pattern of the diffraction is best seen when the intensity is high, such as for a flare, and has been observed with TRACE (Lin, Nightingale, and Tarbell, 2001). The diffraction pattern depends on the wavelength of the light, its intensity, the spacing of the mesh bars, the thickness of the bars, and the rotation angle of the mesh with respect to the focal plane. Each EUV channel has two filters in its aperture. The filters are oriented so that the mesh is rotated approximately 40° and 50°, respectively, relative to the CCD rows and columns. So the diffraction patterns can be separated for each filter and deconvolved from the image, since the diffraction patterns themselves are well known. The omission of this step primarily effects the presentation of flare images that have bright features.

5.2. Level 1.5 Data Processing

The processing of Level 1 data is sufficient to produce calibrated AIA images. However, a key advantage of AIA over previous missions is high temporal cadence, full-solar-disk images in multiple wavelength channels. Thus, it is important to be able to display the AIA images as a movie sequence. Level 1.5 processing accomplishes three things that enable the display of AIA movies.

Level 1.5 processing starts with Level 1 data and applies three corrections in a single processing step. The images are rotated so that solar North is at 0° (at the top of the image array). The four AIA telescopes were aligned prior to launch but no further adjustment is available on orbit, so a small residual roll angle exists between the four AIA telescopes. This residual roll is removed and the spacecraft star-tracker signal is used to place solar North at 0°. The second correction has to do with the plate-scale size of the telescopes. Each telescope has a slightly different focal length; as a result, the number of arcsecs per CCD pixel varies slightly between the telescopes. A plate-scale adjustment is made to set the image scale to exactly 0.6 arcsecs per pixel. If it becomes necessary to refocus a telescope during the mission, the plate-scale calibration will be repeated to ensure that the Level 1.5 processing produces the correct plate scale. Finally, as with the roll angle, the bore-sight pointing of each telescope was coaligned prior to launch to within 20 arcsec. The post-launch measurements indicate that the alignment was not significantly changed by the launch activities. The four telescope bore-sights are coaligned by adjusting on orbit the secondary-mirror offsets. Residual differences are removed during ground processing by interpolating the images onto a new pixel coordinate that places the Sun's center at the middle of each telescope detector. The data processing pipeline implements a damped bi-cubic interpolation algorithm (Park and Schowengerdt, 1983) to make plate-scale, roll, and alignment adjustments. Intensities are rounded off and saved as 16-bit integer values. Those who wish to preserve fractional intensity values may optionally process Level 1 with routines available through SolarSoft (Freeland and Bentley, 2000).

As with the parameters used for the Level 1 processing, small time-dependent changes in the plate-scale and alignment parameters are anticipated during the life of the mission, mostly due to orbital effects that cause variations in the distance to the Sun and thermal variations of the telescopes. Calibration data will be regularly acquired and the Level 1.5 processing will access the calibration parameters appropriate to the epoch of data that is processed.

Along with the processing actions described above, the associated metadata is updated to reflect the final Level 1.5 calibrated state. Because of the nature of these processing steps, the operation of processing Level 1 to Level 1.5 is not reversible. When exported as FITS files, the file header structure for Level 1 and Level 1.5 have exactly the same keywords but with values set to reflect the level of processing. HMI magnetograms are provided along with AIA Level 1.5 data. The magnetograms are rescaled down to the AIA plate scale of 0.6 arcsec (which reduces the HMI resolution slightly) and the roll and alignment are adjusted to be the same as AIA.

Table 5 Useful links for AIA data, events, and analysis documentation.

Item	URL
Analysis user's guide	www.lmsal.com/sdodocs/public_docs/sdo_analysis_guide.html
General documentation	www.lmsal.com/sdodocs
Recent AIA Images	sdowww.lmsal.com/suntoday
Sungate portal	www.lmsal.com/sungate

5.3. Data Distribution

The processed AIA data are made freely available for public access approximately 48 hours after receipt at the JSOC. The large volume of Level 1 and 1.5 data products are transferred to the Smithsonian Astrophysical Observatory to act as a data-distribution center for scientists around the world. Many researchers will make use of the Virtual Solar Observatory (VSO), which has developed a data-distribution architecture that is consistent with the JSOC. The Heliophysics Event Knowledgebase (Hurlburt *et al.*, 2011) will aid users in their search of the SDO event database and also facilitate efficient data retrieval. Additional ancillary databases are made available in the SolarSoft database distribution. Further information about SDO data and database distribution systems are available at the links provided in Table 5.

6. Conclusion

The AIA doors were opened and first images were acquired on 27 March 2010. Since then AIA has returned TRACE-like high-resolution images, but with a field of view that covers the entire solar disk in seven EUV wavelength channels and with an unprecedented cadence. AIA movies reveal a fascinating level of structure and variability in the solar atmosphere. The full-disk, high-resolution observations reveal that even minor flare and filament eruptions and disturbances are related to global magnetic restructuring.

In addition to the seven EUV channels, AIA also observes in two UV channels near 1600 and 1700 Å. A visible-light channel provides a means for coalignment with other observatories. All four telescopes are well coaligned with each other and their individual image-stabilization systems are working as expected, to eliminate jitter produced by the spacecraft. Except for two periods of the year near the Vernal and Autumnal Equinoxes, AIA is able to observe almost continuously (more than 95% of the time) due to the dedicated ground station in White Sands, New Mexico. The regular synoptic observing sequence can be adjusted, but is typically operated a cadence of 10 or 12 seconds. The flight software has flexibility to enable seasonal and long-term refinements to the observing program.

After nearly a year on orbit the spacecraft and the AIA are functionally extremely well. There is an excellent probability that the AIA performance will exceed its five-year prime mission lifetime. All of the SDO data are available to the public without restriction.

Acknowledgements The effort required to build an instrument such as AIA requires a large, skillful, and dedicated team. We wish to acknowledge many individuals who contributed to the success of the AIA: Robert Batista, Roger Chevalier, Dustin Cram, Cliff Evans, Scott Gibb, Dwana Kacensky, Robert Honeycut, Bruce Imai, Alex Price, Lawrence Shing, Edgar Thomas, Shanti Varaich, Ross Yamamoto, Kent Zickhur (Lockheed Martin), Gerald Austin (Smithsonian Astrophysical Observatory), David McKenzie (Montana State Univ.), Eberhard Spiller, Jeff C. Robinson, Sherry L. Baker (Lawrence Livermore National Lab.), Travis Ayers, Heidi

Lopez, Forbes Powell, (Luxel Corp.), and Tom Anderson, Elizabeth Citrin, Julie Lander, Chad Salo, and Mike Scott (NASA Goddard Space Flight Center). Lawrence Livermore National Laboratory's efforts are partially supported by the U.S. Department of Energy under contract DE-AC52-07NA27344. This work is supported by NASA under contract NNG04EA00C and the Lockheed Martin Independent Research Program.

References

Boerner, P., Edwards, C., Lemen, J., Rausch, A., Schrijver, C., Shine, R., Shing, L., Stern, R.A., Tarbell, T., Title, A., Wolfson, C.J., Gullikson, E., Soufli, R., Spiller, E., McKenzie, D., Windt, D., Golub, L., Podgorski, W., Testa, P., Weber, M.: 2011, *Solar Phys.*, accepted.

Cheimets, P., Caldwell, D.C., Chou, C., Gates, R., Lemen, J., Podgorski, W.A., Wolfson, C.J., Wuelser, J.-P.: 2009, *Proc. SPIE* **7483**, 36.

Delaboudinière, J.-P., Artzner, G.E., Brunaud, J., Gabriel, A.H., Hochedez, J.F., Millier, F., Song, X.Y., Au, B., Dere, K.P., Howard, R.A., Kreplin, R., Michels, D.J., Moses, J.D., Defise, J.M., Jamar, C., Rochus, P., Chauvineau, J.P., Marioge, J.P., Catura, R.C., Lemen, J.R., Shing, L., Stern, R.A., Gurman, J.B., Neupert, W.M., Maucherat, A., Clette, F., Cugnon, P., van Dessel, E.L.: 1995, *Solar Phys.* **162**, 291.

Dere, K.P., Landi, E., Mason, H.E., Monsignori Fossi, B.C., Young, P.R.: 1997, *Astron. Astrophys. Suppl.* **125**, 149.

Dere, K.P., Landi, E., Young, P.R., Del Zanna, G., Landini, M., Mason, H.E.: 2009, *Astron. Astrophys.* **498**, 915.

Freeland, S.L., Bentley, R.D.: 2000, *Encyc. Astron. Astrophys.*, IOP Publishing, Bristol. doi:10.1888/0333750888/3390.

Golub, L., DeLuca, E., Austin, G., Bookbinder, J., Caldwell, D., Cheimets, P., *et al.*: 2007, *Solar Phys.* **243**, 63.

Gullikson, E.M., Mrowka, S., Kaufmann, B.B.: 2001, *Proc. SPIE* **4343**, 363.

Handy, B.N., Acton, L.W., Kankelborg, C.C., Wolfson, C.J., Akin, D.J., Bruner, M.E., Caravalho, R., Catura, R.C., Chevalier, R., Duncan, D.W., Edwards, C.G., Feinstein, C.N., Freeland, S.L., Friedlaender, F.M., Hoffmann, C.H., Hurlburt, N.E., Jurcevich, B.K., Katz, N.L., Kelly, G.A., Lemen, J.R., Levay, M., Lindgren, R.W., Mathur, D.P., Meyer, S.B., Morrison, S.J., Morrison, M.D., Nightingale, R.W., Pope, T.P., Rehse, R.A., Schrijver, C.J., Shine, R.A., Shing, L., Strong, K.T., Tarbell, T.D., Title, A.M., Torgerson, D.D., Golub, L., Bookbinder, J.A., Caldwell, D., Cheimets, P.N., Davis, W.N., Deluca, E.E., McMullen, R.A., Warren, H.P., Amato, D., Fisher, R., Maldonado, H., Parkinson, C.: 1999, *Solar Phys.* **187**, 229.

Howard, R.A., Moses, J.D., Vourlidas, A., Newmark, J.S., Socker, D.G., Plunkett, S.P., Korendyke, C.M., Cook, J.W., Hurley, A., Davila, J.M., Thompson, W.T., St Cyr, O.C., Mentzell, E., Mehalick, K., Lemen, J.R., Wuelser, J.P., Duncan, D.W., Tarbell, T.D., Wolfson, C.J., Moore, A., Harrison, R.A., Waltham, N.R., Lang, J., Davis, C.J., Eyles, C.J., Mapson-Menard, H., Simnett, G.M., Halain, J.P., Defise, J.M., Mazy, E., Rochus, P., Mercier, R., Ravet, M.F., Delmotte, F., Auchère, F., Delaboudinière, J.P., Bothmer, V., Deutsch, W., Wang, D., Rich, N., Cooper, S., Stephens, V., Maahs, G., Baugh, R., McMullin, D., Carter, T.: 2008, *Space Sci. Rev.* **136**, 67.

Hurlburt, N., Cheung, M., Schrijver, C., Chang, L., Freeland, S., Green, S., Heck, C., Jaffey, A., Kobashi, A., Schiff, D., Serafin, J., Seguin, R., Slater, G., Somani, A., Timmons, R.: 2011, *Solar Phys.*, doi:10.1007/s11207-010-9624-2.

Lin, A.C., Nightingale, R.W., Tarbell, T.D.: 2001, *Solar Phys.* **198**, 385.

Park, S.K., Schowengerdt, R.A.: 1983, *Comput. Vis. Graph. Image Process.* **23**, 258.

Podgorski, W.A., Cheimets, P.N., Boerner, P., Glenn, P.: 2009, *Proc. SPIE* **7438**, 13.

Schou, J., Scherrer, P.H., Bush, R.I., Wachter, R., Couvidat, S., Rabello-Soares, M.C., *et al.*: 2011, *Solar Phys.*, in preparation.

Soufli, R., Windt, D.L., Robinson, J.C., Baker, S.L., Spiller, E.A., Dollar, F.J., Aquila, A.L., Gullikson, E.M., Kjornrattanawanich, B., Seely, J.F., Golub, L.: 2005, *Proc. SPIE* **5901**, 59010M.

Soufli, R., Baker, S.L., Windt, D.L., Gullikson, E.M., Robinson, J.C., Podgorski, W.A., Golub, L.: 2007, *Appl. Opt.* **46**, 3156.

Thompson, W.T.: 2006, *Astron. Astrophys.* **449**, 791.

Thompson, W.T., Brekke, P.: 2000, *Solar Phys.* **195**, 45.

Underwood, J.H., Gullikson, E.M., Koike, M., Batson, P.J.: 1997, *Proc. SPIE* **3113**, 214.

Waltham, N., Beardsley, S., Clapp, M., Lang, J., Jerram, P., Pool, P., Auker, G., Morris, D., Duncan, D.: 2011, In: *Internat. Conf. on Space Optics*, in press.

Woods, T.N., Eparvier, F.G., Hock, R., Jones, A.R., Woodraska, D., Judge, D., Didkovsky, L., Lean, J., Mariska, J., Warren, H., McMullin, D., Chamberlin, P., Berthiaume, G., Bailey, S., Fuller-Rowell, T., Sojka, J., Tobiska, W.K., Viereck, R.: 2011, *Solar Phys.*, doi:10.1007/s11207-009-9487-6.

Solar Phys (2012) 275:41–66
DOI 10.1007/s11207-011-9804-8

Initial Calibration of the *Atmospheric Imaging Assembly* (AIA) on the *Solar Dynamics Observatory* (SDO)

Paul Boerner · Christopher Edwards · James Lemen · Adam Rausch · Carolus Schrijver · Richard Shine · Lawrence Shing · Robert Stern · Theodore Tarbell · Alan Title · C. Jacob Wolfson · Regina Soufli · Eberhard Spiller · Eric Gullikson · David McKenzie · David Windt · Leon Golub · William Podgorski · Paola Testa · Mark Weber

Received: 8 September 2010 / Accepted: 24 May 2011 / Published online: 22 July 2011

Abstract The *Atmospheric Imaging Assembly* (AIA) instrument onboard the *Solar Dynamics Observatory* (SDO) is an array of four normal-incidence reflecting telescopes that image the Sun in ten EUV and UV wavelength channels. We present the initial photometric calibration of AIA, based on preflight measurements of the response of the telescope components. The estimated accuracy is of order 25%, which is consistent with the results of comparisons with full-disk irradiance measurements and spectral models. We also describe the characterization of the instrument performance, including image resolution, alignment, camera-system gain, flat-fielding, and data compression.

Keywords Instrumentation · EUV · Soft X-ray · Chromosphere · Corona · Transition region

The Solar Dynamics Observatory
Guest Editors: W. Dean Pesnell, Phillip C. Chamberlin, and Barbara J. Thompson

P. Boerner (✉) · C. Edwards · J. Lemen · A. Rausch · C. Schrijver · R. Shine · L. Shing · R. Stern · T. Tarbell · A. Title · C.J. Wolfson
Solar and Astrophysics Laboratory, Dept ADBS Bldg 252, Lockheed Martin Advanced Technology Center, 3251 Hanover St., Palo Alto, CA 94304, USA
e-mail: boerner@lmsal.com

R. Soufli · E. Spiller
Lawrence Livermore National Laboratory, Livermore, CA, USA

E. Gullikson
Lawrence Berkeley National Laboratory, Berkeley, CA, USA

D. McKenzie
Montana State University, Bozeman, MT, USA

D. Windt
Reflective X-ray Optics LLC, New York, NY, USA

L. Golub · W. Podgorski · P. Testa · M. Weber
Smithsonian Astrophysical Observatory, Cambridge, MA, USA

1. Introduction

The *Atmospheric Imaging Assembly* (AIA) instrument onboard the *Solar Dynamics Observatory* (SDO) explores the structure and dynamics of the solar atmosphere with unprecedented cadence and coverage, both spatial and spectral. It has operated essentially continuously since 28 April 2010. AIA is described in detail in Lemen *et al.* (2011); the data processing approach is described by Hurlburt *et al.* (2011). It consists of an array of four dual-channel normal-incidence telescopes, which observe an 41-arcmin field of view in seven EUV and three UV–visible-light channels with 0.6-arcsec pixels on 4096 × 4096 CCDs. In the normal observing mode, each camera records an image every 5 – 6 seconds, so that each of the EUV channels is imaged every 10 – 12 seconds (the UV/visible channels are imaged at a lower cadence). In this article, we describe the initial calibration of AIA, including the photometric calibration that is required to perform quantitative analysis of the AIA data, and characterization of the instrument parameters that affect the data products.

2. Photometric Calibration

The basic AIA data product consists of an image of pixel values $p_i(\mathbf{x})$, where the index i refers to one of the ten wavelength channels and $\mathbf{x}$ refers to a particular position in the image plane. These pixel values are measurements of the solar spectral radiance integrated over the solid angle subtended by the pixel and the wavelength passband of the telescope channel:

$$p_i(\mathbf{x}) = \int_0^\infty \eta_i(\lambda)\,\mathrm{d}\lambda \int_{\text{pixel}\,\mathbf{x}} I(\lambda, \boldsymbol{\theta})\,\mathrm{d}\boldsymbol{\theta}. \tag{1}$$

Here η_i is the efficiency function of the ith channel of the telescope – its passband, in units of digital number [DN] per unit flux at the aperture. It can be calculated as follows:

$$\eta(\lambda, t, \mathbf{x}) = A_{\text{eff}}(\lambda, t)G(\lambda)F(\mathbf{x}),$$

where G is the gain of the CCD-camera system in DN per photon – it combines the standard conversion of photons to detected electrons with the camera gain [g], in DN per electron:

$$G(\lambda) = (12398/\lambda/3.65)g.$$

The field-angle (position in the image) dependent factors, including vignetting, filter grid shadowing, and CCD sensitivity variations, are included in the flat-field function [F].

The effective area [A_{eff}] contains information about the efficiency of the telescope optics, as follows:

$$A_{\text{eff}}(\lambda, t) = A_{\text{geo}} R_{\text{P}}(\lambda) R_{\text{S}}(\lambda) T_{\text{E}}(\lambda) T_{\text{F}}(\lambda) D(\lambda, t) Q(\lambda).$$

The geometrical collecting area [A_{geo}] is multiplied by the reflectance [R] of the primary and secondary mirrors, the transmission efficiency [T] of the entrance and focal-plane filters, the quantum efficiency [Q] of the CCD, and an additional correction [D] to account for the time-varying effect of degradation due to contamination or deterioration of the components.

In practice, calibration of the AIA data proceeds in two directions. Where possible, the data are corrected for instrumental effects, so the vignetting and flat-field variations are divided out of the image data for Level 1 and higher (Hurlburt *et al.*, 2011). However, because the AIA CCDs provide no spectroscopic information, there is no straightforward way to

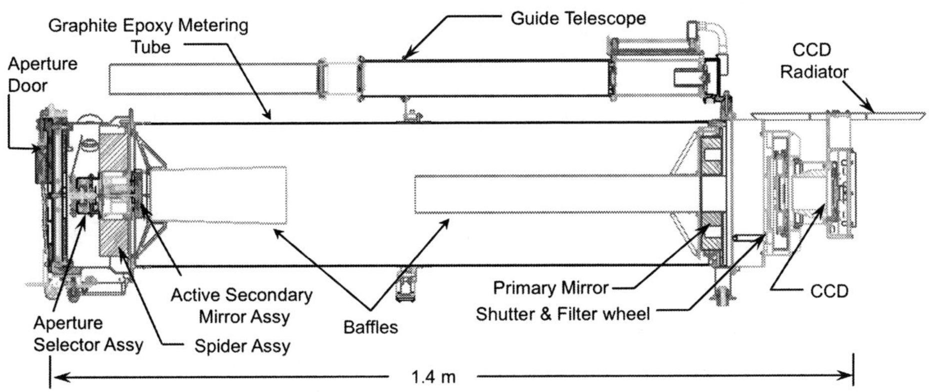

Figure 1 Schematic layout of an AIA telescope assembly.

apply the wavelength-dependent calibration to the data directly. Therefore the instrument-response function $[R(\lambda, t) = A_{\mathrm{eff}}(\lambda, t)G(\lambda)]$ is provided to the data user and can be incorporated into an analysis of the images.

2.1. Measuring the Response Function

Due to the difficulty in obtaining a well-calibrated, narrowband, collimated EUV source, an end-to-end EUV calibration measurement of the AIA response function was not feasible. Instead, the photometric calibration was obtained by making component-level measurements of all the optical elements in the AIA telescopes (mirrors, filters, and CCD), and combining those measurements analytically to produce a model of the system performance. This approach was used in the calibration of TRACE (Handy *et al.*, 1999) and EIT (Dere *et al.*, 2000) as well. The optical elements are shown schematically in Figure 1.

2.1.1. Geometric Area

The AIA telescope assemblies (ATAs) are 20-cm aperture Ritchey–Chrétien optical systems whose apertures have been divided in half by applying different coatings to the top and bottom sectors of the mirrors. For each exposure, one half of the aperture is selected either by blocking the other half with a motorized shutter at the aperture, or by selecting a focal-plane filter that rejects photons at the other channel's wavelength. The coated areas of the mirrors are masked in order to prevent reflections from one half of the primary mirror reaching the opposite half of the secondary. A semi-annular mask on the primary mirror defines the pupil of the system. The geometric collecting area is further limited by obstruction due to the secondary mirror and the structural elements of the telescope spider and entrance-filter supports. The geometric area of each EUV channel is 83.0 cm^2.

The UV channel does not make use of the full semi-annulus of open area in the primary-mirror mask; its aperture is stopped down by a roughly circular mask in front of the UV window. The geometric area of the UV channel is 30.8 cm^2. The filters and mirrors in the UV channel are qualitatively different from the EUV filters and mirrors; therefore, they are discussed in Section 2.1.5 after the discussion of the EUV channel filters and mirrors.

Table 1 Parameters of the models used to fit the measurements of the thin-film metal filters.

Filter	Model
Al thin/Al entrance	1450 Å Al + 87 Å Al_2O_3
Al thick	2450 Å Al + 103 Å Al_2O_3
Zr thin/Zr entrance	2160 Å Zr + 199 Å ZrO_2
Zr thick	3236 Å Zr + 171 Å ZrO_2
Zr on polyimide	4000 Å Poly + 2570 Å Zr + 267 Å ZrO_2

2.1.2. EUV Filters

AIA's EUV channels use thin-film metal filters mounted on an electroformed nickel mesh to reject visible and IR radiation. The mesh transmission is 82 – 84% (this factor is included in the filter transmission and not in the geometric collecting area, although it acts similarly). The filters were manufactured by Luxel Corporation. Most channels use Al filters; the two short-wavelength channels, below the Al L edge (94 Å and 131 Å), use Zr filters instead. All channels have two focal-plane filters available for their use: a thin one (1500 Å Al or 2000 Å Zr), which is used the majority of the time, and a thicker one (2500 Å Al or 3000 Å Zr) which is primarily used as a spare. The thick focal-plane filter for the 131-Å channel has a Zr film on a polyimide substrate (rather than a nickel mesh); this provides enhanced attenuation of the signal in the 131-Å channel (necessary because of the extreme dynamic range of the 131-Å images during flares).

The focal-plane filters were measured using the EUV reflectometer at the Lockheed Martin Solar and Astrophysics Laboratory (LMSAL) (Catura *et al.*, 1987). A Henke X-ray tube or hollow-cathode discharge was used to illuminate a 1-m McPherson grazing-incidence grating monochromator, which selected a single wavelength and sent a pencil beam to illuminate the detector (a proportional counter for $\lambda < 200$ Å, and a microchannel plate cross-calibrated with a National Institute of Standards and Technology (NIST) windowless photodiode for $\lambda > 200$ Å). The ratio of the detected signal with and without the filter in place was measured. The results of the filter calibration measurements were used to constrain a model of the filter composition, which was calculated using optical constants taken from the LLNL CXRO database (Henke, Gullikson, and Davis, 1993). The details of the filter model are described in Table 1; the measured and modeled filter transmittance are shown in Figure 2.

The flight focal-plane filters themselves were not measured, but the filter calibration program included a series of measurements on filter samples and flight-like filters from different processing runs. The scatter in repeated measurements on a single filter, or between measurements on different flight-like filters and test coupons from the same run, was 1% in absolute transmittance. This is slightly larger than the shot noise and is probably due to variation in the source or detector. Measurements on different filter lots with the same prescription agreed to within 1.5%. The focal-plane filter models were also used for the entrance filters, which had the same prescription. The entrance filters were measured for visible-light rejection, but they were not calibrated in-band.

There is some additional uncertainty in applying the model to wavelengths where the optical constants are not well known; in particular, the optical constants for Al change rapidly with wavelength (due to the presence of the Al $L_{2,3}$ absorption edge) in the vicinity of the Fe IX line targeted by the 171-Å channel, so small variations in the assumed Al optical constants can have substantial effects on the predicted intensity of the images in the 171-Å channel. The overall uncertainty in applying the filter model to the transmission of the flight filters is therefore of order 5%.

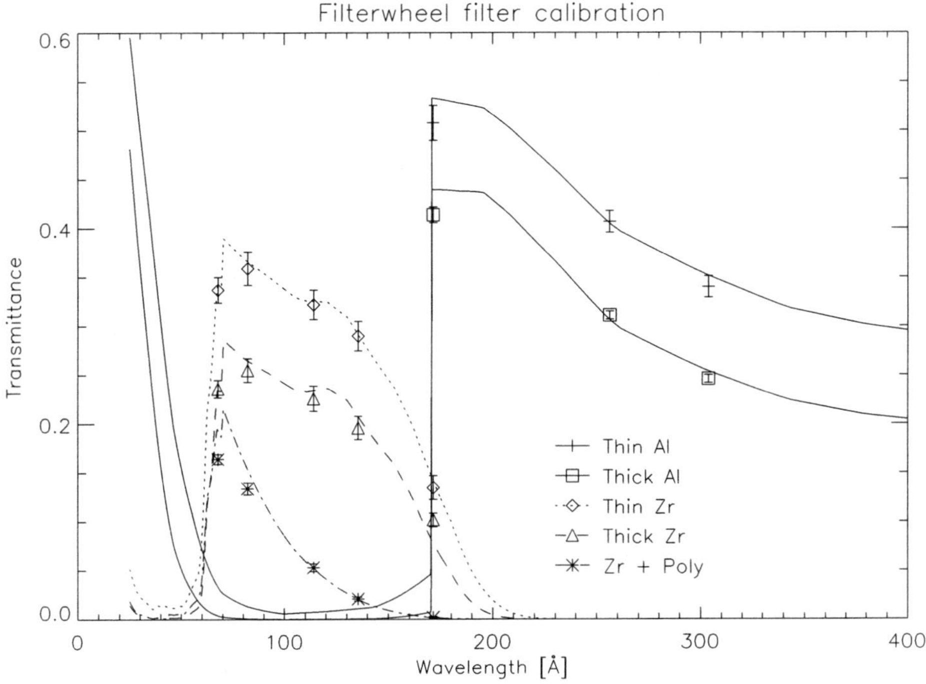

Figure 2 The transmittance of the focal-plane filters, measured at LMSAL, and a model based on optical constants from Henke, Gullikson, and Davis (1993), fit to the measurements. Error bars indicate 1σ scatter in a set of measurements made on different filters and coupons.

2.1.3. Mirrors

The AIA flight-mirror substrates are made of Zerodur™ and were figured and polished independently by two different optics vendors: SAGEM (France) and Tinsley (Richmond, California). The high-spatial-frequency roughness of the mirror substrates, which can significantly affect the EUV reflective performance of the multilayer-coated mirrors, was measured by Atomic Force Microscopy at Lawrence Livermore National Laboratory (LLNL); the results are discussed in Soufli *et al.* (2007). Multilayer coatings are necessary in order to render the mirrors efficient reflectors in the EUV region. The multilayer coating development for the seven EUV AIA channels is discussed in Soufli *et al.* (2005). Multilayer coatings on the AIA flight-mirror substrates were applied at Reflective X-ray Optics (RXO), LLC for the 94 (Mo/Y), 131 (Mo/Si), 304 and 335 (SiC/Si) Å channels and at LLNL for the 171, 193 and 211 (Mo/Si) Å channels. The mirrors were then measured *vs.* wavelength across the peak of their response (in-band reflectance measurements) at beamline 6.3.2. of the Advanced Light Source (ALS) synchrotron at Lawrence Berkeley National Laboratory. In-band and off-band reflectance measurements were also performed at the ALS on AIA multilayer witness samples that were deposited at the same time as the flight-mirror coatings in the wavelength region 43 – 500 Å. These measurements were performed three – four years after the flight coatings were deposited, and the results were used in the analysis of the AIA instrument performance, as discussed later in this section. The in-band and off-band AIA measurements on flight mirrors and witnesses will be described in detail in an upcoming

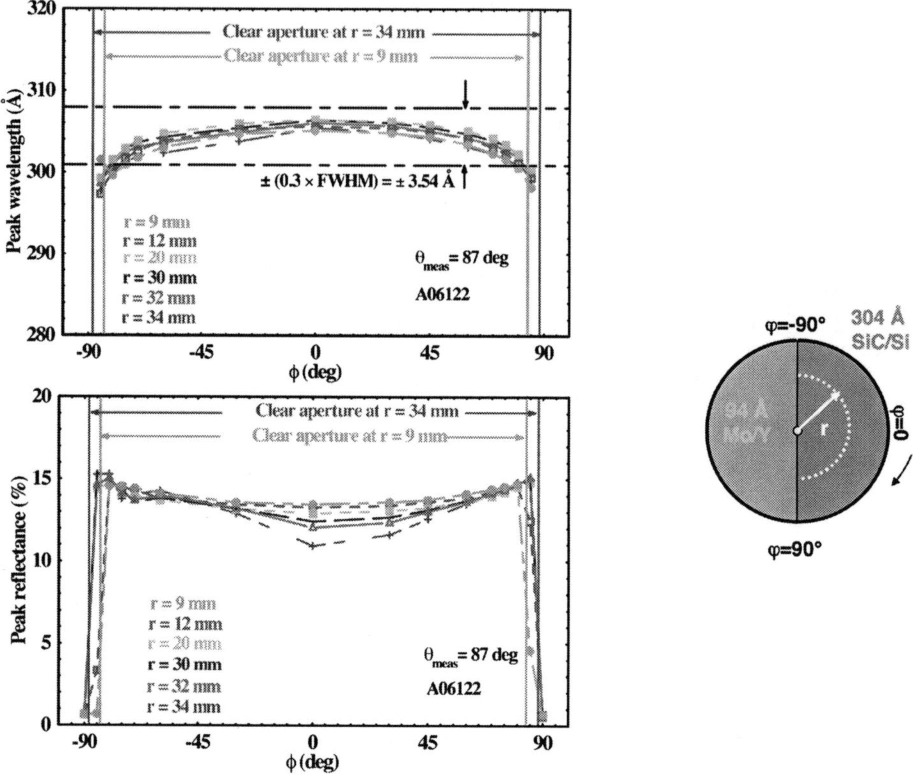

Figure 3 The spatial variation in the peak reflectance and peak wavelength of the SiC–Si multilayer coating on the 304-Å secondary mirror was sampled in detail, and an area-weighted average was constructed. It should be noted that each azimuth [ϕ] scan obtained at radial location [r] corresponds to an area proportional to r^2, per the drawing on the right.

publication. Two examples from these measurements and analysis are shown in Figures 3 and 4.

For the in-band reflectance measurements, each AIA flight mirror was measured at a number of locations within the clear aperture, as shown in Figure 3. At each location, the reflectance was scanned *vs.* wavelength within the bandpass of each EUV channel. These measurements were combined using an area-weighted average (Podgorski *et al.*, 2009) to produce a single composite reflectivity function for each mirror. There is some variation (of order 10%) in the measurements taken at different points on the mirror. Note that this standard deviation reflects a real variation in the multilayer across the mirror's surface and is accurately represented by the weighted average of measurements. The individual measurements that go into the mirror calibration are extremely accurate ($< 0.6\%$ error on the measurement of a given point at a given wavelength). There may be some residual error ($< 5\%$) in the composite reflectivity function due to spatial variations that are not adequately sampled during the mirror calibration. The results of the calibration measurements across a typical mirror are shown in Figure 3.

The other potential error source in the mirror calibration is a change in the mirror performance (due to aging effects such as oxidation of the top layer of the multilayer coating) between the calibration measurement (typically done within a day to roughly one month

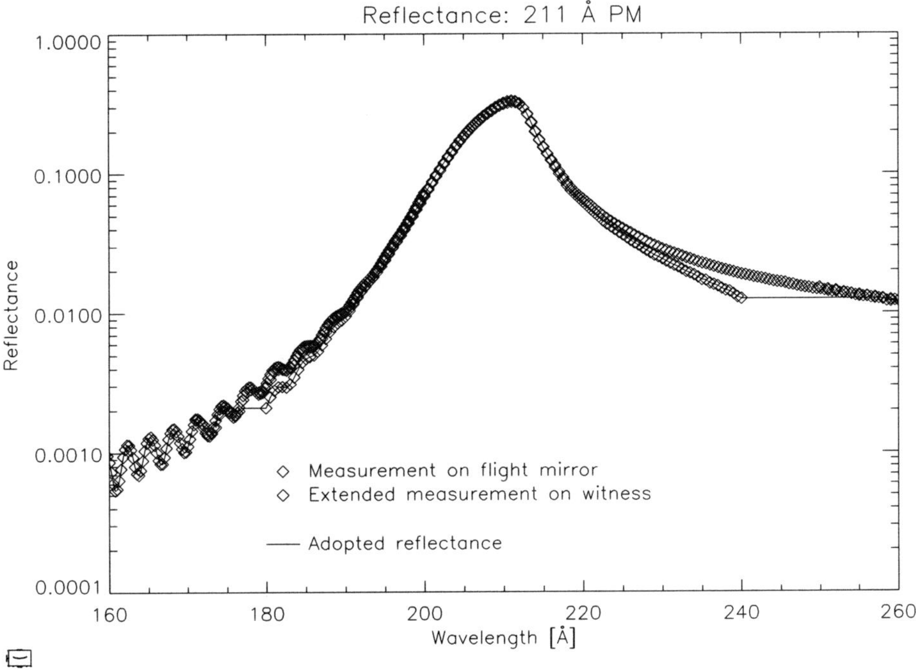

Figure 4 An example of using the full wavelength range measurements from a coating witness to extend the calibration of the flight mirror (in this case, the 211-Å primary).

after the multilayer deposition) and the beginning of on-orbit operations. This change in any case is not expected to exceed 5%, based on trending measurements of coating-witness samples deposited along with the flight mirrors.

The mirrors were sampled on a very fine wavelength grid, so it is not necessary to fit a model to the measurements in order to interpolate the response in wavelength. However, the original measurements of the mirrors did not extend to wavelengths well away from the reflectance peak. In order to sample the wings of the multilayer reflectance, coating-witness samples that were deposited along with each of the flight-mirror coatings were re-measured at the ALS in 2009 over the wavelength range from 43 – 500 Å. The results were combined with the area-weighted average measurements of the core of the mirror-reflectance curves, as depicted in Figure 4. The flight-mirror measurements were used in the wavelength range where they exist; for extended wavelengths, the calibration adopts the lesser of the nearest flight measurement or the witness measurement. The accuracy of the extended-wavelength measurements on flight witness samples (when applied to the flight mirrors) is not as good as that of the measurements on the flight mirrors themselves, but in general the impact of this portion of the multilayer response (or the portion outside of the range of the witness-sample measurements) on the spectral content of the channel's images is not significant. The one notable exception is the second-order peak of the multilayers for the 335-Å channel, which falls at 183.5 Å. This is discussed further in Section 2.2.1.

Figure 5 shows the results of the mirror calibration measurements for the AIA EUV channels.

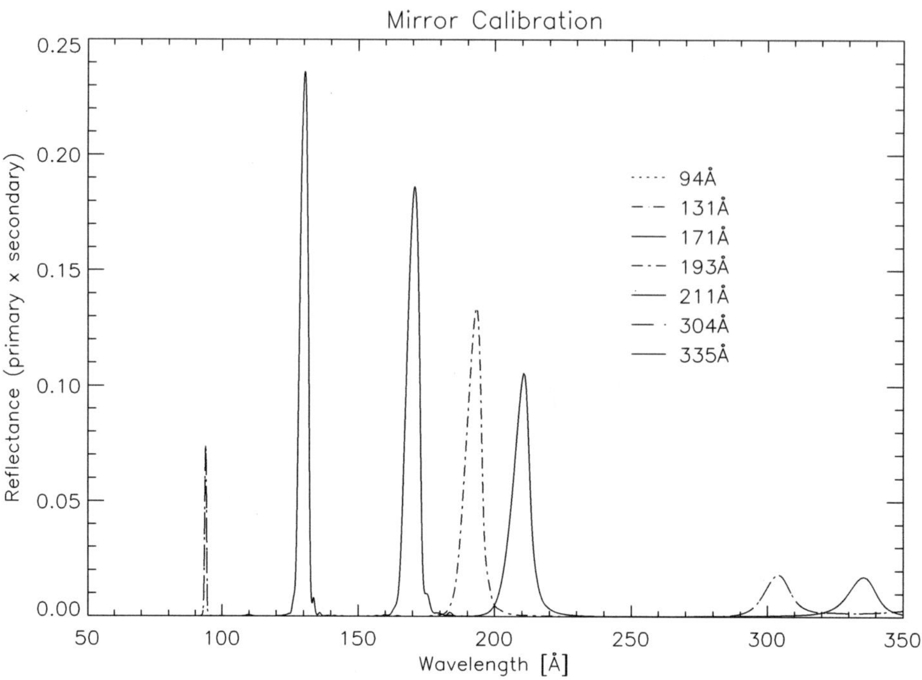

Figure 5 The combined reflectance of the primary and secondary mirrors for the EUV channels.

2.1.4. CCDs

Each AIA telescope uses a 16-megapixel back-illuminated, ion-implanted, and laser-annealed CCD manufactured by e2v Technologies. The CCDs are divided into four 2048 × 2048 quadrants, which are read out separately and simultaneously at a 2 Mpix sec^{-1} rate with camera electronics made by the Rutherford Appleton Laboratory (RAL). A discussion of the CCD and camera electronics design and fabrication is given by Waltham *et al.* (2011). Prior to mounting in the AIA telescopes, the CCDs were screened and calibrated using the LMSAL XUV Calibration facility (Windt and Catura, 1988; Stern *et al.*, 2004) and an ARC Gen II camera system.

Each flight CCD was illuminated in each quadrant at a number of discrete EUV, soft X-ray, and UV wavelengths, and the intensity of the incoming beam was measured using a proportional counter or a microchannel plate (MCP) cross-calibrated with NIST diodes. Unlike the measurements of the filters and mirrors, the CCD calibration measurements rely on the absolute calibration of the reference detector used in the experiment. Variations in the accuracy of the beam normalization due to changes or errors in the calibration of the microchannel plate or proportional counter are the dominant source of uncertainty in the QE measurement (< 5% for the proportional counter, and around 10% for the cross-calibrated MCP). After accounting for the effect of the CCD-camera system gain (see below), the measurements were used to constrain a model of the detector quantum efficiency (Stern *et al.*, 1994, 2004). The measurements of the four flight CCDs all agreed to within 5 – 6%, less than their error bars, so the mean CCD QEs were fit to the same CCD model, which was used for all channels. The results of the CCD calibration are shown in Figure 6. The model shows excellent agreement with the measurements over the range of the AIA EUV channels

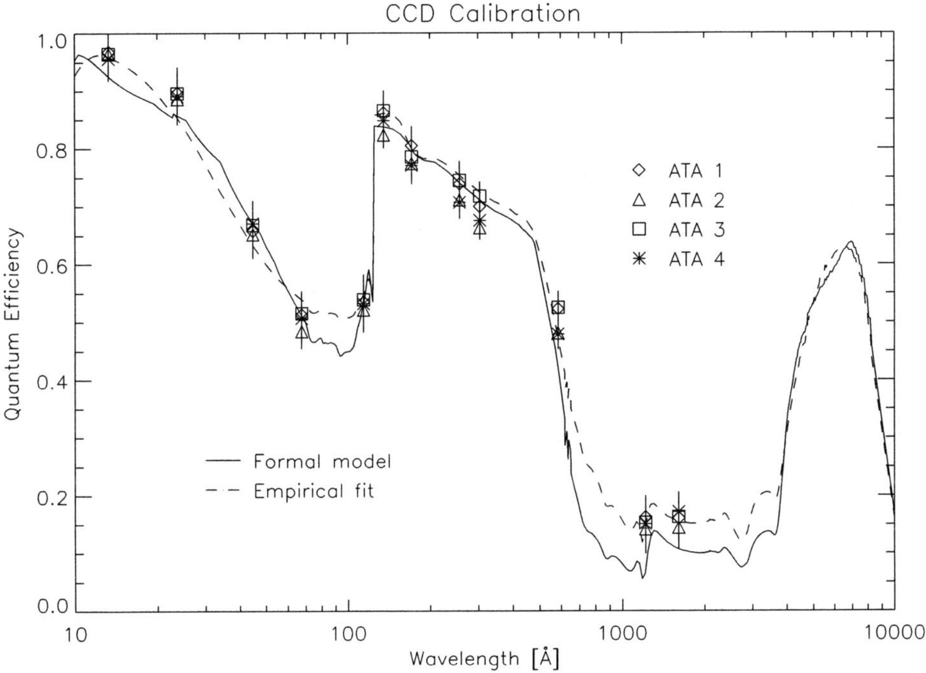

Figure 6 The quantum efficiency of the CCDs measured at LMSAL, and a model fit to the measurements. Agreement between the measurements of the different flight devices and the model is excellent in the EUV range.

and additional data taken in the soft X-ray range; the agreement is not as good for the UV wavelength range. The uncertainties in both the optical constants that go into the model and the QE measurements are larger in the UV range. Therefore, instead of the formal model of the QE that was used for the EUV channels based upon an estimated implant distribution (see Stern, Shing, and Blouke (1994) for the description of the formal and empirical model approaches), we used an empirical fit to the CCD measurements in the calibration of the UV channels. In addition, the CCD response is flat over the bandpass of the AIA UV channels, so the accuracy of the CCD calibration for the UV channels can likely be improved by on-orbit cross-calibration with instruments such as TIMED/SEE (Woods *et al.*, 2005).

2.1.5. UV Optics

Like the EUV channels, the UV channel uses half of the aperture of one of the telescopes, but it can record images in three distinct wavelength channels – 1600 Å, 1700 Å, and 4500 Å – using different focal-plane filters. The mirror coatings for the UV channel are Al with a MgF_2 overcoating, optimized for reflectance at 1600 Å. The mirror coatings were deposited by Acton Research Optics (ARO). The UV channel mirror reflectance was not measured; a model for the UV channel reflectance was calculated using data and code from the internet materials database (IMD: Windt, 1998). The model reflectance agrees well with measurements of similar coatings performed by ARO.

Instead of thin-film filters at the entrance and in the filterwheel, the UV channel uses a MgF_2 window with a metal–dielectric bandpass coating to limit visible – IR throughput and select a narrow range of UV wavelengths. The 1600-Å channel uses a similar MgF_2 optic in the filter wheel to further narrow the bandpass. The 1700-Å channel has a fused-silica filter with metal–dielectric coatings in the filter wheel, while the 4500-Å channel uses a more conventional three-cavity bandpass filter. The UV window and focal-plane filters were made by ARO; the 4500 Å focal-plane filter was made by Andover Corporation. While the UV window and 4500 Å focal-plane filter are plano–plano optics, the 1600-Å and 1700-Å focal-plane filters each has a convex surface in order to keep them parfocal with the 171-Å channel that shares their telescope assembly.

The UV window and filters were calibrated using an ARO FUV spectrophotometer. A Cary spectrophotometer was used to measure the out-of-band (NUV–visible) transmittance.

2.1.6. Contamination

AIA was kept in a clean environment throughout integration and test; cleanliness was strictly maintained even during spacecraft-level integration due to the sensitivity of both AIA and SDO's *Extreme Ultraviolet Variability Experiment* (EVE) to molecular contamination. Nevertheless, it is inevitable that some amount of contamination will accumulate on the optics and (especially) on the detector during instrument operations, and our throughput model reflects this.

A venting analysis of the telescope system suggested that a total effective path length of a few hundred Ångstroms of volatile molecules would likely accumulate in the AIA optical system. We used IMD to model the EUV throughput of a typical contaminant, Triphenyl Phosphate ($C_{18}H_{15}O_4P$), at a thickness of 275 Å, and included this in our preflight estimate of the system throughput. The contaminant transmittance curve is shown in Figure 7. Note that it has the strongest effect in the longer-wavelength channels, especially 304 Å and 335 Å.

There were no preflight measurements that could constrain the thickness of the contaminant layer, so the uncertainty on the contamination function is as large as the correction itself. We expect to adjust this parameter in order to achieve better agreement with reference spectra from EVE and other instruments during on-orbit cross-calibration, and may use the thickness of the contaminant layer as a parameter to optimize the ongoing cross-calibration.

The throughput of contaminants at UV and visible wavelengths is more difficult to model analytically; therefore we did not include an allocation for contamination in the initial calibration functions for the UV channels. Corrections for the effect of contamination will be done empirically by cross-calibration with other instruments.

2.2. Wavelength Response

Table 2 illustrates how the end-to-end wavelength-response functions $[R(\lambda)]$ are calculated from the component-level measurements by showing the efficiency of each component and the total instrument response at the wavelength of the strongest solar emission line in each channel. The effective areas for the seven EUV channels as a function of the wavelength are plotted in Figure 8. The effective areas for the three UV–Vis channels as a function of the wavelength are shown in Figure 9.

2.2.1. Channel Crosstalk

For three of the four AIA telescopes, both channels (both halves of the aperture) are illuminated at all times, and the active channel is selected using the focal-plane filters. (The

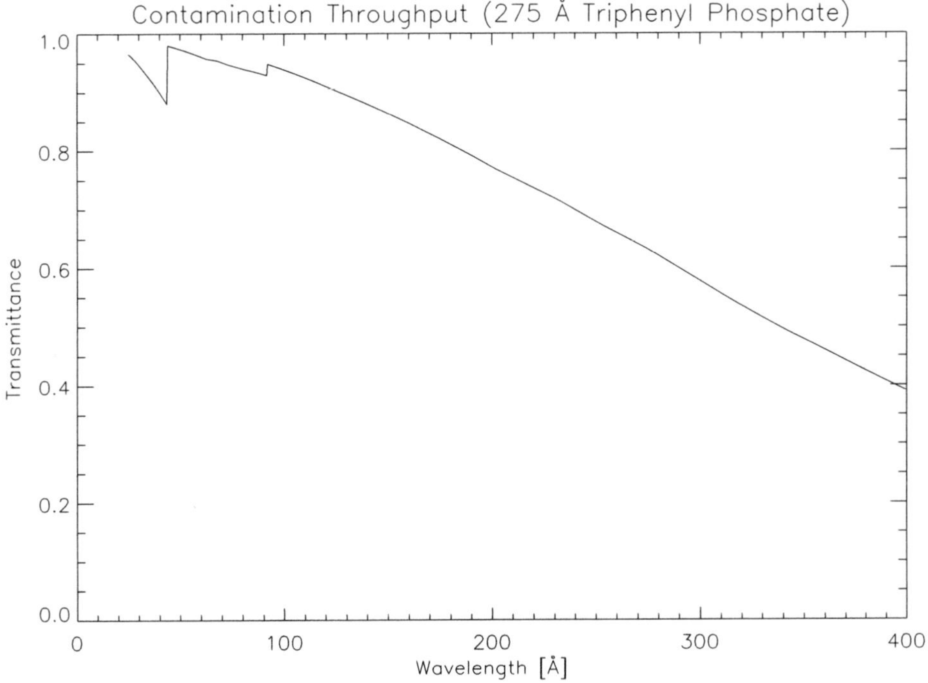

Figure 7 Transmittance for the model contamination layer assumed in the preflight calibration.

exception is telescope 2, which includes the 193-Å and 211-Å channels, where a mechanical aperture selector is used to block half of the aperture in order to select the active channel for an image.) While the wavelength channels were grouped to minimize crosstalk, the focal-plane filters on telescopes 1 (which includes the 131- and 335-Å channels) and 4 (94 and 304 Å) do not perfectly reject light from the opposite channel, so there is some crosstalk between the wavelength channels.

This is most significant for the 335-Å channel: it has a relatively low reflectivity, and is paired with the highly reflective 131-Å channel. The effective area of the 335-Å channel (using thin and thick focal-plane filters) is shown over a wide wavelength range in Figure 10. The contribution of the 131-Å coatings on the opposite half of the telescope aperture, as well as the higher-order peak from the 335-Å multilayer mirror discussed above, are evident. We have estimated the relative contribution of the 131-Å coatings to the 335-Å channel images by folding solar spectra through the effective-area function. This analysis demonstrates that, while the 131-Å channel generally accounts for less than 2% of the total counts in the 335-Å images in quiet Sun or active regions, it can contribute up to 40% during flares (when the hot lines around 131 Å are enhanced substantially compared to the 335-Å Fe XVI line). The crosstalk can be minimized (at the expense of throughput) by using the thick focal-plane filter on that channel. However, given the low level of solar activity thus far in this solar cycle, we have continued to primarily use the thin filter.

2.3. Error Budget

An estimate of the error in the photometric calibration can be calculated by adding the fractional errors on each component-level calibration in quadrature (see Table 3). The calibration

Table 2 The components of the wavelength-response function at the strongest line in each channel.

Component	Wavelength of strongest line [Å]						
	93.9	131.2	171.1	195.1	211.3	303.8	335.4
A_{geo} [cm^2]				83.0			
T_E	0.348	0.306	0.533	0.523	0.497	0.352	0.324
R_P	0.241	0.505	0.424	0.283	0.331	0.117	0.117
R_S	0.308	0.399	0.434	0.303	0.305	0.129	0.125
T_F	0.348	0.306	0.533	0.523	0.497	0.352	0.324
Q	0.442	0.838	0.801	0.779	0.774	0.712	0.696
D	0.946	0.893	0.827	0.782	0.752	0.569	0.504
A_{eff}	0.312	1.172	2.881	1.188	1.206	0.063	0.045
G [$\mathrm{DN\,phot^{-1}}$]	2.128	1.523	1.168	1.024	0.946	0.658	0.596
$R(\lambda)$ [$\mathrm{cm^2\,DN\,phot^{-1}}$]	0.664	1.785	3.365	1.217	1.14	0.041	0.027

Component	Nominal bandpass center [Å]		
	1600	1700	4500
A_{geo} [cm^2]		30.8	
T_E	0.114	0.053	3.80×10^{-5}
R_P	0.9	0.904	0.884
R_S	0.9	0.904	0.884
T_F	0.043	0.191	0.705
Q	0.155	0.153	0.442
A_{eff}	1.92×10^{-2}	3.89×10^{-2}	2.83×10^{-4}
G [$\mathrm{DN\,phot^{-1}}$]	0.125	0.118	0.044
$R(\lambda)$ [$\mathrm{cm^2\,DN\,phot^{-1}}$]	2.4×10^{-3}	4.6×10^{-3}	1.3×10^{-5}

error for each component is the estimated 1σ difference between the value of the component efficiency that has been adopted into the calibration model at a given wavelength and the true efficiency of the component at that wavelength. (Note that the errors in Table 3 are approximate, and do not include subtleties such as the fact that certain errors affect some channels or some wavelengths more than others.) The largest terms are the uncertainty in the contamination thickness and the CCD-QE determination. The overall accuracy of the preflight calibration is estimated to be of order 25%.

2.4. Cross-Calibration with EVE

EVE (Woods *et al.*, 2010) measures the solar spectral irradiance at 1 Å spectral resolution over much more than the full wavelength range of AIA's EUV channels, and at a very similar cadence to AIA. EVE expects to maintain an absolute calibration accuracy $\approx 20\%$ by cross-calibrating with a series of sounding-rocket underflights; the first rocket underflight went off successfully on 3 May 2010. AIA will take advantage of EVE's excellent absolute calibration to refine our estimate of the instrument-response functions.

A rather crude cross-calibration can be obtained by integrating the EVE irradiance measurements [$I(\lambda)$: $\mathrm{phot\,cm^{-2}\,s^{-1}\,Å^{-1}}$] times the AIA response functions

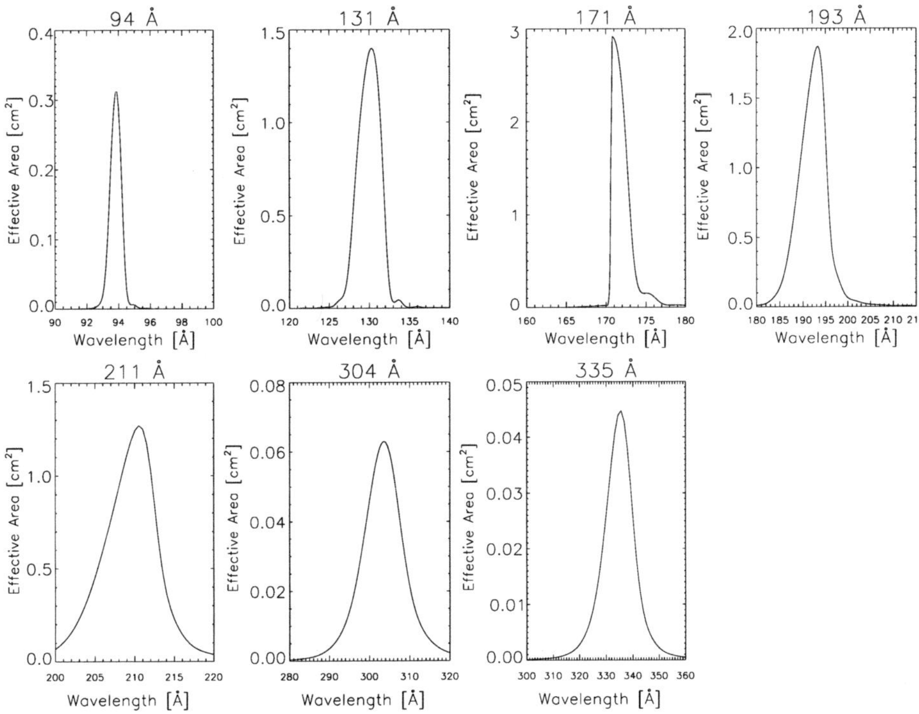

Figure 8 EUV effective-area functions.

Table 3 Calibration error budget. Sources of calibration are (M)odel, experimental (D)ata, or a combination of the two where data at selected wavelengths are used to constrain a model (DM).

Component	Calibration Error [%]	Source of Calibration	Dominant Source of Error
T_E	7	M	Uncertainty in optical constants; differences between measured filter samples and flight filters
R_P	6	D	Degradation in mirror after calibration
R_S	6	D	Degradation in mirror after calibration
T_F	5	DM	Uncertainty in optical constants
D	20	M	Uncertainty in contaminant thickness
Q	15	DM	Absolute calibration of reference detector; uncertainty in optical constants
Total	28		

[$R(\lambda)$: cm^2 DN phot^{-1}] to predict the total count rate that AIA should observe (in DN sec^{-1}), integrated over the full image, and then comparing this prediction with the actual full-frame count rate. This approach is limited in a number of ways – it ignores the fact that AIA's field of view is somewhat smaller than EVE's, and it is subject to errors up to 40% due to the width of EVE's spectral response (particularly for AIA channels with relatively sharp fea-

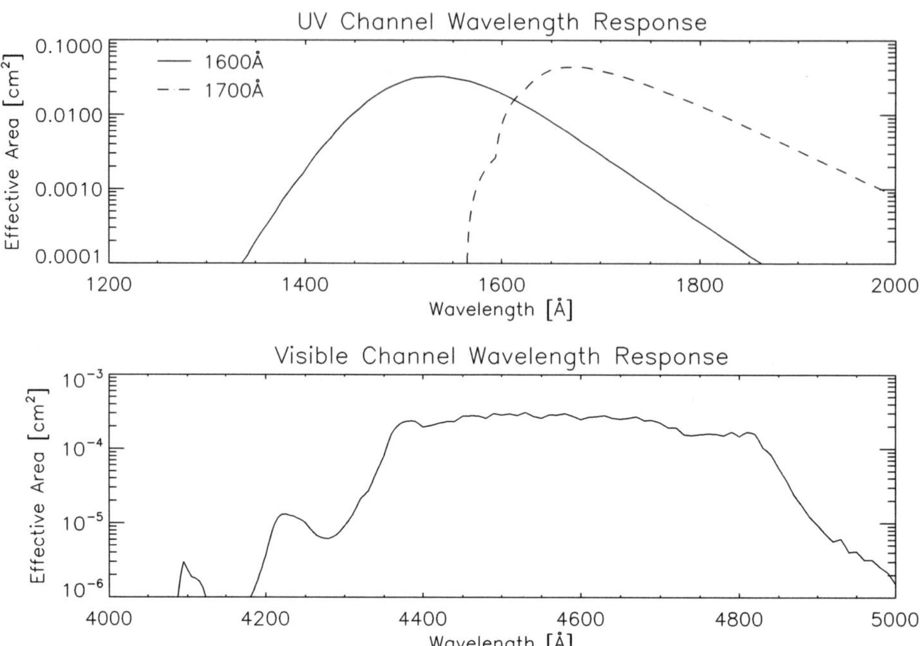

Figure 9 UV–Vis channel effective-area functions.

Table 4 Cross-calibration between the AIA EUV channels and the EVE calibration rocket underflight on 3 May 2010; and between the AIA UV channels and TIMED/SEE observations on the same date.

Channel	Full-disk count rate [DN s^{-1}]		Ratio predicted/observed
	Predicted from EVE rocket data	Observed by AIA	
94 Å	3.3×10^7	3.6×10^7	0.90
131 Å	8.3×10^7	7.3×10^7	1.15
171 Å	1.7×10^9	1.8×10^9	0.91
193 Å	1.7×10^9	1.9×10^9	0.91
211 Å	5.6×10^8	5.6×10^8	1.00
304 Å	3.0×10^8	4.4×10^8	0.69
335 Å	4.3×10^7	5.9×10^7	0.73

Channel	Full-disk count rate [DN s^{-1}]		Ratio predicted/observed
	Predicted from TIMED/SEE	Observed by AIA	
1600 Å	1.2×10^9	5.6×10^8	2.1
1700 Å	5.1×10^9	6.8×10^9	0.75

tures in wavelength, such as the 171-Å edge or the 94-Å channel). However, as a first-order check this approach has much merit.

This cross-calibration was performed using sample data from the EVE calibration rocket flown on 3 May and AIA images taken at the same time. The results are shown in Table 4.

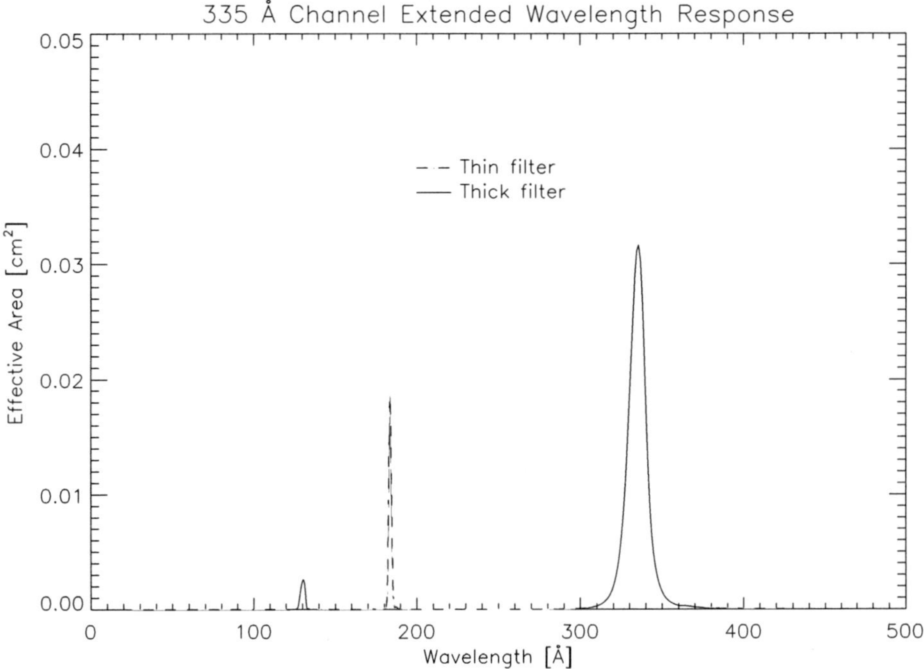

Figure 10 Full wavelength range effective area for the 335-Å channel, showing the effects of crosstalk from the 131-Å channel which shares the aperture of telescope 1, and a second-order peak in the mirror reflectivity around 184 Å. The other EUV channels do not show any significant throughput outside of the range shown in Figure 8.

The agreement is good, and likely can be improved by some straightforward updates to the AIA response functions (such as optimizing the thickness of the contaminant layer, which may be throwing off the predictions for the 304-Å and 335-Å channels).

Cross-calibration with EVE will be performed routinely during the AIA mission. This will allow us to track any degradation in the instrument. Furthermore, by observing how the ratio of observed AIA intensities to those intensities predicted by combining the EVE measurements and the AIA response functions varies with solar activity, we can refine the shape of the AIA wavelength response and improve the accuracy of the temperature-response function.

2.5. Temperature Response

The goal of the instrument calibration is to enable the user to extract useful scientific information from the AIA data. In Equation (1), if all of the effects that go into p and η are well known, then the user can obtain information about the emitted solar intensity [I]. One way to translate the information about I into information about the thermodynamic state of the corona is to express the solar spectral radiance as an integral over temperature:

$$I(\lambda, \mathbf{x}) = \int_0^\infty G(\lambda, T)\mathrm{DEM}(T, \mathbf{x})\, \mathrm{d}T.$$

Here G is the contribution function, which contains information on the plasma and atomic physics governing how material at a given temperature emits radiation. Given a few assump-

Table 5 Count rates predicted using CHIANTI, and observed quiet-Sun count rates.

Channel	Predicted count rate [DN pix^{-1} s^{-1}]				Observed quiet-Sun count rate [DN pix^{-1} s^{-1}]
	Coronal Hole	Quiet Sun	Active Region	Flare	
94 Å	0.0	0.3	14.5	49249.6	0.2
131 Å	2.1	4.6	39.9	368727.1	4.3
171 Å	65.5	272.8	2711.2	63727.1	323.9
193 Å	9.3	106.8	1988.8	475965.5	170.5
211 Å	1.7	28.3	750.5	32006.3	38.5
304 Å	17.4	25.5	126.8	98966.5	125.5
335 Å	0.3	1.4	44.6	18521.5	4.2

tions, it can be calculated using CHIANTI (Dere *et al.*, 1997, 2009). DEM is the differential emission measure function,

$$\mathrm{DEM}(T) = n_\mathrm{e}^2 \frac{\mathrm{d}z}{\mathrm{d}T},$$

which describes the thermal structure of the coronal plasma along the line of sight to the pixel.

It is possible to calculate temperature-response functions $[K(T)]$ for the AIA channels by folding the CHIANTI data and the instrument-wavelength response together:

$$K_i(T) = \int_0^\infty G(\lambda, T) R_i(\lambda)\, \mathrm{d}\lambda.$$

Then the observations can be expressed as an integral over temperature instead of wavelength:

$$p_i(\mathbf{x}) = \int_0^\infty K_i(T)\mathrm{DEM}(T, \mathbf{x})\, \mathrm{d}T$$

and this equation can (in principle) be inverted to obtain a DEM from a given set of observations $[p]$ and an instrument calibration $[K]$.

While the challenges of this problem are discussed in detail in the astrophysical literature (*e.g.*, Pottasch, 1963; Withbroe, 1975; Sylwester, Schrijver, and Mewe, 1980; Judge, Hubeny, and Brown, 1997), it remains a useful approach, as the calculation of the instrument temperature-response functions provides some basic insight into the interpretation of the images from the various channels.

We used CHIANTI to produce a typical spectral contribution function $[G]$. We assumed the coronal abundances of Feldman and Widing (1993), and the ionization balance of Dere *et al.* (2009) with a constant pressure of 10^{15} cm^{-3} K. With this contribution function, we computed temperature-response functions for the AIA EUV channels. The results are shown in Figure 11. We then combined these temperature-response functions with the DEM functions distributed with CHIANTI to predict count rates for typical solar features. The results are shown in Table 5, along with observed count rates from a patch of quiet Sun. Note that we have included an empirical correction factor of 20× on the He II 304-Å intensities. The He II line at 304 Å is not well-modeled by CHIANTI, and generally should not be used in DEM analysis (Warren, 2005).

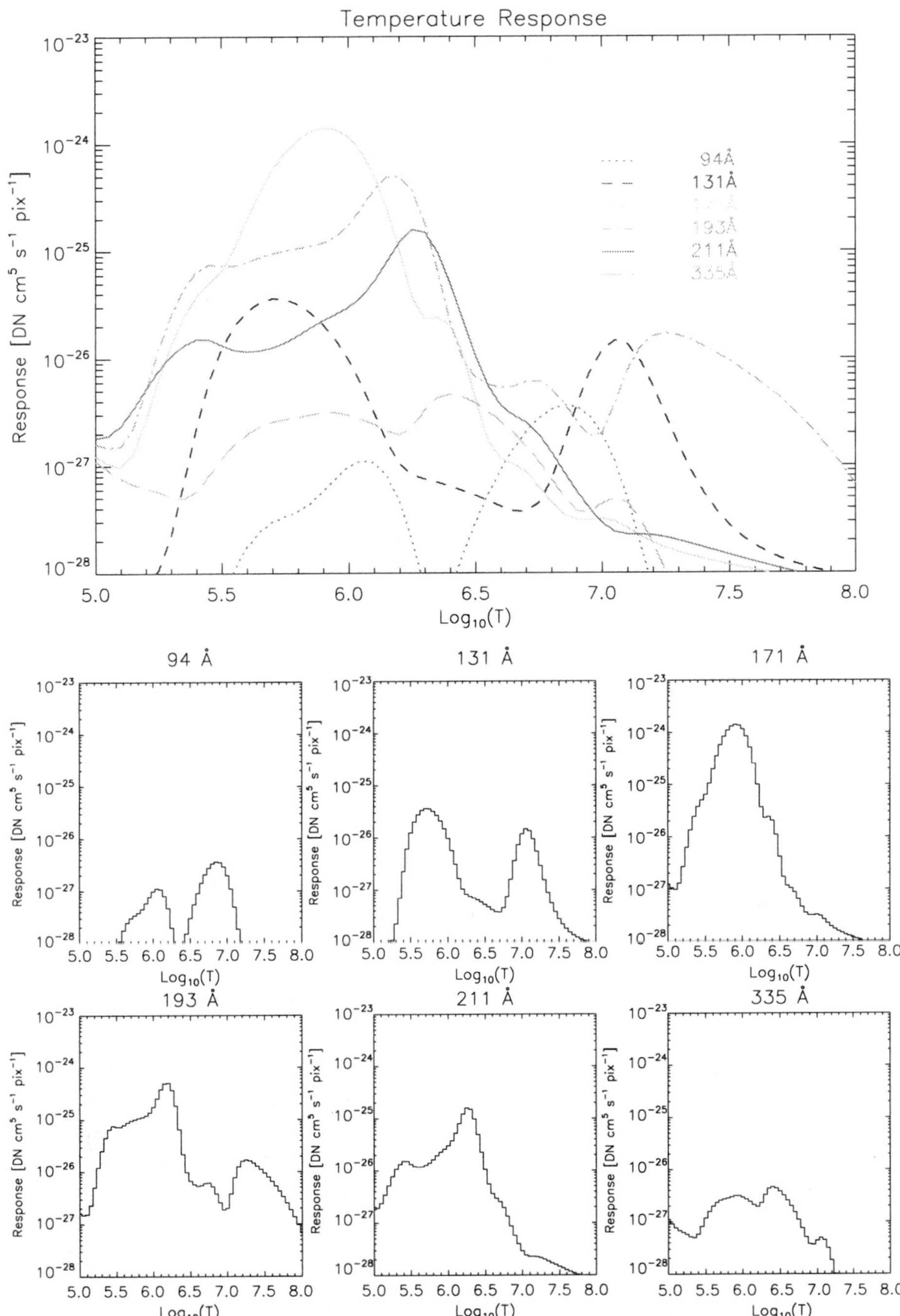

Figure 11 Temperature-response functions for the six EUV channels dominated by Fe emission, calculated from the effective-area functions and a CHIANTI model of solar emissivity.

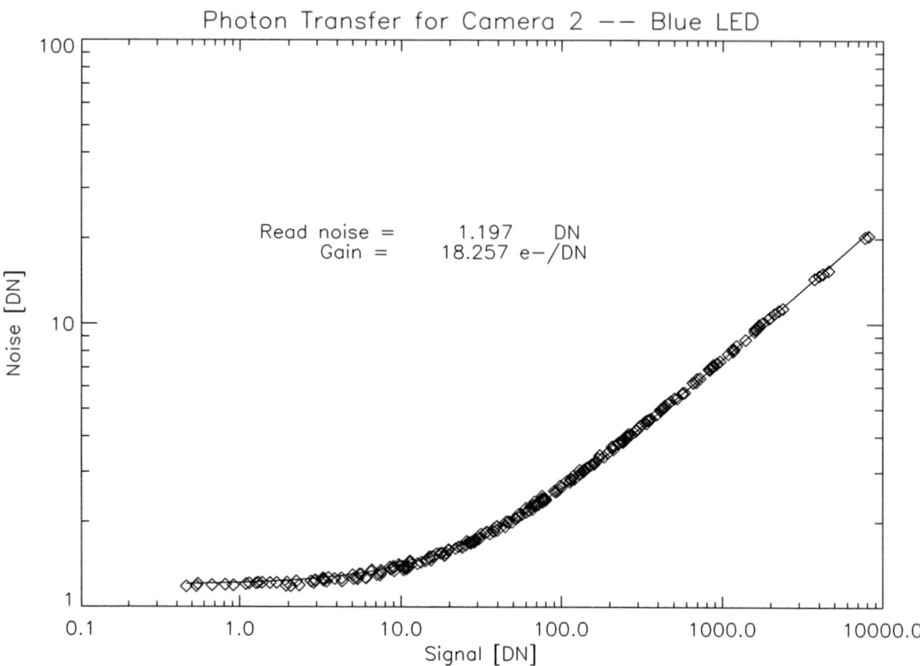

Figure 12 A sample photon-transfer curve for the CCD–camera system on ATA2. The slope of the linear part of the curve gives the gain of the system in electrons per DN; this gain is incorporated into the wavelength-response functions. The level at which the curve flattens out gives the read noise.

The default temperature-response and wavelength-response functions can be accessed through the SolarSoft (SSW) IDL library (Freeland and Handy, 1998), as described in the Appendix.

3. Instrument Characterization

3.1. Camera–CCD System Characterization

The AIA cameras use a 14-bit analog-to-digital converter (ADC) to translate the charge read out from each pixel to a digital number [DN] value. The gain of the camera (in electrons DN^{-1}) directly affects the accuracy of the instrument calibration. The camera gain was measured prior to launch both using ^{55}Fe X-rays and at visible wavelengths. In visible light, the gain is determined by measuring the camera photon-transfer curve. A dark image is taken along with a pair of light images (to eliminate pixel-to-pixel variation), and for a number of subregions within the images the difference between the light images is plotted *vs.* their average value. This is done for a series of increasing integration times, producing a curve like the one in Figure 12. For large-enough signals, the difference between the images is proportional to the shot noise (the square root of the number of photons), so the number of photons can be determined for a given DN value (and thus the number of detected electrons, and the number of electrons DN^{-1}). For small signals, the difference goes to a constant value, which is the read noise of the camera system in DN. Note that, in all cases, the read noise is $\approx 20-22$ e^{-} RMS, corresponding to, *e.g.*, the charge produced and collected by fewer than

Table 6 Camera system gain and read noise.

Camera	Wavelengths [Å]	Gain [$e^- DN^{-1}$]	Read Noise [DN]
1	131,335	17.6	1.18
2	193,211	18.3	1.20
3	171, UV–Vis	17.7	1.15
4	94,304	18.3	1.14

three photons at 304 Å. Thus, for all practical purposes, the images are photon-noise limited except at the faintest signal levels. The gains of the four AIA cameras are listed in Table 6.

The camera–CCD system gain is a weak function of temperature, and may change as the camera electronics age. Therefore, the light-transfer curve will be checked periodically on orbit using solar EUV illumination and visible light from the blue LED built into the telescope spider assemblies.

The light-transfer curve also provides information about the linearity of the camera–CCD system (from the ratio of average signal to exposure time). The AIA camera system is linear to better than 1% for signals up to 11 000 DN.

3.2. Flat-Field

The instrument flat-field has been measured on orbit using a variety of techniques. Most of the data were taken by sending a series of small offsets to the active secondary mirror, displacing the image to ≈ 12 different positions with shifts of up to ≈ 35 arcsecs. Larger scale offsets were obtained during a coordinated spacecraft maneuver that sent SDO to a pattern of offpoints with a scale of about 25 arcmins. The offpoint data were analyzed using a Kuhn algorithm (Kuhn, Lin, and Loranz, 1991) to produce a map of the sensitivity of different parts of the image. The flat-field is normalized to a mean value of one.

A sample flat-field for the 171-Å channel is shown in Figure 13. It consists of several distinctly recognizable components: the quadrant scaling due to slight gain mismatches between the four amplifiers used to read out each CCD; a "tire track" pattern that consists of diagonal stripes with a variation of a few percent, characteristic of the CCD laser-anneal process; the grid of the focal-plane filters, which results in differential obscuration of up to a few percent; pixel-to-pixel variations in the CCD; "blobs" on the scale of a few tens of pixels across that may indicate contamination or may be artifacts of the CCD processing; and large-scale variations in sensitivity due to vignetting in the telescope optical system. So far, these patterns appear to be reasonably stable, and the RMS variation in the flat-field is of order 2%. We will continue to monitor the flat-field through monthly secondary-mirror offpoints, and occasional spacecraft-offpoint maneuvers.

3.3. Optical Performance

The telescope mirrors were integrated into the tube assembly and aligned at SAO using a visible-light interferometer (Podgorski *et al.*, 2009). The focal-plane assembly was integrated to the tube assembly at LMSAL. Optical performance of the integrated telescope assembly was checked at visible wavelengths using an infinite collimator to project a resolution target into the aperture. While this procedure allowed for an accurate determination of the position of best focus and could verify the absence of gross errors in the telescope's alignment or imaging performance, the visible-light measurements (using wavelengths $\approx 30\times$

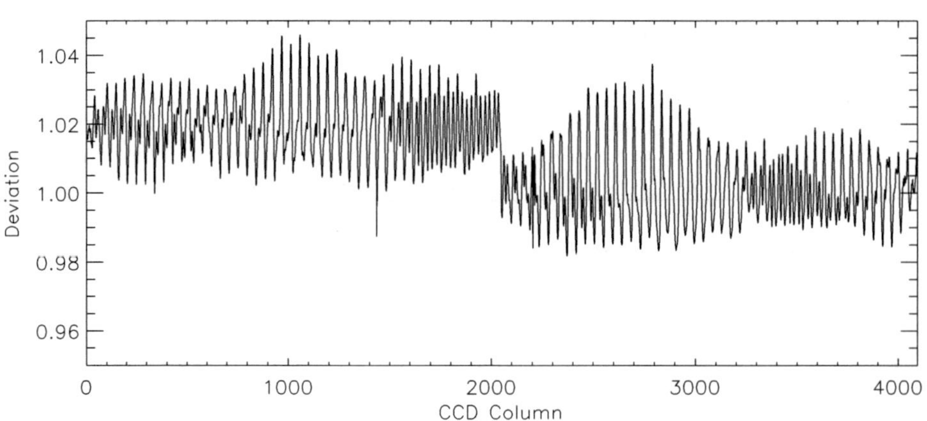

Figure 13 The flat-field of the 171-Å channel. The dark line indicates the position of the cut, plotted below the image.

Table 7 A sample image-resolution error budget for the 171-Å channel. The effects are added in quadrature to give an estimate of the system-level optical performance.

Item	Contribution to RMS Spot diameter [arcsec]
Optical prescription	0.60
Fabrication, alignment, and assembly effects	1.21
Launch shift effects	0.10
On-orbit thermal effects	0.21
Focus error (with on-orbit correction)	0.10
Jitter residual	0.48
Diffraction (for UV–Vis channels)	0.00
Detector Pixelization	0.48
CCD Charge spreading	0.80
On-orbit performance Prediction	1.73

Table 8 Predicted resolution (RMS spot diameter in arcsec) for the AIA images.

Channel	Predicted Resolution
94 Å	1.74
131 Å	1.67
171 Å	1.73
193 Å	1.38
211 Å	1.55
304 Å	1.52
335 Å	1.73
1600 Å	1.65
1700 Å	1.65
4500 Å	2.43

longer than the telescope's operational wavelengths) are limited by diffraction. They did not allow an accurate measurement of the in-band point-spread function.

The telescope spatial resolution was determined analytically by combining component-level interferometric measurements of the mirror figures with a raytrace model of the telescope prescription (including tolerances for mirror misalignment and structural and thermal deformations on orbit). Allocations for jitter, CCD pixelization, and charge spreading, diffraction (significant only for the UV–visible channels), and scatter by the mirror surfaces and the filter-support meshes were included as well. A minimal amount of defocus was assumed; the telescopes all have focus mechanisms that work by translating the secondary mirror along the optical axis, and on-orbit measurements have shown the focus to be stable. The stackup of these allocations for the 171-Å channel is shown in Table 7, and the predicted RMS spot diameters for all channels are listed in Table 8. Note that the metric used in these tables is the RMS spot diameter, averaged over the field of view. This is not the limit of the spatial resolution of the instrument; there is significant contrast transfer at finer angular scales than those listed in Table 8.

3.3.1. Filter Mesh Diffraction

The entrance-filter mesh creates a diffraction pattern, which is particularly noticeable for bright, compact sources such as flare kernels. The entrance filter for each AIA EUV channel consists of two independently-mounted mesh segments. In order to ensure that the diffraction patterns from the two filter segments can be separated from each other and from the solar image, the mesh grid of one segment is oriented at about 40° with respect to the CCD columns, and the other is at about 50°. This results in an eight-legged diffraction pattern. The spacing of the maxima along each leg is given by

$$\sin(\theta) = \frac{m\lambda}{a},$$

where θ is the diffraction angle, m is the order and a is the mesh spacing. The mesh period is 70 lines per inch; the bars of the mesh cells are 29 – 33 microns thick.

The effect of diffraction by an entrance-filter mesh has been analyzed extensively for the TRACE entrance filters (Lin, Nightingale, and Tarbell, 2001; Gburek, Sylwester, and Martens, 2006). Similar techniques will be applied to the AIA images to characterize the mesh diffraction. The focal-plane filters use the same mesh, and therefore diffract light by the same mechanism; however, because they are so much closer to the CCD, the effect of this diffraction is much less significant. Instead, as previously discussed, they produce a screen-like shadow that is removed by flat fielding.

The telescope PSF, including the core (defined by the terms in Table 6) and the wings (which are dominated by the filter diffraction pattern), can be measured on orbit using the technique of DeForest, Martens, and Wills-Davey (2009), or by analysis of bright flare kernels. Once an accurate PSF is obtained, we will produce higher level data products with the PSF deconvolved out of the image.

3.3.2. Stray Light

The EUV entrance filters are routinely monitored for leakage of visible light by taking images without a focal-plane filter in place. The Zr entrance filters on the 94-Å and 131-Å channels transmit enough visible light to form a good-quality image of the photosphere in a 500-msec exposure. The combination of the entrance filters and the focal-plane filters provides excellent rejection of out-of-band light for the EUV channels, and EUV light does not scatter efficiently off the interior surfaces of the telescope, so there is no substantial stray light in the EUV images.

The UV-channel images do show a significant ghost image approximately 2 arcmin to the Southeast of the main image. This is most likely due to internal reflections within the UV filter substrate. It may be possible to correct for the ghost image in the data processing pipeline.

3.4. Guide Telescope–ISS

Each ATA has a guide telescope (GT) rigidly mounted to it, which provides continuous measurements of the offset between the sunvector and the telescope boresight in pitch and yaw. This information is used by the telescope's image-stabilization system (ISS) to control the deflection of the active secondary mirror in order to cancel image motion or blurring due to spacecraft jitter. It is also sent to the spacecraft attitude control system (ACS) and used to maintain the observatory's pointing at the Sun.

The ISS is an open-loop system in the sense that the corrections applied by the active secondary mirror are not directly sensed by the GT. Therefore, it relies on careful calibration of the relationship between the sunvector reported by the GT and the image deflection applied by the ISS electronics. The ISS calibration varies seasonally (due to changes in the apparent size of the solar disk, which changes the amplitude of the GT signal for a given angular deflection of the sunvector). Therefore, we perform monthly calibration maneuvers by offpointing the observatory in a cruciform pattern and measuring the response of the GT signals and the active secondary-mirror deflection at each position.

The image-stabilization system achieves a reduction of jitter by a factor of > 10 for frequencies below ≈ 5 Hz, and provides some attenuation for jitter up to ≈ 20 Hz. This eliminates the impact from most of the dominant sources of jitter on SDO: the slow, thermally-driven drift in spacecraft pointing following eclipses, the stiction due to zero-speed crossings of the reaction wheels, and the torques of the AIA mechanisms themselves. There is some higher-frequency jitter (primarily due to the stepping of the high-gain antennas) which may introduce image blur at the level of a few tenths of an arcsecond for occasional exposures.

Differential shifts between the GT and the science telescope (ST) will cause pointing errors in the science data. This is generally not an issue, as the GT is rigidly coupled to the ST. However, during the first month of operations, the ISS introduced a slow ($\approx$ three – five minutes) oscillation of ≈ 1 arcsec amplitude as the cycling of the telescope heaters drove a differential pointing shift between the GT and ST. After this effect was discovered, the heater operations were tuned to maintain a stable temperature without cycling; since then, the pointing of all AIA channels has been extremely stable (less than 0.1 pixels of image motion on timescales of tens of minutes). Some image shift (up to a few pixels) due to thermal variations on longer time scales (days to months) persists, and there are small changes in pointing whenever the focus position of a channel is adjusted. These drifts will be corrected out as the operations and image processing techniques mature.

3.5. Image Coalignment

During integration, the four telescope assembly boresights were aligned to within 20 arcsecs of a common reference frame, and they maintained that coalignment through launch. The GTs were aligned to within 45 arcsecs of their respective STs; the GT boresights can effectively be shifted by up to 90 arcsecs in the onboard electronics by applying a bias to the signal from the GT limb sensors. Because the SDO ACS uses the GT signals for fine guiding, AIA is able to control the pointing of the observatory using the GT biases. The GT biases were chosen to coalign the four GTs, such that they all produce null signals when the spacecraft aligns the science reference boresight (SRB) (defined as the mean of the extreme science telescope boresights in pitch and yaw) with the Sun vector. This combined system has shown excellent stability, consistently maintaining science pointing to within a fraction of an arcsecond.

DC offsets of up to 15 arcsecs were applied to the active secondary mirrors of the telescopes to coalign each ST with the SRB. This reduced the misalignment of the science telescopes to a few arcseconds – thus, the center of the solar disk is within $\approx$ five pixels of the center of the CCD on all AIA channels before applying corrections in ground software. The CCDs are coaligned in roll to within ± 0.25 degrees.

The residual misalignments between the channels have been measured using limb-fitting software and "manual" comparison of bright features on the disk and at the limb. The alignment information is documented in the image headers at Level 1, and corrected out of the image at Level 1.5.

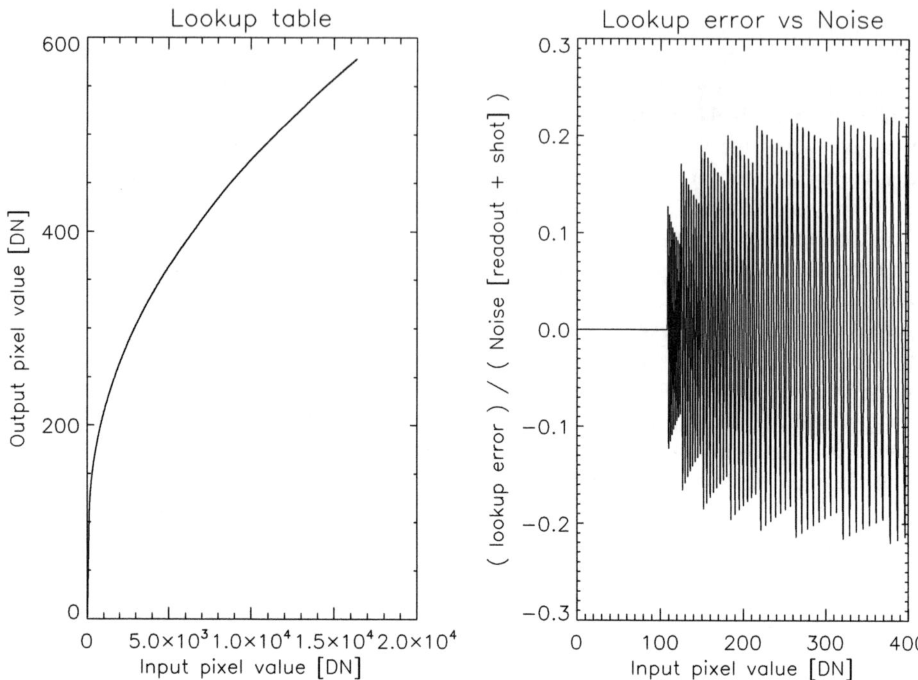

Figure 14 Left: An example onboard lookup table used for lossy compression. Right: The ratio of the error in pixel values due to the lookup table to the intrinsic error in the measurement (due to shot noise and readout noise).

3.6. Compression

AIA downlinks science data at a higher rate than any solar science instrument that has ever flown. Its telemetry allocation is capped at 67 Mbps; the observing sequence is designed to fill 80 – 90% of this bandwidth all the time. In order to maintain a cadence of 10 or 12 seconds for full-frame images from all eight AIA channels while fitting within this bandwidth, it is necessary to perform some compression onboard. Without compression, the eight channel-by-ten second cadence generates $4096 \times 4096 \times 16 \times 8/10 = 215$ Mbps of data, so compression of approximately a factor of four is required (to approximately four bits per pixel). This level of compression cannot be reliably achieved without using slightly lossy compression techniques.

The image data are compressed by an onboard FPGA using a Rice compression algorithm (Rice, 1979) that converts the pixel values into differences between adjacent pixels. Small differences compress extremely efficiently (down to a limit of two bits per pixel for very smoothly varying images), while sharp intensity variations (especially due to cosmic-ray hits or bright coronal features) require much more data to transmit. The Rice algorithm itself is lossless; however, in order to maximize its effectiveness, it is necessary to minimize the impact of large intensity variations. This is accomplished by remapping the image pixel values using a lookup table before applying the Rice compression. A sample lookup table is shown in Figure 14. Note that the difference between the "true" pixel value and the value that it is rounded off to (the error introduced by the compression scheme) is approximately proportional to the square root of the true pixel value – in this case, it is always less than 1/4 of the value. This ensures that the compression error will be small compared to the intrinsic shot noise in the image for all types of observations. The images are decompressed during the production of level 0 data (Lemen *et al.*, 2011), and therefore the data user does not need to make any corrections for the compression.

4. Conclusion

The preflight photometric calibration of the AIA instrument shows good agreement with absolute irradiance measurements by the EVE rocket and with theoretical intensity predictions from CHIANTI. The response functions will be updated within the first year of operations based on cross-calibrations with EVE and closer analysis of the AIA images themselves.

The preliminary on-orbit characterization of the instrument performance is complete, and, where possible, corrections for instrumental effects (such as flat-fielding, image offsets, *etc.*) have been applied to the calibrated data. The instrument is healthy and in good working order, and should continue to provide excellent data for years to come.

Acknowledgements This work is supported by NASA under contract NNG04EA00C.

Appendix: Accessing the Response Functions

The instrument-wavelength response and temperature-response functions are available through the SolarSoft (SSW) IDL library (Freeland and Handy, 1998), using the function aia_get_response. The wavelength-response functions are retrieved as follows: IDL> aia_resp = aia_get_response(/area [, /dn, /full, /uv, /all]) where the optional keyword /dn specifies that the photon-to-DN conversion should be included; /full indicates that the detailed component-level calibration results (including, *e.g.*, the mirror reflectivity, CCD QE, *etc.*) should be returned; /uv selects the UV and visible channels rather than the (default) EUV channels; and /all indicates that responses for both thin and thick focal-plane filter positions should be returned.

The default solar emissivity, calculated using CHIANTI 6.0.1, can be returned as follows: IDL> aia_emiss = aia_get_response(/emissivity [, /full]) where the optional keyword /full will return the detailed line list.

The temperature-response functions of the EUV channels can be accessed as follows: IDL> aia_tresp = aia_get_response(/temp [, /full, /dn, /all]) where the optional keywords /dn and /all have similar meanings to those used with the effective area, and /full in this context returns the response for each channel as a function of both wavelength and temperature. The temperature response is calculated on demand, using effective area and emissivity data that are stored in the SSW database.

A number of lower-level routines (with filenames beginning with aia_bp_) are also available in SSW, allowing the user to adjust or apply the response functions (*e.g.* to calculate count rates given an input spectrum, generate a temperature-response function using alternate assumptions about ionization balance, *etc.*).

References

Catura, R.C., Stern, R.A., Cash, W., Windt, D.L., Culhane, J.L., Lappington, J., Barnsdale, K.: 1987, *Proc. SPIE* **830**, 204.

DeForest, C.E., Martens, P.C.H., Wills-Davey, M.J.: 2009, *Astrophys. J.* **690**, 1264.

Dere, K.P., Landi, E., Mason, H.E., Monsignori Fossi, B.C., Young, P.R.: 1997, *Astron. Astrophys. Suppl.* **125**, 149.

Dere, K.P., Moses, J.D., Delaboudinière, J.-P., Brunaud, J., Carabetian, C., Hochedez, J.-F., Song, X.Y., Catura, R.C., Clette, F., Defise, J.-M.: 2000, *Solar Phys.* **195**, 13.

Dere, K.P., Landi, E., Young, P.R., Del Zanna, G., Landini, M., Mason, H.E.: 2009, *Astron. Astrophys.* **498**, 915.

Feldman, U., Widing, K.G.: 1993, *Astrophys. J.* **414**, 381.

Freeland, S.L., Handy, B.N.: 1998, *Solar Phys.* **182**, 497.

Gburek, S., Sylwester, J., Martens, P.: 2006, *Solar Phys.* **239**, 531.

Handy, B.N., Acton, L.W., Kankelborg, C.C., Wolfson, C.J., Akin, D.J., Bruner, M.E., Caravalho, R., Catura, R.C., Chevalier, R., Duncan, D.W., Edwards, C.G., Feinstein, C.N., Freeland, S.L., Friedlaender, F.M., Hoffmann, C.H., Hurlburt, N.E., Jurcevich, B.K., Katz, N.L., Kelly, G.A., Lemen, J.R., Levay, M., Lindgren, R.W., Mathur, D.P., Meyer, S.B., Morrison, S.J., Morrison, M.D., Nightingale, R.W., Pope, T.P., Rehse, R.A., Schrijver, C.J., Shine, R.A., Shing, L., Strong, K.T., Tarbell, T.D., Title, A.M., Torgerson, D.D., Golub, L., Bookbinder, J.A., Caldwell, D., Cheimets, P.N., Davis, W.N., Deluca, E.E., McMullen, R.A., Warren, H.P., Amato, D., Fisher, R., Maldonado, H., Parkinson, C.: 1999, *Solar Phys.* **187**, 229.

Henke, B.L., Gullikson, E., Davis, J.C.: 1993, *Atom. Data Nucl. Data Tables* **54**(2), 181.

Hurlburt, N., Cheung, M., Schrijver, C., Chang, L., Freeland, S., Green, S., Heck, C., Jaffey, A., Kobashi, A., Schiff, D., Serafin, J., Seguin, R., Slater, G., Somani, A., Timmons, R.: 2011, *Solar Phys.* doi:10.1007/s11207-010-9624-2.

Judge, P.G., Hubeny, V., Brown, J.C.: 1997, *Astrophys. J.* **475**, 275.

Kuhn, J.R., Lin, H., Loranz, D.: 1991, *Publ. Astron. Soc. Pac.* **103**, 1097.

Lemen, J.R., Title, A.M., Akin, D.J., Boerner, P.F., Chou, C., Drake, J.F., Duncan, D.W., Edwards, C.G., Friedlaender, F.M., Heyman, G.F., Hurlburt, N.E., Katz, N.L., Kushner, G.D., Levay, M., Lindgren, R.W., Mathur, D.P., McFeaters, E.L., Mitchell, S., Rehse, R.A., Schrijver, C.J., Springer, L.A., Stern, R.A., Tarbell, T.D., Wuelser, J.-P., Wolfson, C.J., Yanari, C.: 2011, *Solar Phys.* doi:10.1007/s11207-011-9776-8

Lin, A.C., Nightingale, R.W., Tarbell, T.D.: 2001, *Solar Phys.* **198**, 385.

Podgorski, W.A., Cheimets, P.N., Boerner, P.F., Glenn, P.: 2009, *Proc. SPIE* **7438**, 13.

Pottasch, S.R.: 1963, *Astrophys. J.* **137**, 945.

Rice, R.F.: 1979, *Some Practical Universal Noiseless Coding Techniques*, JPL Publication 79-22, Pasadena, CA.

Soufli, R., Baker, S.L., Windt, D.L., Gullikson, E.M., Robinson, J.C., Podgorski, W.A., Golub, L.: 2007, *Appl. Opt.* **46**, 3156.

Soufli, R., Windt, D.L., Robinson, J.C., Baker, S.L., Spiller, E., Dollar, F.J., Aquila, A.L., Gullikson, E.M., Kjornrattanawanich, B., Seely, J.F., Golub, L.: 2005, *Proc. SPIE* **5901**, 173.

Stern, R.A., Shing, L., Blouke, M.: 1994, *Appl. Opt.* **33**, 2521.

Stern, R.A., Shing, L., Catura, P., Morrison, M., Duncan, D., Lemen, J., Eaton, T., Pool, P., Steward, R., Walton, D., Smith, A.: 2004, *Proc. SPIE* **5171**, 77.

Sylwester, J., Schrijver, J., Mewe, R.: 1980, *Solar Phys.* **67**, 285.

Waltham, N., Beardsley, S., Clapp, M., Lang, J., Jerram, P., Pool, P., Auker, G., Morris, D., Duncan, D.: 2011, In: *Internat. Conf. Space Optics*, ESA, Noordwijk, in press.

Warren, H.: 2005, *Astrophys. J. Suppl.* **157**, 147.

Windt, D.: 1998, *Comput. Phys.* **12**, 4 (), 360.

Windt, D., Catura, R.: 1988, *Proc. SPIE* **984**, 82.

Withbroe, G.L.: 1975, *Solar Phys.* **45**, 301.

Woods, T.N., Eparvier, F.G., Bailey, S.M., Chamberlin, P.C., Lean, J., Rottman, G.J., Solomon, S.C., Tobiska, W.K., Woodraska, D.L.: 2005, *J. Geophys. Res.* **110**, 01312.

Woods, T.N., Woods, T.N., Chamberlin, P., Eparvier, F.G., Hock, R., Jones, A., Woodraska, D., Judge, D., Didkosky, L., Lean, J., Mariska, J., McMullin, D., Warren, H., Berthiaume, G., Bailey, S., Fuller-Rowell, T., Sojka, J., Tobiska, W.K., Viereck, R.: 2010, *Solar Phys.* doi:10.1007/s11207-009-9487-6.

Solar Phys (2012) 275:67–78
DOI 10.1007/s11207-010-9624-2

Heliophysics Event Knowledgebase for the *Solar Dynamics Observatory* (SDO) and Beyond

N. Hurlburt · M. Cheung · C. Schrijver · L. Chang · S. Freeland · S. Green · C. Heck · A. Jaffey · A. Kobashi · D. Schiff · J. Serafin · R. Seguin · G. Slater · A. Somani · R. Timmons

Received: 14 May 2009 / Accepted: 9 August 2010 / Published online: 7 December 2010

Abstract The immense volume of data generated by the suite of instruments on the *Solar Dynamics Observatory* (SDO) requires new tools for efficient identifying and accessing data that is most relevant for research. We have developed the Heliophysics Events Knowledgebase (HEK) to fill this need. The HEK system combines automated data mining using feature-detection methods and high-performance visualization systems for data markup. In addition, web services and clients are provided for searching the resulting metadata, reviewing results, and efficiently accessing the data. We review these components and present examples of their use with SDO data.

Keywords Data markup · Solar dynamics observatory · Solar features and events

1. Introduction

The *Atmospheric Imaging Assembly* (AIA: Lemen *et al.*, 2010) and the *Helioseismic and Magnetic Imager* (HMI: Schou *et al.*, 2010) on the *Solar Dynamics Observatory* (SDO) represent a fundamental change in the approach to solar observations. Earlier solar missions were constrained by the technology of their time to follow two operational strategies that are both reflected in the instrumental precursors to AIA. Developers of the *Extreme ultraviolet Imaging Telescope* (EIT) project on SOHO (Delaboudinière *et al.*, 1995), for example, designed their investigations to observe the entire solar disk and adjacent corona at a regular sampling rate. This enables studies of global coronal dynamics and guaranteed complete coverage of all solar events, but it sacrifices spatial and temporal resolution required to make

The Solar Dynamics Observatory
Guest Editors: W. Dean Pesnell, Phillip C. Chamberlin, and Barbara J. Thompson.

N. Hurlburt (✉) · M. Cheung · C. Schrijver · L. Chang · S. Freeland · S. Green · C. Heck · A. Jaffey · A. Kobashi · D. Schiff · J. Serafin · R. Seguin · G. Slater · A. Somani · R. Timmons
Lockheed Martin Advanced Technology Center, Palo Alto, CA 94304, USA
e-mail: hurlburt@lmsal.com
url: http://www.lmsal.com/sungate

detailed studies. As a complementary approach, the TRACE project (Handy *et al.*, 1999) designed their investigation to capture the dynamics of the corona in high temporal and spatial resolution. This limited the field-of-view of TRACE and forced observers to make careful plans to anticipate where and when interesting phenomena might occur.

The design of the AIA program removes the compromises of coverage *versus* resolution by observing the full disk at resolutions comparable to those of TRACE, with even better temporal coverage and improved spectral coverage. While this offers a dramatic improvement to the community, it shifts the operational problem from planning and coverage to one of data management. The data to be returned by AIA dwarf those of TRACE and EIT. In one day, it delivers the equivalent of five years of TRACE data (2TB). To put the problem simply, the constraining factor for successful AIA science investigations is to efficiently find the data needed, and only the data needed for their purpose.

The Heliophysics Event Knowledgebase is designed to alleviate this problem. Its purpose is to catalog interesting solar events and features and to present them to the community members in such a way that it guides them to the most relevant data for their purposes. This is a problem of data markup that is arising in many scientific and other fields. Our approach shares many aspects with those from other fields. For example, the Monterrey Bay Aquarium Research Institute developed a similar system for annotating and cataloging video sequences of ocean fauna and activities (Cline, Edgington, and Mariette, 2007); various sports leagues have systems for cataloging clips of athletic events. The distinguishing factor for SDO is the large image format and complex event types. In the following sections we present the design and implementation of the HEK system and then present a short description of its use.

The design and implementation of the HEK and its interfaces allow expansion beyond its primary function as a searchable database that contains metadata on solar events and on the observation sets from which these are derived. It can, and it does, track temporary data products that users may wish to have online for a limited period. It is also possible to include numerical data sets based on assimilated observations, or even such data sets and events within them that were generated as an observational data set would be. And, as a final example here, it can contain information on papers published based on certain data sets and events within them as associated metadata so that for any given event users can be

1.1. HEK Design Goals

The Heliophysics Events Knowledgebase (HEK) is an integrated metadata system designed with the following goals in mind:

- Help researchers find data sets relevant to their topics of interest.
- Serve as an open forum where solar/heliospheric features and events can be reported and annotated.
- Facilitate discovery of statistical trends/relationships between different classes of features and events.
- Avoid overloading SDO data systems with attempts to download data sets too large to transmit over the internet.

To achieve these goals, HEK consists of registries to store metadata pertaining to observational sequences (Heliophysics Coverage Registry, or HCR), heliophysical events (Heliophysics Events Registry, or HER) and browse products such as movies. Interfaces for communications and querying between the different registries are also provided in terms of web services.

1.1.1. Heliophysical Events

While many think of events as physical processes, it is more appropriate for our purposes to use a more empirical definition. This enables our system to include entries that come from multiple sources and methods, while avoiding ambiguities. Events can be grouped into two general categories: those that are directly found in data, and those that are inferred by other means. In the former case, any report of an event must specify an event class and the method, data, and associated parameters used to identify the event as well as contributed to the event. For example: "I" found a "flare" in "AIA 195 data" using the SolarSoft routine "ssw_flare_locator" with "default parameters". The extensible list of possible event classes, shown in Table 1, is maintained as a controlled vocabulary by the HEK team. For each event class, a unique set of required and optional attributes is also defined by the HEK team. All events are required to have a duration and bounding box that contains the event in space and time. Some assumptions are made about locations dependent upon the event class. For instance, solar events are assumed to be on the solar surface unless explicitly stated otherwise. Similarly, heliospheric events seen from a single perspective, such as CMEs observed by SOHO/LASCO, are taken to be on the plane of the sky containing the solar center.

The second category of events, namely those inferred by other means, can also be captured within the HEK. In this case, the required information is the inference method, the metadata that it operated upon, and the parameters it used for making the inference. For instance, active regions (ARs) are reported and assigned numbers daily by NOAA. Using our empirical definitions, each AR observation is an event bounded within a 24-hour time interval. An obvious inference is that all NOAA active regions with the same active region number are the same active region. In this case, the data source is the Heliophysics Events

Table 1 The current list of events classes supported by the HEK. This is a controlled list that is expandable as new events classes are defined.

Event Class	Description
Active Region	Solar Active Region
Coronal Mass Ejection	Ejection of material from the solar corona
Coronal Dimming	A large-scale reduction in EUV emission
Coronal Jet	A jet-like object observed in the low corona
Coronal Wave	EIT or Morton waves spanning a large fraction of the solar disk
Emerging Flux	Regions of new magnetic flux in the solar photosphere
Filament	Solar Filament or Prominence
Filament Eruption	A sudden launching of a filament into the corona
Filament Activation	A sudden change in a filament without launching
Flare	Solar Flare
Loop	Magnetic loops typically traced out using coronal imagery
Oscillation	A region with oscillating coronal field lines
Sigmoid	S-shaped regions seen in soft X rays; indicator for flares
Spray Surge	Sudden or sustained intrusion of chromospheric material well into the corona
Sunspot	Sunspots on the solar disk
Plage	Bright areas associated with active regions
Other	Something that could not be classified – good candidate for further research
Nothing Reported	Used to indicate that the particular data were examined, but had nothing noteworthy to the observer

Registry and the method is a query to the HER for active regions with a given active region number. Similar higher-level events can be constructed to the point where they can be matched against the "physical" counterpart and these can either by registered back into the HER or derived on the fly.

We have extended the VOEvent schema developed by the IVOA (White *et al.*, 2006) to encapsulate all metadata describing HEK events. This provides a convenient means for exchanging event descriptions between various providers and can be used as a means for distributed processing of events (Hurlburt, Cheung, and Bose, 2008). SolarSoft (Freeland and Bentley, 2000) routines have been developed that support HEK events.

The concept of an "event" can easily be expanded. For instance, the VOEvent schema supports descriptions of observatory configurations at specified times as well. We have used these extended properties to capture metadata associated with instrument operations and incorporated them into a database that describes the type and purpose of the data collected. These "Coverage Events" include information on who collected the data, what wavelengths were observed, and in what regions and time intervals the data were collected. Additional event classes are being defined to support the execution of models and simulations of heliophysical phenomena. In each case, the VOEvent standard permits easy exchange and transport of event descriptions. We have developed service and client software that permits these exchanges.

These descriptions can also be annotated after the fact, resulting in something akin to social-networking site for heliophysical data. It can be integrated with scientific publications by explicit links or references, such as those provided the Astrophysical Data System (http://ads.harvard.edu). Alternatively, they can be referenced implicitly using standard descriptions of events supported by publishers, such as the SOL convention that has been adopted by journals supporting solar-physics publications (Solar Physics Editors, 2010).

1.2. Example Usage

The research community interacts with the HEK by several methods. For example, iSolSearch (Figure 1) is a web application that presents a unified experience with the services provided by the HEK and SolarSoft. It permits web users to search for different event classes with a variety of constraints. Here, a set of common event types is selected, including flares, CMEs, and Active Regions. Events are displayed as icons on a solar disk or Carrington grid and as a list. There are a variety of options for saving these lists for future use. When an individual event is selected in iSolSearch, a summary panel of the event opens with links to more-detailed information including movies and images that help identify events worthy of further study. In addition, iSolSearch finds and displays observational sequences available in the vicinity of the selected event. Users follow these links to download the data containing the events. These links lead directly to the appropriate data sources or, when appropriate, pass the request on to the Virtual Solar Observatory.

An alternative approach to using the HEK is through SolarSoft. This snippet of IDL code uses the HEK-provided routines to find flares within a specified time interval:

```
IDL> t0='1-jan-2000' & t1='1-jan-2001'
IDL> query = ssw_her_make_query(t0,t1,/fl)
IDL> events = ssw_her_query(query)
```

Further analysis of these events can be fed into a search for data:

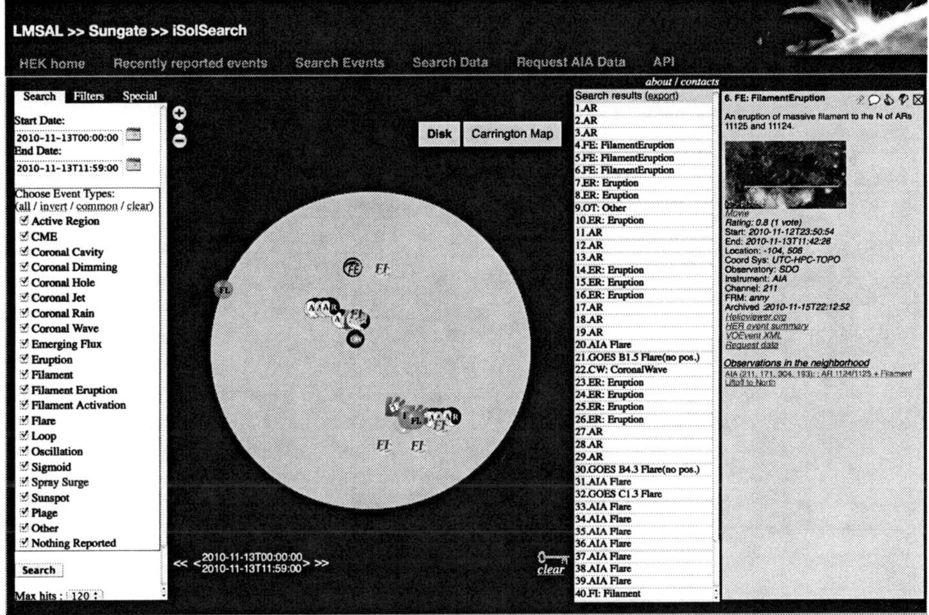

Figure 1 The iSolSearch client, http://lmsal.com/isolsearch, interacts with various HEK registries to present a unified view of events and available data.

```
IDL> t0=events.fl(0).event_starttime &
     t1=events.fl(0).event_endtime
IDL> dataset = ssw_hcr_query(ssw_hcr_make_query(t0,t1))
```

2. HEK Implementation

We have developed a system for extracting events and features from SDO data and for presenting them to users in a convenient fashion. Here we review the system design and operation. The overall design is displayed in Figure 2, with major functional areas color coded for easy identification. These three areas encompass Mission Assets (green), primarily focused on missions such as TRACE and *Hinode* that have an extensive planning component; the Joint Science Operations Center (JSOC) Assets (gray), which are the data-oriented components involved in SDO/HMI and AIA operations; and Public Assets (red), which are primarily involved in presenting the outcomes of the first two to the general user community. Here we discuss each of these in more detail.

2.1. Mission (Coverage) Assets

2.1.1. Capturing Planning Metadata

Metadata associated with TRACE and *Hinode* images and observations consist of detailed image catalogs which describe individual images and higher-level descriptions of image sets, including pre-observation notes from the planner or observer, and post-observation

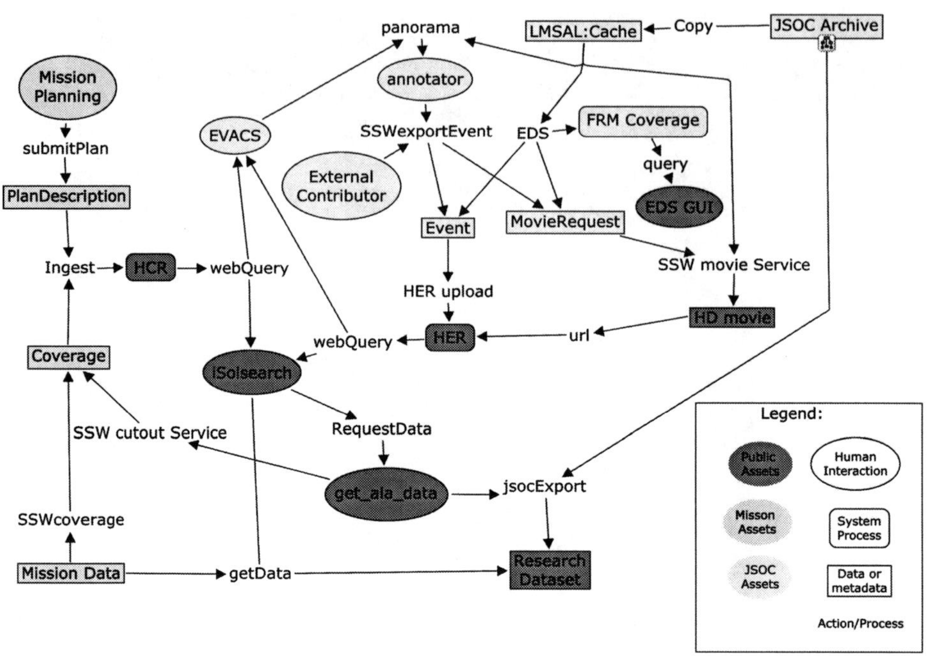

Figure 2 HEK Components include the Event and Coverage Registries (HER, HCR), Inspection and Analysis Tools, Event Identification System, and Movie Processing. Event services enable web clients to interact with the HEK and use the results to request data from the JSOC at Stanford.

annotations by the TRACE, SOT, and XRT science teams. Metadata associated with the observation planner is captured at the time the daily observing plan is committed. A summary of the plan is extracted by automated tools that then registers these plan descriptions into the HCR. After the plans are executed and the resulting data are cataloged, the results are compared and cross-referenced to the plan.

2.2. JSOC Assets

Calibrated SDO image data are accessed from the Joint Science Operations Center Science Data Processing (JSOC/SDP) system at Stanford over a private link to a 100 TB cache at the LMSAL. This cache feeds into three components within the LMSAL facility: the Event Detection System, which operates on a dedicated computer cluster; the AIA Visualization Center, containing a high-performance display system; and an image-extraction and browse-product generation system.

2.2.1. Event Detection System

The Event Detection System (EDS) autonomously orchestrates a variety of feature- and event-detection modules in order to populate the HEK with events from SDO data. The EDS continually acquires incoming data, provides the appropriate data to each module, receives results from modules through a standard API, and uploads the results to the HER. The outputs of the EDS are event descriptions in the VOEvent format (White *et al.*, 2006), including at a minimum the type, time, and location of the detection. Some modules produce ancillary data products and the EDS creates links from the HER to them, and the EDS also sends information to the movie-creation service to create summary clips for its events.

The Smithsonian Astrophysical Observatory-led Feature Finding Team (FFT) is the main source of modules for SDO (Martens *et al.*, 2010). However, the interfaces needed for modules to run in the EDS are published openly, and contributions from the wider solar-physics community are encouraged. Modules may run in a standalone mode or can be "triggered", meaning they subscribe to certain types of VOEvents from other modules in the EDS and run only when and where those types of events are found. This mode is useful for feature-recognition methods that are too computationally expensive to run on the entirety of SDO data.

The EDS consists of a single head node and a number of worker nodes connected via JXTA, a Java-based peer-to-peer networking technology (http://jxta.dev.java.net). At system initialization, the head node assigns modules to nodes. During operation, the head node monitors a cache and distributes announcements of the appropriate wavelengths/types of new data to the modules. Module results are sent back to the head node, which uploads them to the HER. The head node also tracks resource-usage statistics and can reassign modules to other nodes if there is a failure or an overload of one node's computing power. In addition, the head node can receive commands to update a module version or add in new modules during operations without restarting the rest of the system. The head node also maintains a Feature Recognition Method Coverage Registry that records when EDS modules have been run, what parameters and code versions were used by the modules, and which ones were triggered by others at what time. Besides internal record keeping, this registry permits identification of "All Clear" periods, which could form the basis for space-weather prediction (using the HER alone, it would not be clear if an absence of events during a specified period was due to actual solar inactivity, or problems that prevented the modules from running during that time).

Some modules (*e.g.* those detecting flares, emerging flux, and coronal dimmings) produce data that are useful for immediate action. These modules can produce "open" events soon after first detection, denoting the beginning of some phenomena, which form the basis for space-weather alerts or trigger immediate execution of other modules. When the event ends, the modules submit a "closed event" that updates the original in the HEK. These modules run on "quicklook" SDO data that arrive less than an hour after observation, while others may run on the final version that comes approximately a day later. The EDS stores this distinction and other information on data provenance in its VOEvents.

Currently there are two instances of the EDS, one at LMSAL and one at the Smithsonian Astrophysical Observatory (SAO); the latter includes some modules that produce auxiliary science products that are archived separately from the HEK. Although the EDS was built primarily to deal with the large data volume from SDO, it is applicable to other solar data sets as well; both Hα observations and SOHO/LASCO data are planned to be used by FFT modules. We are considering extending it to scientific data-processing applications outside of solar physics as well.

2.2.2. AIA Visualization Center

In parallel with the automated EDS, all AIA data are inspected by the AIA science team. A dedicated inspection facility, with a large high-resolution display and viewing tools, is used to review and annotate the automated detections from EDS and to add events that have no reliable means of automated detection. The results are passed along in the same format as those from the EDS. Similar, but less intensive, data inspection can be made by the science team directly from their workstations, or by external team members. The inspection

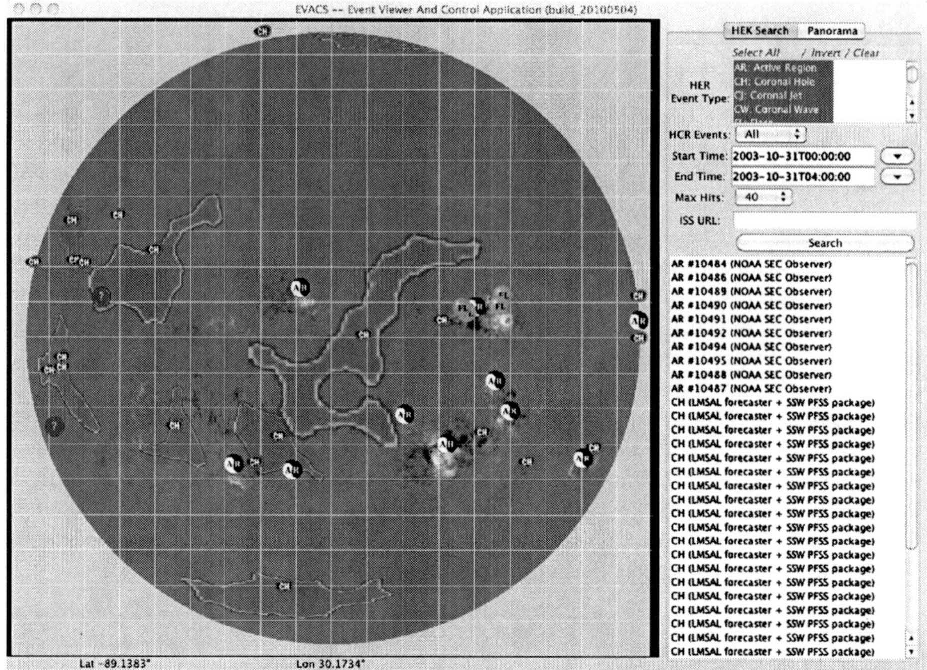

Figure 3 The Event Viewer and Control System control panel.

and browsing system consists of multiple tools that generate images, movies, and metadata which are then uploaded and stored into the HER and the HCR.

The Event Viewer And Control System (EVACS) tool displayed in Figure 3 uses metadata stored in the HER and HCR to display time and channel information for a given observation. EVACS controls a high-performance image-viewing tool called Panorama for use in reviewing the data and an annotation tool for reporting events. Panorama displays multiple channels of time-series of FITS or other image data. It can overlay them, pan and zoom spatially, and select regions in space and in time on a HiPerSpace data wall (DeFanti *et al.*, 2009) such as that displayed in Figure 4. Once a desired sequence and event is identified, the user launches the Annotator tool from within Panorama or EVACS. Panorama takes full advantage of modern hardware-accelerated graphics including the CGLX library (Doerr and Kuester, 2010) for multi-tiled display and the performance available from today's graphics chipsets to achieve high-speed visualization of very large image sets.

The Annotator tool has an extensive user interface to assist AIA data analysts in describing observed events. Once all appropriate values are supplied, the Annotator generates VOEvent XML content and registers it in the HER. If the data being annotated are associated with a coverage event, appropriate linkage and citation between the events in the HER and HCR are created and preserved so that users can browse in both directions from source observations to the derived data products and identified features.

2.2.3. Browse Products

Movie processing takes the inputs from EDS and inspection tools to generate movie clips in a variety of formats and resolutions. The particular combination of wavelength, cadence,

Figure 4 HiPerSpace data wall controlled by EVACS/Panorama.

and resolution are dependent both on the method that found the data and the type of feature reported. The resulting movies and ancillary data are web accessible with references in the HER.

2.3. Public Assets

2.3.1. *The Heliophysics Event and Coverage Registries*

A set of registries store the resulting outputs of all of the analyses described above and form the core of publicly accessible components of the HEK system. The Heliophysics Event Registry accepts the VOEvent files and registers their contents into a PostGIS database. The Heliophysics Coverage Registry contains descriptions of the instrument coverage, that is the "where", "when", and "how" of a particular observation. A variety of coordinate systems are supported for contributions based upon the World Coordinate System (Thompson, 2006).

2.3.2. *HEK Services*

The HEK is presented through a web server in a variety of forms. It is closely tied to the Stanford website (http://jsoc.stanford.edu) and supports an interface for passing the results of event searches and movie browsing directly to the data request services developed at Stanford. Besides the standard web interface – such as presenting users with movies, pre-computed data products, and static forms for interrogating the registries – the web server supports a variety of services that can be used by external teams to develop web applications.

2.3.3. *HEK Clients*

Aside from iSolSearch and SolarSoft clients that have been discussed earlier, clients from other developers are encouraged. Two that are currently under development are those of the

Table 2 A summary of HEK web services. Complete documentation is available on the LMSAL SDO documentation site (http://www.lmsal.com/sdodocs).

Service Area	Service	Description	Result
HER	search	Event search	Event list in a variety of formats
	view-event	Event display	Document describing event
	export-event	Reports back event details	VOEvent file or other format
	view attributes	Returns all values of attributes	List
	home	HER home page	Displays recent events submitted by humans
HEK	Heks	Authenticated Services	https://www.lmsal.com/hek/heks
	login	Login to HEK	Authorization
	comment	Comment on event	Records comment to existing event entry
	rate	Rate event	Records rating of existing entry
	add reference	Create link to event	Adds a URL link to existing entry
HCR	search	Instrument coverage search	Coverage list
	display	Details	Document describing instrument coverage
	ingest	Accepts coverage descriptions	Registers instrument coverage in HCR
SolarSoft	SSWexportEvent	Create and upload event to HER	VOEvent file
	SSWcoverage	Create and upload to HCR	VOEvent file
	SSWcutout	Create cropped image sets	Web summaries
	general	Time and coordinate services	Converted units

Helioviewer.org and jHelioviewer projects (http://helioviewer.org) and the Virtual Solar Observatory (Hill *et al.*, 2009). In addition, both EVACS and Panorama are distributed through SolarSoft in the Panorama package.

2.3.4. Data Request Management

One of the prime motivations for developing the HEK is to assist users in requesting subsets of the full image stream, rather than requesting large chunks of data that then requires winnowing on their part. This reduces both the network load and user overload. The last step in the use of the HEK is to get the data. This requires finding subsets of associated data with HER events and, when needed, ordering data based on field-of-view, time, resolution, and wavelength. A means to share and track these datasets is also useful to prevent repeated regeneration of popular datasets.

The HEK guides researchers to the most useful times and locations and will *suggest* reasonable fields-of-view and durations. With *guidance* from the HEK, researchers can access data by specifying cutouts, wavelengths, and sampling rates in space and time. These research data products use the best flat fields and calibration data available at the time that they are created. They are tagged in the HCR. Subsequent researchers can be directed to existing, local copies of research products through the HEK.

3. Conclusions

We have developed the Heliophysics Event Knowledgebase to address the immediate needs of the *Solar Dynamics Observatory*. However, the underlying motivation of devising a means for coping with petabyte datasets resonates with many other missions and projects throughout modern science. Our system design and implementation were intended to grow into a wider effort, as suggested by the use of "Heliophysics" in its title, when SDO is primarily a solar mission. We hope that our description here will encourage and benefit related efforts throughout the field of heliosphere research and beyond. For more information visit http://lmsal.com/sungate.

Acknowledgements We thank our partners at all of the institutions that have contributed to our efforts, most notably those at the Smithsonian Astrophysical Observatory and at the Royal Observatory of Belgium. This work was supported by NASA through grants NAS5-38099, NNM07AA01C, NNG04EA00C and Lockheed Martin Internal Research Funds.

References

Cline, D., Edgington, D., Mariette, J.: 2007, In: *MTS/IEEE Oceans 2007 Conf.*, IEEE Press, Vancouver, 1.

DeFanti, T., Leigh, J., Renambot, L., Jeong, B., Verlo, A., Long, L., Brown, M., Sandin, D., Vishwanath, V., Liu, Q., Katz, M., Papadopoulos, P., Keefe, J., Hidley, G., Dawe, G., Kaufman, I., Glogowski, B., Doerr, K., Singh, R., Girado, J., Schulze, J., Kuester, F., Smarr, L.: 2009, *Future Gener. Comput. Syst.* **25**(2), 114.

Delaboudinière, J.-P., Artzner, G.E., Brunaud, J., Gabriel, A.H., Hochedez, J.F., Millier, F., Song, X.Y., Au, B., Dere, K.P., Howard, R.A., Kreplin, R., Michels, D.J., Moses, J.D., Defise, J.M., Jamar, C., Rochus, P., Chauvineau, J.P., Marioge, J.P., Catura, R.C., Lemen, J.R., Shing, L., Stern, R.A., Gurman, J.B., Neupert, W.M., Maucherat, A., Clette, F., Cugnon, P., van Dessel, E.L.: 1995, *Solar Phys.* **162**, 291.

Doerr, K.-U., Kuester, F.: 2010, *IEEE Trans. Vis. Comput. Graph.* **99**, RapidPosts. doi:10.1109/TVCG.2010.59.

Freeland, S.L., Bentley, R.D.: 2000, *Encyc. Astron. Astrophys.*, IOP Publishing, Bristol. doi:10.1888/0333750888/3390

Handy, B.N., Acton, L.W., Kankelborg, C.C., Wolfson, C.J., Akin, D.J., Bruner, M.E., Caravalho, R., Catura, R.C., Chevalier, R., Duncan, D.W., Edwards, C.G., Feinstein, C.N., Freeland, S.L., Friedlaender, F.M., Hoffmann, C.H., Hurlburt, N.E., Jurcevich, B.K., Katz, N.L., Kelly, G.A., Lemen, J.R., Levay, M., Lindgren, R.W., Mathur, D.P., Meyer, S.B., Morrison, S.J., Morrison, M.D., Nightingale, R.W., Pope, T.P., Rehse, R.A., Schrijver, C.J., Shine, R.A., Shing, L., Strong, K.T., Tarbell, T.D., Title, A.M., Torgerson, D.D., Golub, L., Bookbinder, J.A., Caldwell, D., Cheimets, P.N., Davis, W.N., Deluca, E.E., McMullen, R.A., Warren, H.P., Amato, D., Fisher, R., Maldonado, H., Parkinson, C.: 1999, *Solar Phys.* **187**, 229.

Hill, F., Martens, P., Yoshimura, K., Gurman, J., Hourclé, J., Dimitoglou, G., Suarez-Sola, I., Wampler, S., Reardon, K., Davey, A., Bogart, R.S., Tian, K.Q.: 2009, *Earth Moon Planets* **104**, 315.

Hurlburt, N., Cheung, M., Bose, P.: 2008, *American Geophysical Union, Fall Meeting 2008*, abstract #SA53A-1580.

Leibacher, J., Sakurai, T., Schrijver, K., van Driel-Gesztelyi, L.: 2010, *Solar Phys.* **263**, 1. doi:10.1007/s11207-010-9553-0

Lemen, J.R., Title, A.M., Akin, D.J., Boerner, P.F., Chou, C., Drake, J.F., Duncan, D.W., Edwards, C.G., Friedlaender, F.M., Heyman, G.F., Hurlburt, N.E., Katz, N.L., Kushner, G.D., Levay, M., Lindgren, R.W., Mathur, D.P., McFeaters, E.L., Mitchell, S., Rehse, R.A., Schrijver, C.J., Springer, L.A., Stern, R.A., Tarbell, T.D., Wuelser, J.-P., Wolfson, C.J., Yanari, C.: 2010, *Solar Phys.*, submitted.

Martens, P.C.H., Attrill, G.D.R., Davey, A.R., Engell, A., Farid, S., Grigis, P.C., Kasper, J., Korreck, K., Saar, S.H., Savcheva, A., Su, Y., Testa, P., Wills-Davey, M., Bernasconi, P.N., Raouafi, N.-E., Delouille, V.A., Hochedez, J.F., Cirtain, J.W., DeForest, C.E., Angryk, R.A., De Moortel, I., Wiegelmann, T., Georgoulis, M.K., McAteer, R.T.J., Timmons, R.P.: 2010, *Solar Phys.*, accepted.

Schou, J., Scherrer, P.H., Bush, R.I., Wachter, R., Couvidat, S., Rabello-Soares, M.C., *et al.*: 2010, *Solar Phys.*, in preparation.

Thompson, W.T.: 2006, *Astron. Astrophys.* **449**, 791.

White, R., Allan, A., Barthelmy, S., Bloom, J., Graham, M., Hessman, F.V., Marka, S., Rots, A., Scholberg, K., Seaman, R., Stoughton, C., Vestrand, W.T., Williams, R., Wozniak, P.: 2006, *Astron. Nach.* **327**, 775.

Solar Phys (2012) 275:79–113
DOI 10.1007/s11207-010-9697-y

Computer Vision for the *Solar Dynamics Observatory* (SDO)

P.C.H. Martens · G.D.R. Attrill · A.R. Davey · A. Engell · S. Farid · P.C. Grigis · J. Kasper · K. Korreck · S.H. Saar · A. Savcheva · Y. Su · P. Testa · M. Wills-Davey · P.N. Bernasconi · N.-E. Raouafi · V.A. Delouille · J.F. Hochedez · J.W. Cirtain · C.E. DeForest · R.A. Angryk · I. De Moortel · T. Wiegelmann · M.K. Georgoulis · R.T.J. McAteer · R.P. Timmons

Received: 3 February 2010 / Accepted: 1 December 2010 / Published online: 7 January 2011

Abstract In Fall 2008 NASA selected a large international consortium to produce a comprehensive automated feature-recognition system for the *Solar Dynamics Observatory* (SDO). The SDO data that we consider are all of the *Atmospheric Imaging Assembly* (AIA) images plus surface magnetic-field images from the *Helioseismic and Magnetic Imager* (HMI). We produce robust, very efficient, professionally coded software modules that can keep up with the SDO data stream and detect, trace, and analyze numerous phenomena, in-

The Solar Dynamics Observatory
Guest Editors: W. Dean Pesnell, Phillip C. Chamberlin, and Barbara J. Thompson.

P.C.H. Martens · G.D.R. Attrill · A.R. Davey · A. Engell · S. Farid · P.C. Grigis · J. Kasper · K. Korreck · S.H. Saar · A. Savcheva · Y. Su · P. Testa · M. Wills-Davey
Harvard-Smithsonian Center for Astrophysics, 60 Garden Street, Cambridge, MA 02138, USA

P.C.H. Martens (✉)
Department of Physics, Montana State University, 247 EPS, Bozeman, MT 59717-3880, USA
e-mail: martens@physics.montana.edu

A. Savcheva
Astronomy Department, Boston University, 725 Commonwealth Avenue, Boston, MA 02215, USA

P.N. Bernasconi · N.-E. Raouafi
Applied Physics Laboratory, Johns Hopkins University, 11100 Johns Hopkins Road, Laurel, MD 20723, USA

V.A. Delouille · J.F. Hochedez
SIDC-Royal Observatory of Belgium, Avenue Circulaire 3, 1180 Brussels, Belgium

J.W. Cirtain
Marshall Space Flight Center-NASA, Mail Code:VP 62, Marshall Space Flight Center, AL 35812, USA

C.E. DeForest
Southwest Research Institute, 1050 Walnut Street Suite 300, Boulder, CO 80302, USA

R.A. Angryk
Department of Computer Science, Montana State University, 362 EPS, Bozeman, MT 59717-3880, USA

cluding flares, sigmoids, filaments, coronal dimmings, polarity inversion lines, sunspots, X-ray bright points, active regions, coronal holes, EIT waves, coronal mass ejections (CMEs), coronal oscillations, and jets. We also track the emergence and evolution of magnetic elements down to the smallest detectable features and will provide at least four full-disk, nonlinear, force-free magnetic field extrapolations per day. The detection of CMEs and filaments is accomplished with *Solar and Heliospheric Observatory* (SOHO)/*Large Angle and Spectrometric Coronagraph* (LASCO) and ground-based Hα data, respectively. A completely new software element is a trainable feature-detection module based on a generalized image-classification algorithm. Such a trainable module can be used to find features that have not yet been discovered (as, for example, sigmoids were in the pre-*Yohkoh* era). Our codes will produce entries in the *Heliophysics Events Knowledgebase* (HEK) as well as produce complete catalogs for results that are too numerous for inclusion in the HEK, such as the X-ray bright-point metadata. This will permit users to locate data on individual events as well as carry out statistical studies on large numbers of events, using the interface provided by the Virtual Solar Observatory. The operations concept for our computer vision system is that the data will be analyzed in near real time as soon as they arrive at the SDO Joint Science Operations Center and have undergone basic processing. This will allow the system to produce timely space-weather alerts and to guide the selection and production of quicklook images and movies, in addition to its prime mission of enabling solar science. We briefly describe the complex and unique data-processing pipeline, consisting of the hardware and control software required to handle the SDO data stream and accommodate the computer-vision modules, which has been set up at the Lockheed-Martin Space Astrophysics Laboratory (LMSAL), with an identical copy at the Smithsonian Astrophysical Observatory (SAO).

Keywords Instrumentation and data management · Solar Dynamics Observatory

1. Introduction

The open and immediate dissemination of solar data from all NASA solar missions is one of the greatest assets in solar physics, and the *Solar Dynamics Observatory* (SDO) represents

I. De Moortel
School of Mathematics & Statistics, University of St Andrews, North Haugh, St Andrews, KY16 9SS, UK

T. Wiegelmann
Max-Planck-Institut für Sonnensystemforschung, Max-Planck Strasse 2, 37191 Katlenburg-Lindau, Germany

M.K. Georgoulis
Research Center for Astronomy and Applied Mathematics, Academy of Athens, 4 Soranou Efesiou Street, Athens 11527, Greece

R.T.J. McAteer
School of Physics, Trinity College Dublin, Dublin 2, Ireland

R.T.J. McAteer
Department of Astronomy, New Mexico State University, P.O. Box 30001, MSC 4500, Las Cruces, NM 88003-8001, USA

R.P. Timmons
Lockheed Martin Advanced Technology Center, Bldg. 252, 3251 Hanover Street, Palo Alto, CA, 94304, USA

a new frontier in quantity and quality of solar data. At about 1.5 TB per day, the data will not be easily digestible by solar physicists using the same methods that have been employed for images from previous missions. The availability of this imagery in a form that is useful for scientists will be crucial to the success of the mission.

In order for solar scientists to use the SDO data effectively they need metadata that will allow them to identify and retrieve data sets that address their particular science questions. Providing this metadata, via the pipeline processes described below, is the core of our computer vision project.

We are building a comprehensive computer-vision post-processing pipeline for SDO, abstracting complete metadata on many of the features and events detectable on the Sun without human intervention. Our feature-finding team will debug, deploy, and support a modular, extensible pipeline framework to augment the existing data system, and comprises experts in each subfield of computer vision. Our project unites more than a dozen individual, existing codes into a systematic tool that can be used by the entire solar community. In addition to static, standards-based codes that detect well-defined and well-studied features, we are developing and deploying a trainable feature recognition system for post-processing. It will enable the systematic study of human-recognizable features that are yet to be identified. This unique capability allows flexible scientific exploration of the SDO data set as new types of features gain interest.

2. Operations Concept and Overview of Modules

The feature-finding codes described here are part of the SDO *Event Detection System* (EDS) at the *Joint Science Operations Center* (JSOC; joint between Stanford and the Lockheed Martin Solar and Astrophysics Laboratory (LMSAL)). The basic purpose of the EDS is to analyze the SDO imagery for events and features as soon as the data reach the JSOC, *i.e.* in near real time. This concept implies that the analysis must be automated. The setup allows for the generation of space-weather alerts, relevant quicklook images and movies, as well as timely alerts for the solar-physics community. The feature-finding modules that are described in the remainder of this article form the core part of the EDS, but there will also be room for other, community-produced tools.

The metadata produced by the EDS software modules are stored in the *Heliophysics Event Knowledgebase* (HEK), which is accessible on-line for the rest of the world directly or via the *Virtual Solar Observatory* (VSO). Solar scientists will be able to use the HEK to select event and feature data to download for science studies. Such a preselection is necessary because it is impossible to retrieve and store the full 1.5 TB per day of SDO imagery. SDO data is available from archives at JSOC, the Smithsonian Astrophysical Observatory (SAO), and European sites. In order for our feature-finding algorithms to survive the throughput of SDO data, we are producing software that is modular (there is an independent, separate program for each feature-finding task), robust, and efficient.

Some of the feature-finding modules described here have been designed to monitor the data stream and extract metadata continuously. Others (such as the flare module) are self-triggered: they carefully monitor part of the data stream for an alert and then spring into full action when such an alert is issued. A third group of modules does not operate continuously, but is triggered by one or more alerts from other modules (*e.g.*, the wave-detection algorithms).

In Tables 1 and 2 we present an overview of the software modules produced by our Feature Finding Team (FFT), specifying their names, modes of operation, scientific purpose,

Table 1 Elements of the SDO computer vision system related to space weather alerts.

Science Target	**Flares**	**Coronal Dimming**	**Magnetic Feature Tracking**	**Filaments**
S/W Module Name	Flare Detective	Dimming Detector	SWAMIS	AAFDCC
Triggered or Continuous?	Triggered	Triggered	Continuous	Continuous
Trigger Source Data	AIA 193	AIA 193		
Trigger Source Module(s)	Self	Self		
Binned Image for Trigger	16 X 16	512 X 512		
Cadence for Trigger	Full	5-6 minutes		
Source Data	All AIA Channels	AIA	HMI Time Averaged Magnetograms	Global Hi-res Ha Network
Source Data Binning	None	512 X 512	None	Full; some smoothing
Source Data Cadence	Full	5-6 minutes	6 minutes	Full (~ 4/day eventually)
S/W Module Data	Flare Detective	Dimming Detector	SWAMIS	AAFDCC
Programming Language	IDL	IDL	Perl/PDL	IDL and C
Module Location	LMSAL	LMSAL	LMSAL	SAO
Computing Requirement		20 minutes CPU/day	8-16 CPUs to operate at 4x realtime	Most < 1 min CPU/Image
Responsible Team Member	Paolo Grigis	Meredith Wills-Davey	Craig DeForest	Pietro Bernasconi
Originating Organization	SAO	SAO	SwRI	JHU-APL
Output Data Volume		10 kB/event, 5 MB/movie	120 MB/day nominal	~ 1.5 MB per Ha image
First Pipeline Installation	June 2010	July 2010	November 2010	March 2010
Heritage	EIT, TRACE, RHESSI	New	MDI	GBO H-alpha
Science Target	**Active Regions**	**Sigmoids**	**PIL Mapping**	**CMEs**
S/W Module Name	SPoCA	Sigmoid Sniffer	PIL Finder	CME Detector/Tracker
Triggered or Continuous?	Continuous	Triggered	Continuous	Triggered
Trigger Source Data		AIA 211		LASCO C2 and C3
Trigger Source Module(s)		Self		Flares, Dimmings, Sigmoids
Binned Image for Trigger		Full Resolution		Full Resolution
Cadence for Trigger		10 min		Full LASCO Cadence
Source Data	AIA 171/195, 211/335, 94	AIA 94, 131	HMI Magnetograms	LASCO C2 and C3
Source Data Binning	None	None	Depends on science specs	None; front smoothing
Source Data Cadence	15 minutes	10-20 s	~5 min	Full (~1 hr)
S/W Module Data	SPoCA	Sigmoid Sniffer	PIL Finder	CME Detector/Tracker
Programming Language	C++	IDL	IDL	IDL
Module Location	LMSAL	LMSAL	SAO	SAO
Computing Requirement	30 sec/pair of full-res AIA	10 min CPU/image	Standard desktop	30 minutes/event
Responsible Team Member	Veronique Delouille	Nour-Eddine Raouafi	Alexander Engell	Meredith Wills-Davey
Originating Organization	ROB	JHU-APL	SAO	SAO
Output Data Volume	6 kB/event	5 MB/day	~ 2 Mb/day	1-8 Mb/day
Scheduled Installation	October 2010	November 2010	December 2010	Spring 2011
Heritage	EIT	SXT, SXI, XRT	Kitt Peak, SOLIS	LASCO

module authors, *etc.* The modules in Table 1 include those that spawn near real-time space weather alerts (flares, coronal dimmings, and emerging flux), and others that are closely related to space weather. The modules in Table 2 will be added during the mission and also run on the already archived data to complete the HEK entries for the entire mission. A detailed individual description of each module is given in the remainder of this paper. At the end, we also describe the hardware setup of the data pipeline in which the modules are operating during the mission. Identical copies of the data pipeline have been installed at LMSAL and SAO. The source codes of the software modules that we are developing will be made available to the scientific community via SolarSoft. With that in mind we list, where known, the CPU and output requirements for each module in the remainder of the article.

3. Flare Detection

The flare detection module of the computer-vision system serves a twofold purpose:

- Provide rapid flare alerts for the space-weather community in near real time (*trigger component*)
- Generate a statistical survey of flares and measure physical parameters relevant for flare science (*analysis component*)

This module is named *flare detective* (FD).

Table 2 Further elements of the SDO computer vision system.

Science Target	**Coronal Holes**	**X-ray Bright Points**	**Sunspots**	**Global NLFFFs**
S/W Module Name	SPoCA	BP Finder	SWAMIS	Optimization Code for Full Disk
Triggered or Continuous?	Continuous	Continuous	Continuous	Continuous
Trigger Source Data				
Trigger Source Module(s)				
Binned Image for Trigger				
Cadence for Trigger				
Source Data	AIA 171/195, 211/335, 94	AIA 171, 195, 211	HMI Magnetograms	HMI+AIA for comparison
Source Data Binning	None	None (TBC)	None	Binned to 256x256 for ARs
Source Data Cadence	15 minutes	Full	TBD (5-60 min)	~5 min
S/W Module Data	SPoCA	BP Finder	SWAMIS	Optimization Code
Programming Language	C++	IDL	PDL	C
Module Location	LMSAL	SAO	LMSAL	MPS
Computing Requirement	Same as for Active Regions		< 1 CPU	2 hours/vector-magnetogram
Responsible Team Member	Veronique Delouille	Steve Saar	Craig DeForest	Thomas Wiegelmann
Originating Organization	ROB	SAO	SwRI	Max Planck Sonnenforschung
Output Data Volume	6 kB/event	600 kB/10 minutes	< 1 MB/day	256x256x256 datacube
First Pipeline Installation	January 2011	February 2011	Spring 2011	When HMI VMG's become routine
Heritage	EIT	EIT	MDI	SOT

Science Target	**Jets**	**Oscillations**	**"EIT Waves"**	**Trainable Feature Recognition**
S/W Module Name	Jet Detector	Oscillation Finder	EIT Wave Tracker	
Triggered or Continuous?	Triggered	Triggered	Triggered	Continuous
Trigger Source Data	TBD AIA Channels	AIA 193	AIA 193	
Trigger Source Module(s)	XRBP & CH	Flares, CMEs, Dimmings	Dimmings	
Binned Image for Trigger	1024x1024 (TBC)	N/A	1024x1024	
Cadence for Trigger	1 min	N/A	Full	
Source Data	AIA 193	Appropriate AIA Channels	AIA 193	All AIA Channels, some HMI Data
Source Data Binning	None	None	None; smoothing applied	128 X 128 subimages
Source Data Cadence	1 min	None	Full	Cadence TBD
S/W Module Data	Jet Detector	Oscillation Finder	EIT Wave Tracker	Trainable Feature Recognition
Programming Language	IDL	IDL	IDL	C/IDL
Module Location	SAO	SAO	SAO	SAO
Computing Requirement	15 seconds per event	60 minutes CPU/event	Very small	
Responsible Team Member	Antonia Savcheva	James McAteer	Meredith Wills-Davey	Rafal Angryk
Originating Organization	SAO	New Mexico State Univ.	SAO	MSU
Output Data Volume	90 MB/jet		2 Mb/event	0.1% of image volume
First Pipeline Installation	February 2011	Spring 2011	Spring 2011	December 2010 for image vectors
Heritage	XRT	TRACE, EIT	TRACE	TRACE

The trigger component of the FD needs to be very fast and works on heavily binned *Atmospheric Imaging Assembly* (AIA) images of 16×16 macropixels, each covering approximately a $150'' \times 150''$ square (see Figure 1 for an example). A peak-detection algorithm is applied to the integrated signal in each macropixel, thus providing the start, peak, end time, and approximate location of the flare. This approach allows detection of simultaneous flares in different active regions (ARs). We use a detection procedure based on the flare-identification algorithm used on *Reuven Ramaty High Energy Solar Spectroscopic Imager* (RHESSI) data by Christe *et al.* (2008), which works well for noisy and background-affected lightcurves. Basically, this algorithm smooths the lightcurves and detects flares as intervals of positive derivative and negative derivative around a local maximum. We are running the trigger in the 193 Å band, because its behavior is well known from past observations. However, if during the mission it is found that some flares are better observed in the new, hotter bands 131 Å and 94 Å, we will use those as well.

The more computationally intensive analysis component runs only on the subset of data close in time and space to the flare (as detected by the trigger component). The key parameters determined for each flare are: flare timing, location, area, peak intensity, plasma temperature, and emission measure, lightcurves in each EUV channel, characteristic time scales, associated NOAA AR number, *Geostationary Operational Environmental Satellite* (GOES) class, peak emissivity in selected EUV and SXR emission lines from SDO's *Extreme ultraviolet Variability Experiment* (EVE), movies of the flare from rise to end, and

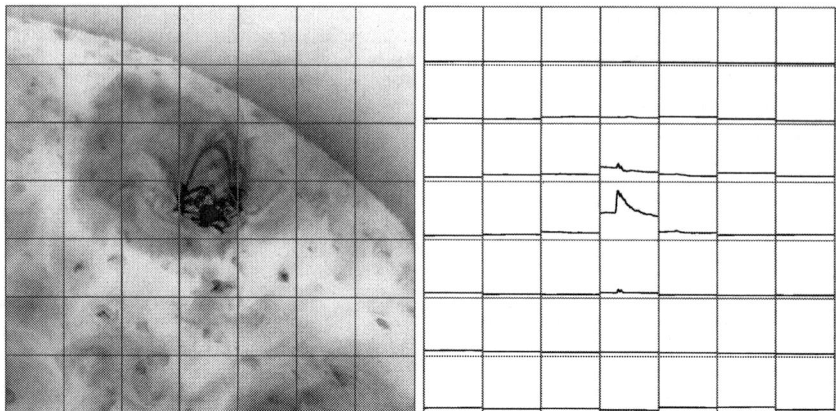

Figure 1 Left: AIA 193 Å image of active region (AR) 11069, taken on 5 May 2010 at 17:02 UT. Each macropixel consists of 256 × 256 AIA pixels, a square with about 150″ sides. Right: Lightcurves for each macropixel over a time interval of one hour. The brightness scale is linear without background subtraction. An M1 flare is clearly visible in the lightcurve corresponding to the macropixel containing the core of the AR.

a classification of the flare in terms of flare type (*e.g.* two-ribbon flares). This classification will be provided by the generic feature-recognition system described in Section 16.

The analysis component determines the flare area and centroid from difference images at peak time *versus* flare onset. A set of background-subtracted lightcurves in all EUV channels is generated for the flare interval. These lightcurves are used to estimate the plasma temperature and to provide approximate emission-measure values. Summary quicklook movies will be produced for each event in the 94 Å, 131 Å, 171 Å, and 193 Å channels. Our approach provides complete coverage at least down to GOES C1-level flares, with possible extension to smaller events.

The set of statistical information about the flares provided by the FD over a substantial fraction of a solar cycle can be used to determine the distribution of flares in space, time, size, and thermal energy. This information is of critical importance to study the energy-release mechanisms in flares.

4. Hα Filaments

Filaments in the solar chromosphere are well-known, large-scale structures of relatively dense and cool plasma suspended in the hot and thin corona. They are particularly well visible in Hα filtergrams. Filaments and their sources, filament channels, are known to align with photospheric magnetic polarity inversion lines (PILs: Martin, 1998). All solar eruptions occur above PILs. In addition, filaments are known to involve helical magnetic fields, twisted beyond their minimum-energy, current-free, magnetic configuration (Martin, Bilimoria, and Tracadas, 1994; Rust and Martin, 1994; Pevtsov, Balasubramaniam, and Rogers, 2003). Non-potential (*i.e.*, helical) magnetic fields are invariably involved in solar eruptions and give rise to coronal mass ejections (CMEs). Filaments themselves often erupt fully or partially into CMEs, leading to a complete or partial filament disappearance from the solar disk (Gilbert *et al.*, 2000; Asai *et al.*, 2003; Jing *et al.*, 2004). If one knows the sense of twist (chirality) in a filament before its disappearance, then one has additional clues about the magnetic helicity of the CME that might be useful in assessing the CME's possible geoeffectiveness (*e.g.*, Yurchyshyn *et al.*, 2001; Rust *et al.*, 2005).

In recent years many algorithms have been developed for filament detection; see, *e.g.*, Gao, Wang, and Zhou (2002), Shih and Kowalski (2003), and Qu *et al.* (2005). The code by Fuller, Aboudarham, and Bentley (2005) was recently implemented as part of the European Grid of Solar Observations (EGSO) project. Most of these codes do not go far beyond the mere detection of filaments. The *Advanced Automated Filament Detection and Characterization Code* (AAFDCC) that we have developed is a step beyond a typical filament identification code. Besides mere identification, *i*) it determines the filament's complete shape, spine, and orientation angle, *ii*) if a filament is broken up into two or more pieces, it correctly identifies them as a single entity, *iii*) it finds the filaments' magnetic chirality (sense of twist), and *iv*) it tracks them from image to image for as long as they are visible on the solar disk. The code was originally developed by Bernasconi, Rust, and Hakim (2005) and is fully functional, tested, and validated. It currently runs daily on an APL server and so far has fully automatically generated a database of daily solar-filament properties from July 2000 until the present with only occasional gaps due to lack of Hα images for specific days. To our knowledge, our code is the only one that accomplishes all of these important tasks at once without any human intervention.

We have described the algorithm extensively in Bernasconi, Rust, and Hakim (2005) so here we only highlight its most important steps. At regular intervals (currently four times per day) the code checks if new (most recent) full-disk Hα images have been posted online by the Global High-Resolution Hα network (http://www.bbso.njit.edu/Research/Halpha). If they have, it downloads the most recent image, performs some preprocessing to clean and standardize it, and begins the analysis by identifying all dimmings in the image (candidate filaments). It then eliminates all very small, or round and very dark features that most likely are not filaments, but sunspots. For each positively identified filament it determines its chain code outlining the shape, and its core skeleton, or spine. The algorithm can also merge filament segments close to each other, thus uncovering the true shape. Once the spine is identified, the code checks for filament barbs whose orientation (left/right-handed) implies the filament magnetic chirality (dextral/sinistral, respectively: Martin, 1998). The helicity of a given filament is determined if a decisive majority of barbs have a common orientation and is left undetermined if the number of right-bearing and left-bearing barbs is evenly split. Finally, the code compares the location of each filament found on the most recent image with the location of filaments detected in an older image. This allows us *i*) to track the evolution of filaments in time; *ii*) to detect the appearance of a new filament; *iii*) to detect if a filament has disappeared from the visible disk, thus indicating a possible filament eruption (this requires a fast cadence which is not yet available). A typical result is shown in Figure 2. We have validated our code by comparing its results with the filament list of Pevtsov, Balasubramaniam, and Rogers (2003). The list was reproduced with an accuracy of 72%. The main difference between our results and those of Pevtsov *et al.* was that in our case a larger number of filaments had undetermined chirality. We were able to identify the orientation of all filament barbs in all cases, so we attribute this discrepancy to our unbiased automated chirality finder, as opposed to the biased chirality determination performed by a human operator (see Bernasconi, Rust, and Hakim, 2005).

For the SDO computer vision project, we have modified the 2005 code to improve its filament identification and characterization performance. We are now applying an adaptive-threshold method that changes the threshold for identification of a filament's outline. This allows the code to identify and characterize thin filaments in active regions that usually are difficult to detect with a single threshold for the entire solar disk, as was done in the 2005

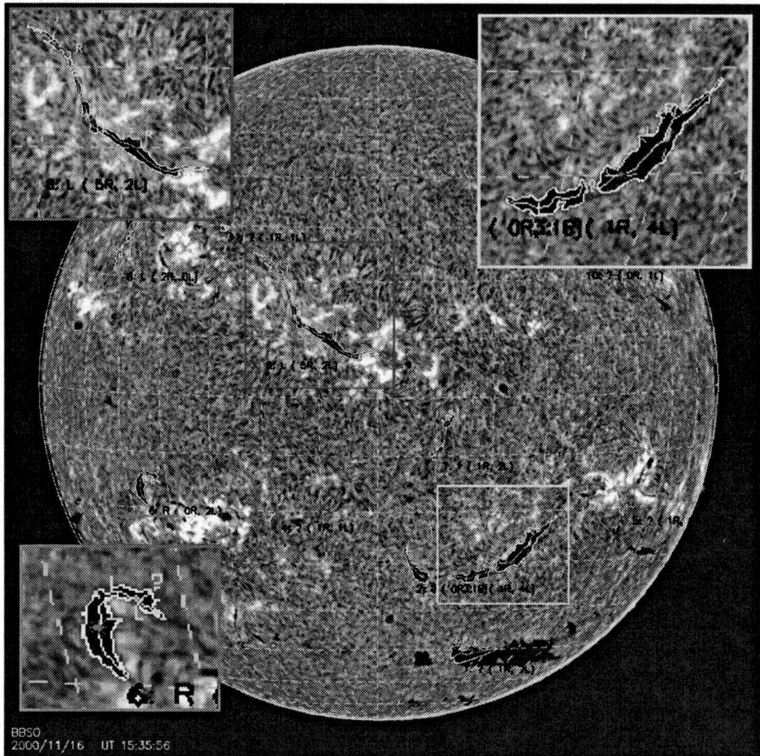

Figure 2 Typical results of the filament finder algorithm of Bernasconi, Rust, and Hakim (2005), applied to an Hα image taken by the Big Bear Solar Observatory (BBSO) Hα telescope at 15:36 UT on 16 November 2000. Insets at the sides show three well-formed filaments magnified: one sinistral (red box) and two dextral (magenta and cyan boxes). Yellow curves outline the filaments, cyan curves indicate their spines, and the red points indicate the intensity-weighted centroid of the filaments.

version. We have also made the detection of merging in a segmented filament more accurate. Another enhancement we have implemented with respect to the 2005 version is adding to the tracking algorithm the capability to determine whether a filament is the result of merging of two or more previously identified filaments, and if an older filament has split into two or more components in a subsequent image. For the SDO feature-finding project we have also produced a new format of module output. Like the current version, it still produces full-disk Hα maps with contours of the identified filaments, location of filament center, filament spine, and barbs, and printed filament ID with chirality (if any is determined). In addition, for each identified filament the module provides a VOEvent entry which contains information about the filament location, outline (chain code), spine skeleton, area and length, number of barbs and in which direction they point, and finally the filament chirality based upon the barb-direction analysis. The module has been deployed since March 2010 and posts results at least twice a day to the HEK. The Big Bear Solar Observatory/New Jersey Institute of Technology and the Global High-Resolution Hα Network provide and maintain the Hα FTP data archive from which we derive our data. The filament code can be used either for statistical study of filament properties over long periods of time spanning a full solar cycle, or if run in near real-time it can provide useful information for space-weather forecasting. When the code detects the disappearance of a large filament, it can deliver a CME warning. Where the code

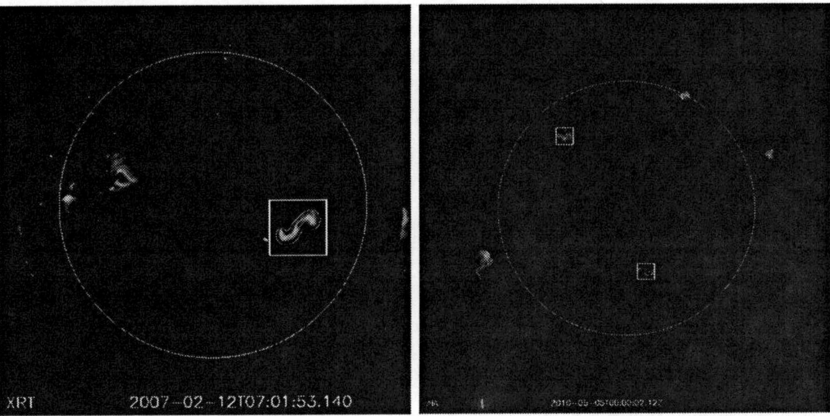

Figure 3 Examples of the results of the sigmoid sniffer module: (left) a right-handed sigmoid is detected in an *Hinode*/XRT full-disk image of 12 February 2007; (right) a left-handed sigmoid (bottom box) detected in a 94 Å image of 5 May 2010 from SDO/AIA. The feature in the top box is a misidentification by the code.

has determined the chirality of the disappeared filament, it provides information about the orientation of the erupting flux-rope and thus the geoeffectiveness of the associated CME.

5. Sigmoids

The solar X-ray corona frequently presents us with bright sigmoidal structures. These *sigmoids* (Rust and Kumar, 1996) are considered to be directly or indirectly related to magnetic structures twisted above their lower energy state (for reviews see Gibson *et al.*, 2006; Green *et al.*, 2007) and, as such, they are broadly recognized as telltale signatures of unstable coronal magnetic-flux systems. This view is supported by the realization that X-ray sigmoids are associated with solar eruptions (Rust and Kumar, 1996). Canfield, Hudson, and McKenzie (1999) found that active regions (ARs) with sigmoidal structures are about 70% more likely to erupt than non-sigmoidal ones, with most ($\approx$ 84%, according to Canfield *et al.*, 2007) triggering solar eruptions within one or two days. Eruptions typically result in the disappearance of an X-ray sigmoid and the launch of a CME that is thought to be associated with the twisted/stressed sigmoidal pre-eruption fields.

Not all CMEs are preceded by sigmoids. Further, not all sigmoids are the same. The most persistent and fainter of them may or may not trigger an eruption, while the brightest, more-transient sigmoids show a much more robust association with eruptions (Gibson *et al.*, 2006). From this viewpoint, although only a fraction of CMEs are associated with sigmoids, it has been accepted that an efficient, automatic, sigmoid recognition offers an unbiased way of identifying short-term progenitors for many CMEs. What is perhaps more important than CME prediction is the physical understanding of what an X-ray sigmoid is and why it occurs. Two model classes are proposed to explain this phenomenon: *i*) The classical view (Rust and Kumar, 1996) treats sigmoids as flux ropes undergoing the helical kink instability. *ii*) More recent views, assisted by numerical simulations, consider upward or downward kinking, inverse bald-patch magnetic geometries, or hyperbolic flux tubes to infer sigmoids with helicity similar or opposite to that of a Titov and Démoulin (1999) flux rope undergoing either the helical kink instability or the torus instability (*e.g.*, Fan and Gibson, 2003, 2004; Kliem, Titov, and Török, 2004; Török and Kliem, 2005, 2007; Gibson *et al.*, 2006;

Green *et al.*, 2007). Regardless of their complicated similarities and differences, all of these models tend to treat sigmoids as *single* flux ropes or current sheets surrounding these flux ropes. Competing with the flux-rope models are models where the flux rope forms before the eruption by a sheared arcade created along a photospheric magnetic PIL (Pneuman, 1983; van Ballegooijen and Martens, 1989, 1990), and models where the sheared arcade gives rise to a flux rope only after the eruption (Antiochos *et al.*, 1994, 1999). Evidence in support of all of these models also exists in recent observations. Of the 107 X-ray sigmoids that Canfield *et al.* (2007) studied using full-resolution observations by the *Soft X-ray Telescope* (SXT: Tsuneta *et al.*, 1992) onboard the *Yohkoh* mission (Ogawara *et al.*, 1991), none appeared to be a single $\mathcal{S}$- or inverse $\mathcal{S}$-shaped loop (see also McKenzie and Canfield, 2008). Instead, they appeared to consist of *multiple* loop patterns arranged such that they are discerned as single sigmoidal structures in limited-resolution images. If these results are confirmed, many sigmoid models may need substantial revision. For the purpose of understanding sigmoids and predicting CMEs, we will use the automatic pattern recognition "sigmoid sniffer" algorithm, first described by LaBonte, Rust, and Bernasconi (2003). A detailed description of the algorithm is given by Bernasconi, Raouafi, and Georgoulis (2010). Full-disk X-ray images are inserted as input to the algorithm, which first uses multiple brightness thresholds to discern persistent bright structures. If one or more candidate sigmoids are identified, the code infers the orientation-angle profile of successive points along the structures' outline and compares it to the expected profile of the theoretical $\mathcal{S}$-shaped curve. From the fit, one recovers the handedness (forward $\mathcal{S}$ [right-handed], or inverse $\mathcal{S}$ [left-handed]), the orientation, and the aspect ratio of the sigmoid. The code further issues a CME warning in case it identifies a sigmoid. This information, along with the location and the total size of the sigmoid, will be provided as output of the module. The sigmoid sniffer algorithm has been applied extensively to full-disk X-ray images from *Yohkoh*/SXT and the GOES/*Solar X-Ray Imager* (SXI) instrument, resulting in the automatic identification of numerous sigmoids. For this project, it will be applied to data from the following instruments.

i) *X-Ray Telescope* (XRT; Golub *et al.*, 2007) on *Hinode* (Kosugi *et al.*, 2007): XRT images have significantly higher spatial resolution (1 arcsec per pixel) than *Yohkoh*/SXT images, so the sigmoid sniffer routinely identifies and delivers crisper, finer sigmoids in *Hinode*/XRT data. The sigmoid sniffer has been implemented to automatically process quicklook *Hinode*/XRT data at SAO. The reported XRT charge-coupled device (CCD) contamination (Narukage *et al.*, 2011, SolarNews, 4 September 2007) covering $< 4\%$ of the XRT field of view is not a major concern for sigmoid detection. The reported irregularities in the *Hinode* downlink have affected the cadence of the incoming XRT images, but even the present cadence of ≈ 1 to a few minutes for the quick-look XRT data is quite sufficient for the purposes of the sigmoid sniffer.

ii) SDO/AIA: Here we will identify sigmoids primarily in the highest-energy channels (94 Å and 131 Å), which correspond to flaring active regions ($10^{6.8-7.2}$ K), but also in some of the lower energy channels (211 Å and 335 Å), which correspond to the (not necessarily flaring) active-region corona ($10^{6.3-6.4}$ K). SDO/AIA data have just become available through JSOC and/or via the Virtual Solar Observatory (VSO), so final preparations for pipeline deployment are being carried out.

An example of the output is depicted in Figure 3. Two sigmoids are detected in *Hinode*/XRT (left) and SDO/AIA (right) full-disk images. Among other characteristics of the detected features, the sigmoid sniffer determines the sigmoid's handedness, which is right for the XRT sigmoid and left for the AIA one in the lower box. Noisy data may result in misidentification, such as the feature shown in the upper box of the AIA image. Reducing misidentifications to an absolute minimum is one of our current development goals.

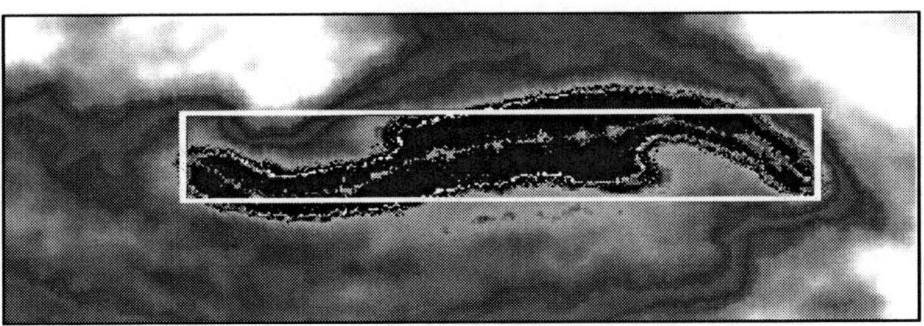

Figure 4 Example of the interaction between the sigmoid sniffer and the filament identification, and the characterization module. The background is a false-color *Yohkoh*/SXT image, with the sigmoid appearing dark. The image box is $51.5'' \times 9.7''$ in size. The white box is drawn by the sigmoid sniffer, while the curved sigmoid outline and the pink sigmoid "spine" stem from the filament-identification code (adapted from Rust and LaBonte, 2005).

The high-resolution *Hinode*/XRT and SDO/AIA observations may help resolve the controversy concerning the morphology of sigmoids as single, monolithic, or multiple structures. In addition, it will be very interesting to see whether the sigmoidal morphology persists in different AIA temperature channels. Serious morphological differences in different temperatures would again constrain the modeling of sigmoids, especially the spatial distribution of temperature enhancements along them. For example, is the transient sigmoid caused by upward kinking (peak temperature on top), giving rise to a classical flare cusp (Masuda *et al.*, 1994), or is it mostly due to a downward kinking toward a possible low-lying bald patch (peak temperature at bottom)? Moreover, what is the physical difference between a *complete* and a *partial* sigmoid eruption, where the sigmoid reappears shortly after its disappearance?

The sigmoid sniffer will be cross-correlated with the filament identification and characterization module of Section 4. In an illustrative example, Rust and LaBonte (2005) first applied the sigmoid sniffer to a *Yohkoh/Soft X-Ray Telescope* (SXT) image to identify a sigmoid and then applied the filament code to it to identify its "spine" and infer its aspect ratio more accurately. The example has been reproduced in Figure 4. Further developments of the sigmoid sniffer based on wavelet filtering aim to improve edge detection of solar features and are expected to enhance the overall efficiency of the package.

6. CME Recognition and Tracking

Over the years, coronal mass ejection (CME) recognition and tracking has become increasingly robust, evolving from observer-dependent methods (*e.g.* Gopalswamy *et al.*, 2009) to fully automated detection. A number of tracking algorithms have been developed, each implementing different methods and criteria; techniques include modified Hough transforms (Robbrecht and Berghmans, 2004), threshold segmentation (Olmedo *et al.*, 2008), multiscale filtering (Byrne *et al.*, 2009), adaptive filtering and segmentation (Boursier *et al.*, 2009), and forward-modeling approximations (Boursier, Lamy, and Llebaria, 2009).

Our algorithms are most comparable to those of Boursier *et al.* (2009) and Boursier, Lamy, and Llebaria (2009) in that CME fronts are tracked organically in two dimensions, using a polar coordinate system. This makes it possible to measure both radial and lateral dynamics, enabling us to follow all aspects of the CME expansion.

By monitoring local intensity changes, events are detected as they emerge from behind the *Large Angle and Spectrometric Coronagraph* (LASCO) C2 and C3 coronagraph disks.

After initial identification, our algorithm tracks the front edge of the CME, identified as the outermost strong positive gradient with respect to the background. Lateral expansion is incorporated by measuring intensity increases along adjacent radial coordinates. Each point along the leading front is then tracked across the field of view, recording the two-dimensional structure of the leading edge as a function of time.

To determine the most energetically favorable course of propagation, the measured fronts are used to create a smoothed "time surface." To do this, the points along each front are splined and assigned a time-equivalent value (usually assuming units of seconds); they are then simultaneously mapped onto a null background and subjected to a recursive-smoothing kernel. The result is a two-dimensional array in which each pixel value corresponds to a point in time. This "time surface" is then subdivided into contours separated by constant time intervals (typically about five minutes) with the original measured fronts also included.

With the data processed in this way, it becomes possible to find two-dimensional velocity trajectories by using Huygens's method to track points along the CME front smoothly forward in time. As a result, we can calculate plane-of-sky velocity and acceleration at any point along the CME front as it propagates.

In cases where the CME is isolated, meets a sufficient brightness threshold, and has an obvious near-side source region (such as a flare or a dimming region), one can use the central axis of the associated "ice cream cone" model (Fisher and Munro, 1984; Leblanc *et al.*, 2001; Michałek, Gopalswamy, and Yashiro, 2003; Xie, Ofman, and Lawrence, 2004; Xue, Wang, and Dou, 2005) to estimate its three-dimensional propagation direction. From this, we can calculate a three-dimensional velocity vector, a quantity far more valuable than the oft-used plane-of-sky CME vectors. This three-dimensional velocity, when combined with the CME mass, will allow for the calculation of the CME kinetic energy.

Determining the CME mass requires a knowledge of the volume and density of the event. By fitting the "ice cream cone" model to the angular width and leading CME front, one can calculate a rough volume for the CME. Since the plasma is optically thin and any LASCO intensity signal is linear in electron column density, the observed Thomson-scattered intensity of a CME is proportional to its column density brightness. Using a base – difference imaging technique, we can find the excess heliospheric density due to the CME to within a factor of two (DeForest, Plunkett, and Andrews, 2001). We note that Webb (2000) and Bemporad *et al.* (2007) have demonstrated that the mass of a CME is not constant, but increases over time. Therefore, mass and probably kinetic energy are calculated as a dynamic quantity.

The algorithm that we use for extracting these metadata is based on a well-established tracking and measurement code (Wills-Davey, 2006). This code is run on calibrated, background-subtracted *Solar and Heliospheric Observatory* SOHO/LASCO coronagraph data. The tracking module monitors the data stream continuously, in order to catch the appearance of a CME from behind the coronagraph disk. In cases of isolated, well-defined CMEs, metadata outputs are extracted and converted to VOEvent format (*e.g.* CME_AngularWidth and CME_Mass) for dissemination via the HEK.

7. Coronal Dimming Regions

The coronal-dimming detection and metadata-extraction algorithm, together with the flare-detection and flux-emergence algorithms, is designed to provide space-weather alerts in near real time. The relationship between coronal dimmings and CMEs, acknowledged since the 1970s (*e.g.*, Rust and Hildner, 1976), has recently been confirmed statistically by Bewsher,

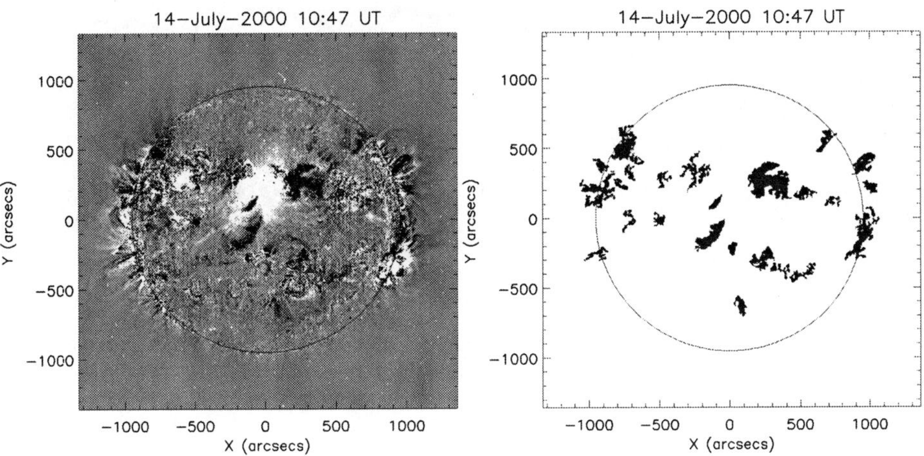

Figure 5 Left: base – difference image of EIT 195 Å data (10:47 – 09:24 UT), showing the coronal dimmings (black regions) during the 14 July 2000 Bastille Day event. White regions indicate increased intensity compared to the pre-event (base) image at 09:24 UT. Regions that do not show any significant change in intensity with respect to the pre-event image appear gray. Right: the result of the coronal dimming algorithm. Only the dimming regions are extracted.

Harrison, and Brown (2008). Coronal dimmings have been noted as reliable indicators of front-side (halo) CMEs, which can be difficult to detect in white-light coronagraph data. In the absence of a coronagraph (such as with SDO), coronal dimmings can act as an important indicator of the launch of a CME. The algorithm described in detail by Attrill and Wills-Davey (2010) is designed to detect and extract coronal-dimming signatures in the low corona, which are associated with eruptive events, such as CMEs.

This algorithm has two main components: *i*) detection and *ii*) metadata extraction. The detection relies on the analysis of statistical properties of EUV images in which coronal dimmings occur. This method was first introduced by Podladchikova and Berghmans (2005). Our implementation differs in that it is adapted to run on non-differenced images, thus reducing the required computer resources. The metadata extraction is implemented with the detection of a coronal dimming. To extract the coronal dimmings, we adopt the Reinard and Biesecker (2008) threshold of more than one σ below the mean pre-event difference image value. Figure 5 shows an example of our algorithm, in which the coronal dimmings associated with the 2000 Bastille Day event are extracted from the corresponding base-difference data. Once the coronal-dimming regions have been identified, our algorithm analyzes intensity changes over time using non-differenced data. Further metadata outputs of the algorithm include area, lightcurves, location coordinates, volume, and mass of the dimmings. Such outputs can make an important contribution to the study of coronal dimmings and their interplanetary CME counterparts. This algorithm is discussed in more detail by Attrill and Wills-Davey (2010).

8. Jets

An extensive study of polar coronal-hole (CH) X-ray jets was conducted with the *X-Ray Telescope* (XRT) onboard the *Hinode* satellite in early 2007. Coronal-hole jets provide grounds for testing models of reconnection and may prove to be one of the sources of the solar wind. During several days of observation, 44 jets were identified in the 171 Å TRACE

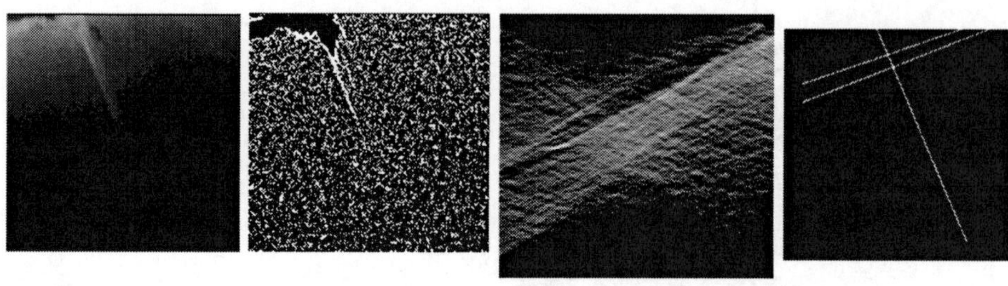

Figure 6 Sequence of images illustrating the steps taken by the jet-finder algorithm, as described in the text. The AIA 193 Å image on the left was taken on 16 June 2010, at 00:04:08 UT. The cutout of the full-disk image represents a location on the limb, inside the South pole coronal hole, just west of the pole.

images. AIA, similar to TRACE in resolution and wavelengths, will have full-Sun field of view and consistent cadence. Therefore, AIA will be able to observe a large number of jets. The coronal jets found in coronal holes are the easiest to identify. Here, the coronal jet (CJ) stands for a jet inside a coronal hole.

The CJ detection calculations will take place at SAO, once triggered by both a bright point (BP) and CH identification by our system (see Sections 11 and 12). The CH boundary is necessary to ensure that the BP is inside the CH. After identifying the boundary of the CH, the BP finder runs on every other image, and the pixels within BPs are identified. The CJ detection and parameter-determination algorithms work on data cubes covering a box enclosing the BP and extending forward in time. Our methods for determining the CJ parameters are described in detail by Savcheva *et al.* (2007). The detection algorithm has been tested on XRT images. The algorithm has been proven to work in the general case for low irregularity in the background, as is the case with XRT images. Further refinements handle the more dynamic background in corresponding EUV images that are taken with AIA. TRACE images have been used to implement this step. Figure 6 illustrates the sequence of steps in the algorithm behind the jet finder part of the jet module for AIA data. The first plot shows the original jet as seen in AIA 193 Å. The second one shows the same portion of the image after a Sobel and a Gaussian filter ($\sigma = 15$ pixels) and linear scaling have been applied – the resultant image consisting only of pixels with values 0 or 1. The third image shows the part of the $\rho - \theta$ space that the scaled image spans after the linear Hough transform has been applied to it. The brightest spot in this image represents the most prominent line in the image in terms of distance from the origin and inclination. The last image shows a reverse Hough transform of the three brightest spots in the Hough space, which is represented by three lines: one for the jet and two for the limb.

The final result from the algorithm is an entry into the HEK containing the following VO Parameters (see http://www.lmsal.com/helio-informatics/hpkb/VOEvent_Spec.html): a jet ID and the corresponding BP ID, the position of the origin of the jet in heliospheric coordinates, duration, length, width, line-of-sight velocity, and BP size. The inclination with respect to the N – S direction and transverse velocity are external parameters that are saved in a text file with the other parameters to create a metadata cube.

9. Oscillations

The study of large data sets (both in area and duration) is needed to improve the statistics of current results regarding coronal oscillations as well as localized, short-lived events. Certain dynamical events such as CMEs, flares, and coronal dimmings are known to be triggers for wave processes, and hence programs for wave analysis will be run on data sets associated with such events, which comprise only a small fraction of the observations ($< 0.1\%$). Other potential wave triggers, such as flux emergence, will also be studied. The codes have been optimized for robustness and accuracy, even at the expense of speed. Wavelet analysis plays an important role in this process, as it is capable of identifying periodicity in a given data set and can also provide both spatial and time localization (De Moortel and Hood, 2000; De Moortel, Hood, and Ireland, 2002; Ireland and De Moortel, 2002). This ability to localize a periodicity in time and space within a given data set can be exploited to "train" the wavelet code to identify certain events.

The adopted code was originally developed by McAteer *et al.* (2004) and De Moortel and McAteer (2004). For a given data cube, it performs a wavelet analysis of the time series in each pixel and outputs the periodicities that are present, the duration (number of cycles) of the detected periodicities, and their start and end times. These data will be entered into the VOEvent catalog. This information allows the team to optimize the wavelet code to detect a variety of events and features found in the solar atmosphere. Using such an automated detection technique on specified observations (*i.e.* observations associated with CMEs, flares, and coronal dimming events) facilitates a thorough investigation into the link between dynamical (explosive) events and observed solar oscillations. Indeed, De Moortel, Munday, and Hood (2004) showed that various wavelet parameters can be altered to favor either the time or frequency resolution. Additionally, by making a detailed study of the wavelet transform of the observed oscillations, it is possible to infer some coronal-plasma properties, as demonstrated by De Moortel and Hood (2000) and De Moortel, Hood, and Ireland (2002).

We are currently exploring a variety of options to improve performance of the code, for example, by binning (which does not seem to affect wave detection). The code will be made available to users for searches with user-specified optimization of search parameters.

10. EIT Wave Tracking

As part of the SDO computer vision project, EIT waves will be tracked and their metadata extracted using algorithms based on the work of Wills-Davey (2006) and Podladchikova and Berghmans (2005). Both sources rely on calibrated, derotated EUV data. Wills-Davey (2006) derives her output from TRACE 171 Å and 195 Å percentage base – difference images, while Podladchikova and Berghmans (2005) use data from the SOHO/*Extreme ultraviolet Imaging Telescope* (EIT) 195 Å "CME Watch" and *Solar Terrestrial Relations Observatory* (STEREO)/EUVI 171 Å and 195 Å data. The Wills-Davey (2006) method assumes that the EIT wave is a coherent front of intensity enhancement and measures intensity increases across the field of view. Podladchikova and Berghmans (2005) instead find EIT waves by measuring the first four moments of a given full-disk EUV image; in particular, the global kurtosis offers a way to find EIT waves that may displace structures while producing only dim fronts. This new algorithm takes aspects of both of these methods, enabling us to find a larger range of events.

Observations suggest that both EIT waves and coronal-dimming regions are highly correlated with CMEs, and many models assume that these events are a product of CME initiation, with EIT waves and dimming regions originating cospatially. Because EIT waves are tracked using a polar coordinate system, the origin of these coordinates is determined by cross-referencing with the coronal-dimming region tracking module, and finding the starting location of the dimming.

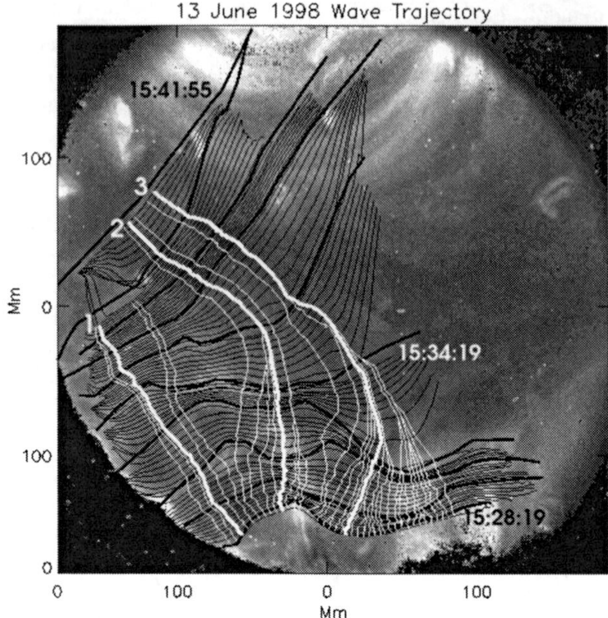

Figure 7 Automated tracking of an EIT wave observed on 13 June 1998.

Two methods are used for finding the EIT wave location. No assumptions are made about the wave's propagation characteristics, other than that it will propagate roughly away from the eruptive region. In cases where a bright-front is observed, the position of the intensity maximum is determined from one frame to the next in the polar coordinate frame. Figure 7 shows an example of this automated bright-front tracking. This analysis is combined with measurements of local kurtosis across overlapping 25 × 25 pixel sections. This use of local kurtosis allows us to track subtle motion at a resolution of order $10''$. Since EIT waves are global structures, such a resolution is more than sufficient for determining structural displacement due to coherent wave propagation. Velocity is calculated using the same Huygens plotting technique discussed for the CME tracker (Section 6).

When intensity measurements show the existence of a bright front, we record the intensity enhancements due to the wave as a function of time. Intensity cross sections can either be calculated as a three-dimensional contoured array, or can be considered as "slices" along the Huygens-plotted velocity trajectories. As EIT waves are defined as propagating intensity increases, only the positive values from the difference images are kept. These data will also be used to find the center-of-front amplitude as a function of time.

By combining intensity enhancement measurements from several overlapping EUV passbands, it is possible to measure the density increase above the pre-event background due to the wave front. Intensity measurements also allow us to calculate the entrained *energy* of a given wave event as follows: $E = \int \mathrm{d}^3 V (P_0 \Delta n_\mathrm{e}^{2\gamma} n_\mathrm{e}^{-2\gamma})$, where E is the entrained energy, V is the volume of interest, P_0 is the equilibrium pressure, n_e is the electron density, and $\gamma = 5/3$.

All but the strongest EIT waves, while easy to pick out by eye in running-difference movies, are nonetheless extremely difficult to find and track *a priori* because noise tends to dominate the difference images. Additionally, studies by Rachmeler and Wills-Davey (2005) show that false-positive detections may be more likely than previously expected. Because of this possibility of false positives, as well as the excessive resource requirements for wave detection from the full data stream, the wave detection program will be conducted *post facto*, triggered by other events detected by our system, particularly flares and coronal dimming regions. Output parameters will be input into the HEK, using the appropriate VOEvent format. With these triggers we expect to observe 100 – 300 events year^{-1}.

11. Detection and Analysis of Active Regions and Coronal Holes

The *Spatial Possibilistic Clustering Algorithm* (SPoCA) we have developed produces a segmentation of EUV solar images into regions that we call "classes" corresponding to active regions (AR), coronal holes (CH), and the quiet Sun (QS); see Barra *et al.* (2005), Barra, Delouille, and Hochedez (2008), Barra *et al.* (2009). Other segmentation methods have been proposed; an overview is given in Barra *et al.* (2009). We have selected SPoCA because of the maturity and flexibility of the program. SPoCA uses a multichannel, fuzzy-logic, clustering procedure. It has been applied successfully to a series of EIT image pairs (171 and 195 Å) spanning almost a full solar cycle; see Barra *et al.* (2009). The classes are determined by minimization of intra-class variance. The method is generic and therefore portable to other instruments, and in particular to SDO/AIA. SPoCA involves a preprocessing by which the limb brightness discontinuity is attenuated. It can take transformed EUV images as input, such as differential emission measure (DEM) maps obtained from AIA images (using software supplied by AIA investigators). This might help address the problem of line-of-sight confusion.

The level-1 product of the procedure is a set of maps giving the membership value to each class. Several higher-level products can be generated from these maps of memberships:

- A segmentation map, attributing a class to the pixel according to its highest membership value;
- A probability-density function giving, for each pixel intensity value, the probability that this pixel belong to a particular class;
- Computation of quantities such as area, integrated intensity, and first statistical moments for each class.

From the maps of, *e.g.* ARs, connected AR pixels are then gathered by means of a region-growing technique; see Figure 8 for an example of our results with AIA data. It provides the instantaneous location of the barycenter, the area, the coordinates of the bounding box, and a mask for each AR.

These elements are computed over time and handled as dynamical quantities. The current tracking method uses an optical-flow algorithm to locate the barycenter of the AR in the next fuzzy map, where the same region-growing technique updates the parameters of the AR being tracked. A starting date for an AR is defined when a new set of connected pixels is identified. An AR "end date" is recorded either when the tracking algorithm can no longer find a connected set, or when the AR disappears over the west limb. The algorithm also handles merging of ARs.

Fuzzy CH maps can be treated exactly as the AR maps, producing area, location of barycenter, and location of boundary for a connected element of the CH map. However, filament channels seen in coronal EUV passbands are often erroneously classified as "coronal holes" in the segmentation. Using Hα images from ground-based observatories, the filament-detection module (see Section 4) separates filaments from the CHs.

On AIA images we will identify AR, QS, and CH using our multichannel segmentation algorithm on pairs of 171 and 193 Å images, 211 and 335 Å images, as well as on 94 Å

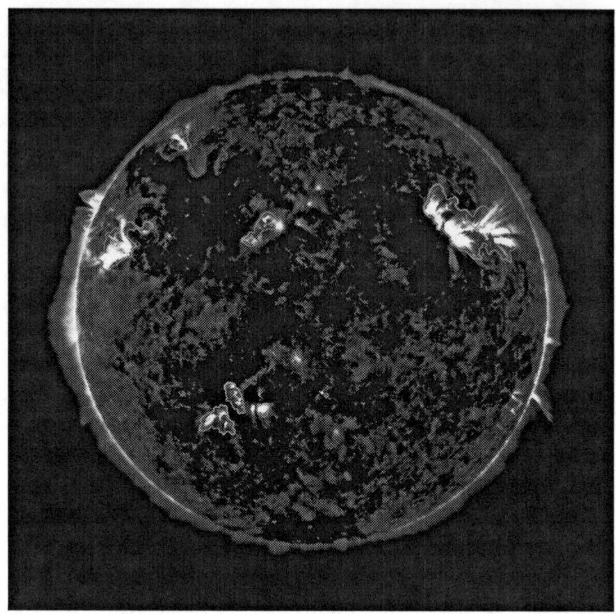

Figure 8 AIA 171 Å image from 29 April 2010, 00:00:42 UT, together with overlays of segmented ARs.

images. When similar morphological information is available in two bandpasses, a multi-channel approach provides a more robust segmentation than a mono-channel one.

12. X-ray Bright Points

AIA delivers a set of full-frame images every ten seconds in each of its seven EUV pass-bands. The bright point (BP) detection algorithm will run on the three coolest coronal pass-bands at wavelengths 171 Å, 193 Å, and 211 Å.

We use a modified version of the BP finder developed by McIntosh and Gurman (2004) to successfully identify and track bright points in SOHO/EIT 171 Å, 193 Å, and 284 Å over an entire solar cycle. The algorithm, written in IDL, uses full-Sun synoptic images to define areas of interest based on levels of intensity above the local background, and number of contiguous pixels that define shape and size (see Figure 9). This process is described in detail in Davey and McIntosh (2007); see also McIntosh and Gurman (2005), McIntosh *et al.* (2009). The bright-point detection algorithm works by applying an $N \times N$ boxcar smooth function to determine the local background intensity. The noise level (σ_{noise}) is also estimated. The smoothed background is subtracted from the original image, and intensity enhancements (BP candidates) are classified by their separation from the background in units of σ_{noise}, with some restrictions on size and shape.

We are implementing several steps to improve the existing code:

- We have included an upper and lower limit on the size of the BPs, weighted by limb angle.
- The McIntosh code can pick up strings of "BPs" which are actually the brighter portions of longer ribbon/loop-like features. After initial passes of BP detection, we slightly re-duce the noise threshold to see if spatially close BPs merge. If so, the underlying ribbon structure is then reassigned.

Figure 9 Full-disk 193 Å AIA image, scaled logarithmically. The pixels found by the BP finder module have been identified and highlighted in the original image. The image was taken on 29 July 2010 at 15:27:56 UT.

- The McIntosh code can pick up patches of locally enhanced quiet Sun (QS) pixels, often with complex boundaries, which are clearly not BPs. We have placed an additional constraint on the ratio of the two major axes of the BP and its perimeter/area ratio to prevent the detection of such QS conglomerations. This also eliminates long ribbon-like structures which are typically remnant loops of decaying active regions.

Once the bright points are defined by heliographic coordinates and time of observation, obtaining statistical information is straightforward. The total number, intensity (mean count rate), size, perimeter, major/minor axes, and area of bright points are determined for each of the regions in each bandpass. To determine lifetimes, the feature-finding program tracks the intensity-weighted center of mass and area in any of the AIA passbands, with the position compared with projected rotation rates as a function of latitude. The lifetime of BPs is found from their appearance and disappearance, or from their rotating on and off the disk. Figure 9 shows an early example of the application of the BP-finder algorithm to a full-disk 193 Å AIA image.

The output of the region finder, made publicly available through the HEK and a full catalog at SAO, includes time of observation, heliospheric coordinates, total area, integrated intensity for each wavelength, and environment associated with each bright point (QS, AR, CH). Also associated with a select number of BPs are 128 × 128 pixel images of the region that can be made into QuickTime movies, accessible, *e.g.* through the VSO interface.

The data-processing time for finding and tracking bright points in three full-disk EIT 171 Å images is four to five seconds using a 2 × 3 GHz Dual-Core Intel Xeon Macintosh HD. The actual processing time will depend on the AIA image-preparation software. Detection parameters (background, dark subtraction, sigma enhancement in each filter) will be adjusted after in-flight calibrations when the AIA image preparation software becomes available. Algorithms that correlate the feature-finding database with other databases will be developed within the first year after launch depending on when the level-2 algorithms are implemented.

12.1. Correlation with Other Databases

The heliographic coordinates determined from the feature-detection program will be correlated to other databases within the pipeline to create other level-3 data products. The magnetic-flux tracker SWAMIS identifies and tracks magnetic-flux concentrations above a particular threshold (see Section 13). Using coordinates of bright points identified in the AIA 171 Å channel, the magnetic-flux database is cross-referenced for footpoints within an area of the bright-point coordinates. The magnetic flux for each of the bright points can then be calculated by summing over the total area in each of the footpoints. The lifetime and unsigned magnetic flux of the region are then appended to the bright-point catalog.

12.2. Correlation with Jets

The algorithm for jet detection uses the location of coronal holes and bright points from the feature-finding program (see Section 8). This algorithm is run every few hours. After jets are detected, the bright-point catalog is updated with a keyword [BPJ] associating the bright-point number to the jet number. The algorithms that correlate the feature-finder database with other databases will be developed after launch, and can be executed retroactively independent of the near real-time pipeline.

Because the basic McIntosh code uses levels of σ_{noise} above background to identify features, with a few modifications the program can also be used to define and track not only bright points, but also active regions and coronal holes. These features will then be used to cross-check the features identified with the segmentation algorithm (see Section 11).

13. Magnetic Feature Tracking and Sunspots

We will provide comprehensive data on flux emergence, interaction, and cancellation over the whole solar disk. Two main types of magnetic-tracking data are available: feature-level data, which represent the motion of resolved line-of-sight magnetic fields over the entire surface of the Sun within 45° of the sub-Earth point; and large-scale emergence data, which identify and highlight emerging-flux regions associated with ephemeral active regions on the supergranular scale and above.

The feature-level data provide comprehensive motion and history information on small magnetic details. They enable correlative statistical studies of flux origin, interaction with coronal features, and diffusion over the solar surface. For example, it will be possible to extract and search on the simplified magnetic geometry, and hence number of coronal null points and field complexity, in the neighborhood of any feature or event in the entire SDO data set. Further, every emergence event observed by the SDO/*Helioseismic and Magnetic Imager* (HMI) over the entire mission life will be cataloged and sorted by position, time, and event size and strength, permitting statistical surveys of the interaction of emerging flux with existing magnetic structures in the chromosphere and corona.

Magnetic feature tracking is easy to prototype but difficult to perform reproducibly and reliably. The pipeline uses a variant of the SWAMIS code described by DeForest *et al.* (2007). For feature-level tracking, we use the "downhill" quasi-watershed method that they discuss; for large-scale emergence tracking we are prototyping both the multiresolution method pioneered by Hagenaar and Cheung (2008) and a post-track feature clustering method. Large-scale flux emergence is challenging to detect with a simple feature finder operating at high spatial resolution; this has led to studies of several methods, including post-track clustering analysis of feature centers, unsigned-flux counting in large image "macropixels" (Hagenaar and Cheung, 2008), and multiple passes through the image data at different spatial resolutions. We identify large-scale flux emergence using the multiresolution approach, in which the data are tracked at multiple spatial resolutions. Larger-scale emergences are identified in degraded copies of the original data. The multiresolution technique is not expected to significantly impact computing resources, as the degraded copies of the data require far less memory and CPU power than the full-resolution stream.

SWAMIS has been used to study flux the emergence patterns with SOHO/MDI (*e.g.*, Lamb *et al.*, 2008) and the *Hinode/Solar Optical Telescope* (SOT) – *Narrowband Filter Imager* (NFI) (see, *e.g.*, DeForest *et al.*, 2008; Lamb *et al.*, 2007). Our basic tracking cadence is six minutes; lifetime studies using SOHO/MDI data show that this is fast enough to capture evolution at the ≈ 1 arcsec resolution of SDO/HMI. Simple feature location and evolution data are the most basic products of every feature recognition code. In addition, we provide origin and demise information on each feature detected. SWAMIS classifies birth events into five categories: Isolated Appearances (a large fraction of new flux features appear in isolation, far from any other feature), as described by Lamb *et al.* (2008), Fragmentations (in which a feature "calves" from a like-sign feature that is undergoing shredding), Emergences (in which a new feature appears in a manner that approximately conserves flux, either due to a new opposing feature appearing nearby or to growth of an existing nearby opposing feature), Errors (which do not appear to conserve flux), and Complex events, where more than two features appear to be interacting directly. Less than 1% of events greater than $10\times$ the detection threshold result in Error events. In each case, SWAMIS also identifies the associated opposing region. Cancellation is classified using the same scheme, reversed in time, to yield Disappearances, Mergers, Cancellations, Errors, and Complex cancellations.

Identification of every feature-interaction event is crucial for the new field of "event-selected ensemble imaging" (ESEI), which allows deep-field study well below the noise floor of an individual observation. ESEI, which uses a combination of magnetic and other feature detections, is a unique and powerful tool to distinguish models of small-scale activity that cannot be resolved any other way.

The exact SWAMIS code used in the SDO pipeline will be made available for download, inspection, and local modification as free software.

14. Polarity Inversion Line Mapping

Identifying the location of *Polarity Inversion Lines* (PILs) – often also called neutral lines – can be of great importance for phenomenological and theoretical studies. Historically, neutral lines in active regions have been useful tools for predicting the locations of flares and CMEs (Falconer, Moore, and Gary, 2002). They also can be used to map out coronal structures (McIntosh, 1994) and are associated with filaments and filament channels (*e.g.*, Martin, Bilimoria, and Tracadas, 1994; Chae *et al.*, 2001).

Until recently, little work appears to have been done to develop automated PIL mapping. Inversion lines have typically either been drawn by hand or determined by contouring algorithms within larger programming packages. The use of a PIL mapping algorithm not only allows for user-independent inclusion in a pipeline, but standardizes the mapping with prespecified, explicit resolution scales.

We have successfully developed a code that identifies PILs based on a well-established code previously developed for National Solar Observatory (NSO)/*Kitt Peak Vacuum Telescope* (KPVT) magnetogram data (Jones, 2004). This method invokes techniques such as

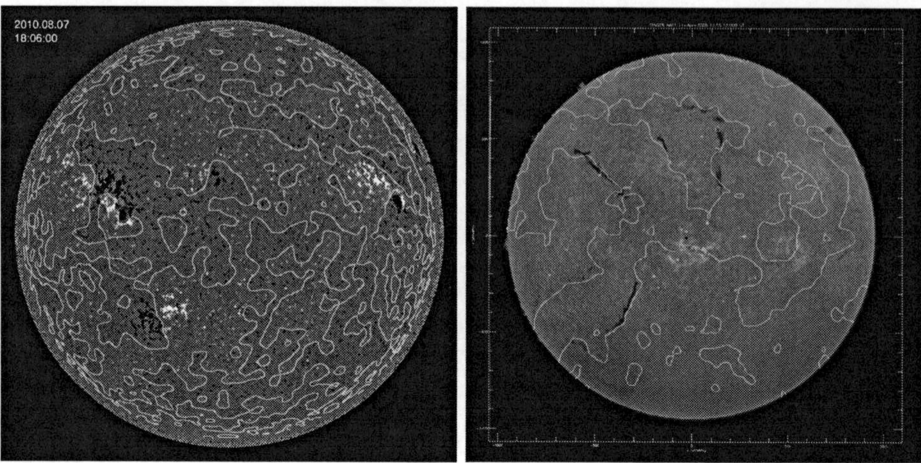

Figure 10 Left: Polarity-inversion contours (in yellow), determined by the PIL module, overlaid on the source SDO/HMI magnetogram. Red dots indicate flare-prone regions of high magnetic gradient. Right: Polarity-inversion lines in blue from a SOLIS magnetogram, calculated with the same code, overlaid on a nearly simultaneous Hα image from BBSO.

edge detection, smoothing, and block averaging to determine the appropriate scale at which to measure the neutral lines. The main difference, and the one that makes this code more desirable and sophisticated, is a multiple-smoothing method. There are three ranges of flux corresponding with the values of the pixels, each of which receives a different amount of smoothing. If the flux is above the upper threshold, little smoothing is applied; if it is below the lower threshold, a lot of smoothing is applied. For all other fluxes between the upper and lower thresholds, a medium amount of smoothing is applied. See Figure 10 (left) for a recent example of PIL code output overlaid on an HMI magnetogram. Areas of high magnetic gradients are identified and indicated in the figure in red. Projection has been taken into account.

To verify our results early on, we overlaid a PIL map determined from a *Synoptic Optical Long-term Investigations of the Sun* (SOLIS) image on an Hα image from BBSO (Figure 10, right). We tested on SOLIS magnetograms because their resolution is closest to that of HMI. Parameters in the PIL routine may be varied in accordance with the user requirements, allowing mapping to be done on a variety of length scales. The output from this code is converted to VOEvent format as a mask data type for the HEK.

15. Nonlinear Force-free Field Extrapolations

The solar magnetic field is key to understanding the physical processes in the solar atmosphere. Unfortunately, we can measure the magnetic field vector routinely with high accuracy only in the photosphere. These measurements are extrapolated into the corona under the assumption that the field is force free, because the magnetic pressure is several orders of magnitudes higher than the plasma pressure. We have to solve the equations $(\nabla \times \mathbf{B}) \times \mathbf{B} = \mathbf{0}$ and $\nabla \cdot \mathbf{B} = 0$.

We solve by minimizing the functional proposed in Wheatland, Sturrock, and Roumeliotis (2000) and effectively encoded in Cartesian and spherical geometry by Wiegelmann

(2004, 2007).

$$L = \int_V \left[B^{-2}\left|(\nabla \times \mathbf{B}) \times \mathbf{B}\right|^2 + |\nabla \cdot \mathbf{B}|^2\right] \mathrm{d}^3x, \tag{1}$$

where an observed preprocessed vector magnetogram specifies the photospheric boundary conditions. Preprocessing of the measured photospheric vector magnetograms is necessary, because nonmagnetic forces such as plasma pressure and gravity are present in the photosphere, and consequently the measured magnetic-field data are not consistent with force-free consistency criteria as defined in Aly (1989). We developed a minimization procedure that uses the measured photospheric field vectors as input to approximate a more chromospheric-like field (Wiegelmann, Inhester, and Sakurai, 2006). The procedure includes force-free consistency integrals and spatial smoothing. Direct chromospheric observations can be taken into account for an improved match to the field direction, as inferred from Hα fibrils by Wiegelmann *et al.* (2008). Recently, the preprocessing routine has also been implemented and tested in spherical geometry by Tadesse, Wiegelmann, and Inhester (2009).

A recent comparison of different nonlinear force-free extrapolation codes has revealed that minimizing the functional (1) is the fastest and most accurate currently available method (Schrijver *et al.*, 2006; Metcalf *et al.*, 2008).

The Cartesian version of the code has been applied to vector magnetographs from space-borne data (*e.g.*, from *Hinode*/SOT by Jing *et al.* (2008)) and ground-based data (*e.g.*, from SOLIS in Thalmann, Wiegelmann, and Raouafi (2008)). The computed data cubes have been analyzed regarding the content of free magnetic energy and the magnetic topology. An ongoing effort is the application of the spherical code to SOLIS data. We use the large and high-resolution HMI field-of-view (FOV) vector-magnetograms to produce estimates several times per day for the free magnetic energy for the ARs on disk, and produce 3D magnetic-field maps outlining the general coronal magnetic-field topology. These data will be cross-referenced with the AR and CH data obtained through other methods (see Section 11).

Further complications in using current vector magnetograms (*e.g.*, from *Hinode*/SOT) are a limited FOV, missing data, and high noise in the transverse B-field measurements (DeRosa *et al.*, 2009). For meaningful extrapolations, it is important to use large and high-resolution FOV vector-magnetograms – as provided by HMI – and to deal with measurement errors and nonmagnetic forces by preprocessing. An updated version of the nonlinear force-free field (NLFFF) code takes the measurement errors into account by adding another term to Equation (1) with a Lagrangian multiplier; see Wiegelmann and Inhester (2010) for details.

For reliable estimates of the free magnetic energy and topology from the extrapolated coronal magnetic-field model, one should validate the model field by additional coronal observations, *e.g.* compare projected field lines with AIA images. This comparison can be done with a newly developed tool to extract coronal loops from EUV images (Inhester, Feng, and Wiegelmann, 2008). This tool has been applied so far to images taken from the STEREO/SECCHI/EUVI instruments, and an application to AIA is straightforward. Wiegelmann, Inhester, and Feng (2009) proposed an algorithm [an additional term in Equation (1)] to incorporate extracted coronal information into the computation of nonlinear force-free coronal magnetic fields, and we will study the use of this tool for our purpose here.

16. Trainable Feature Recognition and Retrieval

Humans have an amazing generic feature-recognition ability that has been hard to match in computers. For example, a solar scientist can instruct student in less than an hour to

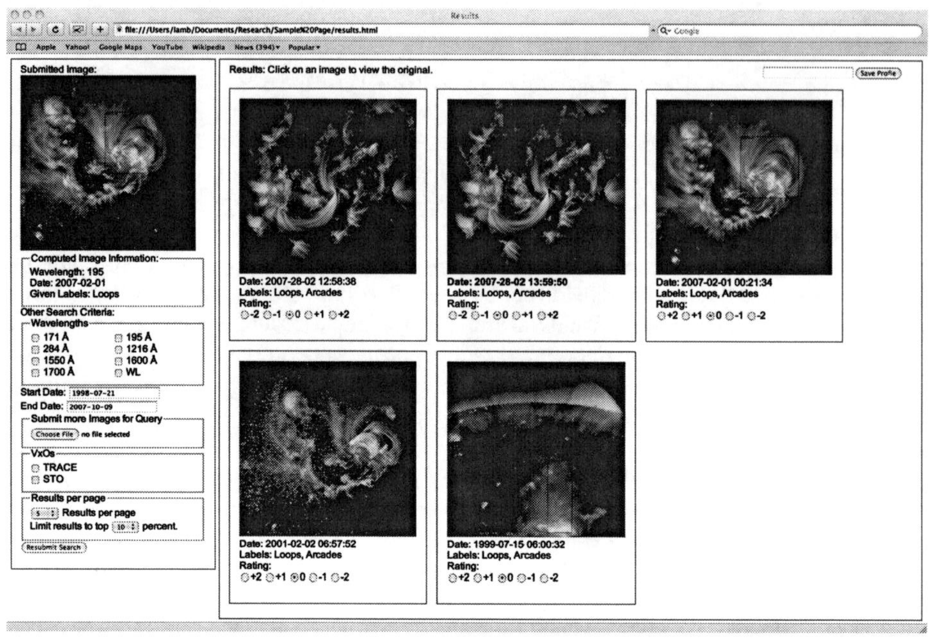

Figure 11 An example of the querying interface showing sample results with the option to rate the returned images and resubmit the image-based query.

recognize sunspots, filaments, loops, and arcades in solar imagery, and the student can then easily produce a catalog of these features from a given set of images. However, a computer feature-finding algorithm takes months to years to develop, and the development must be repeated almost from scratch for every new feature.

Motivated by the successful development and implementation of a generic, more human-like, feature detection method for mammography (*e.g.*, Yang *et al.*, 2007), we are in the process of creating an automated solar-feature-retrieval system that is generic in nature; *i.e.*, the software can detect features of any kind, and rank the returned images based on their similarity to an image provided by the user (see Figure 11 for a sample of the user interface). Rather than developing a new task-specific application to identify each separate feature, our generic feature-recognition software can detect and catalog a wide range of solar features, even involving serendipitous discovery. This work will benefit the solar community in three ways. It will deliver *i*) a benchmark data set that includes balanced representations of common types of phenomena and that can be used to perform comparative evaluation of different image recognition systems, which will be made freely available. (Our benchmark data set for TRACE images is available at http://www.cs.montana.edu/angryk/SDO/data/.) It will also deliver *ii*) catalogs for common types of phenomena with database indices, speeding up data search and retrieval, and *iii*) a content-based image retrieval system with a ranking mechanism that returns images based on their similarity to the image provided as a query, and that can learn from users' feedback.

16.1. Method

There are two steps to our method. The first is implemented at the end of the metadata pipeline. The second is totally separate from it and can be performed by any user at any

Table 3 Texture parameters used for analyzing TRACE images in Lamb (2008).

Name	Equation
Mean	$m = \frac{1}{L}\sum_{i=0}^{L-1} z_i$
Standard deviation	$\sigma = \sqrt{\frac{1}{L}\sum_{i=0}^{L-1}(z_i - m)^2}$
Third moment	$\mu_3 = \sum_{i=0}^{L-1}(z_i - m)^3 p(z_i)$
Fourth moment	$\mu_4 = \sum_{i=0}^{L-1}(z_i - m)^4 p(z_i)$
Relative smoothness	$R = 1 - \frac{1}{1+\sigma^2(z)}$
Entropy	$E = -\sum_{i-0}^{L-1} p(z_i)\log_2 p(z_i)$
Uniformity	$U = \sum_{i=0}^{L-1} p^2(z_i)$

time, even on a laptop, without requiring SDO JSOC resources. Step one consists of calculating for each AIA image the texture parameters, which are then stored in the image texture catalog. Specifically, each AIA image is subdivided in 1024 128×128 sections, and for each of those a set of texture parameters such as entropy, average intensity, standard deviation, kurtosis, *etc.* is calculated. The parameters that we have used for analyzing TRACE images are given in Table 3. Assuming that we derive 16 parameters per image section, each 16 bits, the information compression from AIA image to catalog entry is roughly a factor of 1000. Therefore, the size of the AIA image-texture catalog will not be prohibitive. For the SDO mission we are investigating the option of even further compressing our catalog entries by discretization of our texture parameters (Banda and Angryk, 2009). We estimate that the calculation of statistics for a single image entry into the catalog requires one second of computing time for the whole grid-segmented image.

Suppose now that a user wants to build a catalog of loop arcades by teaching the algorithm how to detect them. The user then downloads a selection, say several dozens, of AIA full-disk images. In those images the user identifies arcades via the simple point-and-click interface shown in Figure 11; there is no need to identify all arcades on the disk. The program then calculates the texture parameters for each of the image segments containing the identified arcades. Using these parameters as a feature definition, the entire AIA image-texture catalog can be quickly searched to filter out irrelevant images and then fine-tune the search for image segments with similar parameters. These segments will contain arcades if the image-texture parameters are adequate for the type of image one is analyzing.

Our work on TRACE images (Lamb, 2008; Lamb, Angryk, and Martens, 2008; Banda and Angryk, 2009) has focused on the crucial step of determining the correct texture parameters: those that have good properties, first from the perspective of computational costs to keep up with our pipeline, and second from the perspective of distinguishing between different types of solar phenomena. Table 4 summarizes the results from Lamb (2008). Here the receiver operating characteristic (ROC) curve plots the true-positive rate on the y-axis and the false-positive rate on the x-axis. The area under the curve represents the overall accuracy of the classifier. The closer the area under the curve is to 1.0, the more accurate the classifier; at 0.5 the classifier would not be any better than randomly picking a class from an evenly distributed set of phenomena. Random under sampling (RUS) and random over sampling (ROS) are different sampling methods to treat data sets with unbalanced numbers of different phenomena (*e.g.*, relatively common images of quiet Sun and coronal loops, *versus* relatively rare occurrences of flares and filaments). The column headers represent different classification techniques. (The method C4.5 performs entropy-based classification, SVM

Table 4 Average area under an ROC curve using two different sampling techniques (RUS and ROS), and two different classification algorithms (C4.5 and SVM) for different types of solar phenomena (Lamb, Angryk, and Martens, 2008).

Phenomenon	C4.5		SVM		AdaBoost C4.5		AdaBoost SVM	
	RUS	ROS	RUS	ROS	RUS	ROS	RUS	ROS
Quiet Sun	0.795	0.872	**0.940**	0.923	0.920	0.912	0.939	0.920
Coronal loop	0.897	0.905	**0.932**	0.922	0.911	0.917	0.912	0.912
Sunspot	0.890	0.917	0.922	**0.958**	0.901	0.932	0.944	0.943
Filament	0.838	**0.960**	0.832	0.848	0.875	0.783	0.898	0.897
Flare	0.977	0.970	**0.988**	0.980	0.977	0.967	0.977	0.976

stands for support vector machines, and AdaBoost is a commonly used boosting technique that emphasizes misclassified samples during classifier training.)

We continue to evaluate our texture parameters and improve our classifiers using TRACE images. The TRACE results will carry over easily to AIA because their imagery is similar. We emphasize again that step two is completely separate from the data pipeline, and that the AIA data archive need not be accessed in its entirety; only a small number of images must be downloaded to provide a training sample to our software. Since the algorithm is further trainable, a user can fine-tune the selection, *e.g.* to distinguish between left-skewed and right-skewed arcades.

Our current list of solar features for which we intend to generate VOEvent entries (time and location at a minimum), using this method, includes: cusps, arcades, null-geometries, flare ribbons (starting with the flare catalog), keyholes (previously detected by *Skylab* and *Yohkoh*/SXT), circular filaments (starting with the filament catalog), faculae, pores, surges, arch filaments, δ-spots (from magnetograms), plumes, and anemones (also called rosettes).

16.2. Discovery of New Features

A promising advantage of generic feature-recognition methods over task-specific ones is the potential for rapid discovery and cataloging, even serendipitously, of new features. An

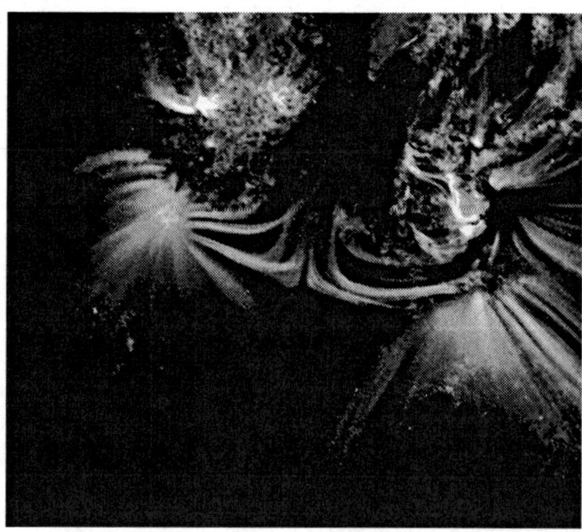

Figure 12 TRACE 171 Å AR image showing a loop geometry that is consistent with the presence of a magnetic null. Field is 3.8×4.2 arcmins.

example is given in Figure 12, a TRACE image of an active region having a special type of loop geometry. The user can start with this image to try to find similar ones in the TRACE or AIA catalog. Not many such images have been observed with TRACE, probably because they are rare, and for an instrument such as AIA, searching the database to find more of such images is a daunting task. Our image-texture characterization makes such a search possible and even rather simple without the user needing to know how to specify the parameters of the search. The scientific interest of Figure 12 is clear: this is what one would expect to see in the neighborhood of a magnetic null, making it a good candidate for inclusion in a study of such topologies.

This then points to a more general application for the image-texture catalog. A user may see a number of images that look very similar and which pique interest without even being able to define exactly what is exceptional about these images (Figure 12 would be an example for someone with limited knowledge of MHD). Our feature-recognition method can then be used to find similar images without an excessive effort, even though the users do not even know exactly what they are looking for!

The image texture catalog, together with a manual and supporting software, will be made available online as a complement to the VOEvent catalog.

17. A Pipeline for the Generation of Feature and Event Metadata

17.1. Computing Facilities at LMSAL

LMSAL has provided a computer infrastructure to enable the running of appropriate feature and event modules. This includes a cluster of Linux servers consisting of either four- or eight-core 2.2 GHz processors with up to 16 GB of memory per server. LMSAL has available > 100 TB of storage which is Network File System (NFS) mounted to these servers. Data transfer between Stanford and LMSAL is achieved by means of a 10 Gbps fiber link, giving LMSAL almost instant access to the SDO data. As part of their commitment to implementing the *Heliophysics Events Knowledgebase* (HEK: Hurlburt *et al.*, 2010), LMSAL has provided for interacting with the *Data Record Management System/Storage Unit Management System* (DRMS/SUMS) and extracting data needed by the feature and event detection modules. LMSAL has created a control system for running and monitoring these modules and tools that enables the easy contribution of events to the HEK. In Section 2.1.1 of their paper, Hurlburt *et al.* (2010) describe the *Event Detection System* (EDS), which continually acquires incoming data and makes it available to a variety of feature and event modules. The same article provides flow charts for the various processes.

Because the available resources on which to run the various feature and event detection modules are finite, SAO provides additional computing power required to successfully run the modules. LMSAL has ported their event pipeline control software to SAO to help with this process.

17.2. SAO Computing Facilities

SAO has deployed an 11-node computer cluster consisting of two dual-core 3 GHz processors with between 16 and 32 GB of memory per node. This has been integrated with a 220 TB storage array from SUN, available to nodes via NFS, which has shown to be more than capable of delivering data at the required rates. SAO uses the storage array to construct a rolling archive of the most recent months of SDO/HMI and AIA level 1.0 data. SAO has implemented and tested a remote version of the DRMS/SUMS setup that is in use at Stanford.

Stanford uses SLONY-I (a master – slave replication system that replicates large databases to a number of slave systems, see htttp://www.slony.info) to replicate meta-information to a limited number of external sites, including SAO. SAO transfers and stores SDO data on the local system as it becomes available. This gives SAO rapid access to SDO data, and enables local processing and easy comparison between SDO data and data from other sources (such as Hα data in the filament detection module).

17.3. Interfacing to DRMS

LMSAL has provided a dedicated Java/IDL-based pipeline framework to interface DRMS/SUMS with the IDL-based feature- and event-detection modules. SAO makes use of the same code locally for any IDL module processed there. For non-IDL modules, *e.g.* the magnetic feature-tracking module, written in *Perl Data Language* (PDL), the code still uses the pipeline software, but with an additional interface layer providing the communication between module and pipeline software.

SAO runs algorithms that either detect events not relevant to the SDO/HEK effort at LMSAL or report exhaustive findings (*e.g.* bright point locations), and so the SAO computer-vision center will port the EDS software to one of their clusters. SAO will also run detection algorithms that do not use SDO data, such as the filaments and CME modules, as well as modules that key off from detection events created by other detection algorithms (*e.g.* flare-detection events will spawn the process to look for oscillations). Initially, there remains a need for human interaction for some event types, *e.g.* EIT waves and oscillations, to ensure that the detections represent actual events.

17.4. Module Testing

The facilities available at SAO have enabled us to create a framework similar to that available for running the feature and event detection modules at LMSAL. We utilize this framework for end-to-end testing and evaluation of the modules before they are deployed at LMSAL, and for prototyping algorithmic or coding improvements to the modules. Part of the evaluation will include investigation of standardization opportunities, particularly in the case of the IDL codes, which will have to be modified to work within the pipeline framework and provide output in VOEvent format, using the environment provided by LMSAL.

17.5. Community Access to Metadata at SAO

The primary target for the feature and event detection modules' output is the SDO/JSOC feature metadata catalog (*i.e.* HEK at LMSAL). In addition to supplying the HEK with features and events, SAO will utilize the newly redesigned VSO catalog infrastructure to make features and events available, with search capabilities that complement those currently proposed for the HEK. This includes the ability to use the catalogs as datasets for science research. In addition, the VSO catalog infrastructure will allow users to comment on data and catalog entries. This information can be used to enhance the data being provided. Additionally, some of the feature- and event-detection modules run at SAO (*e.g.* for non-SDO data), requiring SAO to host any secondary data products generated as the module runs. By utilizing the VSO infrastructure, LMSAL and SAO can easily make data available to other virtual observatories of the heliosphere (such as the Virtual Heliospheric Observatory and the Virtual Wave Observatory) and provide enhanced data integration for all of heliophysics. Dissemination of the pipeline capabilities will be accomplished through the use of a guidebook, two annual workshops, and a web interface; see http://helio.cfa.harvard.edu/sdosc/.

18. Conclusions

The original motivation for NASA to submit an announcement of opportunity for SDO automated feature-finding projects was that the continuous SDO data stream would be simply overwhelming for the solar community: an AIA image every ten seconds in each of the eight channels, an HMI magnetogram every 45 seconds, and a vector magnetogram every 90 seconds, all with no interruption. This would be simply too much to inspect for any solar research group, and for the community, even with the best coordination of efforts. Automated event alerts, from near real-time space-weather warnings to automated cataloging of features and events, must provide at least part of the solution here. We emphasize that automated feature recognition has found wide application in other areas where the amount of imagery has become overwhelming for human observers, *e.g.* elementary particle physics with particle colliders, diagnosis of medical X-ray images, monitoring of urban traffic, and security surveillance in public areas such as airports and sport stadiums. Hence, application to solar physics seems a natural development.

Thus motivated, our group set out to develop a computer-vision system for SDO. We consulted the SDO instrument PIs to select the events and features that should have the highest priority, and began developing a suite of 16 modules, described here, that would meet these priorities best. For most of the modules we were able to find development teams with proven track records of their modules for pre-SDO data, thus minimizing the risk of unwelcome surprises in the little time that we had for development pre-launch. From the beginning we did not want to restrict our computer-vision project to SDO data; for example, it is obvious that for scientific and space-weather uses automated monitoring of filaments and CMEs is required, and hence we incorporated those modules.

As the computer-vision project is taking shape it has become increasingly evident that the production of very high accuracy catalogs of solar features and events is desirable for the field of solar physics in its own right. Due to the fragmented nature of solar observations – different observatories, different time coverage, limited image cadence – for too long solar physicists have limited themselves to the study of single or a few events, often finding that physical models that seem to apply very well in a number of cases turn out to be irrelevant for others.

Examples of the potential unleashed by the computer vision project are: *i*) Draw a butterfly diagram for Active Regions, *ii*) Find all filaments that coincide with sigmoids, and then correlate sigmoid handedness with filament chirality, *iii*) Correlate EUV jets with small-scale flux emergence in coronal holes, *iv*) Draw PIL maps with regions of high shear and large magnetic field gradients overlayed, to pinpoint potential flaring regions. Then correlate with actual flare occurrence. All of these tasks will be accomplished with great ease; the power of this method is limited merely by the imagination of the researcher. In addition, the space-weather alerts generated by our modules allow for the automation of the production and online publication of flare lightcurves and other basic characteristics, quicklook movies, and also automated email alerts to interested parties, from space-weather forecasters to instrument planners. Correlating phenomena will be merely a matter of a few line commands on cataloged metadata.

We are convinced that the presence of automated feature recognition will facilitate the paradigm shift of solar physics from a discipline mostly focused on the analysis of single events, or very limited sets of events, to a discipline capable of the analysis of very large representative sets of events. This, in turn, will enable the discovery of statistical patterns, leading to the recognition of the underlying physical mechanisms, as well as the prediction of the probability of space-weather events.

The multipurpose, trainable, feature-recognition module that we are developing has, as discussed in Section 16, the potential of dealing with solar phenomena for which no task-specific modules exist (*e.g.* cusps, arcades, faculae, pores, arch filaments, *etc.*), as well as quickly identifying newly discovered phenomena (such as sigmoids were in the *Yohkoh* era), and even phenomena that the user is not quite able to define yet (see Section 16 for the example of coronal nulls). We anticipate that the trainable module will find wide applications is heliophysics, not just limited to coronal imagery and magnetograms. Thus, the trainable module has the potential of saving considerable amounts of programming time and effort in comparison with developing task-specific modules for every phenomenon.

The trainable module will also be used to help quantify the accuracy of the task-specific modules. In this article we have not discussed the issue of accuracy in detail, since we are still in the initial phase of the computer-vision project. However, we have developed plans to quantify the accuracy and precision of all our modules in the second phase of our work, which we shall briefly review here. Each module will be tested against human observers. For that we will identify an appropriate time period (say, a week or a month) for which human volunteers will annotate all the AIA and relevant HMI images for the features identified by the task-specific modules. Comparison of the discrepancies between the human-produced data and the module data will allow us to correct in part for the misidentifications by the human observers and produce a dataset that we will consider "true". Discrepancies between the module data and this set can then be used to pinpoint the accuracy and precision of the module-produced metadata. In addition, we will apply the trainable module to the same dataset, and cross-compare the results. Comparison with the output of other automated feature detection algorithms, such as the GOES flare catalog, and the output from modules developed by other groups will allow for further analysis. These quantitative results then define error bars for statistical analysis of events and correlations. Only very limited quantitative verification of feature-recognition algorithms has been carried out so far; one exception is the verification of an early version of our filament detection algorithm AAFDCC, using the filament list by Pevtsov, Balasubramaniam, and Rogers (2003); see Section 4 for more details.

We have some final comments on further feature-recognition modules produced by the solar-physics community. The Feature Finding Team has produced the modules that were deemed most urgent for the SDO mission, but there is absolutely no reason or intent to limit SDO feature finding to these modules. We are aware of multiple efforts by groups in Europe and the US to produce additional SDO feature-recognition algorithms. For the modules of this article we have developed general data input and metadata output subroutines that are available to any group intent on developing feature-finding routines. In addition, we point out that any modules that run on the continuous data stream, even if for only a single AIA passband, must be able to operate in a semi-autonomous manner in a pipeline setup, because of the data volume. This requires a different approach to module programming than is customary in the solar-physics community. A near real-time pipeline exists at LMSAL, but pipelines operating with a lag time of the order of a day can be set up by any institution that receives at least part of the data stream. As described above, SAO operates a pipeline identical to that at LMSAL that runs several of our modules and that is open, in principle, to community-produced algorithms. The metadata produced by additional modules can be delivered to the HEK in the appropriate format, but there is no reason to limit delivery to the HEK. Some of our modules (the bright-point module and the trainable module) do not deliver all their metadata to the HEK because the former produces a volume of metadata that is too large, while the latter produces metadata that are completely incompatible with the HEK.

In closing, we present our vision for the near and medium future of helioinformatics. We foresee a situation in which heliospheric virtual observatories, such as the HEK and the VSO, provide a simple and seamless interface between data, metadata, and computer-vision software systems. Heliospheric virtual observatories and computer-vision systems will work together to monitor the Sun constantly, provide space weather warnings, populate catalogs of metadata, analyze trends, and produce real-time online imagery of current events. Supported by this extensive apparatus, the field of solar physics will transition from an event-driven research mode to a system-oriented approach and will thus develop better predictive capabilities.

Acknowledgements This research and development project is supported by NASA Grant NNX09AB03G to the Smithsonian Astrophysical Observatory with subcontracts to the Southwest Research Institute, Montana State University, the Applied Physics Laboratory of Johns Hopkins University, and the Lockheed Martin Solar and Astrophysics Laboratory, and with the participation of the NASA Marshall Space Flight Center. This project would be impossible without the major contributions from our European collaborators. The institutions involved are the Royal Observatory of Belgium, the Academy of Athens, Trinity College Dublin, the University of St Andrews, and the Max-Planck-Institut für Sonnensystemforschung.

We thank the referee for helpful and constructive comments, and Trae Winter for carefully proofreading the manuscript.

References

Aly, J.J.: 1989, On the reconstruction of the nonlinear force-free coronal magnetic field from boundary data. *Solar Phys.* **120**, 19 – 48.

Antiochos, S.K., Dahlburg, R.B., Klimchuk, J.A.: 1994, The magnetic field of solar prominences. *Astrophys. J.* **420**, 41 – 44. doi:10.1086/187158.

Antiochos, S.K., DeVore, C.R., Klimchuk, J.A.: 1999, A model for solar coronal mass ejections. *Astrophys. J.* **510**, 485 – 493. doi:10.1086/306563.

Asai, A., Ishii, T.T., Kurokawa, H., Yokoyama, T., Shimojo, M.: 2003, Evolution of conjugate footpoints inside flare ribbons during a great two-ribbon flare on 2001 April 10. *Astrophys. J.* **586**, 624 – 629. doi:10.1086/367694.

Attrill, G.D.R., Wills-Davey, M.J.: 2010, Automatic detection and extraction of coronal dimmings from SDO/AIA data. *Solar Phys.* **262**, 461 – 480. doi:10.1007/s11207-009-9444-4.

Banda, J.M., Angryk, R.: 2009, On the effectiveness of fuzzy clustering as a data discretization technique for large-scale classification of solar images. In: Feng, G.G. (ed.) *Proceedings of the 18th IEEE International Conference on Fuzzy Systems, FUZZ-IEEE '09*, IEEE, New York, 2019 – 2024. doi:10.1109/FUZZY.2009.5277273.

Barra, V., Delouille, V., Hochedez, J.F.: 2008, Segmentation of extreme ultraviolet solar images *via* multichannel fuzzy clustering. *Adv. Space Res.* **42**, 917 – 925. doi:10.1016/j.asr.2007.10.021.

Barra, V., Delouille, V., Hochedez, J.F., Chainais, P.: 2005, Segmentation of EIT images using fuzzy clustering: a preliminary study. In: Danesy, D., Poedts, S., De Groof, A., Andries, J. (eds.) *The Dynamic Sun: Challenges for Theory and Observations*, **SP-600**, ESA, Noordwijk, 71 – 80.

Barra, V., Delouille, V., Kretzschmar, M., Hochedez, J.: 2009, Fast and robust segmentation of solar EUV images: algorithm and results for solar cycle 23. *Astron. Astrophys.* **505**, 361 – 371. doi:10.1051/0004-6361/200811416.

Bemporad, A., Raymond, J., Poletto, G., Romoli, M.: 2007, A comprehensive study of the initiation and early evolution of a coronal mass ejection from ultraviolet and white-light data. *Astrophys. J.* **655**, 576 – 590. doi:10.1086/509569.

Bernasconi, P.N., Rust, D.M., Hakim, D.: 2005, Advanced automated solar filament detection and characterization code: description, performance, and results. *Solar Phys.* **228**, 97 – 117. doi:10.1007/s11207-005-2766-y.

Bernasconi, P.N., Raouafi, N.E., Georgoulis, M.K.: 2011, The sigmoid sniffer. *Solar Phys.*, submitted.

Bewsher, D., Harrison, R.A., Brown, D.S.: 2008, The relationship between EUV dimming and coronal mass ejections. I. Statistical study and probability model. *Astron. Astrophys.* **478**, 897 – 906. doi:10.1051/0004-6361:20078615.

Boursier, Y., Lamy, P., Llebaria, A.: 2009, Three-dimensional kinematics of coronal mass ejections from STEREO/SECCHI-COR2 observations in 2007 – 2008. *Solar Phys.* **256**, 131 – 147. doi:10.1007/s11207-009-9358-1.

Boursier, Y., Lamy, P., Llebaria, A., Goudail, F., Robelus, S.: 2009, The ARTEMIS catalog of LASCO coronal mass ejections. Automatic recognition of transient events and Marseille inventory from synoptic maps. *Solar Phys.* **257**, 125 – 147. doi:10.1007/s11207-009-9370-5.

Byrne, J.P., Gallagher, P.T., McAteer, R.T.J., Young, C.A.: 2009, The kinematics of coronal mass ejections using multiscale methods. *Astron. Astrophys.* **495**, 325 – 334. doi:10.1051/0004-6361:200809811.

Canfield, R.C., Hudson, H.S., McKenzie, D.E.: 1999, Sigmoidal morphology and eruptive solar activity. *Geophys. Res. Lett.* **26**, 627 – 630. doi:10.1029/1999GL900105.

Canfield, R.C., Kazachenko, M.D., Acton, L.W., Mackay, D.H., Son, J., Freeman, T.L.: 2007, *Yohkoh* SXT full-resolution observations of sigmoids: structure, formation, and eruption. *Astrophys. J. Lett.* **671**, 81 – 84. doi:10.1086/524729.

Chae, J., Martin, S.F., Yun, H.S., Kim, J., Lee, S., Goode, P.R., Spirock, T., Wang, H.: 2001, Small magnetic bipoles emerging in a filament channel. *Astrophys. J.* **548**, 497 – 507. doi:10.1086/318661.

Christe, S., Hannah, I.G., Krucker, S., McTiernan, J., Lin, R.P.: 2008, RHESSI microflare statistics. I. Flare-finding and frequency distributions. *Astrophys. J.* **677**, 1385 – 1394. doi:10.1086/529011.

Davey, A.R., McIntosh, S.: 2007, The SoHO/EIT brightpoint database: mining the database for science. *Bull. Am. Astron. Soc.* **38**, 327.

De Moortel, I., Hood, A.W.: 2000, Wavelet analysis and the determination of coronal plasma properties. *Astron. Astrophys.* **363**, 269 – 278.

De Moortel, I., McAteer, R.T.J.: 2004, Waves and wavelets: an automated detection technique for solar oscillations. *Solar Phys.* **223**, 1 – 11. doi:10.1007/s11207-004-0806-7.

De Moortel, I., Hood, A.W., Ireland, J.: 2002, Coronal seismology through wavelet analysis. *Astron. Astrophys.* **381**, 311 – 323. doi:10.1051/0004-6361:20011659.

De Moortel, I., Munday, S.A., Hood, A.W.: 2004, Wavelet analysis: the effect of varying basic wavelet parameters. *Solar Phys.* **222**, 203 – 228. doi:10.1023/B:SOLA.0000043578.01201.2d.

DeForest, C.E., Plunkett, S.P., Andrews, M.D.: 2001, Observation of polar plumes at high solar altitudes. *Astrophys. J.* **546**, 569 – 575. doi:10.1086/318221.

DeForest, C.E., Hagenaar, H.J., Lamb, D.A., Parnell, C.E., Welsch, B.T.: 2007, Solar magnetic tracking. I. Software comparison and recommended practices. *Astrophys. J.* **666**, 576 – 587. doi:10.1086/518994.

DeForest, C.E., Lamb, D.A., Berger, T., Hagenaar, H., Parnell, C., Welsch, B.: 2008, The small-scale field measured with *Hinode*/SOT and feature tracking: where is the mixed-polarity flux? *AGU Spring Meeting Abstracts*, 1.

DeRosa, M.L., Schrijver, C.J., Barnes, G., Leka, K.D., Lites, B.W., Aschwanden, M.J., Amari, T., Canou, A., McTiernan, J.M., Régnier, S., Thalmann, J.K., Valori, G., Wheatland, M.S., Wiegelmann, T., Cheung, M.C.M., Conlon, P.A., Fuhrmann, M., Inhester, B., Tadesse, T.: 2009, A critical assessment of nonlinear force-free field modeling of the solar corona for active region 10953. *Astrophys. J.* **696**, 1780 – 1791. doi:10.1088/0004-637X/696/2/1780.

Falconer, D.A., Moore, R.L., Gary, G.A.: 2002, Correlation of the coronal mass ejection productivity of solar active regions with measures of their global nonpotentiality from vector magnetograms: baseline results. *Astrophys. J.* **569**, 1016 – 1025. doi:10.1086/339161.

Fan, Y., Gibson, S.E.: 2003, The emergence of a twisted magnetic flux tube into a preexisting coronal arcade. *Astrophys. J. Lett.* **589**, 105 – 108. doi:10.1086/375834.

Fan, Y., Gibson, S.E.: 2004, Numerical simulations of three-dimensional coronal magnetic fields resulting from the emergence of twisted magnetic flux tubes. *Astrophys. J.* **609**, 1123 – 1133. doi:10.1086/421238.

Fisher, R.R., Munro, R.H.: 1984, Coronal transient geometry. I – The flare-associated event of 1981 March 25. *Astrophys. J.* **280**, 428 – 439. doi:10.1086/162009.

Fuller, N., Aboudarham, J., Bentley, R.D.: 2005, Filament recognition and image cleaning on Meudon Hα spectroheliograms. *Solar Phys.* **227**, 61 – 73. doi:10.1007/s11207-005-8364-1.

Gao, J., Wang, H., Zhou, M.: 2002, Development of an automatic filament disappearance detection system. *Solar Phys.* **205**, 93 – 103.

Gibson, S.E., Fan, Y., Török, T., Kliem, B.: 2006, The evolving sigmoid: evidence for magnetic flux ropes in the corona before, during, and after CMES. *Space Sci. Rev.* **124**, 131 – 144. doi:10.1007/s11214-006-9101-2.

Gilbert, H.R., Holzer, T.E., Burkepile, J.T., Hundhausen, A.J.: 2000, Active and eruptive prominences and their relationship to coronal mass ejections. *Astrophys. J.* **537**, 503 – 515. doi:10.1086/309030.

Golub, L., Deluca, E., Austin, G., Bookbinder, J., Caldwell, D., Cheimets, P., Cirtain, J., Cosmo, M., Reid, P., Sette, A., Weber, M., Sakao, T., Kano, R., Shibasaki, K., Hara, H., Tsuneta, S., Kumagai, K., Tamura, T., Shimojo, M., McCracken, J., Carpenter, J., Haight, H., Siler, R., Wright, E., Tucker, J., Rutledge, H., Barbera, M., Peres, G., Varisco, S.: 2007, The X-ray telescope (XRT) for the *Hinode* mission. *Solar Phys.* **243**, 63 – 86. doi:10.1007/s11207-007-0182-1.

Gopalswamy, N., Yashiro, S., Michalek, G., Stenborg, G., Vourlidas, A., Freeland, S., Howard, R.: 2009, The SOHO/LASCO CME catalog. *Earth Moon Planets* **104**, 295 – 313. doi:10.1007/s11038-008-9282-7.

Green, L.M., Kliem, B., Török, T., van Driel-Gesztelyi, L., Attrill, G.D.R.: 2007, Transient coronal sigmoids and rotating erupting flux ropes. *Solar Phys.* **246**, 365 – 391. doi:10.1007/s11207-007-9061-z.

Hagenaar, H., Cheung, M.: 2008, Magnetic flux emergence on different scales. In: *12th European Solar Physics Meeting* **2**, 2.53. http://espm.kis.uni-freiburg.de.

Hurlburt, N., Cheung, M., Schrijver, C., Chang, L., Freeland, S., Green, S., Heck, C., Jaffey, A., Kobashi, A., Schiff, D., Serafin, J., Seguin, R., Slater, G., Somani, A., Timmons, R.: 2010, Heliophysics event knowledgebase for the solar dynamics observatory and beyond. *Solar Phys.* doi:10.1007/s11207-010-9624-2.

Inhester, B., Feng, L., Wiegelmann, T.: 2008, Segmentation of loops from coronal EUV images. *Solar Phys.* **248**, 379 – 393. doi:10.1007/s11207-007-9027-1.

Ireland, J., De Moortel, I.: 2002, Application of wavelet analysis to transversal coronal loop oscillations. *Astron. Astrophys.* **391**, 339 – 351. doi:10.1051/0004-6361:20020643.

Jing, J., Yurchyshyn, V.B., Yang, G., Xu, Y., Wang, H.: 2004, On the relation between filament eruptions, flares, and coronal mass ejections. *Astrophys. J.* **614**, 1054 – 1062. doi:10.1086/423781.

Jing, J., Wiegelmann, T., Suematsu, Y., Kubo, M., Wang, H.: 2008, Changes of magnetic structure in three dimensions associated with the X3.4 flare of 2006 December 13. *Astrophys. J. Lett.* **676**, 81 – 84. doi:10.1086/587058.

Jones, H.P.: 2004, Counting magnetic bipoles on the Sun by polarity inversion. In: Negoita, M.G., Howlett, R.J., Jain, L.C. (eds.) *Knowledge-Based Intelligent Information and Engineering Systems: 8th International Conference, KES 2004, Lecture Notes in Computer Science* **3215**, 433 – 439.

Kliem, B., Titov, V.S., Török, T.: 2004, Formation of current sheets and sigmoidal structure by the kink instability of a magnetic loop. *Astron. Astrophys.* **413**, 23 – 26. doi:10.1051/0004-6361:20031690.

Kosugi, T., Matsuzaki, K., Sakao, T., Shimizu, T., Sone, Y., Tachikawa, S., Hashimoto, T., Minesugi, K., Ohnishi, A.A., Yamada, T., Tsuneta, S., Hara, H., Ichimoto, K., Suematsu, Y., Shimojo, M., Watanabe, T., Shimada, S., Davis, J.M., Hill, L.D., Owens, J.K., Title, A.M., Culhane, J.L., Harra, L.K., Doschek, G.A., Golub, L.: 2007, The *Hinode* (Solar-B) mission: an overview. *Solar Phys.* **243**, 3 – 17. doi:10.1007/s11207-007-9014-6.

LaBonte, B.J., Rust, D.M., Bernasconi, P.N.: 2003, An automated system for detecting sigmoids in solar X-ray images. *Bull. Am. Astron. Soc.* **35**, 814.

Lamb, D.A., DeForest, C.E., Hagenaar, H.J., Parnell, C.E., Welsch, B.T.: 2008, Solar magnetic tracking. II. The apparent unipolar origin of quiet-sun flux. *Astrophys. J.* **674**, 520 – 529. doi:10.1086/524372.

Lamb, D., Deforest, C.E., Hagenaar, H.J., Parnell, C.E., Welsch, B.T.: 2007, Feature tracking of *Hinode* magnetograms. *AGU Fall Meeting Abstracts*, 1066.

Lamb, R.: 2008, An information retrieval system for images from the TRACE satellite. Master's thesis, Montana State University, Bozeman, MT, USA. http://www.cs.montana.edu/techreports/2008/Lamb.pdf.

Lamb, R., Angryk, R., Martens, P.: 2008, An example-based image retrieval system for the TRACE repository. In: Ejiri, M., Kasturi, R., Sanniti di Baja, G. (eds.) *Proceedings of the 19th International Conference on Pattern Recognition (ICPR '08)*, IEEE, New York, 1 – 4. doi:10.1109/ICPR.2008.4761078.

Leblanc, Y., Dulk, G.A., Vourlidas, A., Bougeret, J.L.: 2001, Tracing shock waves from the corona to 1 AU: Type II radio emission and relationship with CMEs. *J. Geophys. Res.* **106**, 25301 – 25312. doi:10.1029/2000JA000260.

Martin, S.F.: 1998, Conditions for the formation and maintenance of filaments (invited review). *Solar Phys.* **182**, 107 – 137. doi:10.1023/A:1005026814076.

Martin, S.F., Bilimoria, R., Tracadas, P.W.: 1994, Magnetic field configurations basic to filament channels and filaments. In: Rutten, R.J., Schrijver, C.J. (eds.) *Solar Surface Magnetism*, Kluwer, Dordrecht, 303 – 338.

Masuda, S., Kosugi, T., Hara, H., Tsuneta, S., Ogawara, Y.: 1994, A loop-top hard X-ray source in a compact solar flare as evidence for magnetic reconnection. *Nature* **371**, 495 – 497. doi:10.1038/371495a0.

McAteer, R.T.J., Gallagher, P.T., Bloomfield, D.S., Williams, D.R., Mathioudakis, M., Keenan, F.P.: 2004, Ultraviolet oscillations in the chromosphere of the quiet Sun. *Astrophys. J.* **602**, 436 – 445. doi:10.1086/380835.

McIntosh, P.S.: 1994, *YOHKOH* X-ray image interpretation with overlays of Hα neutral lines. In: Uchida, Y., Watanabe, T., Shibata, K., Hudson, H.S. (eds.) *X-ray Solar Physics from Yohkoh*, Universal Academy Press, Tokyo, 271 – 272.

McIntosh, S.W., Gurman, J.B.: 2004, EIT EUV brightpoints over the SOHO mission so far. In: Walsh, R.W., Ireland, J., Danesy, D., Fleck, B. (eds.) *SOHO 15 Coronal Heating, ESA Special Publication* **575**, 235.

McIntosh, S.W., Gurman, J.B.: 2005, Nine years of EUV bright points. *Solar Phys.* **228**, 285 – 299. doi:10.1007/s11207-005-4725-z.

McIntosh, S.W., Sitongia, L., Markel, R., Judge, P.G., Davey, A.R.: 2009, Mining a massive brightpoint database for science. *Bull. Am. Astron. Soc.* **41**, 839.

McKenzie, D.E., Canfield, R.C.: 2008, *Hinode* XRT observations of a long-lasting coronal sigmoid. *Astron. Astrophys.* **481**, 65 – 68. doi:10.1051/0004-6361:20079035.

Metcalf, T.R., Derosa, M.L., Schrijver, C.J., Barnes, G., van Ballegooijen, A.A., Wiegelmann, T., Wheatland, M.S., Valori, G., McTtiernan, J.M.: 2008, Nonlinear force-free modeling of coronal magnetic fields. II. Modeling a filament arcade and simulated chromospheric and photospheric vector fields. *Solar Phys.* **247**, 269 – 299. doi:10.1007/s11207-007-9110-7.

Michałek, G., Gopalswamy, N., Yashiro, S.: 2003, A new method for estimating widths, velocities, and source location of Halo coronal mass ejections. *Astrophys. J.* **584**, 472 – 478. doi:10.1086/345526.

Narukage, N., Sakao, T., Kano, R., Hara, H., Shimojo, M., Bando, T., Urayama, F., DeLuca, E., Golub, L., Weber, M., Grigis, P., Cirtain, J., Tsuneta, S.: 2011, Coronal-temperature-diagnostic capability of the *Hinode*/X-Ray Telescope based on self-consistent calibration. *Solar Phys.* doi:10.1007/s11207-010-9685-2.

Ogawara, Y., Takano, T., Kato, T., Kosugi, T., Tsuneta, S., Watanabe, T., Kondo, I., Uchida, Y.: 1991, The solar-A mission – an overview. *Solar Phys.* **136**, 1 – 16. doi:10.1007/BF00151692.

Olmedo, O., Zhang, J., Wechsler, H., Poland, A., Borne, K.: 2008, Automatic detection and tracking of coronal mass ejections in coronagraph time series. *Solar Phys.* **248**, 485 – 499. doi:10.1007/s11207-007-9104-5.

Pevtsov, A.A., Balasubramaniam, K.S., Rogers, J.W.: 2003, Chirality of chromospheric filaments. *Astrophys. J.* **595**, 500 – 505. doi:10.1086/377339.

Pneuman, G.W.: 1983, The formation of solar prominences by magnetic reconnection and condensation. *Solar Phys.* **88**, 219 – 239. doi:10.1007/BF00196189.

Podladchikova, O., Berghmans, D.: 2005, Automated detection of EIT waves and dimmings. *Solar Phys.* **228**, 265 – 284. doi:10.1007/s11207-005-5373-z.

Qu, M., Shih, F.Y., Jing, J., Wang, H.: 2005, Automatic solar filament detection using image processing techniques. *Solar Phys.* **228**, 119 – 135. doi:10.1007/s11207-005-5780-1.

Rachmeler, L.A., Wills-Davey, M.J.: 2005, Observations of unusual "EIT Wave" dynamics. *AGU Spring Meeting Abstracts*, 9.

Reinard, A.A., Biesecker, D.A.: 2008, Coronal mass ejection-associated coronal dimmings. *Astrophys. J.* **674**, 576 – 585. doi:10.1086/525269.

Robbrecht, E., Berghmans, D.: 2004, Automated recognition of coronal mass ejections (CMEs) in near-real-time data. *Astron. Astrophys.* **425**, 1097 – 1106. doi:10.1051/0004-6361:20041302.

Rust, D.M., Hildner, E.: 1976, Expansion of an X-ray coronal arch into the outer corona. *Solar Phys.* **48**, 381 – 387.

Rust, D.M., Kumar, A.: 1996, Evidence for helically kinked magnetic flux ropes in solar eruptions. *Astrophys. J. Lett.* **464**, 199 – 202. doi:10.1086/310118.

Rust, D.M., LaBonte, B.J.: 2005, Observational evidence of the kink instability in solar filament eruptions and sigmoids. *Astrophys. J. Lett.* **622**, 69 – 72. doi:10.1086/429379.

Rust, D.M., Martin, S.F.: 1994, A correlation between sunspot whirls and filament type. In: Balasubramaniam, K.S., Simon, G.W. (eds.) *Solar Active Region Evolution: Comparing Models with Observations* **CS-68**, Astron. Soc. Pac., San Francisco, 337.

Rust, D.M., Anderson, B.J., Andrews, M.D., Acuña, M.H., Russell, C.T., Schuck, P.W., Mulligan, T.: 2005, Comparison of interplanetary disturbances at the NEAR spacecraft with coronal mass ejections at the Sun. *Astrophys. J.* **621**, L524 – L536. doi:10.1086/427401.

Savcheva, A., Cirtain, J., Deluca, E.E., Lundquist, L.L., Golub, L., Weber, M., Shimojo, M., Shibasaki, K., Sakao, T., Narukage, N., Tsuneta, S., Kano, R.: 2007, A study of polar jet parameters based on *Hinode* XRT observations. *Publ. Astron. Soc. Japan* **59**, 771 – 778.

Schrijver, C.J., Derosa, M.L., Metcalf, T.R., Liu, Y., McTiernan, J., Régnier, S., Valori, G., Wheatland, M.S., Wiegelmann, T.: 2006, Nonlinear force-free modeling of coronal magnetic fields, Part I: A quantitative comparison of methods. *Solar Phys.* **235**, 161 – 190. doi:10.1007/s11207-006-0068-7.

Shih, F.Y., Kowalski, A.J.: 2003, Automatic extraction of filaments in Hα solar images. *Solar Phys.* **218**, 99 – 122. doi:10.1023/B:SOLA.0000013052.34180.58.

Tadesse, T., Wiegelmann, T., Inhester, B.: 2009, Nonlinear force-free coronal magnetic field modeling and preprocessing of vector magnetograms in spherical geometry. *Astron. Astrophys.* **508**, 421 – 432.

Thalmann, J.K., Wiegelmann, T., Raouafi, N.E.: 2008, First nonlinear force-free field extrapolations of SOLIS/VSM data. *Astron. Astrophys.* **488**, 71 – 74. doi:10.1051/0004-6361:200810235.

Titov, V.S., Démoulin, P.: 1999, Basic topology of twisted magnetic configurations in solar flares. *Astron. Astrophys.* **351**, 707 – 720.

Török, T., Kliem, B.: 2005, Confined and ejective eruptions of kink-unstable flux ropes. *Astrophys. J. Lett.* **630**, L97 – L100. doi:10.1086/462412.

Török, T., Kliem, B.: 2007, Numerical simulations of fast and slow coronal mass ejections. *Astron. Nachr.* **328**, 743 – 746. doi:10.1002/asna.200710795.

Tsuneta, S., Hara, H., Shimizu, T., Acton, L.W., Strong, K.T., Hudson, H.S., Ogawara, Y.: 1992, Observation of a solar flare at the limb with the *YOHKOH* soft X-ray telescope. *Publ. Astron. Soc. Japan* **44**, 63 – 69.

van Ballegooijen, A.A., Martens, P.C.H.: 1989, Formation and eruption of solar prominences. *Astrophys. J.* **343**, 971 – 984. doi:10.1086/167766.

van Ballegooijen, A.A., Martens, P.C.H.: 1990, Magnetic fields in quiescent prominences. *Astrophys. J.* **361**, 283 – 289. doi:10.1086/169193.

Webb, D.F.: 2000, Understanding CMEs and their source regions. *J. Atmos. Solar-Terr. Phys.* **62**, 1415 – 1426.

Wheatland, M.S., Sturrock, P.A., Roumeliotis, G.: 2000, An optimization approach to reconstructing force-free fields. *Astrophys. J.* **540**, 1150 – 1155. doi:10.1086/309355.

Wiegelmann, T.: 2004, Optimization code with weighting function for the reconstruction of coronal magnetic fields. *Solar Phys.* **219**, 87 – 108. doi:10.1023/B:SOLA.0000021799.39465.36.

Wiegelmann, T.: 2007, Computing nonlinear force-free coronal magnetic fields in spherical geometry. *Solar Phys.* **240**, 227 – 239. doi:10.1007/s11207-006-0266-3.

Wiegelmann, T., Inhester, B.: 2010, How to deal with measurement errors and lacking data in nonlinear force-free coronal magnetic field modelling? *Astron. Astrophys.* **516**, 107. doi:10.1051/0004-6361/201014391.

Wiegelmann, T., Inhester, B., Sakurai, T.: 2006, Preprocessing of vector magnetograph data for a nonlinear force-free magnetic field reconstruction. *Solar Phys.* **233**, 215 – 232. doi:10.1007/s11207-006-2092-z.

Wiegelmann, T., Inhester, B., Feng, L.: 2009, Solar stereoscopy – where are we and what developments do we require to progress? *Ann. Geophys.* **27**, 2925 – 2936.

Wiegelmann, T., Thalmann, J.K., Schrijver, C.J., Derosa, M.L., Metcalf, T.R.: 2008, Can we improve the preprocessing of photospheric vector magnetograms by the inclusion of chromospheric observations? *Solar Phys.* **247**, 249 – 267. doi:10.1007/s11207-008-9130-y.

Wills-Davey, M.J.: 2006, Tracking large-scale propagating coronal wave fronts (EIT waves) using automated methods. *Astrophys. J.* **645**, 757 – 765. doi:10.1086/504144.

Xie, H., Ofman, L., Lawrence, G.: 2004, Cone model for halo CMEs: application to space weather forecasting. *J. Geophys. Res. (Space Phys.)* **109**(A18), 3109 – 3121. doi:10.1029/2003JA010226.

Xue, X.H., Wang, C.B., Dou, X.K.: 2005, An ice-cream cone model for coronal mass ejections. *J. Geophys. Res. (Space Phys.)* **110**(A9), 8103 – 8114. doi:10.1029/2004JA010698.

Yang, L., Jin, R., Sukthankar, R., Zheng, B., Mummert, L., Satyanarayanan, M., Chen, M., Jukic, D.: 2007, Learning distance metrics for interactive search-assisted diagnosis of mammograms. In: *SPIE* **CS-6514**. doi:10.1117/12.710076.

Yurchyshyn, V.B., Wang, H., Goode, P.R., Deng, Y.: 2001, Orientation of the magnetic fields in interplanetary flux ropes and solar filaments. *Astrophys. J.* **563**, 381 – 388. doi:10.1086/323778.

Solar Phys (2012) 275:115–143
DOI 10.1007/s11207-009-9487-6

Extreme Ultraviolet Variability Experiment (EVE) on the *Solar Dynamics Observatory* (SDO): Overview of Science Objectives, Instrument Design, Data Products, and Model Developments

T.N. Woods · F.G. Eparvier · R. Hock · A.R. Jones · D. Woodraska · D. Judge · L. Didkovsky · J. Lean · J. Mariska · H. Warren · D. McMullin · P. Chamberlin · G. Berthiaume · S. Bailey · T. Fuller-Rowell · J. Sojka · W.K. Tobiska · R. Viereck

Received: 5 October 2009 / Accepted: 19 November 2009 / Published online: 12 January 2010

The Solar Dynamics Observatory
Guest Editors: W. Dean Pesnell, Phillip C. Chamberlin, and Barbara J. Thompson

T.N. Woods (✉) · F.G. Eparvier · R. Hock · A.R. Jones · D. Woodraska
Laboratory for Atmospheric and Space Physics, University of Colorado, 1234 Innovation Dr., Boulder, CO 80303, USA
e-mail: tom.woods@lasp.colorado.edu

D. Judge · L. Didkovsky
Space Sciences Center, University of Southern California, Los Angeles, CA 90089, USA

J. Lean · J. Mariska · H. Warren
Naval Research Laboratory, 4555 Overlook Ave. S.W., Washington, DC 20375, USA

D. McMullin
Space Systems Research Corporation, 1940 Duke St., Alexandria, VA 22314, USA

P. Chamberlin
NASA Goddard Space Flight Center, Greenbelt, MD 20771, USA

G. Berthiaume
Lincoln Laboratory, Massachusetts Institute of Technology, Lexington, MA 02420, USA

S. Bailey
Electrical and Computer Engineering Department, Virginia Tech, Blacksburg, VA 24061, USA

T. Fuller-Rowell
CIRES University of Colorado and NOAA Space Weather Prediction Center, 325 Broadway, Boulder, CO 80305, USA

J. Sojka
Center for Atmospheric and Space Sciences, Utah State University, Logan, UT 84322, USA

W.K. Tobiska
Space Environment Technologies, 1676 Palisades Dr., Pacific Palisades, CA 90272, USA

R. Viereck
NOAA Space Weather Prediction Center, 325 Broadway, Boulder, CO 80305, USA

Abstract The highly variable solar extreme ultraviolet (EUV) radiation is the major energy input to the Earth's upper atmosphere, strongly impacting the geospace environment, affecting satellite operations, communications, and navigation. The *Extreme ultraviolet Variability Experiment* (EVE) onboard the NASA *Solar Dynamics Observatory* (SDO) will measure the solar EUV irradiance from 0.1 to 105 nm with unprecedented spectral resolution (0.1 nm), temporal cadence (ten seconds), and accuracy (20%). EVE includes several irradiance instruments: The *Multiple EUV Grating Spectrographs* (MEGS)-A is a grazing-incidence spectrograph that measures the solar EUV irradiance in the 5 to 37 nm range with 0.1-nm resolution, and the MEGS-B is a normal-incidence, dual-pass spectrograph that measures the solar EUV irradiance in the 35 to 105 nm range with 0.1-nm resolution. To provide MEGS in-flight calibration, the *EUV SpectroPhotometer* (ESP) measures the solar EUV irradiance in broadbands between 0.1 and 39 nm, and a *MEGS-Photometer* measures the Sun's bright hydrogen emission at 121.6 nm. The EVE data products include a near real-time space-weather product (Level 0C), which provides the solar EUV irradiance in specific bands and also spectra in 0.1-nm intervals with a cadence of one minute and with a time delay of less than 15 minutes. The EVE higher-level products are Level 2 with the solar EUV irradiance at higher time cadence (0.25 seconds for photometers and ten seconds for spectrographs) and Level 3 with averages of the solar irradiance over a day and over each one-hour period. The EVE team also plans to advance existing models of solar EUV irradiance and to operationally use the EVE measurements in models of Earth's ionosphere and thermosphere. Improved understanding of the evolution of solar flares and extending the various models to incorporate solar flare events are high priorities for the EVE team.

Keywords EVE · SDO · Solar EUV irradiance · Space weather research

1. Introduction

The *Solar Dynamics Observatory* (SDO) is the first spacecraft in NASA's Living With a Star (LWS) program and is scheduled for a nominal five-year mission starting in February 2010 (or later). The goal of the SDO mission is to understand solar variability and its societal and technological impacts. SDO will investigate how the Sun's magnetic field is generated and structured and how this stored energy is converted and released into the heliosphere and geospace environment through the solar-wind, energetic-particles, and photon output. An underlying theme of SDO is scientific research to enable improved space-weather predictive capabilities, thus transitioning research to operations.

The *EUV Variability Experiment* (EVE) is one of three instruments onboard SDO. EVE will measure the solar extreme ultraviolet (EUV) and soft X-ray (XUV) spectral irradiance in order to better understand how solar magnetic activity is manifest in the ultraviolet wavelength ranges that drive the terrestrial upper atmosphere. The *Helioseismic and Magnetic Imager* (HMI) is a vector magnetograph designed to understand magnetic activity, which is the dominant source for solar variability that EVE will measure. HMI also provides high spatial resolution helioseismic data that can provide forecasting capability for the EUV irradiance, both from subsurface magnetic flux and from far-side imaging. The *Atmospheric Imaging Assembly* (AIA) will make full-disk solar images at multiple wavelengths to link magnetic changes on the surface and interior to those in the solar atmosphere. Most of the AIA wavelengths overlap with EVE, so integration over the solar disk of the AIA images can be compared to EVE irradiance results. The SDO measurements from EVE, HMI, and AIA will facilitate improved understanding of irradiance variations, flares, and coronal mass ejections (CMEs), for use in ionosphere and thermosphere models for space-weather operations, to better track satellites, and to manage communication and navigation systems. Furthermore, the EVE observations provide an important extension of the solar EUV irradiance record currently being made by the *Solar and Heliospheric Observatory* (SOHO) and *Thermosphere, Ionosphere, Mesosphere, Energetics and Dynamics* (TIMED) spacecraft.

The EVE instrument development coincidently has helped to advance technology in a few areas as briefly noted here. The Charge Coupled Device (CCD) cameras in EVE use radiation-hard, back-illuminated 1024 × 2048 CCDs from MIT Lincoln Laboratory and achieve high sensitivity in the EUV range, excellent stability over time, and extremely low noise of about one electron when operated colder than −70°C (Westhoff *et al.*, 2007). The grazing-incidence spectrograph in EVE implements a new, innovative grating design whereby the CCD is near normal incidence, instead of the traditional grazing incidence, so that this spectrograph achieves much higher sensitivity (Crotser *et al.*, 2007). The *EUV Spectrophotometer* (ESP) instrument in EVE is an expanded version of the SOHO *Solar EUV Monitor* (SEM) with multiple channels and with a quadrant photodiode that can provide in realtime the location of flares on the solar disk (Didkovsky *et al.*, 2010). As a final technology example, the EVE *Solar Aspect Monitor* (SAM) channel is a pinhole camera to provide not only alignment information for EVE in visible light but also solar X-ray spectra and images through detecting individual X-ray photon events (Hock *et al.*, 2010).

This paper describes the science objectives, instrumentation, data products, and models for the EVE program. Hock *et al.* (2010) and Didkovsky *et al.* (2010) provide more complete instrument description and details of pre-flight calibration results.

2. EVE Science Plan

The EUV photons that EVE measures originate in the Sun's chromosphere, transition region, and corona and deposit their energy in the Earth's ionosphere and thermosphere, thus directly connecting the Sun and the Earth in just eight minutes. The solar output in the EUV (10 – 121 nm) and XUV (0.1 – 10 nm) spectrum varies with solar activity from a factor of two to several orders of magnitude depending on wavelength, and on timescales from seconds and minutes (flares) to days and months (solar rotation) to years and decades (solar sunspot or magnetic cycle) (Woods and Rottman, 2002). The EUV and XUV irradiance is the primary energy input to the Earth's upper atmosphere: heating the thermosphere, creating the ionosphere, and initiating many complex photochemical reactions and dynamical motions. Fluctuating EUV and XUV irradiance drives variability in the atmosphere, affecting satellite operations, navigation systems, and communications. For example, the heating of the thermosphere by solar EUV radiation increases the neutral density with higher levels of solar activity and thus increases satellite drag (Picone, Emmert, and Lean, 2005; Marcos, Bowman, and Sheehan, 2006; Woodraska, Woods, and Eparvier, 2007; Bowman *et al.*, 2008a, 2008b). As shown in Figure 1, the neutral density changes by more than a factor of ten at 500 km over the 11-year solar cycle, while the thermospheric temperature changes by a factor of 1.5. A small number of previous missions have measured the solar EUV irradiance and produced daily-averaged irradiance spectra (Woods *et al.*, 2004), but none have done so with the accuracy and high time cadence of EVE. EVE will undertake the first comprehensive study of the solar EUV irradiance variability on the time scale of flares.

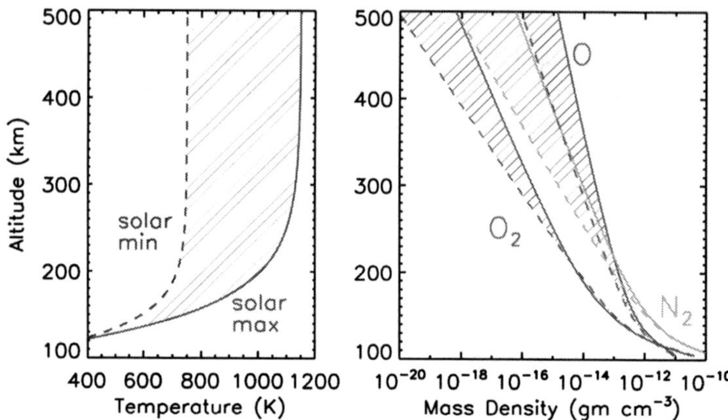

Figure 1 Solar-cycle variation of the thermospheric temperature and neutral density. These are global estimates from the NRLMSIS model, illustrating the large increases in average conditions from solar-cycle minimum to maximum.

2.1. EVE Science Objectives

EVE's measurements, modeling activities, and collaborations with the other SDO instruments pursue the following four scientific objectives:

i) specify the solar EUV irradiance and its variability on multiple times scales from seconds to years,
ii) advance current understanding of how and why the solar EUV spectral irradiance varies,
iii) improve the capability to predict (both nowcast and forecast) the EUV spectral variability, and
iv) understand the response of the geospace environment to variations in the solar EUV spectral irradiance and the impact on human endeavors.

2.1.1. EVE Objective 1 – Specify Solar EUV Irradiance

The first objective of EVE, and its highest priority, is the acquisition of a suitable database to characterize the solar EUV irradiance spectrum and its variations during flares, active-region evolution, and the solar cycle. Without such observations, the NASA Living With a Star (LWS) program will not be able to develop reliable space environment climatologies. For thermosphere and ionosphere energy-input comparisons, Knipp *et al.* (2005) demonstrated the dominance of EUV energy *versus* Joule heating and particle precipitation for Solar Cycles 22 and 23. Figure 2 illustrates modeled variations in the total EUV energy (Lean *et al.*, 2003), compared with that of solar-wind particles, during recent solar cycles.

The solar EUV spectrum (Figure 3) is comprised of thousands of emission lines and a few continua whose irradiances span more than four orders of magnitude. This radiation is emitted from the Sun's chromosphere, transition region, and corona at temperatures in the 10^4 to 10^6 K range. EUV radiation varies continuously at all wavelengths with hotter emissions generally varying more. Because lines at similar wavelengths can originate from sources on the Sun with different temperatures and densities, the EUV irradiance variability displays complex time-dependent wavelength dependences (see, *e.g.*, Woods and Rottman, 2002). The current TIMED *Solar EUV Experiment* (SEE) measurements (Woods *et al.*, 2005) have improved the absolute accuracy of the solar EUV spectral irradiances to about 20%, and the

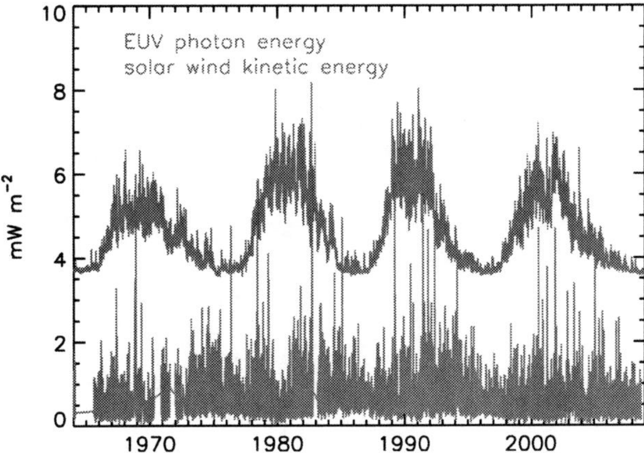

Figure 2 Key solar-energy input for Earth's ionosphere and thermosphere. Shown are variations in the daily total energy of EUV photons at wavelengths less than 120 nm, compared with the kinetic energy in solar-wind particles (mainly protons). (Updated from Lean, 2005).

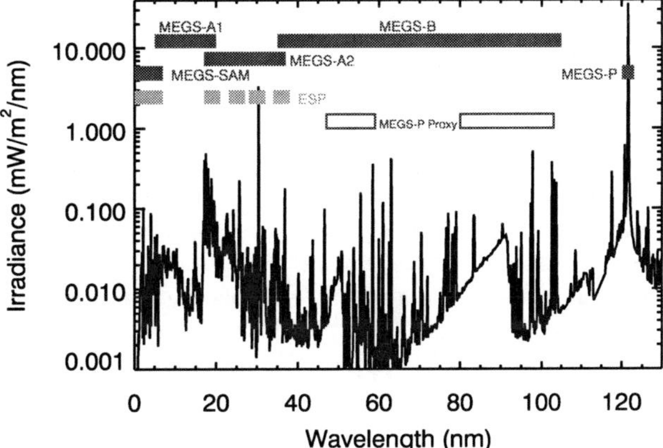

Figure 3 EVE instrument spectral ranges. The wavelength coverage of the various EVE instruments plotted over the solar minimum spectrum obtained with the prototype EVE instrument on 14 April 2008 (Chamberlin *et al.*, 2009).

EVE measurements will further increase this accuracy at some wavelengths to about 10% over the SDO mission and with much higher spectral resolution of 0.1 nm. The TIMED/SEE measurements have also achieved the first systematic characterization of the EUV spectral irradiance variations during solar rotation and the descending phase of a solar cycle, as shown in Figure 4 for SEE Level 3 product (Version 10) in 1-nm intervals. EVE will expand these observations to longer time scales, a factor of ten higher spectral resolution and much higher temporal cadence.

The historical solar EUV irradiance database, including the TIMED/SEE measurements, consists mainly of daily measurements, albeit with many large gaps such as from 1979 to 2002. Observations of the solar EUV irradiance with higher temporal cadence, as needed to study flares, are severely limited. The SOHO *Solar EUV Monitor* (SEM, Judge *et al.*, 1998) and GOES/XRS (Garcia, 1994) measurements do provide high time cadence but only in a few broad wavelength bands. EVE's spectrally-resolved observations with a temporal cadence of ten seconds will greatly advance the specification and understanding of the spectral variations during flare events throughout the XUV and EUV spectrum.

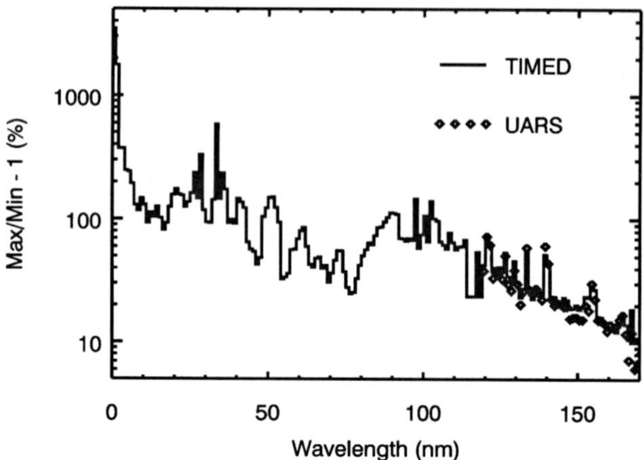

Figure 4 Solar EUV irradiance variations of the 11-year solar cycle. The TIMED/SEE results (line) agree very well with the UARS results (diamonds) in the far ultraviolet range (Woods *et al.*, 2005).

2.1.2. EVE Objective 2 – Understand Why Solar EUV Irradiance Varies

Extensive, multi-faceted investigations are planned to advance understanding of the sources of EUV irradiance variations within a physical framework. This understanding is needed to develop a predictive capability for past and future space-environment climatologies, and for verifying the direct EUV irradiance observations. Models will be developed to account for the observed EUV irradiances and their variations, with traceability to magnetic-flux emergence and the solar dynamo. This approach will be accomplished in collaboration with complementary SDO efforts, including those of the HMI and AIA instruments and other space missions. Auxiliary data from space and from the ground, *e.g.* by the *Synoptic Optical Long-term Investigations of the Sun* (SOLIS), Big Bear Solar Observatory programs, and the National Science Foundation's *Precision Solar Photometric Telescope* (PSPT), will also be used.

EUV radiation emerges from the outer layers of the Sun's atmosphere, which are sufficiently hot to excite highly-ionized species of gases; however, the solar atmosphere is not homogeneous, as the images in Figures 5 and 6 clearly illustrate. The EUV brightness inhomogeneities on the Sun are regions where magnetic fields alter the solar atmosphere. The fundamental determinant of the EUV spectrum is therefore the solar dynamo. There is a broad conceptual appreciation of the connections that relate the solar dynamo to surface magnetic-flux emergence, upward field propagation and expansion, active-region and coronal-hole formation, and EUV irradiance modulation, but the physical links remain largely undetermined. A self-consistent, end-to-end formulation that quantitatively relates the net EUV radiation to subsurface magnetic fields does not yet exist.

The recent approach of using differential-emission measures to model solar EUV irradiance variations (Warren, Mariska, and Lean, 2001) is a first step to quantify crucial physical processes. Emission-measure distributions are derived from spatially and spectrally resolved measurements of line intensities, and describe the temperature and density structure of the solar atmosphere. A quantitative estimate of the EUV radiation spectrum is possible by combining intensities calculated from emission-measure distributions of representative EUV sources, determinations from full-disk images (such as in Figures 5 and 6) of the fractional coverage of these sources, limb brightening, atomic properties and abundances. The EVE investigation will utilize the AIA full-disk images to quantify and track the fractional occurrence of EUV irradiance sources (*e.g.* active regions, active network, coronal holes)

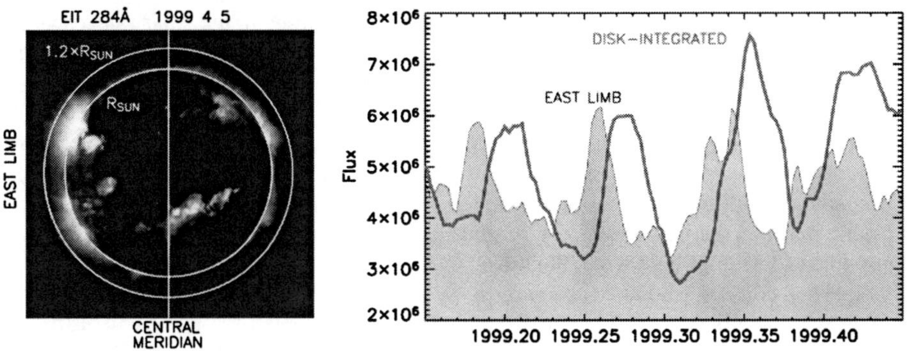

Figure 5 Forecasting technique for solar irradiance. Shown on the left are aspects of the solar disk utilized to develop a predictive capability. On the right, two time series derived from two different non-intersecting subsets of the EIT 28.4 nm images are compared. The east-limb flux (shaded region) is the signal from only those pixels between 1 and $1.2 \times R_{\mathrm{Sun}}$ in the eastern half of the image. The disk-integrated flux (solid line) is the signal from all pixels on the disk. This forecast technique is described in more detail by Lean, Picone, and Emmert (2009).

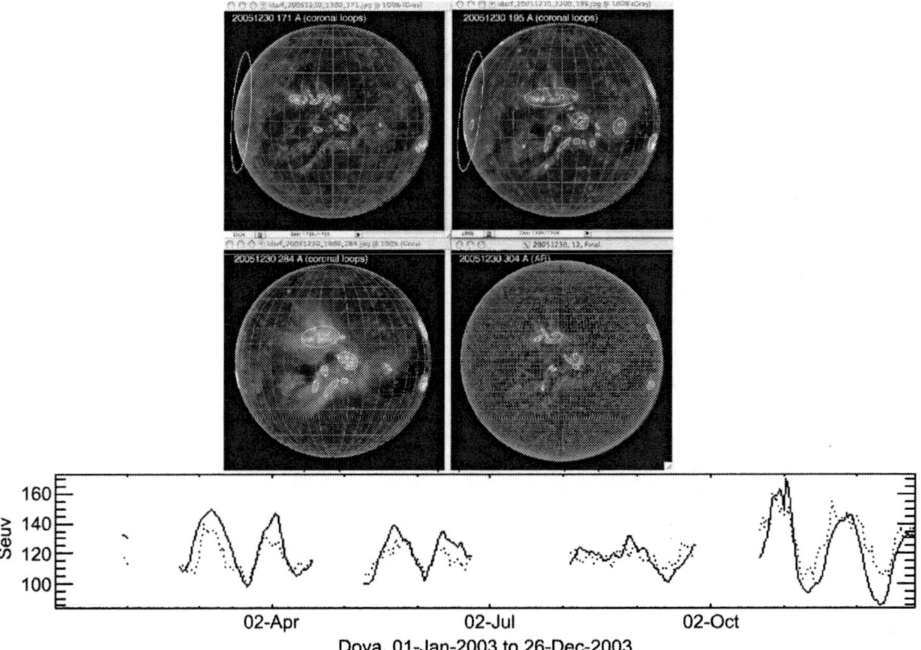

Figure 6 SOHO images are processed by the SET IDAR algorithm to extract active regions and coronal loops for three to seven days operational solar-irradiance forecasting. The 30.4 nm active-region and plage data are extracted to forecast the one-day prediction of the F10.7 index (SEUV, dotted line) for use in the JB2006 and JB2008 models. The actual F10.7 index is shown as the solid line.

throughout the SDO mission. Comparisons of the EUV irradiance modeled from AIA observations with the direct EVE measurements will reveal how well, or how poorly, this physical theory accounts for irradiance variations.

With quantitative relationships between magnetic structures in the solar images and the measured EUV irradiance established, subsequent exploration of the physical connections between EUV irradiance and surface magnetic fields becomes possible. SDO's HMI (and also ground-based magnetograms) map surface magnetic fields. Combining the HMI and AIA images (made at different temperatures) provides empirical tracing of magnetic fields from their relatively compact photospheric footprints to the more expanded EUV sources. Comparisons with modeled extrapolations of surface fields will help characterize the relationship between the fields and EUV brightness. Initial simulations have elucidated the primary role of footprint field strength and the less important role of loop length in accounting for observed coronal irradiance changes during solar rotation and the solar cycle. Analogous studies will combine data and results from the EVE instruments and other SDO instruments to quantify the relationship between surface magnetic structures and the EUV irradiance.

Once the coronal and chromospheric magnetic-field configurations that cause bright and dark EUV sources are properly connected to surface magnetic fields, the EUV irradiance variations become amenable to study in terms of source emergence, meridional transport and diffusion, as encapsulated in current flux-transport models (Wang, Lean, and Sheeley, 2000; Wang, Sheeley, and Lean, 2000). For example, the NRL flux transport model (Wang and Sheeley, 1991) may be used to simulate the evolution of the large-scale surface magnetic field for comparison with synoptic changes in the chromosphere and corona deduced from the AIA images. Relationships between subsurface signatures of flux emergence derived from HMI images, and their surface manifestations ultimately link the solar dynamo and the EUV radiation. New understanding of the solar dynamo and its possible long-term evolution and intermittent behavior (such as during and since the Maunder Minimum) will then be directly applicable to studying the EUV irradiance variations on longer time scales, beyond those accessible to space-based observations (see, *e.g.*, Wang, Lean, and Sheeley, 2005).

Section 5 describes in more detail the plans to update the physics-based NRLEUV model (Warren, Mariska, and Lean, 2001), the empirical Flare Irradiance Spectral Model (FISM) (Chamberlin, Woods, and Eparvier, 2007, 2008), and the hybrid Solar Irradiance Platform (SIP) system (Tobiska, 2004, 2008; Tobiska and Bouwer, 2006; Tobiska *et al.*, 2000). Physics-based models provide more insight into the causes of the solar variability and better links to the solar image results. Empirical models that relate variability of proxies to the solar variability at other wavelengths can provide accurate and easy-to-calculate results for space-weather research and operations when direct measurements are not available. Hybrid systems, such as SIP, combine real-time data, reference spectra, empirical, and physics-based models to report, specify, and forecast the solar irradiances.

2.1.3. EVE Objective 3 – Forecast Solar EUV Variations

The application of the physical understanding and specification of the solar EUV irradiance developed in EVE Objectives 1 and 2 will facilitate a unique, and hitherto unavailable, capability for EUV irradiance predictions on multiple time scales associated with the solar cycle, rotation, and flares. Predictions for the various time scales ranging from the long-term solar cycle to the short-term flare eruptions require different techniques.

Solar-activity cycles are currently predicted in terms of sunspot numbers or 10.7 cm radio fluxes. NASA uses such predictions for planning shuttle and space station activities, for example. Most previous predictions use either statistical regression techniques based on average properties and mean behavior of sunspot patterns or the geophysical precursor technique, which recognizes the extended nature of the solar cycle (Hathaway, Wilson, and Reichmann, 1999). Were the EUV irradiance and its variability better known, these predictions could be transformed empirically to much improved equivalent EUV energies. EVE measurements, with high accuracy and repeatability, will greatly improve the specification of the relationship between EUV irradiance and sunspot numbers (or 10.7 cm fluxes) to facilitate useful nowcasts and forecasts of present and future long-term irradiance cycles.

Predicting the details of day-to-day EUV irradiance variations associated with solar rotations requires a different approach. The US Air Force has an operational requirement for predicting space weather on a three-hour time granularity to 72 hours and a daily-forecast requirement to seven days or beyond while NOAA provides public predictions of radio flux to 45 days. Commercial aerospace organizations use commercially provided irradiance predictions out to five solar rotations. Because solar EUV radiation is a primary driver of geospace fluctuations on these time scales, the predictions ultimately require reliable forecasts of future solar EUV irradiance. A new capability will be developed to predict day-to-day EUV irradiances by using HMI far-side images and AIA limb emission to quantify emerging flux and active regions soon to rotate onto the face of the Sun visible at the Earth. The physical associations between surface fields and active regions in the chromosphere and corona (developed in EVE Objective 2) then permit quantitative forecasts of EUV irradiance using the far-side and east-limb active regions. The far-side active regions are currently predicted using helioseismology with SOHO/MDI data (Lindsey and Braun, 2000) and backscattered H I Lyman-α emissions with SOHO/SWAN data (Bertaux *et al.*, 2000). Fontenla *et al.* (2009) provide examples of forecasting the solar EUV irradiance using these far-side techniques.

As Figures 5 and 6 demonstrate, the emission on the east limb of the solar disk is a useful predictor of full-disk flux (irradiance) from three to nine days later. Separately, EVE's direct observations acquired in Objective 1 provide a database to quantify and validate this association using AIA images, and to develop, as well, an independent statistical predictive capability from accumulated characterizations of EUV irradiance variations at particular phases of the solar cycle. Together, the complementary physical and statistical approaches enable state-of-the-art near-term EUV irradiance forecasts. The database of ten-second EUV irradiance spectra will be particularly useful for nowcasting space weather.

The ability to forecast short-term solar eruptions has progressed in the past few years as a result of empirical associations between certain magnetic-field configurations (such as helicity features, dimmings, and shadows) and subsequent flares and coronal mass ejections. With its high time cadence, EVE will record the increase in EUV irradiance that accompanies each and every flare throughout the SDO mission. This unique database will facilitate classifications of the types and nature of geoeffective flares, ones that have sufficient brightness to dominate the disk-integrated signal during an eruptive event. This database will enable empirical associations with flare precursors detected in HMI and AIA images. EVE will develop, in collaboration with the SDO imaging teams, physical descriptions of the associations of precursor magnetic field configurations and subsequent EUV irradiance enhancements. These studies will contribute to more reliable short-term forecasts of EUV irradiance levels and hence of abrupt space-weather phenomena.

2.1.4. EVE Objective 4 – Understand Response of Geospace Environment

EVE will provide reliable knowledge of the solar EUV spectrum and its variability that the geophysics community has sought for decades, and without which LWS cannot fully succeed. Variations in EUV irradiance initiate space-weather phenomena through both direct and indirect processes and are, consequently, crucial inputs for many geospace models. The direct terrestrial effects of solar EUV electromagnetic radiation are well recognized (National Space Weather Plan, 2000). The EUV irradiances specified by EVE will enable progress in understanding, specifying, and forecasting myriad space-weather phenomena, including spacecraft drag, communications, and navigation.

Through both direct and collaborative methods, the EVE team will support modeling and interpretation of the neutral atmosphere and ionosphere responses to EUV irradiance variations. Examples of such studies include simulations using the NOAA three-dimensional Coupled Thermosphere Ionosphere Plasmasphere electrodynamics model (CTIPe; Fuller-Rowell *et al.*, 1996) and the Utah State University Time-Dependent Ionospheric Model (TDIM; Sojka, 1989; Schunk *et al.*, 2004) with a hierarchy of inputs ranging from direct high time-cadence observations of flares, modeled flare spectra, active-region time scale and near-term forecast responses, solar-cycle-induced variations and extreme conditions of solar activity. EVE will enable the solar component in the day-to-day variability of ionospheric electron density and total electron content to be quantified for the first time. Such a comprehensive study of atmosphere-ionosphere responses to EUV radiation has never been conducted. Prior studies, undertaken only infrequently because of their interdisciplinary nature, have suffered from the quality of their adopted EUV irradiance inputs. Only very recently, for example, were sufficient interdisciplinary resources brought together to simulate the ionospheric response to EUV spectrum changes during a flare (see, *e.g.*, Pawlowski and Ridley, 2008; Smithtro and Solomon, 2008).

2.2. EVE Science Team

The EVE Principal Investigator is Tom Woods at the University of Colorado's Laboratory for Atmospheric and Space Physics (LASP). The EVE scientists include Frank Eparvier and Andrew Jones at LASP, Darrell Judge and Leonid Didkovsky at the University of Southern California (USC), Judith Lean, Don McMullin, John Mariska, and Harry Warren at the Naval Research Laboratory (NRL), Greg Berthiaume at the Massachusetts Institute of Technology Lincoln Laboratory (MIT-LL), and Scott Bailey at Virginia Tech.

LASP provides overall EVE project management, instrument design, fabrication, calibration, instrument operations, and data processing software development. USC contributes a portion of the flight hardware and expertise in solar EUV irradiance measurements. MIT leads the development of CCD detectors (Westhoff *et al.*, 2007). NRL undertakes solar spectral-irradiance modeling, in particular improvements to the NRLEUV semi-empirical irradiance variability model (Warren, Mariska, and Lean, 2001; Lean *et al.*, 2003).

The EVE Science Team also includes several collaborators whose participation is vital to EVE's success: Tim Fuller-Rowell and Rodney Viereck (NOAA Space Weather Prediction Center, SWPC), Jan Sojka (Utah State University, USU), and Kent Tobiska (Space Environment Technologies, SET) participate in various aspects of the space-weather and operations effort through geospace and solar operational modeling. They will assist in transitioning EVE research to operations. In particular, EVE data will be used to improve, validate, or constrain solar irradiance models, such as SIP (Tobiska, 2004, 2008; Tobiska and Bouwer, 2006; Tobiska *et al.*, 2000), and atmospheric models, such as CTIPe (Fuller-Rowell *et al.*, 1996), Global Assimilation of Ionospheric Measurements (GAIM; Schunk *et al.*, 2004), NRLMSIS (Picone *et al.*, 2002), JB2006 (Bowman *et al.*, 2008a), JB2008 (Bowman *et al.*, 2008b), and a newly developed improved density specification model that takes into account an extensive density database and includes direct formulation with the Mg index (Emmert, Picone, and Meier, 2008). These atmospheric models are described in more detail in Section 6.

2.3. EVE Measurement Requirements

To help meet the EVE objectives, the EVE instrument suite will measure the spectral irradiance from 0.1 to 5 nm at 1-nm resolution, from 5 to 105 nm with a resolution of 0.1 nm, plus the hydrogen Lyman-α line at 121.6 nm with 1-nm resolution. The full spectral range will be measured every ten seconds, continuously (except during satellite eclipse periods and planned calibration activities). The absolute accuracy of EVE's spectral irradiance measurements will be better than 25% throughout the nominal five-year mission. The pre-flight tests and calibrations verify that EVE meets these requirements, and at many wavelengths EVE greatly exceeds the 25% accuracy. The measurement requirements represent significant improvements over previous measurements, especially noteworthy for the spectral resolution, time cadence, and duty cycle.

3. EVE Instrumentation

To meet the measurement and accuracy requirements, the EVE suite is composed of several instruments as listed in Table 1. The wavelength coverage of all instruments is shown in Figure 3, along with the solar-cycle minimum spectrum obtained with the prototype EVE onboard a sounding rocket flight on 14 April 2008 (Chamberlin *et al.*, 2009). Eparvier *et al.* (2004), Crotser *et al.* (2004, 2007), Chamberlin *et al.* (2007), Didkovsky *et al.* (2007, 2010), and Hock *et al.* (2010) provide details of the optical designs for the EVE instruments, but a brief overview of the instruments is included here.

The primary, high-spectral-resolution irradiance measurements are made by the *Multiple EUV Grating Spectrographs* (MEGS) (Crotser *et al.*, 2004, 2007), which have heritage from the TIMED/SEE *EUV Grating Spectrograph* (EGS) (Woods *et al.*, 2005). The MEGS is composed of two spectrographs: MEGS-A is a grazing-incidence spectrograph covering the 5 to 37 nm range, and MEGS-B is a two-grating, cross-dispersing spectrograph covering the 35 to 105 nm range. Both MEGS-A and B have 0.1-nm spectral resolution. Included as part of the MEGS-A package is a pinhole camera for use as a solar aspect monitor (MEGS-SAM) to provide a pointing reference for the EVE instruments. MEGS-SAM will also make a spectral measurement of the solar irradiance in the 0.1 to 5 nm wavelength range at approximately 1-nm resolution. In addition, a photodiode with a filter to isolate Lyman-α at 121.6 nm (MEGS-P) is part of the MEGS-B. This measurement is used to track potential changes in the sensitivity of the MEGS on timescales of weeks and months. Annual sounding rocket underflights of similar instruments will track longer-term changes in the sensitivity of the EVE instruments. Eparvier *et al.* (2004), Crotser *et al.* (2004, 2007), Chamberlin *et al.* (2007), and Hock *et al.* (2010) provide more details about LASP's MEGS and its calibration results.

Additional onboard short-term calibration tracking is achieved by redundant, lower-spectral-resolution measurements at select bandpasses, made by the *EUV Spectrophotometer* (ESP). The ESP is a transmission grating and photodiode instrument similar to the SOHO/SEM (Judge *et al.*, 1998). ESP has four channels centered on 18.2, 25.7, 30.4, and 36.6 nm that are each approximately 4 nm in spectral width. The ESP also has a central, zeroth-order diode with a filter to make the primary irradiance measurement in the 0.1 to 7 nm range. The ESP measurements are made at a high time cadence (0.25 seconds) and so are useful as quick indicators of space-weather events such as flares. Didkovsky *et al.* (2007, 2010) provide more details about USC's ESP and its calibration results.

3.1. MEGS-A Instrument

The MEGS-A instrument is an 80° grazing-incidence, off-Rowland circle spectrograph with a CCD detector to measure the solar spectrum between 5 and 37 nm at a resolution just

Table 1 EVE solar irradiance instruments. $\Delta\lambda$ is the spectral resolution, Δt is the normal operations integration time (cadence), R is the grating radius of curvature, d is the grating line spacing, α is the grating incidence angle, β is the grating diffraction angle, 1 and 2 for MEGS-B refer to its two different gratings, 2nd is the filter used for second- (and higher-) order sorting calibrations, BI is Back-Illuminated (back-thinned) for the CCD type, AXUV is XUV grade (n-on-p) Si photodiodes from IRD.

Instrument (slit, mm^2)	λ Range (nm)	$\Delta\lambda$ (nm)	Δt (sec)	Filters	Grating	Detector	Description
MEGS-A					$R = 600$ mm	1024 × 2048	Grazing Incidence, Off-Rowland Circle Spectrograph.
Slit A1	5 – 18	0.1	10	Zr/C	$d = 1304$ nm	BI CCD	Grating and Detector are shared with both slits and SAM.
(0.02 × 2)				2nd: Zr/Si/C	$\alpha = 80°$		
Slit A2	17 – 37	0.1	10	Al/Ge/C	$\beta = 73 – 79°$		
(0.02 × 2)				2nd: Al/Mg/C			
MEGS-B	35 – 105	0.1	10	None	$R_1 = 200$ mm	1024 × 2048	Normal Incidence, Double-Pass, Cross-Dispersing Rowland Circle Spectrograph.
(0.035 × 3.5)				2nd: Al/Mg/C	$d_1 = 1111$ nm	BI CCD	Entrance Slit and First Grating are shared with MEGS-P.
					$\alpha_1 = 1.8°$		
					$\beta_1 = 4 – 7°$		
					$R_2 = 200$ mm		
					$d_2 = 467$ nm		
					$\alpha_2 = 14°$		
					$\beta_2 = 19 – 28°$		
MEGS-SAM					None	1024 × 2048	Pinhole Camera.
(26 µm Dia.)	250	80	10	Acton 250W		BI CCD	X-ray Photon Counting.
	1 – 7	1	10	Al/Ti/C			Uses one corner of MEGS-A CCD.

Table 1 (*Continued*)

Instrument (slit, mm^2)	λ Range (nm)	Δλ (nm)	Δt (sec)	Filters	Grating	Detector	Description
MEGS-P	121.6	10	0.25	Acton 122XN	MEGS-B1	AXUV-100	H I Lyman-α Photometer.
(0.035 × 3.5)	Dark	N/A	0.25	Thick Ta		(10 × 10 mm^2)	Uses MEGS-B Slit and Grating.
ESP					$d = 400$ nm	AXUV-SP2	Normal Incidence,
(1 × 10) Ch 1:	36.6	4.7	0.25	Al	$\alpha = 0°$	(6 × 16 mm^2)	Transmission Grating Spectrograph.
Ch 2:	25.7	4.5	0.25	Al	$\beta_{\pm 1} = 3-5°$		Entrance Slit, Al Filter, and
Ch 3:	Dark	N/A	0.25	Thick Ta		Quad Diode	Grating shared with all ESP channels.
Ch 8:	18.2	3.6	0.25	Al		AXUV-PS5	Quad Diode (QD) is at 0th
Ch 9:	30.4	3.8	0.25	Al	$\beta_0 = 0°$	(6 × 16 mm^2)	order, other channels are at
Ch 4 – 7 QD:	0.1 – 7	6	0.25	Al + Ti/C			±1st order.

less than 0.1 nm. MEGS-A has two entrance slits, each 20 microns wide and 2 mm high, oriented top-to-bottom. The off-Rowland circle design is an innovative approach to having the CCD at near-normal incidence, instead of grazing incidence, so that MEGS-A has much higher sensitivity (Crotser *et al.*, 2007). In front of the slits is a filter-wheel mechanism with bandpass-limiting thin foil filters (made by Luxel Corp.). The primary science filters are Zr (280 nm) / C (20 nm) for slit 1 to isolate 5 to 18 nm and Al (200 nm) / Ge (20 nm) / C (20 nm) for slit 2 to isolate 17 to 37 nm. Secondary filters are available to further limit the bandpasses of each slit to provide an occasional check on higher orders: Zr (230 nm) / Si (120 nm) / C (20 nm) for slit 1 to pass 13 to 18 nm, and Al (180 nm) / Mg (300 nm) to pass 25 to 37 nm for slit 2. The filter-wheel mechanism also has a blanked-off position for dark measurements. The grating, produced by Jobin Yvon (JY), is a spherical holographic grating with a radius of curvature of 600 mm, platinum coating, and 767 grooves mm^{-1} with a laminar groove profile to suppress even orders.

The detector for MEGS-A is a back-thinned, back-illuminated, split-frame transfer CCD with 1024×2048 pixels, built by MIT-LL (Westhoff *et al.*, 2007). The pixel size is $15~\mu\mathrm{m} \times 15~\mu\mathrm{m}$. The top and bottom halves, each 512×2048 pixels, are read out through separate, redundant signal chains (charge amps) in about eight seconds. The gain on the CCD signal is about two electrons per data number (DN), and its noise is about two electrons once its temperature is below about $-70^{\circ}\mathrm{C}$. The CCD is maintained at $-95^{\circ}\mathrm{C} \pm 5^{\circ}\mathrm{C}$ during flight to suppress noise and to minimize radiation effects in the geosynchronous environment. Furthermore, the CCDs are heavily shielded by about 25 mm Al to reduce the radiation dose to less than 10 kRad over the SDO mission.

3.2. MEGS-B Instrument

The MEGS-B instrument is a normal-incidence, double-pass, cross-dispersing Rowland circle spectrograph with a CCD detector to measure the solar spectrum between 35 to 105 nm at a resolution just less than 0.1 nm. MEGS-B has a single entrance slit, 35 microns wide and 3.5 mm high. There are no known reliable bandpass filters for the MEGS-B wavelength range, so a two-grating design isolates the entire 35 to 105 nm range. A filter-wheel mechanism is also included with clear positions for primary science measurements, a Al (180 nm) / Mg (300 nm) foil filter for higher-order checks, and a blanked-off position for dark measurements. Both MEGS-B gratings are also produced by JY and are spherical holographic gratings with platinum coating and laminar groove profiles to suppress even orders. The first grating has 900 grooves mm^{-1} and the second has 2140 grooves mm^{-1}. The detector for MEGS-B is identical to the MEGS-A detector and permits about three pixels across the optical resolution of 0.1 nm and over 100 pixels along the slit image.

3.3. MEGS-SAM Instrument

The MEGS-SAM instrument is a pinhole camera within the MEGS-A housing, using a separate aperture, but producing an image of the Sun on a portion of the MEGS-A CCD where the bandpass filter for slit 2 blocks essentially all light. The SAM aperture has a separate filter-wheel mechanism that allows for three modes. In aspect-monitor mode, a UV filter is in place and the resultant image of the Sun is centroided to give pointing information for all of EVE relative to the boresights established during pre-flight calibrations to roughly one arcminute accuracy. In XUV photon-counting mode, a Ti (300 nm) / Al (150 nm) / C (40 nm) foil filter is in place to isolate 0.1 to 7 nm. This filter is designed to be the same as the ESP Al and Ti/C filter combination used with ESP's zeroth-order photometer. The pinhole and filter are optimized so that single photon events can occur during the ten-second CCD integration. The energy (or wavelength) is determined from the magnitude of each photon event, thus binning photon events from over the entire image of the Sun gives a low (≈ 1 nm) spectral resolution for the SAM XUV bandpass. Summing consecutive integrations over a few minutes generates XUV images of the Sun. The third mode for SAM has the filter wheel in a blanked-off position for dark measurements. The SAM is normally in its XUV mode.

3.4. MEGS-P Instrument

The MEGS-P instrument is an International Radiation Detectors (IRD) silicon photodiode placed at the minus first order of the first MEGS-B grating. In front of the diode is an Acton interference filter to isolate the solar hydrogen Lyman-α line at 121.6 nm. The filter has a bandwidth of 10 nm, but the solar spectrum is such that greater than 99% of the signal is due to Lyman-α. Next to the primary MEGS-P diode is an identical diode that is masked to give simultaneous dark information, used to correct the MEGS-P measurements for background noise induced by particle radiation.

3.5. ESP Instrument

The ESP instrument is a non-focusing, broadband spectrograph with a transmission grating and IRD silicon photodiodes similar to the SOHO/SEM instrument (Judge *et al.*, 1998). In front of the entrance slit is an Al (150 nm) foil filter made by Luxel Corp. to limit the out-of-band light that enters the instrument. The free-standing transmission grating, made by X-Opt. Inc., is essentially a set of thin bars with no substrate, spaced so that there are 2500 lines mm^{-1}. Silicon photodiodes are placed at both plus and minus first orders and positioned so that the bandpass centers are at 18.2, 25.7, 30.4, and 36.6 nm. Masks in front of the diodes are sized to give approximately 4-nm bandpasses centered on each of these wavelengths. The central, zeroth-order position has a silicon quadrant photodiode with an additional thin foil Ti (300 nm) / C (40 nm) filter to isolate 0.1 to 7 nm. The sum of the quadrants gives the solar irradiance in this bandpass. Differencing the quadrants allows for determination of the pointing of the ESP and can also provide the location of flares on the solar disk (Didkovsky *et al.*, 2010). The ESP has a filter-wheel mechanism with Al foil filters for solar measurements, a blank position for dark measurements, and a fused-silica window to check for visible stray light. The ESP has the fastest measurement cadence of all of the instruments in the EVE suite at 0.25 seconds.

3.6. Other EVE Subsystems

The EVE instruments are packaged together onto the EVE Optical Package (EOP) as shown in Figure 7. The EOP is mounted to the SDO spacecraft deck using three Ti-flex structures. The microprocessor, interface electronics, control electronics, and most of the power conditioning and regulation electronics are housed in the EVE Electronics Box (EEB). The EEB resides behind the EOP and directly mounts to the SDO spacecraft deck. The interfaces to the SDO spacecraft include unregulated 28 V DC for power, 1553 for commands and housekeeping telemetry, and High Speed Bus (HSB) for science telemetry. Several radiators are also part of the EVE package, to passively remove heat by radiating to deep space. The resources for EVE include mass of 54 kg, orbit-average power of 44 W, housekeeping telemetry of two kilobits per second (kbps), and science telemetry of seven Megabits $\sec^{-1}$ (Mbps).

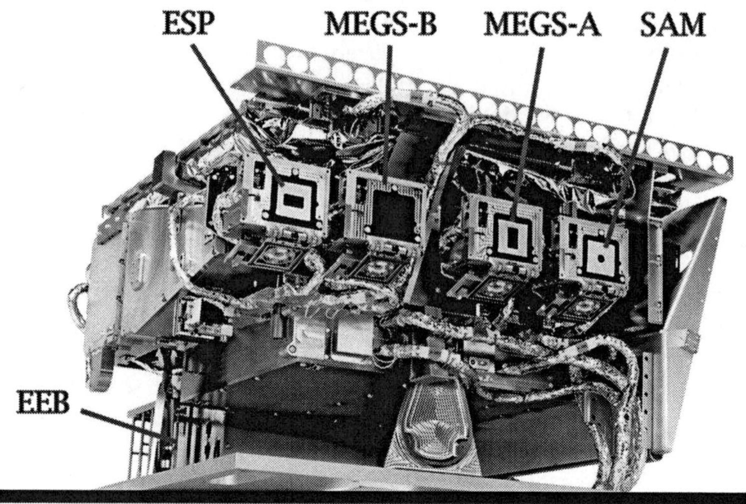

Figure 7 The EVE instruments are mounted onto the EVE Optical Package (EOP), and the EVE Electronics Box (EEB) provides the electrical interfaces to the SDO spacecraft. The entrance baffle in the door mechanisms are indicated for the various MEGS and ESP instruments.

4. EVE Data Products

The primary EVE data products are solar EUV irradiances at 0.1-nm and 1-nm resolution at a ten-second time cadence and as daily averages. In addition, specific solar emission lines and broadband irradiances will be extracted and provided at both time cadences. These data products will be available within a day or so of receipt on the ground. Near real-time data products for space-weather operations will also be produced and available within approximately fifteen minutes of ground receipt. The space-weather products will not be as fully corrected or processed as the primary data products, and may consist of solar indices and spectra with preliminary calibration. The EVE space-weather products are intended for use in near real-time applications, such as in ionospheric and thermospheric models and for "nowcasting" short-time-scale solar events such as flares. The following sections describe the EVE data-processing system and data products (Table 2). The raw telemetry records are called Level 0A, and raw data sorted by channel are called Level 0B; neither of these products is discussed below. The algorithms for converting the raw data into irradiance are provided in the instrument calibration papers by Hock *et al.* (2010) for MEGS and Didkovsky *et al.* (2010) for ESP.

4.1. EVE Level 0C – Space Weather Product

The EVE space-weather products, called Level 0C and Level 0CS, provide the solar EUV irradiance as soon as possible after the observation. To reduce the processing time, only simple forms of the calibration parameters are applied. The solar irradiance is provided as spectra from 6 to 105 nm in 0.1-nm intervals and for several different spectral bands, each with one-minute cadence and with less than 15-minute latency. The spacecraft and ground systems impose a 15-minute latency, which is adequate for using ionosphere and thermosphere models for space-weather operations (see Section 6). Some of the bands are also provided with about one-minute latency for the most time-critical space-weather operations, namely flare events and predictions for Solar Energetic Particles (SEPs).

Table 2 EVE data products.

Level	Description	Spectral Range / Intervals (nm)	Lines / Bands	Temporal Sampling	File Time Span	Latency of availability
0C	Space Weather Products from Ka-Band data: 0C-Bands 0C-latest-Bands (15-min files) 0C-Spectrum 0C-latest-Spectrum (15-min files)	6 – 105 nm / 0.1 nm	ESP, MEGS-P, integrated bands from MEGS, ESP Quad, Various Proxies	one-minute averages	24 hours (15 min)	$<$ 15 minutes
0CS	Space Weather Products from S-Band data: 0CS-Bands 0CS-latest-Bands (15-min files)	Spectra are not available.	ESP, MEGS-P, integrated bands from MEGS, ESP Quad, Various Proxies	one-minute averages	24 hours (15 min)	$\approx$ one minute
1	Fully calibrated/corrected irradiance, no averaging, full science potential: 1-MEGS-A (not public, use L2) 1-MEGS-B (not public, use L2) 1-MEGS-P (121.5 nm) 1-ESP 1A-SAM (photon events) 1B-SAM (spectra from 1A-SAM)	6 – 105 nm / 0.02 nm	ESP, MEGS-P, integrated bands from MEGS, ESP Quad	Spectra at 10 sec, Bands at 0.25 sec	Spectra in 1 hour, Bands in 24 hours	Next Day
2	Fully calibrated irradiance, combined MEGS-A and MEGS-B, no averaging, full science potential: 2-Spectrum 2-Lines	6 – 105 nm / 0.02 nm	Extracted lines from MEGS	ten second	24 hours	Next Day
3	Daily-averaged, combined irradiance spectrum from SAM, MEGS-A and MEGS-B, plus MEGS-P and ESP. Spectra available in three different intervals.	0.1 – 105 nm / 0.02, 0.1, 1 nm	Subset of L1 Bands and L2 Lines	UTC day averages	1 year	Next Day

The space-weather products are all provided in ASCII files to support cross-platform compatibility and allow for human readability. The top of each file contains information (metadata) about the file contents and data formatting. Comment lines begin with the semicolon character. All space-weather products are generated on little-endian x86-based computers, so line endings in hexadecimal representation are 0x0a (LF). Windows-based computers must convert this to 0x0d (CR), 0x0a (LF). Each line represents a unique instant of time. We anticipate a latency of less than ten minutes for the routine, Ka-band space-weather products. The ground system for SDO was not designed to accommodate real-time space-weather operations, so the high-rate data lag real-time by about three minutes. This is the time that it takes for data collected on the spacecraft to reach the science-processing operations center before any processing can begin.

The space-weather products are produced as rapidly as possible. Due to the desire to produce a faster product, the quality is lower and these products cannot meet the full science potential. These products are roughly calibrated using recent values for calibration parameters, and these products are not subject to accuracy requirements of the higher-level products, where latency is not a high priority.

Both Level 0C and 0CS space-weather products contain the ESP and MEGS-P broadband photometer measurements averaged to one-minute sampling. Each line of data corresponds to a one-minute average, and the file will be updated once per minute. These files will span one UT day. A smaller version of this file will also be made available which contains only the last 15 minutes. This reduced-size file will serve most real-time space-weather requests. A separate version of these EVE photometer bands is also available through the SDO S-band communication link, and these EVE Level 0CS products are expected to have less than one-minute latency.

The Level 0C space-weather product also contains one-minute averaged solar spectra at 0.1-nm sampling. Similar to the broadband file, this file will span one UT day with updates every minute. A reduced-size file containing only the most recent 15 minutes will also be produced to reduce download times for real-time space-weather needs.

4.2. EVE Level 1

The full science potential is achieved in the MEGS and ESP Level 1 science products. There is a Level 1 product for each of the EVE instruments: MEGS-A, MEGS-B, MEGS-SAM, MEGS-P, and ESP. The Level 1 products include spectra in 0.02 nm intervals and ten-second cadence and photometer bands with 0.25-second cadence that has better than ten millisecond accuracy. For those who need composite spectra, where MEGS-A and MEGS- B spectra are combined, one can more conveniently use EVE's Level 2 and Level 3 data products that contain the full spectral coverage in a single product.

The Level 1 products contain every measurement collected with no temporal averaging, and the best calibration applied. After 14:00 UT each day the routine science processing can begin for the prior UT day. This provides time for the Mission Operations Center to ensure that the Level 0 files containing the ancillary science data packet information are complete. After this period the 48-node EVE computing cluster will begin Level 1 processing. The Level 1 processing is separated for each science component to allow parallel tasks to run simultaneously. MEGS-A slit 1 and slit 2 are stored together in one-hour files. MEGS-B is stored in separate one-hour duration files as well. However, MEGS-P is such a small amount of data that it is stored as a daily file, as is ESP. The remaining science instrument, SAM, is also divided into one-hour files. SAM will have a Level 1a event list (time, energy, location), from which the Level 1b spectral product is derived.

4.3. EVE Level 2

The Level 2 processing produces a combined set of merged spectra from 6 to 105 nm using the data from MEGS-A and MEGS-B, extracts solar emission features of special interest, and provides time-averaged photometer measurements to simplify comparisons with the spectra. This is done through a pair of files that will be produced: one file contains the spectra for one hour, and the other contains the lines and bands.

The spectrum files span one hour each. The individual ten-second spectra are sampled at a common, uniform wavelength spacing of 0.02 nm. This provides five resolution samples per instrument resolution element (0.1 nm). The MEGS-A slit 1 (6 – 16 nm) and slit 2 (16 – 37 nm) spectra are spliced together and then spliced together with MEGS-B (37 – 105 nm). The Level 2 spectrum files provide full instrument resolution at the full measurement cadence.

Table 3 lists the lines and bands planned for the EVE data products. The lines have been identified to include many plasma temperatures with good sampling between 20 000 K and 10 MK. For example, this list includes the ionization states of Fe VIII through Fe XX. The extracted emission lines are total irradiances over the lines including the background. There are several broadbands including EVE's photometer measurements and several wavelength bands extracted from the MEGS spectra that simulate bands from other instruments, such as AIA, SOHO/*Solar EUV Monitor* (SEM) (Judge *et al.*, 1998), and GOES/*EUV Sensor* (EUVS) (Viereck *et al.*, 2007). For these other instrument bands, their normalized responsivity profiles are convolved with the MEGS spectrum so that the EVE equivalent bands can be more directly compared for calibration and validation purposes. Some additional bands may also be produced using the band definitions for the PROBA2/*Large Yield Radiometer* (LYRA) instrument (Hochedez *et al.*, 2006), the EUVAC solar irradiance model (Richards, Fennelly, and Torr, 1994), JB2006 and JB2008 atmospheric models (Bowman *et al.*, 2008a, 2008b), and TIM-GCM model solar input (Solomon and Qian, 2005).

4.4. EVE Level 3

The Level 3 products are merged files spanning one year of daily averages of the spectra. At this level, the daily-averaged SAM spectrum will also be provided from 0.1 to 6 nm with spectral resolution ranging from about 0.1 nm at 0.1 nm to about 1 nm at 6 nm. The Level 3 product retains the high-resolution 0.02-nm sampling. An additional pair of products, integrated into 0.1-nm and 1.0-nm bins, will also be provided. The daily average is derived as the median value at each wavelength and thus normally represents the daily solar activity without flare contributions.

4.5. EVE Data System

The EVE data system was designed to be robust and fault tolerant to support near real-time operational product creation and distribution. The interface to the Data Distribution System (DDS) is comprised of twin (identical) redundant Solaris 10 (sparc) servers. This minimizes data interruptions for routine, planned system maintenance. Approximately one-minute duration data files received from the DDS are then passed to the processing cluster. A 48-node, high-availability, GNU-Linux cluster verifies the telemetry files, and performs the distributed data processing tasks. The head node runs a custom scheduler that assigns tasks to the cluster nodes. The tasks performed on each node are independent, so computer capacity may be increased by adding nodes. Furthermore, individual node failures have a

Table 3 EVE line / band list. The first and middle columns are the isolated, dominant lines from MEGS-A and MEGS-B, respectively. The last columns are the ESP and MEGS-P bands and other bands derived from the MEGS spectra.

Line / Wavelength (nm)	log(T)	Line / Wavelength (nm)	log(T)	Band / Wavelength (nm)	Description
Fe XVIII 9.39	6.8	Mg IX 44.37	6.0	ESP-Quad: 0.1 – 7	ESP zeroth order (Quad Diode)
Ne VII 12.77	5.7	Ne VII 46.52	5.6	ESP171: 15 – 20	ESP Fe IX channel
Fe VIII 13.12	5.6	Si XII 49.94	6.3	ESP256: 21 – 29	ESP He II channel
Fe XX 13.29	7.0	O III 52.58	4.9	ESP304: 29 – 32	ESP He II channel
Ne V 14.87	5.5	O IV 55.44	5.3	ESP366: 34 – 39	ESP Fe XVI / Mg IX channel
Fe IX 17.11	5.8	He I 58.4	4.3	MEGSP: 121 – 122	MEGS-P H I Lyman-α
Fe X 17.72	6.0	O III 59.96	4.9	AIA94: 9.35 – 9.44	AIA Fe XVIII channel
Fe XI 18.04	6.1	Mg X 62.49	6.1	AIA133: 13.06 – 13.51	AIA Fe XX channel
Fe XII 19.51	6.1	O V 62.97	5.4	AIA171: 16.87 – 17.34	AIA Fe IX channel
Fe XIII 20.20	6.2	O II 71.85	4.8	AIA195: 19.05 – 19.65	AIA Fe XII channel
Fe XIV 21.13	6.3	N IV 76.51	5.2	AIA211: 20.78 – 21.48	AIA Fe XIV channel
He II 25.63	4.8	Ne VIII 77.04	5.8	AIA304: 29.74 – 31.01	AIA He II channel
Fe XV 28.42	6.3	O IV 79.02	5.3	AIA335: 32.72 – 34.37	AIA Fe XVI channel
He II 30.38	4.8	H I 97.25	4.7	SEM304: 25-34	SOHO/SEM He II channel
Fe XVI 33.54	6.4	C III 97.70	4.7	QEUV: 0.1 – 45	TIMED/GUVI Photoelectrons
Fe XVI 36.08	6.4	H I 102.57	4.7	GOES-A: 5 – 15	GOES-13 EUVS Band A
Mg IX 36.81	6.0	O VI 103.19	5.4	GOES-B: 25 – 34	GOES-13 EUVS Band B
				GOES-C: 42 – 63	GOES-13 EUVS Band C
				GOES-D: 17 – 81	GOES-13 EUVS Band D
				GOES/XRS-B: 0.1 – 0.8	GOES/XRS Band B

minimal impact on overall processing performance. Serialized processing is performed on each file as soon as it is verified to be complete and free of transmission errors. The Level 0B processing separates the packets from each science instrument, and in the case of MEGS-A and MEGS-B efficiently de-multiplexes and decodes each pixel (twos-complemented and shifted), and then assembles the CCD images into time-ordered integrations. The Level 0B products are then used as input to generate space-weather products. The Level 0B code is written in C.

The largest source of data latency in the EVE space-weather processing is the approximately three-minute delay that was designed into the ground station at the White Sands Complex. This is needed since the DDS core system receives data from the antenna sites on a one-minute cadence, then separates the Virtual Channel Data Units (VCDUs) by instrument, stores data in a temporary 30-day disk storage repository, and performs statistics monitoring of the data and data files.

In addition to providing a repository of data products, the EVE Science Operations Center (SOC) is responsible for archiving the EVE data, and disseminating the processed data products. The EVE web site is http://lasp.colorado.edu/eve/.

5. Solar Irradiance Models

The EVE team plans to advance the understanding of EUV irradiance modulation by using magnetic features and their connections to surface magnetic-flux emergence. Physical processes will be developed, parameterized, and modeled, working toward the overall goal of a self-consistent, end-to-end formulation of EUV irradiance variations that arise from dynamo-driven magnetic fields. Although the primary focus is physics-based modeling, empirical approaches will also be employed to relate EUV irradiance variations to various proxies of solar activity that extend over long time periods, and into the future. These proxy models, once calibrated properly with the SDO measurements, are simpler to use for routine space-weather operations as long as the proxies have long-term stability (see, *e.g.*, Lilensten *et al.*, 2008).

A fundamental, new, and unique aspect of EVE-based modeling is physics-based simulations of EUV spectra in flares, for which detailed variations are not well known over the full EUV range. A preliminary attempt has been made to use differential-emission measures (DEMs) and fractional areas from TRACE, GOES, and SXT observations to model the increase in the EUV spectral irradiance during the Bastille Day 2000 flare (Meier *et al.*, 2002). For operations since 2005, the SOLARFLARE model (Tobiska and Bouwer, 2005; Tobiska, 2007), which is part of SIP, uses GOES XRS 0.1 to 0.8 nm data, transformed into an effective temperature for the Mewe model, to create a 0.1-nm, one-minute flare spectrum from 0.1 to 30.0 nm. Additionally, a few flares have been observed by TIMED/SEE and some empirical relationships have been developed for the FISM (Chamberlin, Woods, and Eparvier, 2008). According to these results, the EUV irradiance flare increase is comparable in magnitude to the solar-cycle increase, but has a remarkably different spectral shape. How realistic are these preliminary flare models? The new EVE and AIA data will enable the development of more accurate models of EUV flare variations. This effort will afford unique insights into the processes of flaring plasma, and significantly advance the capability for EUV irradiance specification on short time scales crucial for space-weather modeling.

EVE's solar modeling effort will also include quantitative studies of how the magnetic fields in the chromosphere and corona, which produce EUV irradiance sources, relate to surface flux emergence, transport, and rotation. This will be undertaken using the AIA and

HMI images with a global scale focus. Relationships between coronal brightness and surface magnetic-field strengths will be investigated using potential-field extrapolation models. When combined with flux-transport models, these efforts will lead to new physics-based parameterizations between EUV irradiance and the solar dynamo that generates the surface flux. Furthermore, forecasting of the EUV irradiance is planned using helioseismic predictions of far-side active regions using HMI data.

These improved solar EUV irradiance models will be especially useful for space-weather operations after the SDO mission. The GOES/EUVS observations are expected to be operational during the SDO mission and to continue on past the SDO mission using a series of GOES satellites. The GOES/EUVS will only observe a few key EUV emissions (Viereck *et al.*, 2007; Eparvier *et al.*, 2009), so the EVE-established relationships of these GOES EUV bands to all other EUV wavelengths will be an important component for the future EUV irradiance-modeling effort.

5.1. NRLEUV Model

One approach to linking the global view of solar soft X-ray and EUV irradiance variability provided by EVE to the spatially-resolved AIA and HMI observations is the NRLEUV irradiance model (Warren, Mariska, and Lean, 2001; Warren, 2005). This model is currently based on four components of irradiance variability: coronal holes, quiet Sun, active network, and active regions. For the corona-hole, quiet-Sun, and active-region components, representative spectra derived from *Skylab* measurements are used to construct differential-emission-measure distributions. The DEM provides an empirical description of the density and temperature structure in the solar transition region and corona. Differential-emission measures for the active network component were determined by interpolating between the quiet-Sun and active-region distributions at transition region temperatures. These differential-emission-measure distributions were combined with databases of atomic physics parameters to compute the line intensity for any optically thin emission line (for, *e.g.*, CHIANTI; Dere *et al.*, 1997; Landi *et al.*, 2006). Observed intensities are used in the model for optically thick emission lines. Once the line intensities are determined, the solar irradiance is computed by using full-disk solar images to estimate the contribution of each surface feature and simple limb-brightening curves to account for the increased path length near the limb. The original version of this model uses full-Sun coronal images from *Yohkoh*/SXT and chromospheric Ca II K line images from the Big Bear Solar Observatory. The NRLEUV formalism successfully reproduces much of the EUV irradiance during solar minimum conditions as well as the irradiance variability on solar-rotational time scales. The model tends to underestimate the solar-cycle variability for many wavelengths.

New observations from SDO will provide for significant improvements to the NRLEUV irradiance model. The broad temperature coverage, multiple channels, and absolute calibration of AIA will allow for accurate differential-emission measures to be computed at every position on the Sun at high cadence. These DEMs will be used to calculate the spectral radiance in each AIA pixel and eliminate both the need for crudely partitioning the Sun into broad intensity ranges and the use of limb-brightening curves at many wavelengths. The AIA 13.1-nm Fe XX/XXIII and 9.4-nm Fe XVIII channels will also provide spatially resolved images at very high temperatures. This will allow the NRLEUV model to be extended to flare temperatures and allow the model to capture very transient contributions to the solar irradiance. The NRLEUV model based on SDO will also incorporate the latest version of the CHIANTI atomic database (version 6), which contains more accurate atomic parameters. With these improvements the model will more accurately reproduce the irradiance variability co-temporally observed with EVE. The primary objective of the NRLEUV modeling effort will be to link changes in the solar soft X-ray and EUV irradiance with the evolution of the magnetic field over many different temporal and spatial scales. For example, the spatially resolved model will allow us to track the irradiance variability driven by the fragmentation and dispersal of active-region magnetic flux.

5.2. FISM Model

The Flare Irradiance Spectral Model (FISM) is an empirical model of the solar-irradiance spectrum from 0.1 to 190 nm at 1-nm spectral resolution and on a one-minute time cadence (Chamberlin, Woods, and Eparvier, 2007, 2008). The goal of FISM is to provide accurate solar spectral irradiances over the aforementioned wavelength range as input for ionospheric and thermospheric models during times when no actual solar measurements are available. FISM is currently based on over seven years of measurements in the EUV (0.1 – 119 nm) from TIMED/SEE and the first five years of FUV observations from UARS/SOLSTICE (119 – 190 nm) for the daily components that model the solar-cycle and solar-rotational irradiance variations. TIMED/SEE also provides the FISM flare estimates for the entire wavelength range, but due to its 3% duty cycle there are only 30 flares (11 impulsive and 19 gradual phase observations) that make up the FISM basis data set.

EVE will contribute greatly in advancing the FISM model. Along with the improved accuracy of EVE over the existing SEE measurements, it will also provide more data from which to derive the FISM relationships of the measurements to the proxies. EVE will expand the base data set through Solar Cycle 24 improving the solar-cycle and solar-rotation components, but most importantly EVE will make available a much larger number of flare observations, as well as continuously observe the flares throughout their evolution due to its 100% duty cycle. EVE will also provide measurements at 0.1-nm spectral resolution, and with the UARS/SOLSTICE and SORCE/SOLSTICE measurements this will allow for the entire FISM range to be enhanced to 0.1-nm resolution once a statistically significant amount of data is collected.

5.3. SIP Hybrid System of Data and Models

Space Environment Technologies (SET) has developed the hybrid system Solar Irradiance Platform (SIP) as a superset of real-time data streams, reference spectra, models (empirical and physics-based) to provide irradiance product services that operationally mitigate space-weather adverse effects (Tobiska, 2008). For example, SET now operationally provides the JB2006, JB2008, F10.7, S10.7, M10.7, and Y10.7 solar proxies and indices that reduce the one-σ uncertainty by up to 50% in atmosphere-density calculations for satellite orbit determination. SET operationally provides high time- and spectral-resolution solar irradiances that capture solar-flare effects on transionospheric communications for the Ionosphere Forecast Model (IFM) that is part of the Communication Alert and Prediction System (CAPS; http://spacewx.com, CAPS link under Products). These solar irradiance products have been developed and tested for *i*) daily time resolution for historical, nowcast, and intermediate-term forecast periods with one-day granularity, one-hour cadence, and one-hour latency extending 4.5 months; *ii*) high time resolution for recent, nowcast, and short-term forecast periods with three-hour granularity, one-hour cadence, and one-hour latency extending 96 hours; and *iii*) precision time resolution for recent, current epoch, and near-term forecast periods with one-minute granularity, two-minute cadence, and five-minute latency extending to six hours. The SIP incorporates empirical and physics-based solar irradiance models such as SOLAR2000 (Tobiska, 2004, 2008; Tobiska and Bouwer, 2006;

Tobiska *et al.*, 2000) and SOLARFLARE (Tobiska and Bouwer, 2005; Tobiska, 2007) along with reference rocket measurements and real-time satellite data stream systems such as APEX for SOHO/SEM data as well as GOES/XRS and TIMED/SEE data.

The SOLARFLARE (SFLR) model has been coupled with SOLAR2000 in order to produce high time- and high spectral-resolution irradiances. SFLR is a mature model at the Technology Readiness Level (TRL) 9 that incorporates the Mewe model to produce 0.1-nm spectral lines based on solar coronal abundances. The Mewe model subroutine requires an input coronal electron temperature (in MK) to produce a spectrum. The GOES/XRS real-time data is collected from NOAA SWPC to predict coronal electron temperatures for use in the Mewe model, whose results are adjusted based on a calibration based on SORCE XPS measurements. The SFLR spectrum in the 0.05 to 30-nm range, calculated at a two-minute cadence, is combined with the daily SOLAR2000 high-resolution spectrum longward of 30 nm.

The solar-irradiance products from SIP will be improved with the new EVE measurements, notably important for the solar flare spectral variability and higher time cadence. In addition, solar reference spectra derived from EVE observations will be incorporated as products available through SIP.

6. Earth's Atmospheric Models

One of the four core objectives for the EVE program is to improve the understanding of how the solar EUV radiation affects the geospace environment, namely Earth's atmosphere above 50 km. This research will entail incorporation of the solar EUV irradiance measurements into modeling the Earth's ionosphere and thermosphere and validation of the model results with atmospheric measurements during periods of interesting solar activity. Ultimately, the near real-time use of the EVE Level 0C space-weather product in some of these ionospheric and thermospheric models will represent an important milestone in the transition from research to operations for this aspect of the EVE program. These planned efforts are supported by EVE collaborators using the Time Dependent Ionospheric Model (TDIM; Schunk, 1988; Sojka, 1989), Ionosphere Forecast Model (IFM; Schunk and Sojka, 1996), and Global Assimilation of Ionospheric Measurements (GAIM; Schunk *et al.*, 2004) at the Utah State University (USU), the Coupled Thermosphere Ionosphere Plasmasphere electrodynamics model (CTIPe) at NOAA (Fuller-Rowell *et al.*, 1996, 2006), the NRLMSIS atmospheric model (Picone *et al.*, 2002) and SAMI ionospheric model (Huba, Joyce, and Fedder, 2000) at NRL, and the JB2006 and JB2008 models (Bowman *et al.*, 2008a, 2008b). We also encourage the involvement of the broader space-weather community in applying the EVE results towards advancing space-weather operations into more reliable forecasts.

6.1. TDIM, IFM, and GAIM Models

The Time Dependent Ionospheric Model (TDIM; Schunk, 1988; Sojka, 1989) uses a framework of physical processes to describe the global F-region ionosphere. This model includes time-dependent ion continuity and momentum equations, convection electric fields and particle precipitation, ion thermal conduction, and diffusion thermal heat flow in solving for the ion and electron densities and temperatures in the F region ionosphere. A derivative model is the Ionosphere Forecast Model (IFM; Schunk and Sojka, 1996), which is used as a physics-based core in the Global Assimilation of Ionospheric Measurements (GAIM). The TDIM will be used to benchmark the ionospheric response to EVE solar irradiances. A similar study with TIMED/SEE solar EUV irradiances was completed early in the EVE development to help refine the EVE measurement requirements. The first EVE-based ionospheric studies will quantify the sensitivity of the ionosphere to the EUV spectrum, for example, determining the responses of the electron density and dynamics of the ionosphere to changes in the solar EUV spectrum. The model will then be used to determine which aspects of the solar spectrum are the most important and also to determine the temporal scales on which the ionosphere responds to solar EUV variability. This work will provide the foundation for EVE observations being used to drive other ionospheric models.

Models under development for operational space-weather forecasting, such as the Utah State University's GAIM model (Schunk *et al.*, 2004), will rely heavily on accurate solar EUV measurements. GAIM is based on a background ionospheric model and then adjusts the electron densities using a Gauss – Markov Kalman filter of observations, such as from GPS ground receivers and Defense Meteorological Satellite Program (DMSP) satellites. A USU-USTAR initiative will run GAIM continuously with a temporal cadence of 15 minutes for routine space-weather operations as well as an evaluation test bed for GAIM models run in real-time at centers such as AFWA and NOAA. This 15-minute operational cadence of GAIM is a driving factor on providing EVE Level 0C spectra with a latency of less than 15 minutes.

6.2. CTIPe Model

The Coupled Thermosphere Ionosphere Plasmasphere electrodynamics (CTIPe) model at NOAA and CIRES (Fuller-Rowell *et al.*, 1996; Millward *et al.*, 1996; Codrescu *et al.*, 2008) is a physical model that solves the dynamic differential equations for momentum, energy, and composition for the neutral and ionized medium, together with the electrodynamics at mid and low latitudes. Physical models such as CTIPe offer an advantage over climatology-based empirical models in that they provide a more detailed, time-dependent structure that is not available in a more statistical-type empirical solution. The physical model is also well suited for use in a data assimilation system in defining the current state and allowing for a more physically realistic evolution of the state (Fuller-Rowell *et al.*, 2006).

CTIPe solves the equations for the thermosphere and ionosphere but requires external input of solar EUV radiation, solar-wind velocity, and interplanetary magnetic field to prescribe the magnetospheric input, and wave forcing from the lower atmosphere. The largest energy source on a day-to-day basis is the solar EUV irradiance so it is one of the key inputs for global modeling of the thermosphere and ionosphere system, as in CTIPe, to provide heating, ionization, and dissociation rates for the upper atmosphere. The accurate, high-time cadence EVE measurements of the solar EUV irradiance provide the opportunity to improve the description of solar forcing in the atmosphere rather than using the daily F10.7 to estimate the highly variable radiation at all EUV wavelengths. A real-time version of CTIPe is currently running at NOAA in a test operational mode to explore the potential application for space-weather forecasting. The EVE real-time observations will be used to drive CTIPe for real-time specification and forecasting of the ionosphere and thermosphere, and the EVE Level 0C space-weather products, once validated after the SDO launch, will be part of this effort.

6.3. NRLMSIS Model

NRLMSIS is an empirical, upper-atmosphere density specification model (Picone *et al.*, 2002). The geospace community uses this model extensively, both for direct specification of neutral densities in operational activities and to provide the neutral-density component of other geospace models, for example the SAMI ionospheric model (Huba, Joyce, and Fedder, 2000). Given solar and geomagnetic inputs, the NRLMSIS estimates temperature, total mass density, and composition within the original MSIS framework (Hedin, 1991) at altitudes up to 1000 km for specified geographical location, time, and day of year. NRLMSIS utilizes daily F10.7, and time-centered 81-day running means, F10.7A, to simulate solar EUV radiation (adjusted from 1 AU to the appropriate Sun – Earth distance). The Ap index is a daily proxy for solar-wind-induced perturbations of the Earth's magnetic field. The total mass densities, temperatures, and composition shown in Figure 1 are global estimates from the NRLMSIS model, illustrating the large increases in average conditions from solar-cycle minimum and maximum (Figure 4).

The advent of extensive new databases of thermospheric densities derived from orbital drag, mass spectrometers, and UV remote sensing by the Global UV Imager (GUVI) on TIMED have afforded systematic, detailed validation and assessment studies of NRLMSIS (see, *e.g.*, Emmert *et al.*, 2006). Rarely is the model able to specify mass densities to better than 15%, and more generally the uncertainties are larger, and often global in scale. A notable limitation of the NRLMSIS is recognized to be its use of F10.7 as an index for solar EUV irradiance variations. Whereas the F10.7 index mainly reflects changes in solar emissions from the corona, the strongest EUV emissions that dominate the Sun's energy input to the thermosphere and ionosphere originate in the solar chromospheres and transition region. This is motivating the development of improved density-specification models formulated using EUV irradiances (Emmert, Picone, and Meier, 2008). Lacking an adequate long-term EUV irradiance database over the past 30 years (the duration of the drag-derived density database), the new models utilize instead EUV irradiances calculated by an empirical model constructed from SEE observations in which each 1-nm irradiance bin is parameterized by a wavelength-dependent combination of the coronal F10.7 index and the chromospheric Mg index. Since the new density specification models are formulated for direct EUV irradiance inputs, they will ingest the EVE measurements directly, permitting rapid and effective incorporation of SDO observations to geospace applications.

6.4. JB2006 and JB2008 Models

The JB2006 empirical atmospheric density model (Bowman *et al.*, 2008a) is developed using the CIRA72 (Jacchia, 1971) model as the basis for the diffusion equations. Solar indices based on on-orbit sensor data are used for the solar irradiances in the extreme and far ultraviolet wavelengths. Exospheric temperature and semiannual density equations are employed to represent the major thermospheric density variations. Temperature correction equations are also developed for diurnal and latitudinal effects, and finally density-correction factors are used for model corrections required at high altitude (1500 – 4000 km). The model is validated through comparisons of accurate daily density drag data previously computed for numerous satellites. For 400 km altitude, the standard deviation of 16% for the standard Jacchia model is reduced to 10% for the new JB2006 model for periods of low geomagnetic storm activity.

The JB2008 empirical atmospheric-density model (Bowman *et al.*, 2008b) is developed as an improved revision to the JB2006 model. The solar indices for the solar irradiances have been extended to include X-ray and H I Lyman-α wavelengths. New exospheric temperature equations are developed to represent the thermospheric EUV and FUV heating. New semiannual density equations based on multiple 81-day average solar indices are used to represent the variations in the semiannual density cycle that result from EUV heating.

Geomagnetic storm effects are modeled using the Dst index as the driver of global density changes. The model is validated through comparisons with accurate daily density drag data previously computed for numerous satellites in the altitude range of 175 to 1000 km. Model comparisons have been computed for the JB2008, JB2006, Jacchia 1970, and NRLMSIS 2000 models. Accelerometer measurements from the CHAMP and GRACE satellites are also used to validate the new geomagnetic-storm equations.

7. Summary

SDO/EVE will measure solar EUV spectral irradiance with unprecedented spectral and temporal resolution and accuracy. These measurements will contribute to space-weather operations, and solar and atmospheric physics research, particularly solar flares and their effects on the geospace environment. EVE research will help validate and improve empirical and first-principle models of the solar irradiance variability and of the geospace environment. The EVE data will also extend the record of continuous solar EUV irradiance measurements from the SOHO/SEM and TIMED/SEE instruments, which started in 1996 and 2003 respectively. The EVE instrument suite completed its final pre-flight calibration in 2007 and is exceeding all of its instrument requirements. The SDO spacecraft integration and test (I&T) activities have been completed at NASA's Goddard Space Flight Center (GSFC), and SDO currently awaits its launch in February 2010 (or later) into geosynchronous orbit for a nominal five-year mission.

Acknowledgements This research is supported by NASA contract NAS5-02140 to the University of Colorado. The authors gratefully acknowledge the many people who have contributed to the success of this new instrument throughout concept, design, fabrication, and testing. Special thanks to Vanessa George for her support in preparing this manuscript.

References

Bertaux, J.-L., Quemerais, E., Lallement, R., Lamassoure, E., Schmidt, W., Kyrölä, E.: 2000, *Geophys. Res. Lett.* **27**, 1331.

Bowman, B.R., Tobiska, W.K., Marcos, F.A., Valladares, C.: 2008a, *J. Atmos. Solar Terr. Phys.* **70**, 774.

Bowman, B.R., Tobiska, W.K., Marcos, F.A., Huang, C.Y., Lin, C.S., Burke, W.J.: 2008b, In: *AIAA 2008-6438*, AIAA-AAS Astrodynamics Specialist Conference.

Chamberlin, P.C., Woods, T.N., Eparvier, F.G.: 2007, *Space Weather J.* **5**, S07005. doi:10.1029/2007SW000316.

Chamberlin, P.C., Hock, R.A., Crotser, D.A., Eparvier, F.G., Furst, M., Triplett, M.A., Woodraska, D., Woods, T.N.: 2007, *SPIE Proc.* **6689**, 66890N. doi:10.1117/12.734116.

Chamberlin, P.C., Woods, T.N., Eparvier, F.G.: 2008, *Space Weather J.* **6**, S05001. doi:10.1029/2007SW000372.

Chamberlin, P.C., Woods, T.N., Crotser, D.A., Eparvier, F.G., Hock, R.A., Woodraska, D.L.: 2009, *Geophys. Res. Lett.* **36**, L05102. doi:10.1029/2008GL037145.

Codrescu, M.V., Fuller-Rowell, T.J., Munteanu, V., Minter, C.F., Millward, G.H.: 2008, *Space Weather J.* **6**, S09005. doi:10.1029/2007SW000364.

Crotser, D.A., Woods, T.N., Eparvier, F.G., Ucker, G., Kohnert, R., Berthiaume, G., Weitz, D.: 2004, *SPIE Proc.* **5563**, 182.

Crotser, D.A., Woods, T.N., Eparvier, F.G., Triplett, M.A., Woodraska, D.L.: 2007, *SPIE Proc.* **6689**, 66890M. doi:10.1117/12.732592.

Dere, K.P., Landi, E., Mason, H.E., Monsignori Fossi, B.C., Young, P.R.: 1997, *Astron. Astrophys. Suppl.* **125**, 149.
Didkovsky, L., Judge, D., Wieman, S., Woods, T., Chamberlin, P., Jones, A., Eparvier, F., Triplett, M., Woodraska, D., McMullin, D., Furst, M., Vest, R.: 2007, *SPIE Proc.* **6689**, 66890P. doi:10.1117/12.732868.
Didkovsky, L., Judge, D., Wieman, S., Woods, T., Jones, A.: 2010, *Solar Phys.* doi:10.1007/s11207-009-9485-8.
Emmert, J.T., Picone, J.M., Meier, R.R.: 2008, *Geophys. Res. Lett.* **35**, L05101. doi:10.1029/2007GL032809.
Emmert, J.T., Meier, R.R., Picone, J.M., Lean, J.L., Christensen, A.B.: 2006, *J. Geophys. Res.* **111**, A10S16. doi:10.1029/2005JA011495.
Eparvier, F.G., Woods, T.N., Crotser, D.A., Ucker, G.J., Kohnert, R.A., Jones, A., Judge, D.L., McMullin, D., Berthiaume, G.D.: 2004, *SPIE Proc.* **5660**, 48.
Eparvier, F.G., Crotser, D., Jones, A.R., McClintock, W.E., Snow, M., Woods, T.N.: 2009, *SPIE Proc.* **7438**, 743804.
Fontenla, J.M., Quemerais, E., Gonzalez Hernandez, I., Lindsey, C., Haberreiter, M.: 2009, *Adv. Space Res.* **44**, 457.
Fuller-Rowell, T.J., Rees, D., Quegan, S., Moffett, R.J., Codrescu, M.V., Millward, G.H.: 1996, In: Schunk, R.W. (ed.) *A Coupled Thermosphere – Ionosphere Model (CTIM), STEP Handbook*, 217.
Fuller-Rowell, T.J., Codrescu, M.V., Minter, C.F., Strickland, D.: 2006, *Adv. Space Res.* **37**, 401.
Garcia, H.: 1994, *Solar Phys.* **154**, 275.
Hathaway, D.H., Wilson, R.M., Reichmann, E.J.: 1999, *J. Geophys. Res.* **104**, 22375.
Hedin, A.: 1991, *J. Geophys. Res.* **96**, 1159.
Hochedez, J.-F., Schmutz, W., Stockman, Y., Schühle, U., Benmoussa, A., Koller, S., Haenen, K., Berghmans, D., Defise, J.-M., Theissen, A., Delouille, V., *et al.*: 2006, *Adv. Space Res.* **37**, 303.
Hock, R.A., Chamberlin, P.C., Woods, T.N., Crotser, D., Eparvier, F.G., Furst, M., Woodraska, D.L., Woods, E.C.: 2010, *Solar Phys.*, in press.
Huba, J.D., Joyce, G., Fedder, J.A.: 2000, *J. Geophys. Res.* **105**, 23035.
Jacchia, L.G.: 1971, *Smithsonian Astrophys. Spec. Rep.* **332**, May.
Judge, D.L., McMullin, D.R., Ogawa, H.S., Hovestadt, D., Klecker, B., Hilchenbach, M., Mobius, E., Canfield, L.R., Vest, R.E., Watts, R.N., Tarrio, C., Kuhne, M., Wurz, P.: 1998, *Solar Phys.* **177**, 161.
Knipp, D.J., Welliver, T., McHarg, M.G., Chun, F.K., Tobiska, W.K., Evans, D.: 2005, *Adv. Space Res.* **36**, 2506.
Landi, E., Del Zanna, G., Young, P.R., Dere, K.P., Mason, H.E., Landini, M.: 2006, *Astron. Astrophys. Suppl.* **162**, 261.
Lean, J.: 2005, *Phys. Today* **58**, 32.
Lean, J.L., Picone, J.M., Emmert, J.T.: 2009, *J. Geophys. Res.* **114**, A07301.
Lean, J.L., Warren, H.P., Mariska, J.T., Bishop, J.: 2003, *J. Geophys. Res.* **108**, 1059.
Lilensten, J., Dudok de Wit, T., Kretzschmar, M., Amblard, P.O., Moussaoui, S., Aboudarham, J., Auchere, F.: 2008, *Ann. Geophys.* **26**, 269.
Lindsey, C., Braun, D.C.: 2000, *Science* **287**, 1799.
Marcos, F.A., Bowman, B.R., Sheehan, R.E.: 2006, In: *AIAA 2006-6167*, AIAA ASM.
Meier, R.R., Warren, H.P., Nicholas, A.C., Bishop, J., Huba, J.D., Drob, D.P., Lean, J., Picone, J.M., Mariska, J.T., Joyce, G., Judge, D.L., Thonnard, S.E., Dymond, K.F., Budzien, S.A.: 2002, *Geophys. Res. Lett.* **29**, 99.
Millward, G.H., Moffett, R.J., Quegan, S., Fuller-Rowell, T.J.: 1996, In: Schunk, R.W. (ed.) *Handbook of Ionospheric Models, STEP Report*, 239.
National Space Weather Program: 2000, *The Implementation Plan*, 2nd edn., Office of the Federal Coordinator for Meteorology, FCM-P31-2000.
Pawlowski, D.J., Ridley, A.J.: 2008, *J. Geophys. Res.* **113**, A10309.
Picone, J.M., Emmert, J.T., Lean, J.L.: 2005, *J. Geophys. Res.* **110**, A03301. doi:10.1029/2004JA010585.
Picone, J.M., Hedin, A.E., Drob, D.P., Aikin, A.C.: 2002, *J. Geophys. Res.* **107**(A12), 1468. doi:10.1029/2002JA009430.
Richards, P.C., Fennelly, J.A., Torr, D.G.: 1994, *J. Geophys. Res.* **99**, 8981.
Schunk, R.W.: 1988, *Ionospheric Modeling*, Birkhäuser, Basel, 255.
Schunk, R.W., Sojka, J.J.: 1996, In: Schunk, R. (ed.) *Solar-Terr. Energy Program: Handbook of Ionospheric Models, SCOSTEP*, 153.
Schunk, R.W., Scherliess, L., Sojka, J.J., Thompson, D.C., Anderson, D.N., Codrescu, M., Minter, C., Fuller-Rowell, T.J., Heelis, R.A., Hairston, M., Howe, B.M.: 2004, *Radio Sci.* **39**, RS1S02.
Smithtro, C.G., Solomon, S.C.: 2008, *J. Geophys. Res.* **113**, A08307.
Solomon, S.C., Qian, L.: 2005, *J. Geophys. Res.* **110**, A10306.
Sojka, J.J.: 1989, *Rev. Geophys.* **27**, 37.

Tobiska, W.K.: 2004, *Adv. Space Res.* **34**, 1736.

Tobiska, W.K.: 2007, *AIAA 2007-0495*.

Tobiska, W.K.: 2008, *AIAA Proceedings*, AIAA-2008-0453.

Tobiska, W.K., Bouwer, S.D.: 2005, In: Goodman, J.M. (ed.) *2005 Ionospheric Effects Symposium*, MG Associates, Alexandria, 76.

Tobiska, W.K., Bouwer, S.D.: 2006, *Adv. Space Res.* **37**, 347.

Tobiska, W.K., Woods, T., Eparvier, F., Viereck, R., Floyd, L., Bouwer, D., Rottman, G., White, O.R.: 2000, *J. Atmos. Solar Terr. Phys.* **62**, 1233.

Viereck, R., Hanser, F., Wise, J., Guha, S., Jones, A., McMullin, D., Plunket, S., Strickland, D., Evans, S.: 2007, *SPIE Proc.* **6689**, 66890K-10.

Wang, Y.-M., Sheeley, N.R. Jr.: 1991, *Astrophys. J.* **375**, 761.

Wang, Y.-M., Lean, J., Sheeley, N.R. Jr.: 2000, *Geophys. Res. Lett.* **27**, 505.

Wang, Y.-M., Sheeley, N.R. Jr., Lean, J.: 2000, *Geophys. Res. Lett.* **27**, 621.

Wang, Y.-M., Lean, J.L., Sheeley, N.R. Jr.: 2005, *Astrophys. J.* **625**, 522.

Warren, H.P.: 2005, *Astrophys. J. Suppl.* **157**, 147. doi:10.1086/427171.

Warren, H.P., Mariska, J.T., Lean, J.: 2001, *J. Geophys. Res.* **106**, 15745.

Westhoff, R.C., Rose, M.K., Gregory, J.A., Berthiaume, G.D., Seely, J.F., Woods, T.N., Ucker, G.: 2007, *SPIE Proc.* **6686**, 668604. doi:10.1117/12.734371.

Woods, T.N., Rottman, G.J.: 2002, In: Mendillo, M., Nagy, A., Waite, J.H. Jr. (eds.) *Comparative Aeronomy in the Solar System*, *Geophys. Monogr. Ser.* **130**, AGU, Washington, 221.

Woods, T., Acton, L.W., Bailey, S., Eparvier, F., Garcia, H., Judge, D., Lean, J., McMullin, D., Schmidtke, G., Solomon, S.C., Tobiska, W.K., Warren, H.P.: 2004, In: Pap, J., Fröhlich, C., Hudson, H., Kuhn, J., McCormack, J., North, G., Sprig, W., Wu, S.T. (eds.) *Solar Variability and Its Effect on Climate*, *Geophys. Monogr. Ser.* **141**, AGU, Washington, 127.

Woods, T.N., Eparvier, F.G., Bailey, S.M., Chamberlin, P.C., Lean, J., Rottman, G.J., Solomon, S.C., Tobiska, W.K., Woodraska, D.L.: 2005, *J. Geophys. Res.* **110**, A01312. doi:10.1029/2004JA010765.

Woodraska, D.L., Woods, T.N., Eparvier, F.G.: 2007, *J. Spacecr. Rockets* **44**(6), 1204. doi:10.2514/1.28639.

Solar Phys (2012) 275:145–178
DOI 10.1007/s11207-010-9520-9

Extreme Ultraviolet Variability Experiment (EVE) *Multiple EUV Grating Spectrographs* (MEGS): Radiometric Calibrations and Results

R.A. Hock · P.C. Chamberlin · T.N. Woods · D. Crotser · F.G. Eparvier · D.L. Woodraska · E.C. Woods

Received: 6 October 2009 / Accepted: 20 January 2010 / Published online: 24 February 2010

Abstract The NASA *Solar Dynamics Observatory* (SDO), scheduled for launch in early 2010, incorporates a suite of instruments including the *Extreme Ultraviolet Variability Experiment* (EVE). EVE has multiple instruments including the *Multiple Extreme ultraviolet Grating Spectrographs* (MEGS) A, B, and P instruments, the *Solar Aspect Monitor* (SAM), and the *Extreme ultraviolet SpectroPhotometer* (ESP). The radiometric calibration of EVE, necessary to convert the instrument counts to physical units, was performed at the National Institute of Standards and Technology (NIST) Synchrotron Ultraviolet Radiation Facility (SURF III) located in Gaithersburg, Maryland. This paper presents the results and derived accuracy of this radiometric calibration for the MEGS A, B, P, and SAM instruments, while the calibration of the ESP instrument is addressed by Didkovsky *et al.* (*Solar Phys.*, 2010, doi:10.1007/s11207-009-9485-8). In addition, solar measurements that were taken on 14 April 2008, during the NASA 36.240 sounding-rocket flight, are shown for the prototype EVE instruments.

Keywords SDO · EVE · Solar EUV irradiance · Calibration · Synchrotron

1. Introduction

The *Solar Dynamics Observatory* (SDO) is the first mission in NASA's Living With a Star (LWS) program. The goal of SDO is to quantify and understand the causes of solar variability and its effects on Earth. The *Extreme Ultraviolet Variability Experiment* (EVE), one

The Solar Dynamics Observatory
Guest Editors: W. Dean Pesnell, Phillip C. Chamberlin, and Barbara J. Thompson.

R.A. Hock (✉) · T.N. Woods · D. Crotser · F.G. Eparvier · D.L. Woodraska
Laboratory for Atmospheric and Space Physics, 1234 Innovation Drive, Boulder, CO 80303, USA
e-mail: rachel.hock@lasp.colorado.edu

P.C. Chamberlin
Solar Physics Laboratory, Code 671, NASA Goddard Space Flight Center, Greenbelt, MD 20771, USA

E.C. Woods
Rhodes College, 2000 North Parkway, Memphis, TN 38112, USA

of the three experiments on SDO, will measure the variability of the Sun's irradiance in the X-ray ultraviolet (XUV) and extreme ultraviolet (EUV) wavelengths from 0.1 to 106 nm plus the hydrogen Lyman-α emission line at 121.6 nm. The EVE irradiance measurements will be made with better than 25% accuracy over the mission lifetime.

Details about EVE can be found in Woods *et al.* (2010), which includes the science (measurements and modeling) goals, as well as the details of the EVE instrument design and data products. The calibration and algorithms of the *Multiple EUV Grating Spectrographs* (MEGS) A, B, P, and *Solar Aspect Monitor* (SAM) instruments are the focus of this paper. The results shown are from the final EVE calibration performed at the National Institute of Standards and Technology (NIST) Synchrotron Ultraviolet Radiation Facility III (SURF III) in Gaithersburg, Maryland, during August 2007. Also included are results from a January 2009 calibration of the prototype EVE instrument that will be flown as an underflight calibration three months after SDO launches. A brief overview of the EVE instruments is given in Section 2, followed by a discussion of EVE's calibration heritage in Section 3. The SURF III radiometric calibration facilities are detailed in Section 4, then the calibration algorithms and results of the MEGS A and B instruments are presented in Section 5, followed by a discussion of the MEGS P and SAM calibrations in Sections 6 and 7, respectively. The solar EUV spectral irradiance from a sounding-rocket flight on 14 April 2008 using the prototype EVE instrument is then presented in Section 8, followed by concluding remarks in Section 9.

2. EVE Instruments

The EVE instrument suite will measure the solar spectral irradiance from 0.1 to 6 nm with 1-nm spectral resolution, 6 to 106 nm with 0.1-nm resolution, and the hydrogen Lyman-α line at 121.6 nm with 1-nm resolution. Solar observations will be taken continuously with a ten-second cadence except during satellite eclipse periods and planned calibration activities, which are expected to total less than 1% of the mission lifetime. To cover the entire spectral range and meet accuracy requirements, EVE is composed of multiple instruments. The EVE instrument design is described in detail by Woods *et al.* (2010), so only a brief overview of the instruments is given here.

The primary, high-spectral-resolution irradiance measurements are made by the MEGS A and B instruments. MEGS A is an off-Rowland-circle grazing-incidence spectrograph covering the 6 to 37 nm range. The entrance aperture of MEGS A includes two slits: denoted A1 and A2. Light illuminating these slits shares the same grating and CCD detector, but has light paths that are offset in the cross-dispersion direction of the grating and therefore do not overlap on the CCD. In addition, MEGS A1 and A2 have separate primary filters with different spectral bandpasses to isolate different portions of the MEGS A wavelength range and reduce the effects from higher orders. MEGS A1 is optimized for the 6 to 18 nm range, while MEGS A2 covers 16 to 37 nm. MEGS B is a two-grating, cross-dispersing spectrograph covering the 36 to 106 nm range. The two orthogonal gratings are designed to eliminate the out-of-band light without the use of a bandpass filter and to disperse the higher-order spectra off the primary first-order spectrum.

Both MEGS A and B use the same type of 2048 × 1024 CCD. Westhoff *et al.* (2007) provide more information about the CCDs used with MEGS. Ray-tracing results for the MEGS spectrometers are found in Crotser *et al.* (2004, 2007). The MEGS CCDs are two-dimensional arrays. One direction, the dispersion direction, covers the wavelength range of the instrument over 2048 pixels. The other direction, the cross-dispersion direction, images the slit at each wavelength over 1024 pixels. As a result, the solar irradiance for any given wavelength is spread over many pixels in the cross-dispersion direction of the CCD. We create our final science products by adding together all pixels with a certain wavelength.

MEGS P measures the Lyman-α emission with 0.25-second cadence using a Si photodiode. Lyman-α is isolated by placing the diode in the MEGS B optical path at the point where 121.6 nm photons are dispersed after MEGS B's first grating. A limiting bandpass Lyman-α filter is also used to eliminate potentially significant out-of-band light.

Included as part of the MEGS A package is SAM: a pinhole camera that observes the short XUV wavelengths (0.1 – 7 nm). SAM is designed to record single photon events that can be used to reconstruct a spectrum with 1-nm spectral resolution. In addition, SAM has the ability to generate images with a coarse spatial resolution of approximately 15 arcseconds per pixel.

The *EUV SpectroPhotometer* (ESP) instrument uses a transmission grating to disperse EUV light onto a set of broadband EUV photometers. This instrument will not only get higher temporal resolution measurements (0.25 seconds), but will also, along with underflight calibration rockets, help quantify and track the long-term degradation of the MEGS instruments. As ESP's diodes are expected to degrade slowly over time, they can help to re-calibrate MEGS results if MEGS undergoes a sudden change in responsivity from a CCD bakeout or science-filter failure. While the absolute calibration of EVE will be tied to the planned annual rocket underflights, ESP and MEGS P can help track degradation trends between rocket flights. The details of the ESP instrument are presented by Didkovsky *et al.* (2010).

All instruments on EVE have a rotatable filter wheel in front of their entrance apertures. These filter wheels house redundant science filters, which may be necessary in the event that a primary science filter develops pinholes. There are also blank filter positions that allow for dark measurements on orbit. One filter position in each of the MEGS A and B filter holders contains a "second-order filter" that blocks the short-wavelength portion of the primary filter and quantifies the effects of higher grating orders. Other positions have filters that transmit visible light to check for out-of-band scattered light. Triplett *et al.* (2007) give calibration results for the EVE filters.

3. EVE MEGS Calibration Heritage

The MEGS instruments on EVE are a follow-on to the *Solar EUV Experiment* (SEE) that has been making measurements since February 2002 of the solar spectral irradiance from 0.1 to 195 nm onboard the *Thermosphere, Ionosphere, Mesosphere Energetics and Dynamics* (TIMED) satellite (Woods *et al.*, 2005). Although EVE has a limited spectral range compared to SEE, EVE will improve upon the temporal and spectral resolution, accuracy, and duty cycle. Currently, SEE is planned to overlap with EVE in time to compare the absolute values of the two experiments and create a continuous record of the solar EUV irradiance.

SEE has provided experience not only in designing and building the EUV spectrographs and photometers, but also in determining the radiometric calibration. For example, the *EUV Grating Spectrograph* (EGS), an instrument on SEE and a copy for underflight rocket payloads, has been calibrated at SURF more than eight times. This includes two calibrations of the flight EGS before launch as well as several calibrations of the rocket underflight copy. Another instrument, the *X-ray Photometer System* (XPS) has been flown as different versions on the *Student Nitric Oxide Experiment* (SNOE; Bailey *et al.*, 2006), TIMED (Woods *et al.*, 2005), and the *Solar Radiation and Climate Experiment* (SORCE; Woods and Rottman, 2005) as well as a prototype version that has flown on various sounding rocket flights. As a result, it has been calibrated at SURF numerous times. The TIMED/SEE calibrations and algorithms are described by Woods *et al.* (2005), while the SORCE/XPS calibrations and results are given by Woods, Rottman, and Vest (2005).

4. SURF Calibration Setup

The primary EVE calibrations are performed in the vacuum chamber at the end of beam line 2 (BL-2) at the SURF III calibration facility in Gaithersburg, Maryland. SURF provides a primary radiometric source that is accurate to 1%, one of the most accurate UV sources available (Arp *et al.*, 2000). There are certain advantages of using this primary source over using secondary, transfer standards that were first calibrated at NIST. Calibrating at SURF not only eliminates the uncertainties associated with using a transfer standard but also provides an adjustable source in both intensity and wavelength that provides flexibility to optimize many of the calibrations discussed throughout this paper. The key calibration parameters for the SURF beam are the beam energy, which determines the peak wavelength, and the beam current, which determines the intensity.

At SURF, the EVE instruments are coherently illuminated by the synchrotron radiation from the electrons that travel around the SURF ring with a frequency of 57 MHz. The diffraction effects from BL-2 baffles are minimized to less than 1% for the EVE calibrations. First, by having the EVE entrance slit smaller than the SURF illuminated beam size and also having the slit located in the center of the beam, the beam line baffle diffraction patterns do not enter the MEGS slits. Furthermore, the EVE grating is much larger than the slit diffraction pattern incident on the grating; thereby, the majority of the radiation reaches the detector. Consequently, no corrections for diffraction are made in the MEGS algorithms.

EVE is mounted in a vacuum chamber that allows the entire system, from electron ring to instrument CCDs, to be in a vacuum environment, thus eliminating any influence or absorption of the EUV photons by air or glass windows. Inside the vacuum chamber, EVE is mounted to a gimbal system that provides pitch and yaw movements. This allows the field-of-view (FOV), or the angle of the SURF beam relative to the normal of the entrance slit, to vary. As the SURF beam is much smaller that the angular size of the Sun, it is important to vary the FOV to characterize the resulting changes in the optical responses. The vacuum chamber also allows for x- and y-translations in order to center the SURF beam in the middle of the aperture for each EVE instrument.

Both the beam energy and current determine the photon flux from the SURF beam. The beam energy determines the spectral shape, while the current determines the intensity. Figure 1 shows the photon flux per milliamp of SURF current for each beam energy used in the EVE calibrations. Lower energies shift the peak intensity to longer wavelengths and have fewer photons per unit current. As a result, using lower energies requires higher currents to provide a similar photon flux into the instrument. Most calibrations were done using the 380 MeV beam energy. The current was adjusted to get high counts but not so high as to saturate any of the CCD's pixels. Correcting for the effects of higher orders requires using multiple energies; the details of this process are discussed in Section 5.4.4.

In general, to improve the counting statistics and reduce uncertainties, at least 20 ten-second images are co-added for each calibration test. Furthermore, each instrument is calibrated separately to allow us to optimize the beam current and energy. Once all of the calibration data have been taken, these data need to be processed to obtain the final responsivity for each of the EVE instruments.

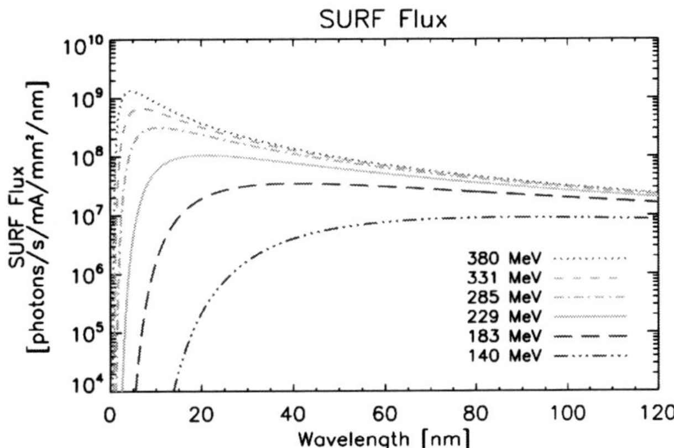

Figure 1 The SURF flux for beam energies from 140 MeV to 380 MeV.

5. MEGS A and B Calibration

As with any spectrograph, the many instrument parameters require calibration and/or analysis in order to use the spectrograph's data for scientific results in meaningful units. In the case for MEGS A and B, the primary result is the solar spectral irradiance, which we plan to report in units of W m^{-2} nm^{-1}. Passing through the light path, the following instrument parameters are briefly introduced: Firstly, the entrance-slit area is needed to provide the m^{-2} part of the measured irradiance. Then the incoming light is transmitted through a filter, diffracted off a grating into individual wavelengths, and absorbed in a CCD pixel where electron-hole pairs are created.

The efficiency of each of these optical elements can be calibrated separately, but in the case of the instrument-level calibrations performed at SURF, the net effect of all of the optics is calibrated simultaneously to derive the MEGS responsivity. A critical part of this responsivity is accurately knowing the incoming irradiance, or flux, of the light source; in this case, the SURF synchrotron irradiance is known with a relative accuracy of about 1%. For this analysis, the responsivity [R] will be given in units of data number [DN] per photon. The detector's signal, in DN, is actually the conversion of the number of electrons detected in each pixel over the integration period. This electronic conversion, enabled by charge amplifier and analog-to-digital converter [ADC], is referred to as gain. The CCD detector's gain per pixel varies slightly with temperature and is dependent on which charge amplifier chain or readout mode is used in reading the CCD pixels. Therefore, the MEGS calibration measurements at SURF are taken at different temperatures and using all readout modes to determine the variations in gain. Another important parameter for the CCD readout is the detector background (dark) signal, which also varies with temperature and readout mode. The angle of the incoming radiation affects where the light falls on the optical elements (filter, grating, CCD), so the responsivity as a function of incoming angle requires careful calibration. The mapping of responsivity over the instrument field-of-view (FOV) is an important part of the SURF calibrations, so that the SURF measurements, which are made with a small beam of light, can be properly averaged for the larger angular size of the Sun and in order to characterize how the responsivity changes with any offset pointing of the Sun from the ideal MEGS optical center point. The final part of the MEGS calibration needed for the spectral-irradiance calculation is specifying the spectral bandpass per CCD pixel, which is accomplished with wavelength-scale calibrations and raytrace analysis of the spectrograph design.

While MEGS A and B have different optical designs as discussed in Section 2 and in more detail by Woods *et al.* (2010), the data from these instruments are calibrated and processed in the same way, as both instruments have the same type of CCDs. This section presents the algorithms and calibration results for MEGS A1, A2, and B. The first subsection, Section 5.1, discusses the algorithms necessary to convert CCD detector counts in DN to irradiance in $\mathrm{W\,m^{-2}\,nm^{-1}}$. Section 5.2 discusses the determination of the wavelength scale and spectral resolution, while Sections 5.3 and 5.4 present the details of the algorithms and calibration results. While the focus of this paper is on the calibration of the EVE instruments to be flown on SDO, final calibration results for the prototype EVE instrument flown on a NASA sounding rocket in 2008 are included for comparison. Some early calibration results for MEGS are presented by Chamberlin *et al.* (2007, 2009).

5.1. MEGS A and B Algorithms

There are two aspects of the MEGS algorithms: One is the responsivity algorithm using the SURF calibration measurements, and the other is the irradiance algorithm for the solar observations. These two algorithms are in concept identical, but inverted, equations. For the SURF calibrations, the responsivity is calculated as the measured signal divided by the known synchrotron irradiance. For the solar observations, the solar spectral irradiance is calculated as the measured signal divided by the instrument responsivity derived from SURF calibrations. In actuality, the algorithms are more detailed as the responsivity parameters are dependent on temperature and field-of-view angle, as well as other effects such as grating higher-order contributions to the measured signals. The following describes the algorithms for the MEGS signal, responsivity, and solar irradiance. The many parameters in these algorithms require characterization, from direct calibration and/or from analysis. The subsequent sections describe the calibration techniques and results for each of the primary instrument parameters.

For MEGS A and B, the responsivity is calculated for each individual pixel on the 2048×1024 CCD detectors, denoted by the indices i and j in the following equations. This pixel-based calibration allows us to bypass the flatfield correction sometimes used for array detectors. Furthermore, the calibration is not based on calibrations of each individual optical element of the instrument, but it is for the complete end-to-end optical system, including filter transmissions as well as grating and detector efficiencies.

The algorithms for the rocket instrument (prototype EVE) are essentially the same as those for the flight EVE. As the rocket version of MEGS does not have a filter wheel, calibrations were done for the primary (and only) filter, which is the same type as the primary science filters on the EVE flight instrument.

The first step for processing all MEGS A and B images, taken either at SURF or on orbit for solar observations, is to correct the raw images to get the corrected count rate $[C']$ in DN second^{-1}:

$$C'(i,j,t) = \left[\frac{C(i,j,t,T_{\mathrm{CCD}},\mathrm{tap})}{\Delta t} - D(i,j,T_{\mathrm{CCD}},C,\mathrm{tap})\right] \times G(i,j,T_{\mathrm{CCD}},\mathrm{tap})\mathrm{Mask}(i,j,C,\mathrm{tap}) \quad (1)$$

The raw counts $[C]$ in each pixel are corrected for the integration time $[\Delta t]$, dark offset $[D]$, and the dependence of the CCDs on the temperature and readout mode [tap] of the electronics $[G]$. In addition, invalid pixels are identified and set to zero by applying a binary mask [Mask], allowing them to be ignored in later steps. While there are very few defective pixels on the CCDs, there are several pixels contaminated in each image by cosmic-ray particle hits during ground calibration and also by energetic particle hits during flight. The details of how each of these corrections are calculated will be discussed in Section 5.3.

Images taken at SURF are further processed to determine the SURF response [R_{SURF}] in DN photon^{-1}:

$$R_{\mathrm{SURF}}(i, j, E_{\mathrm{beam}}, \mathrm{filter}, \alpha, \beta) = \frac{\frac{1}{n}\sum_{k=1}^{n} \frac{C'_k(i,j,t,E_{\mathrm{beam}},\mathrm{filter},\alpha,\beta)}{I_{\mathrm{SURF}}(t)}}{F_{\mathrm{SURF}}(i, j, E_{\mathrm{beam}}, \alpha, \beta) A_{\mathrm{slit}} \Delta\lambda(i, j)} \tag{2}$$

Each corrected image [C'] is normalized by the SURF beam current [I_{SURF}] in mA. Then, all images taken with the same beam energy [E_{beam}], filter, and field-of-view pointing [α, β] are co-added to reduce the uncertainty. Finally, this co-added image is divided by the SURF beam flux [F_{SURF}] in photons s^{-1} mA^{-1} mm^{-2} nm^{-1}, area of the limiting aperture [A_{slit}] in mm^2, and bandpass [$\Delta\lambda$] in nm to derive the SURF response [R_{SURF}].

The corrected counts used to determine the SURF response in Equation (2) can contain both first order and higher orders. As shown later, MEGS B SURF calibrations have essentially no higher-order contributions, so Equation (2) can be directly used for MEGS B. MEGS A responsivities do have contributions from higher orders so R_{SURF} needs to be adjusted to include only the first-order response. Equation (5) shows the relationship between R_{SURF} and true first-order response [R_1] as derived from:

$$R_{\mathrm{SURF}}(i, j, E_{\mathrm{beam}}, \mathrm{filter}, \alpha, \beta) = \sum_{k=1}^{m} \frac{1}{k} \frac{F^k_{\mathrm{SURF}}(i, j, E_{\mathrm{beam}}, \alpha, \beta)}{F^{k=1}_{\mathrm{SURF}}(i, j, E_{\mathrm{beam}}, \alpha, \beta)} R_k(i, j, \mathrm{filter}, \alpha, \beta) \tag{3}$$

where k denotes the order and

$$F^k_{\mathrm{SURF}}(\lambda) = F_{\mathrm{SURF}}\left(\frac{\lambda}{k}\right) \tag{4}$$

The contributions of the higher orders are found and separated by a proven method that utilizes multiple beam energies at SURF (Saloman, 1975; Rottman, Woods, and Sparn, 1993; Chamberlin, Woods, and Eparvier, 2002). This multiple-beam-energy method exploits the differences in the spectral distribution of SURF fluxes for different beam energies to vary R_{SURF}. Using Equation (3), a system of linear equations can be written and then solved numerically to determine each R_k. By using two different SURF beam energies, first and second order can be separated; using three energies, it is possible to calculate the contributions from first, second, and third orders. The results of this order-sorting correction are applied as a scalar correction to R_{SURF}:

$$f_{\mathrm{OS}}(i, j, E_{\mathrm{beam}}, \mathrm{filter}, \alpha, \beta) = \frac{R_1(i, j, \mathrm{filter}, \alpha, \beta)}{R_{\mathrm{SURF}}(i, j, E_{\mathrm{beam}}, \mathrm{filter}, \alpha, \beta)} \tag{5}$$

Determining R_{SURF} and the higher-order correction will be discussed in Section 5.4.

For processing solar data, the SURF response is used to calculate the flight responsivity [R_{flight}] in DN s^{-1} (W m^{-2} nm^{-1})$^{-1}$:

$$\begin{aligned} &R_{\mathrm{flight}}(i, j, \mathrm{filter}) \\ &\quad = \frac{\lambda(i, j)}{hc}\left[\sum_{\alpha,\beta} w(\alpha, \beta) R_{\mathrm{SURF}}(i, j, E_{\mathrm{beam}}, \mathrm{filter}, \alpha, \beta) f_{\mathrm{OS}}(i, j, E_{\mathrm{beam}}, \mathrm{filter}, \alpha, \beta)\right] \\ &\qquad \times A_{\mathrm{slit}} \Delta\lambda(i, j) \end{aligned} \tag{6}$$

Table 1 Example FOV map weights for solar observations. These weights are appropriate for solar observations when the on-orbit pointing is at the optical center.

		$\alpha[°]$		
		−0.5	0.0	+0.5
$\beta[°]$	+0.5	0.0249	0.1455	0.0249
	0.0	0.1455	0.3180	0.1455
	−0.5	0.0249	0.1455	0.0249

where hc/λ converts from photon units to energy units, and R_{SURF}, f_{OS}, A_{slit}, and $\Delta\lambda$ are the SURF response, order-sorting correction, slit area, and bandpass as described above. The weight of each SURF FOV point, given by w, is used to average the SURF FOV map over the approximately 0.5° FOV of the Sun. As mentioned in Section 4, the SURF beam is significantly smaller in angular diameter than the full-disk Sun. By correctly weighting SURF responses taken at different angles, it is possible to combine the SURF responses to determine the appropriate weighted response for the solar measurements. Table 1 gives the values of w for the expected on-orbit pointing. The FOV maps are discussed in Section 5.4.3.

Next, the corrected count rate [C': Equation (1)] for solar data is converted to irradiance:

$$I(i,j,t) = \frac{C'(i,j,t,\mathrm{filter})}{R_{\mathrm{flight}}(i,j,\mathrm{filter})} f_{\mathrm{degrad}}(i,j,t,\mathrm{filter}) f_{1\ \mathrm{AU}}(t) \tag{7}$$

where R_{flight} is the flight responsivity defined in Equation (6), f_{degrad} is the degradation correction, and $f_{1\ \mathrm{AU}}$ is the correction to normalize the irradiance to 1 AU to account for the eccentricity of Earth's orbit. The final step is to convert the irradiance image (as all calibrations are applied on the pixel level) to get a spectrum of irradiance *versus* wavelength. This is done by adding together all pixels whose wavelength falls within a 0.02 nm wavelength bin of the final spectrum; note that there are multiple wavelength bins within the spectral resolution of 0.1 nm.

We must carefully track and propagate the uncertainties in all of the algorithms. The details of the uncertainties for each correction will also be discussed in the appropriate sections. Here, we simply present the equations used to propagate the uncertainties that determine the irradiance accuracy. The uncertainty in C' [Equation (1)] is given by:

$$\sigma_{C'}^2 = (C')^2 \left[\frac{(\frac{C}{\Delta t})^2 (\frac{\sigma_C^2}{(C)^2} + \frac{\sigma_{\Delta t}^2}{(\Delta t)^2}) + \sigma_D^2}{(\frac{C}{\Delta t} - D)^2} + \frac{\sigma_G^2}{(G)^2} \right] \tag{8}$$

while the uncertainty in R_{SURF} [Equation (2)] is given by

$$\sigma_{R_{\mathrm{SURF}}}^2 = (R_{\mathrm{SURF}})^2 \left[\frac{\frac{1}{n^2} \sum_{k=1}^{n} (\frac{C_k'}{I_{\mathrm{SURF}}})^2 [\frac{\sigma_{C'}^2}{(C_k')^2} + \frac{\sigma_{I_{\mathrm{SURF}}}^2}{(I_{\mathrm{SURF}})^2}]}{(\frac{1}{n} \sum_{k=1}^{n} \frac{C'}{I_{\mathrm{SURF}}})^2} + \frac{\sigma_{F_{\mathrm{SURF}}}^2}{(F_{\mathrm{SURF}})^2} \right] \tag{9}$$

Slit area [A_{slit}] and bandpass [$\Delta\lambda$] do not contribute to the uncertainty of R_{flight} as the terms cancel out in Equations (2) and (6). The uncertainty for the order-sorting correction (Equation (5)) is found by comparing order-sorting results using different combinations of beam energies. This is discussed in more detail in Section 5.4.4. Then, the uncertainty in R_{flight} [Equation (6)] is given by:

$$\sigma_{R_{\mathrm{flight}}}^2 = (R_{\mathrm{flight}})^2 \left[\frac{\sigma_\lambda^2}{(\lambda)^2} + \frac{\sum_{\alpha,\beta} (w R_{\mathrm{SURF}} f_{\mathrm{OS}})^2 [\frac{\sigma_w^2}{(w)^2} + \frac{\sigma_{R_{\mathrm{SURF}}}^2}{(R_{\mathrm{SURF}})^2} + \frac{\sigma_{f_{\mathrm{OS}}}^2}{(f_{\mathrm{OS}})^2}]}{(\sum_{\alpha,\beta} w R_{\mathrm{all}} f_{\mathrm{OS}})^2} \right] \tag{10}$$

Finally, the irradiance accuracy using Equation (7) is given by:

$$\sigma_I^2 = I^2\left[\frac{\sigma_{C'}^2}{(C')^2} + \frac{\sigma_{R_{\text{flight}}}^2}{(R_{\text{flight}})^2} + \frac{\sigma_{f_{\text{degrad}}}^2}{(f_{\text{degrad}})^2} + \frac{\sigma_{f_{1\,\text{AU}}}^2}{(f_{1\,\text{AU}})^2}\right] \tag{11}$$

The following subsections describe the calibration technique and results for determining each of these parameters in the MEGS algorithms.

5.2. Wavelength Scale, Bandpass, and Spectral Resolution

Before any radiometric calibrations are performed, the wavelength scale for each pixel on the CCDs must be determined. The wavelength scale or the wavelength associated with each pixel on the CCD is used to determine the SURF flux illuminating each pixel. It is also used when converting an irradiance image into a spectrum.

Because the primary calibration facility, SURF, is a continuum source, measuring a definitive wavelength scale is not straightforward. Instead, a more accurate method was employed prior to going to SURF by illuminating MEGS A and B with well-known emission spectra from various gases using a hollow-cathode lamp.

5.2.1. LASP Calibration Setup

The Multiple Optical Beam Instrument (MOBI) vacuum chamber at the Laboratory for Atmospheric and Space Physics (LASP) in Boulder, Colorado incorporates a hollow-cathode lamp capable of generating emission-line spectra from the visible down to $\approx$25 nm using various gases including neon, argon, helium, and nitrogen. These spectral lines provide focus and alignment information, as well as provide a method to determine the wavelength scale and spectral resolution of MEGS A and B.

The hollow-cathode lamp is mounted to the tank with a large, flexible vacuum bellows and is differentially pumped to provide an unimpeded path for the EUV light, which would otherwise be absorbed by air or glass windows. One side of the bellows is mounted to an external $x-y$ translation system that allows $\pm 4^\circ$ off-axis illumination in both directions for focus and alignment activities. EVE is mounted to a vacuum x-stage inside MOBI, which allows for additional off-axis maneuvers as well as illumination of all EVE instruments without breaking vacuum and repositioning the experiment.

5.2.2. Wavelength Scale

With EVE in the MOBI vacuum chamber, emission spectra are observed for each of the instruments. For each spectrum, a subset of bright lines is selected whose wavelengths are known. The profile of each line is fit to a Gaussian to determine the pixel location of the peak. Then a high-order polynomial is used to fit wavelength *versus* pixel location in the dispersion direction. Each row (cross-dispersion direction) is fit independently to allow for any curvature of the slit image on the detector. This process is repeated for different pointing angles to determine how the wavelength scale varies with field-of-view. These fits of the wavelength scale are verified by ray tracing the optical system (*e.g.* see Crotser *et al.* 2004, 2007).

Once on orbit, the solar-irradiance spectrum will be used to verify the pre-flight wavelength scale. This is particularly important for MEGS A, as the shortest wavelength seen in the MOBI spectra is around 25 nm. Therefore the extrapolation of the wavelength fit to the MEGS A spectral range down to 6 nm can have increased uncertainties.

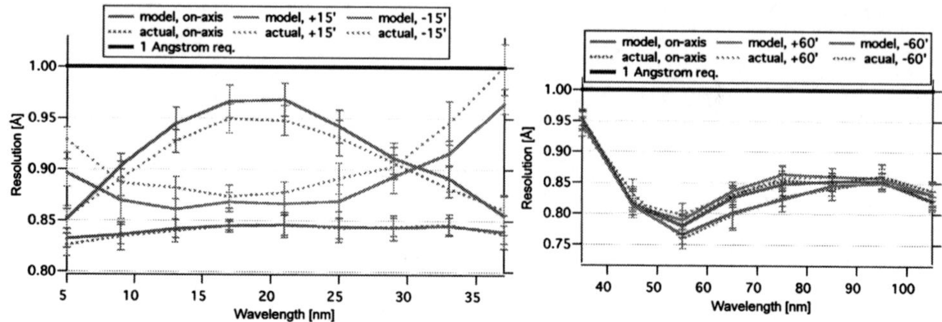

Figure 2 The expected (modeled) and measured spectral-resolution performance for MEGS A (left) and MEGS B (right). There results are for both on-axis and off-axis illumination. The measured results are based on instrument performance using the spectra produced by various hollow-cathode gases. These results indicate that the spectral resolution is better than 0.1.

5.2.3. *Bandpass*

The bandpass [$\Delta\lambda$] in Equations (2) and (6) is the spectral width of each pixel. The bandpass is not related to spectral resolution, which is the width of the spectral lines, but is instead related to the step size along the wavelength scale. The bandpass for pixel (i, j) is defined as the absolute value of the difference between wavelength of the pixel on the left $(i - 1, j)$ and the pixel on the right $(i + 1, j)$ divided by two:

$$\Delta\lambda(i, j) = \frac{|\lambda(i + 1, j) - \lambda(i - 1, j)|}{2} \tag{12}$$

The absolute value is necessary as the wavelength of pixel does not necessary increase as the pixel indices increase.

5.2.4. *Spectral Resolution*

The spectral resolution as a function of wavelength and off-axis pointing can be determined using the MOBI spectra. The measured spectral resolution is the full width at half maximum [FWHM] of the Gaussian fits of the line profiles in the dispersion direction. Figure 2 shows the measured spectral resolution for MEGS as a function of wavelength as well as the modeled solar spectral resolution. These results, shown for on-axis and off-axis illumination, indicate that the measured spectral performance is comparable to the modeled performance. Furthermore, the results indicate that the spectral resolution requirement of 0.1 nm has been exceeded.

5.3. Corrected Count Rate

The first step in processing the raw image is to convert the images to raw count rates by dividing the raw counts by the integration time. The uncertainty of the raw counts, used in Equation (8), is about two data number [DN] per pixel, or four electrons per pixel. The standard integration time for MEGS images is ten seconds with an uncertainty of 0.001 second. The following sections describe generating the corrections to this count rate for MEGS A and B as shown in Equation (1).

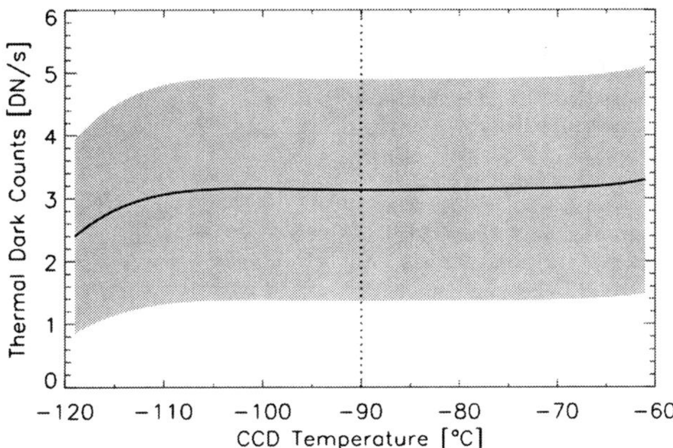

Figure 3 The thermal dark as a function of CCD temperature for one pixel on the MEGS A CCD. The shaded region shows the uncertainty in the thermal dark correction. The vertical dotted line indicates the flight operating temperature for the CCDs.

5.3.1. *Dark Count Correction*

There are two dark-rate components that need to be considered: one due to the bias in the electronics, and one due to the thermal noise of the CCDs. The electronic bias is determined by averaging the four "virtual" columns, being the first four columns in the image that read the charge amplifier signal (bias) prior to reading actual CCD pixels. In data processing, the electronic dark component is the average of the four virtual columns for each half of the CCD. Each half of the CCD is read out through different charge amplifier chains, so each CCD half has unique electronic-bias characteristics. The uncertainty of the electronic dark is the standard deviation of these pixels.

The second dark component is the thermal dark. This dark is due to the thermal noise of the CCDs and cannot be directly measured for each solar image. Instead, a fit to the thermal dark is determined by using dark images taken over a range of temperatures and repeated during several cycles as part of environmental thermal vacuum tests. The temperature behavior of each pixel is well characterized by a polynomial fit after subtracting the electronic dark in each image. Figure 3 shows the thermal dark component for one pixel in MEGS A as a function of temperature. The uncertainty in the thermal dark is determined by the goodness of the polynomial fit, and an example is given by the shaded region in Figure 3. Each CCD pixel has its own thermal dark function. The thermal dark and uncertainty are very small (1 DN is about two electrons) for these cold operating temperatures.

5.3.2. *Scattered Light*

Light scattering, primarily from the diffraction gratings but also possible from reflections off other surfaces (*e.g.* baffle edges, optics masks, limiting apertures), can be a significant uncertainty in spectrometers. With EVE, multiple proven techniques were used early in the design to reduce the amount of scattered light, such as using a holographically ruled grating, machining knife edges on all apertures and baffles, and inserting bandpass filters to reject out-of-band light from entering the optical cavity. As a result, there is very little measured scattered light, amounting to less that 1% of the total counts. The scattered light is therefore insignificant, and no correction for it is applied in data processing.

5.3.3. *Linearity*

An important feature of the MEGS CCDs is the high degree of linearity of their response, as expected for silicon devices. For a linear detector, the dark-corrected counts are a linear

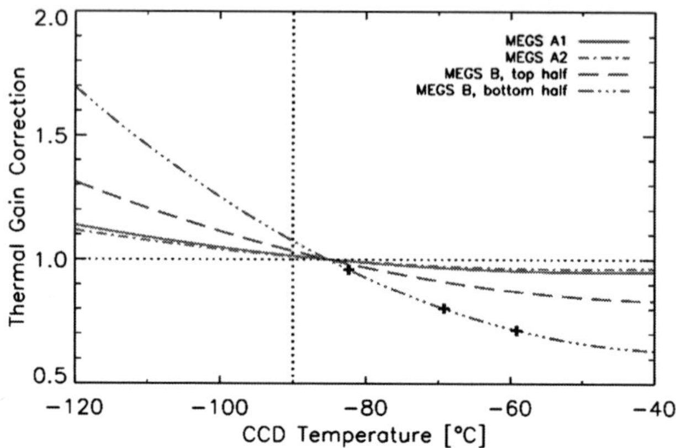

Figure 4 Thermal gain correction for each half of the CCDs as a function of CCD temperature. The gain correction is forced to be unity at −85°C. The vertical line at −90°C shows where we expect to operate on orbit. As an example, the data points used for the bottom half of MEGS B are shown as black crosses.

function of the input signal. The linearity of the MEGS CCDs is determined at SURF by measuring the count rate *versus* SURF beam current for a range of beam currents. By varying the SURF beam current by two orders of magnitude, we measure the linearity over the full dynamic range of the CCDs, which is limited by its 14-bit ADC. The residuals of a linear fit to the corrected count rate and beam current are much less than 1% for the full dynamic range of the CCDs. As the residuals are so small, there is no need to apply a linearity correction.

5.3.4. Gain Correction

There are two gain components that account for changes in the CCD responsivity due to temperature and readout mode of the electronics. Both of these gain components are relative, *i.e.* we normalize the MEGS images to a default state (CCD temperature at −85°C and default readout mode). The default temperature is chosen to be −85°C to balance the warmer temperatures at SURF and the colder temperatures expected on orbit. While these gain parameters are unity at the default state, the true gain of the electronics is about two electrons per DN.

The first gain component is the temperature gain. This is determined by taking data at SURF at multiple temperatures. The images are dark-corrected and normalized by the SURF beam current. These corrected counts are then divided by the corrected counts at −85°C to get the normalized counts for each temperature. Then a polynomial is fit to the inverse of the normalized counts to get the relative change as a function of CCD temperature for each half of the CCD. This polynomial is the multiplicative temperature-gain correction as seen in Figure 4. The coefficients of the polynomial fits for both the flight and rocket instruments are given in Table 2. The uncertainty in the thermal gain is 1%, which is the average residual of the fit.

The second gain component is associated with the readout mode of the electronics and is necessary as there are redundant charge amplifiers that have different gain factors. The default readout mode for the top CCD half is the left charge amplifier and the right amplifier is redundant. For the bottom CCD half, the default readout is the right charge amplifier and the left is redundant. The readout mode gain is calculated by averaging the ratio of co-added corrected images taken with the default readout mode to co-added corrected images taken with the redundant readout mode. For the default readout mode, this gain is unity. For MEGS A, the difference in the gain from unity for other readout modes is up to 7%; for

Table 2 MEGS CCD temperature gain coefficients. The temperature-gain function is a polynomial of the form $G = a + b(T + 85) + c(T + 85)^2$. Instances where no data are available to fit the gain are marked with an asterisk (*).

Instrument	Readout mode	a	b	c
Flight MEGS A	Left (top half)	1.028	3.363×10^{-3}	3.572×10^{-5}
	Right (top half)	1.046	3.801×10^{-3}	3.832×10^{-5}
	Left (bottom half)	1.068	3.869×10^{-3}	3.612×10^{-5}
	Right (bottom half)	1.044	3.285×10^{-3}	3.251×10^{-5}
Flight MEGS B	Left (top half)	0.904	4.422×10^{-3}	6.526×10^{-5}
	Right (top half)	0.842	2.674×10^{-3}	5.327×10^{-5}
	Left (bottom half)	0.774	9.350×10^{-3}	1.409×10^{-4}
	Right (bottom half)	0.814	1.046×10^{-3}	1.484×10^{-4}
Rocket MEGS A	Left (top half)	1.007	8.288×10^{-4}	8.826×10^{-6}
	Right (top half)*	1	0	0
	Left (bottom half)*	1	0	0
	Right (bottom half)	1.003	6.739×10^{-4}	7.550×10^{-6}
Rocket MEGS B	Left (top half)	0.977	1.171×10^{-3}	1.705×10^{-5}
	Right (top half)*	1	0	0
	Left (bottom half)*	1	0	0
	Right (bottom half)	0.940	3.474×10^{-4}	1.235×10^{-5}

MEGS B, it is a correction of up to 12%. The uncertainty for the default readout mode is zero. For other readout modes, it is 5%: the standard deviation of the gain ratio.

5.3.5. Invalid Pixel Masking

The invalid pixel mask [Mask in Equation (1)] is a combination of several binary masks that returns one if the pixel contains valid data and zero if the data are invalid. As the mask process simply identifies unusable pixels there is no uncertainty associated with it. The number of pixels eliminated by this mask is small (<1%), although it is expected to increase for large solar storms.

The first mask is to eliminate particle hits. MEGS A and B are sensitive to cosmic-ray hits, which can be seen on the CCDs as bright dots or streaks. Particle hits are not expected to persist from one image to the next so by comparing an image to the previous image, it is possible to identify pixels with counts that are much brighter. These are identified as particle hits and are masked out.

The next type of invalid pixels is saturated pixels. It is possible that during extremely large flares some pixels on MEGS A and B may become saturated. Including these pixels would underestimate the final solar irradiance so they are masked out and excluded from processing. The columns of virtual pixels used to determine the electronic dark as discussed in Section 5.3.1 are also masked off as they do not contain valid data.

Finally, pixels that behave differently from the other pixels are masked out. These pixels are identified using the flatfield images. Both MEGS A and B have flatfield lamps that uniformly illuminate the detector. The flatfield lamps, blue and violet LEDs, were used daily during SURF calibrations and will also be used for daily in-flight calibrations. While flatfield images are usually used to correct for the pixel-to-pixel variations in order to combine

Table 3 Uncertainty for SURF beam current. This relative uncertainty is dependent on the SURF beam configuration, such as the use of 380 MeV and 183 MeV beam energies, for each MEGS instrument. This uncertainty is calculated using Equation (13).

Instrument	380 MeV	183 MeV
MEGS A1	0.001%	0.4%
MEGS A2	0.003%	0.2%
MEGS B	0.1%	0.4%

the pixels of similar wavelengths (Chamberlin, Woods, and Eparvier, 2002), this is not necessary for MEGS A and B as the responsivity is found at SURF for each individual pixel. However, by comparing images taken with different flatfield-lamp intensity levels, we can identify pixels that do not behave the same way as the majority and define a flatfield mask to exclude these defective pixels.

5.4. SURF Responsivity

The next step in processing the SURF data is to take the corrected count rate and determine the SURF responsivity. The following sections detail this calculation as shown in Equation (2). Also included are the results of the FOV maps and higher-order corrections shown in Equations (3 – 6).

It is important to note that after normalizing the corrected count rate by the SURF beam current, multiple images are co-added to reduce the uncertainty. In general, four minutes of ten-second integrations are co-added for each point in the FOV map. This reduces the uncertainty by a factor of almost five. For the higher-order corrections, ten minutes of data are used to reduce the uncertainty by a factor of almost eight and are necessary as some of the higher energies produce very low count rates due to the current limitations of SURF.

5.4.1. SURF Beam Current

For the August 2007 SURF calibrations, beam currents were recorded every five seconds. For processing the SURF data, these measurements are linearly interpolated to the middle of the ten-second CCD integration.

The uncertainty in the SURF beam current is proportional to the rate of change of the current and to the uncertainty in the timing between the SURF control computer and EVE microprocessor:

$$\sigma_{I_{\mathrm{SURF}}} = \sigma_t \frac{\mathrm{d}I_{\mathrm{SURF}}}{\mathrm{d}t} \tag{13}$$

Higher beam currents, which are needed with the lower SURF beam energies, decay faster so there is more uncertainty in those current measurements. Furthermore, the beam current is recorded on a separate computer from EVE. While the EVE microprocessor and the SURF control computer are synchronized to a reference time, the relative time between the computers can and does vary. The uncertainty in the timing of the SURF beam current [σ_t] is estimated to be less than one second. Table 3 shows the uncertainties for both 380 MeV and 183 MeV beam energies for MEGS A1, A2, and B.

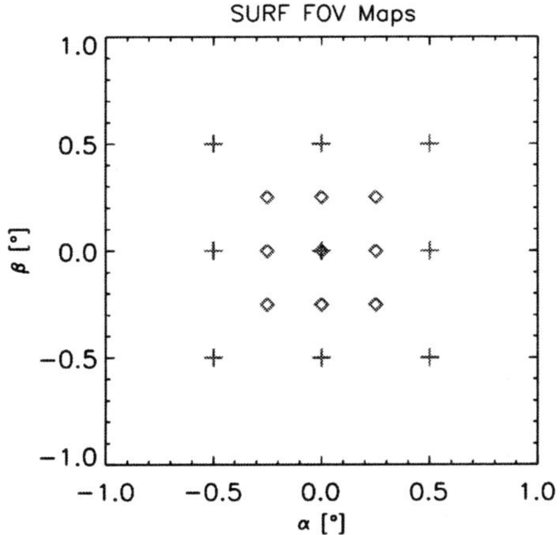

Figure 5 The primary (red diamonds) and secondary (blue crosses) 3 × 3 FOV maps. The dispersion direction is the α axis, while the cross-dispersion direction is the β axis. Both maps share a common center point.

5.4.2. SURF Flux

The photon flux provided by SURF is provided as flux *versus* wavelength as seen in Figure 1. As the calibrations are calculated on a pixel basis, the provided SURF flux is interpolated onto the wavelength scale. Aided by the ability of EVE to view the primary radiometric source of SURF, the relative uncertainty in the SURF flux is 1% (Arp *et al.*, 2000).

5.4.3. Field-of-View (FOV) Maps

An important characterization is determining the responsivity of the instrument as a function of FOV as the SURF beam is significantly narrower than the Sun. The SURF beam FOV is a few arcminutes wide in the horizontal plane but is less than one arcminute vertically. The SURF responsivity is calculated for each angle in the primary and secondary 3 × 3 FOV maps. The primary map is a 0.5° × 0.5° FOV range with 0.25° steps in each direction, with α defined as the dispersion direction and β the cross-dispersion direction. The secondary map is a 1.0° × 1.0° FOV range with 0.5° steps in each direction. The primary map allows for averaging over the solar FOV of about 30-arcminutes diameter and the secondary map allows for offsets from the ideal optical center. In Figure 5, the primary map angles are shown as the red diamonds, while the secondary map points are the blue crosses. Both maps share a common center so that different maps can be compared. While the primary map is sufficient if the pointing of EVE once on orbit is exactly centered, the secondary map allows EVE to make valid science measurements even if the pointing is off-center.

Figures 6, 7, and 8 show the SURF responsivity for the primary FOV map for MEGS A1, A2, and B. For each figure, the top plot shows the responsivity for each of the nine FOV angles. The red curve is the center. The bottom plot shows the relative difference between each FOV angle and the center. Overall, the differences due to FOV are less than 20%. As mentioned in Section 5.2.2, the wavelength scale for MEGS A1 and A2 has higher uncertainty than for MEGS B; therefore, there may be wavelength shifts that have not been accounted for in this instrument that would contribute to these larger than expected uncertainties. MEGS B, where the wavelength scale is well understood, has even smaller FOV differences. Once on orbit, in-flight FOV maps will be obtained to validate these pre-flight maps.

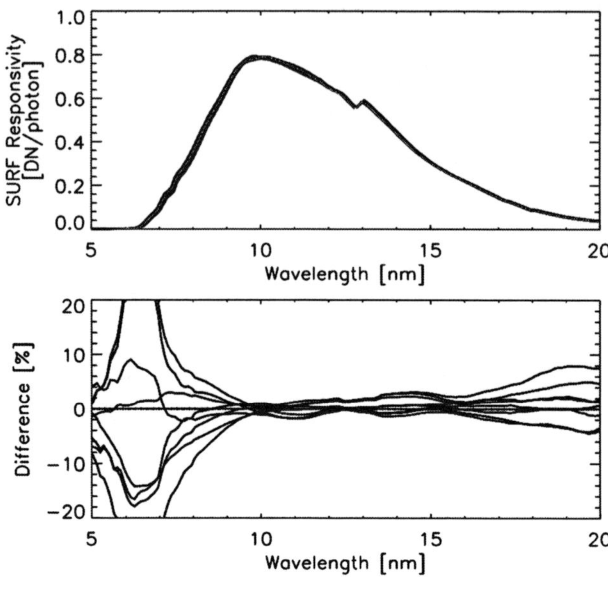

Figure 6 Primary FOV map results for MEGS A1. The top panel shows the responsivity for each of the nine FOV points. The red curve is the center point. The bottom panel shows the relative difference between each FOV point and the center point.

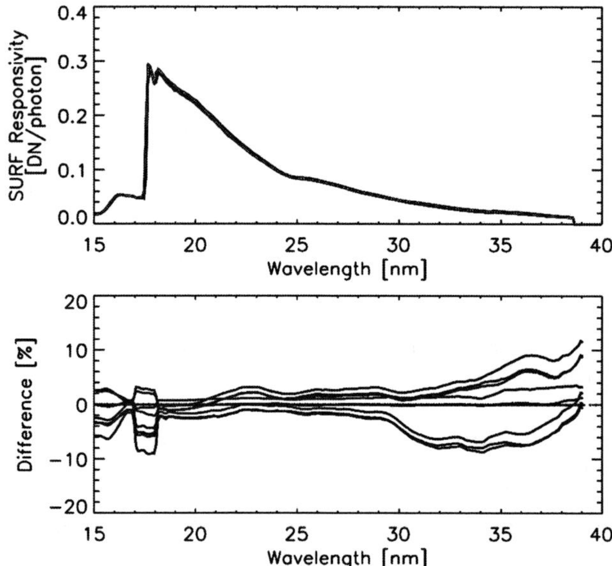

Figure 7 Primary FOV map results for MEGS A2. The top panel shows the responsivity for each of the nine FOV points. The red curve is the center point. The bottom panel shows the relative difference between each FOV point and the center point.

5.4.4. Higher-Order Correction

Correcting for contributions from higher orders is critical for meeting our uncertainty requirements. As discussed in Section 5.1, higher orders for SURF calibrations are separated by using multiple beam energies. Figures 9 and 10 show the order-sorting results for MEGS A1 and A2. We did not find any higher-order contributions for MEGS B. These results from MEGS A1, A2, and B all agree with what we expect if we calculate the higher-order contributions using the theoretical grating efficiencies, filter transmissions (A1 and A2 only), and CCD responsivity. Given these results, the optical design was successful in

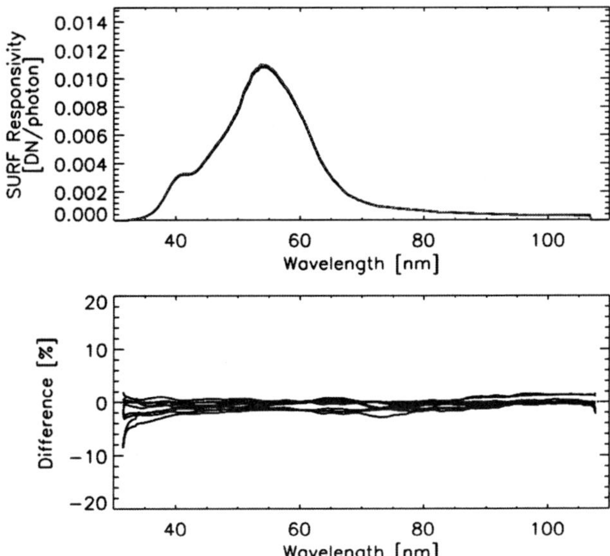

Figure 8 Primary FOV map results for MEGS B. The top panel shows the responsivity for each of the nine FOV points. The red curve is the center point. The bottom panel shows the relative difference between each FOV point and the center point.

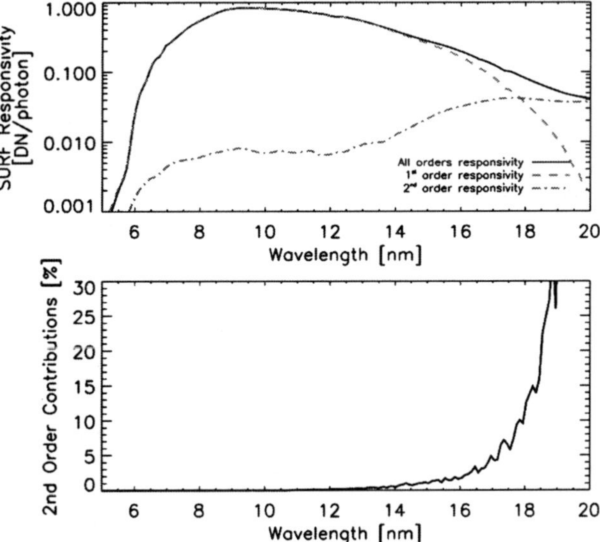

Figure 9 Results of the multiple-beam-energy method for MEGS A1. The top panel shows the SURF responsivity without any correction (black), the true first-order responsivity (red) and the second-order responsivity (blue). The bottom panel shows the expected contribution to the first-order spectrum from the second-order spectrum at half the wavelength. MEGS A1 is intended for solar measurements below 17 nm.

eliminating, or at least significantly reducing, the higher orders in these instruments using holographically ruled gratings that suppress even orders and in MEGS B using the dual orthogonal gratings.

The top panels of Figures 9 and 10 show the SURF responsivity without any correction (black), the true first-order responsivity (red), and the second-order responsivity (blue). These were determined using two beam energies. However, multiple combinations of beam energies were tested to verify the results.

Figure 10 Results of the multiple-beam-energy method for MEGS A2. The top panel shows the SURF responsivity without any correction (black), the true first-order responsivity (red) and the second-order responsivity (blue). The bottom panel shows the expected contribution to the first-order spectrum from the second-order spectrum at half the wavelength.

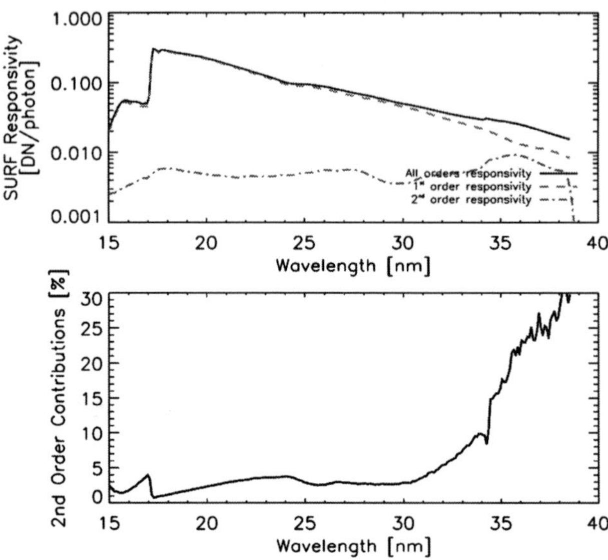

The bottom panels of Figures 9 and 10 show the expected contribution to the first-order spectrum from second order at half the wavelength:

$$f_{\mathrm{2nd}} = \frac{\frac{1}{2} R_2(\frac{\lambda}{2})}{R_1(\lambda)} \times 100\% \qquad (14)$$

For MEGS A1, there is very little contribution below 15 nm. For MEGS A2, the higher-order contribution increases sharply near 34 nm, which is the second-order rise due to the Al filter edge near 17 nm.

5.5. Flight Responsivity

The final step in calibrating MEGS A and B is calculating the flight responsivity, defined in Equation (6). The irradiance from a solar MEGS image is the corrected count rate for that image divided by the flight responsivity, being the SURF response averaged over the solar disk field-of-view.

Figure 11 shows the flight responsivity for MEGS A1 and A2 (top) and the associated uncertainty (bottom), while Figure 12 shows the flight responsivity (top) and uncertainty (bottom) for MEGS B. The uncertainties are for individual pixels. For the final irradiance spectrum, pixels along the slit are added together into 0.02 nm wavelength bins, so the irradiance uncertainty is reduced when assuming random errors. Figures 13 and 14 show the ratio of the flight responsivity for the rocket MEGS instruments to the flight MEGS instruments. Both sets of MEGS instruments have similar responsivity. The MEGS A results might indicate a wavelength shift between these different calibrations; these possible wavelength shifts are expected to be resolved once solar measurements have been made. For MEGS B, the differences in responsivity are mostly related to different CCDs selected for the flight and rocket instruments.

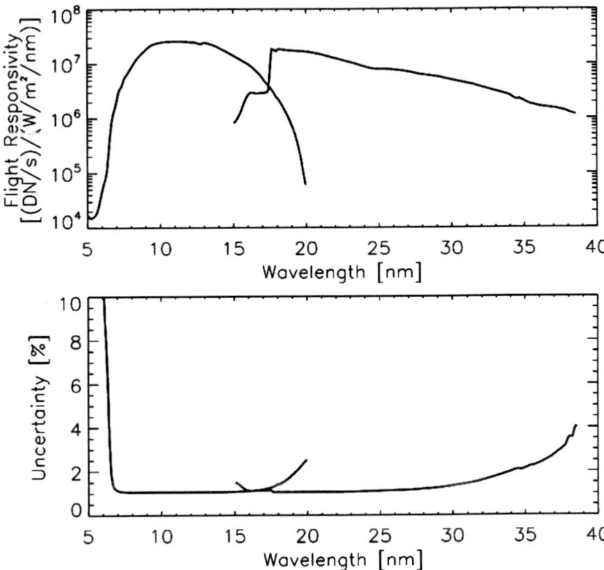

Figure 11 Flight responsivity for MEGS A1 and A2 (top) and the associated relative uncertainty (bottom).

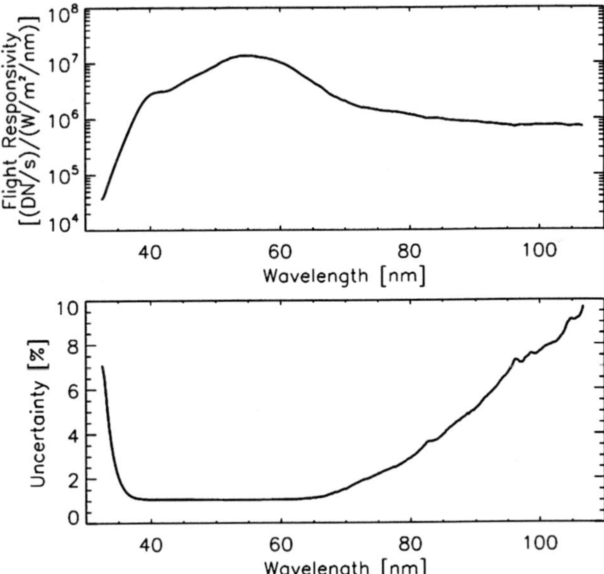

Figure 12 Flight responsivity for MEGS B (top) and the associated relative uncertainty (bottom).

6. MEGS P Calibration

The MEGS P photometer is an integrated part of the MEGS B spectrograph and consists of a Si photodiode and Acton 122 nm filter located after the first MEGS B grating and positioned to measure the bright H I Lyman-α emission at 121.6 nm. The MEGS P photodiode is not at the focal plane of this concave grating due to space limitations within the MEGS B housing. By not being at the focal plane, the spectrum across the diode is a fairly broadband of about 10 nm. Even so, the Lyman-α emission still contributes about 98% of the total solar signal

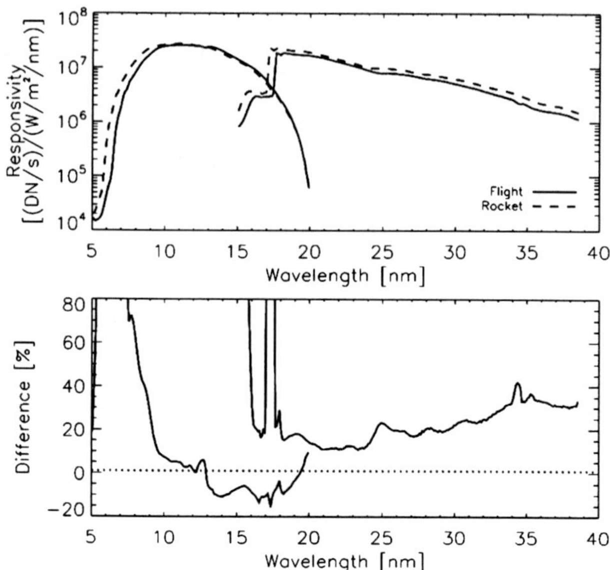

Figure 13 Responsivity for MEGS A1 and A2 (top) for flight and rocket instruments. We believe the differences are mostly due to a wavelength shift in the flight instrument. The bottom panel shows the percent differences between the flight responsivity for the rocket MEGS A channel and the flight MEGS A channel.

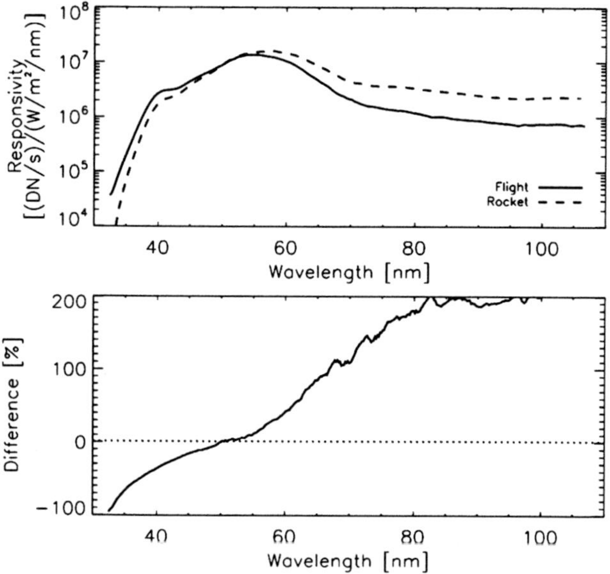

Figure 14 Responsivity for MEGS B (top) for flight and rocket instruments. We believe the differences are mostly due to variations in the CCDs. The bottom panel shows the percent differences between the flight responsivity for the rocket MEGS B channel and the flight MEGS B channel.

on the MEGS P photodiode. More details on the MEGS P instrument are given by Woods *et al.* (2010).

The MEGS P calibration uses a set of algorithms that draws on the heritage of many previous instruments using similar Si photodiodes with a bandpass filter as discussed in Section 3. This is also similar to the algorithms discussed by Didkovsky *et al.* (2010) for the ESP instrument on EVE.

6.1. MEGS P Algorithm

The responsivity algorithm for MEGS P is a simplified version of that for MEGS A and B. As with MEGS A and B, the raw signal is dark-corrected to get the corrected count rate $[C']$:

$$C'(t) = \frac{C(t) - D(t)}{\Delta t} \tag{15}$$

where C is the data number [DN] from each integration from the photodiode electrometer that consists of a current amplifier and voltage-to-frequency converter [VFC]. The dark (background counts) is given by D, which is about 160 DN second^{-1} during SURF calibrations and ground testing. The dark correction is expected to be much more complicated in-flight due to radiation background effects, so an identical Si photodiode, but fully enclosed in an Al frame, is included as part of MEGS P to be a proxy of the background signal for the Lyman-α photodiode. The integration time $[\Delta t]$ is 0.25 seconds.

The SURF responsivity is then:

$$R_{\mathrm{SURF}}(E_{\mathrm{beam}}, \mathrm{filter}, \alpha, \beta) = \frac{C'(t, E_{\mathrm{beam}}, \mathrm{filter}, \alpha, \beta)}{I_{\mathrm{SURF}}(t) F_{\mathrm{SURF}}(E_{\mathrm{beam}}, \alpha, \beta)} \tag{16}$$

As with MEGS A and B, the corrected count rate is divided by the SURF beam current $[I_{\mathrm{SURF}}]$ and the SURF beam flux $[F_{\mathrm{SURF}}]$. The slit area of the MEGS B entrance slit is not included in the MEGS P responsivity as it is the same in the SURF calibration and solar observations.

The MEGS P responsivity needs to be corrected for the differences in the bandpass between SURF and solar measurements. As MEGS P is not in focus, it has a fairly broad bandpass and its responsivity is dependent on the spectrum being measured. The bandpass-corrected responsivity is given by:

$$R_{\mathrm{P}}(\mathrm{filter}, \alpha, \beta) = R_{\mathrm{SURF}}(E_{\mathrm{beam}}, \mathrm{filter}, \alpha, \beta) \frac{\Delta\lambda_{\mathrm{Sun}}}{\Delta\lambda_{\mathrm{SURF}}(E_{\mathrm{beam}})} \tag{17}$$

where $\Delta\lambda_{\mathrm{Sun}}$ is the bandpass of MEGS P for a solar spectrum, while $\Delta\lambda_{\mathrm{SURF}}$ is the bandpass using SURF fluxes. These bandpasses are defined in Equations (18) and (19).

$$\Delta\lambda_{\mathrm{SURF}}(E_{\mathrm{beam}}) = \int_{\mathrm{all}\ \lambda} \frac{F_{\mathrm{SURF}}(\lambda, E_{\mathrm{beam}})}{F_{\mathrm{SURF}}(121.6\ \mathrm{nm}, E_{\mathrm{beam}})} P_{\mathrm{G}}(\lambda) P_{\mathrm{F}}(\lambda) P_{\mathrm{D}}(\lambda)\, \mathrm{d}\lambda \tag{18}$$

$$\Delta\lambda_{\mathrm{Sun}} = \int_{\mathrm{all}\ \lambda} \frac{F_{\mathrm{Sun}}(\lambda)}{F_{\mathrm{Sun}}(121.6\ \mathrm{nm})} P_{\mathrm{G}}(\lambda) P_{\mathrm{F}}(\lambda) P_{\mathrm{D}}(\lambda)\, \mathrm{d}\lambda \tag{19}$$

The bandpass is calculated by convolving the normalized spectrum, SURF flux, or the solar spectral irradiance, with the spectral profiles of the MEGS P optical components, which are the grating $[P_{\mathrm{G}}]$, filter $[P_{\mathrm{F}}]$, and diode $[P_{\mathrm{D}}]$. These profiles are not measured for MEGS P, but are estimated based on raytracing for the grating profile and from measurements on similar Acton filters and International Radiations Detectors (IRD) Si photodiodes. The detector profile is flat across the MEGS P range, but the filter and grating profile are strongly peaked near 121.6 nm. These spectral profiles are normalized by their peak efficiencies, and the spectrum is normalized to its value at 121.6 nm. Table 4 lists the spectral bandpasses for the different SURF beams and solar observations. As the SURF flux is a broad continuum, $\Delta\lambda_{\mathrm{SURF}}$ is more representative of the full width of the MEGS P bandpass (≈10 nm). The solar spectrum is dominated by the H I Lyman-α emission, so the $\Delta\lambda_{\mathrm{Sun}}$ is narrower (≈1 nm).

Table 4 MEGS P spectral bandpass widths.

Irradiance source	Bandpass width ($\Delta\lambda$)
SURF 183 MeV	10.641 nm
SURF 380 MeV	10.618 nm
Sun – Solar-Cycle Min	1.023 nm
Sun – Solar-Cycle Max	1.016 nm
Sun – average	1.020 nm

6.2. Uncertainty Algorithm

The uncertainty for R_{P} is derived from Equation (15) and is as follows.

$$\sigma_{R_{\mathrm{P}}}^2 = R_{\mathrm{P}}^2\left[\frac{\sigma_C^2+\sigma_D^2}{(C-D)^2}+\left(\frac{\sigma_{\Delta t}}{\Delta t}\right)^2+\left(\frac{\sigma_{F_{\mathrm{SURF}}}}{F_{\mathrm{SURF}}}\right)^2+\left(\frac{\sigma_{I_{\mathrm{SURF}}}}{I_{\mathrm{SURF}}}\right)^2+\left(\frac{\sigma_{\Delta\lambda_{\mathrm{SURF}}}}{\Delta\lambda_{\mathrm{SURF}}}\right)^2+\left(\frac{\sigma_{\Delta\lambda_{\mathrm{Sun}}}}{\Delta\lambda_{\mathrm{Sun}}}\right)^2\right] \tag{20}$$

The relative uncertainties for C, D, F_{SURF}, and I_{SURF} are each 1% or less. There is an unknown amount of uncertainty for the spectral bandpass (profile) at this time, but this could be the largest uncertainty for the MEGS P calibration. For now, the uncertainty of each spectral bandpass is assumed to be 10%. Combining these as in Equation (16), the relative uncertainty for the MEGS P responsivity [R_{P}] is estimated to be 14%.

6.3. Calibration Results

SURF calibration results for MEGS P are obtained simultaneously while doing the MEGS B calibrations, which primarily use 140 MeV SURF beams. In addition, special MEGS P calibrations at 183 MeV and at 380 MeV are done at the center point ($\alpha = 0$, $\beta = 0$) to minimize the MEGS P responsivity uncertainty. Using SURF beam energy of 183 MeV and a beam current of 300 mA, the quantity C' in Equation (15) is about 150 DN second^{-1}. The center point R_{P} is 1421 DN (W m^{-2})$^{-1}$ for the flight EVE and is 1922 DN (W m^{-2})$^{-1}$ for the protoflight (rocket) EVE, each having relative uncertainty of about 14%. The largest contribution to the uncertainty is the bandpass uncertainty. A future improvement for MEGS P calibrations is to measure the bandpass directly instead of estimating the bandpass by analysis.

The variation of the responsivity across the solar FOV (0.5°) is less than 1% and has a sensitivity of about 0.02% per arcminute. With solar pointing from SDO expected to vary much less than one arcminute, a FOV correction is not currently included in the MEGS P solar-irradiance algorithm.

6.4. Application for MEGS In-flight Calibrations

In addition to the scientific reasons for measuring Lyman-α, another purpose for the MEGS P instrument is to be a calibration proxy for other H and He emission lines in the MEGS A and B spectra. In particular, if MEGS P has little degradation, which is expected to be less than 1% per year, then the MEGS P can serve as reference proxy for degradation of MEGS A and B. From TIMED/SEE observations, the relationship of the H I Lyman-α (121.6 nm) emission to H I emission lines and continuum as well as He I and He II emission

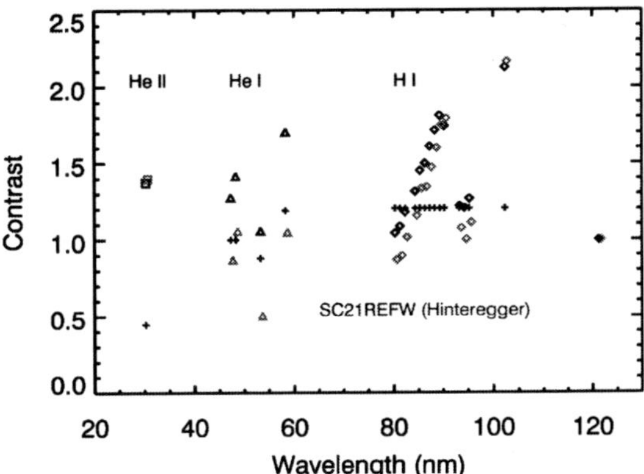

Figure 15 Contrast ratios for H I, He I, and He II emissions relative to H I Lyman-α emission. The black triangles are the short-term contrast ratios, and the green diamonds are the long-term contrast ratios, both derived using TIMED/SEE Level 3 data. The red plus symbols are the contrast ratios in the SC21REFW solar reference spectrum/model (Hinteregger, Fukui, and Gilson, 1981).

lines in the EUV are all accurately known. The contrast ratio, being the ratio of variability between one emission and a proxy, is expected to be the same for short-term variability (<81 days) and long-term variability (>81 days) if the emission and the proxy are from the same solar atmospheric layer (Woods *et al.*, 2000). This expectation was established for the solar far ultraviolet irradiances from *Upper Atmosphere Research Satellite* (UARS)/*SOLar STellar Irradiance Comparison Experiment* (SOLSTICE).

To see if this relationship holds for the EUV wavelengths, the comparison of short-term and long-term variability is shown in Figure 15 using SEE data. As expected, the short-term and long-term contrast ratios are about the same for the H I, He I, and He II emissions; however, there are larger differences for the He I emissions. The uncertainty in the SEE degradation trends in the 45 to 70 nm range appear to be the cause of this difference for the long-term contrast ratios for the He I emissions. The short-term contrast ratios are much more accurate than the long-term contrast ratios because instrument degradation over 81 days is much less of a concern than that over six years. Therefore, the short-term contrast ratios, as shown in Figure 15 and listed in Table 5, will be used with the MEGS P solar Lyman-α measurements to help establish the degradation trends of MEGS A and B. These contrast ratios are derived using the SEE Level 3 (Version 10) data given in 1-nm intervals. The underflight rocket EVE calibrations provide the ultimate long-term accuracy, so the MEGS P proxy estimates for the H and He emissions between 25 and 103 nm are intended only for the time between the calibration rocket flights.

An interesting result from generating these contrast ratios for the H and He emissions in the EUV relative to the solar Lyman-α is that the contrast ratios are not constant values as shown for the Hinteregger, Fukui, and Gilson (1981) results (proxy model). The SEE data are in 1-nm intervals, so contamination of a few weaker lines is only slightly affecting these results, but there are systematic trends, notably in the H I continuum, that likely indicate intrinsic solar-variability relationships. Existing solar empirical models that use the Hinteregger, Fukui, and Gilson (1981) results should be updated with the TIMED/SEE results. Because these relationships are based on short-term variations, we can independently verify them during the first year of the SDO mission.

7. MEGS SAM Calibration

The *Solar Aspect Monitor* (SAM) instrument is a simple pinhole-camera optical design that provides a low-resolution (≈15 arcseconds pixel^{-1}) solar image onto an unused part of the

Table 5 Contrast ratios for H I, He I, and He II emissions relative to MEGS P Lyman-α.

Emission	Wavelength (nm)	Short-term contrast ratio
H I	80.5	1.045
H I	81.5	1.088
H I	82.5	1.180
H I	84.5	1.315
H I	85.5	1.451
H I	86.5	1.500
H I	87.5	1.610
H I	88.5	1.717
H I	89.5	1.813
H I	90.5	1.743
H I Ly-ε	93.8	1.221
H I Ly-δ	95.0	1.239
H I Ly-β	102.6	2.126

Emission	Wavelength (nm)	Short-term contrast ratio
He I	47.5	1.277
He I	48.5	1.416
He I	53.7	1.059
He I	58.4	1.706
He II	30.4	1.373

MEGS A CCD. SAM has two operating modes depending on which filter is used. The first uses a visible-light filter for alignment purposes for both ground and in-flight calibrations to produce better than one-arcminute alignment (pointing) information for the MEGS spectrographs. The second mode uses an X-ray filter, equivalent to the ESP zeroth-order filter consisting of mostly Al and Ti, to provide solar images and solar spectra for the 0.1 to 7 nm range. The X-ray spectrum is possible because the filter and pinhole size are designed so that only single photon events can be detected in a ten-second integration in any given pixel. For the Si-based CCD, each X-ray photon produces a few hundred electron-hole pairs within a couple pixels. The number of electrons is expected to be the energy of the photon divided by the Si band-gap energy (3.65 eV). This technique is advantageous because not only is an image produced but also the energy of each photon event, and therefore the photon wavelength, can be determined. The current plan for SAM data processing with the X-ray filter is to produce a spectrum in 1-nm intervals between 0.1 and 7 nm with a cadence of one minute and an X-ray image with a cadence of five minutes. More information about SAM and MEGS A CCD is provided by Woods *et al.* (2010).

The solar aspect mode (visible-light filter) will be used only once a day to verify the pointing of the MEGS spectrographs, while the X-ray filter will be used most of the time as part of SAM's normal science observations. The visible solar images are dominated by diffraction, as the pinhole size has only a 26 micron diameter, and thus are more useful for solar position information than for scientific research. The solar diameter and diffraction pattern are estimated to provide a Sun-center position with an accuracy to better than two pixels (0.5 arcminute). The diffraction is not an issue with X-ray photons so the X-ray solar images are useful for scientific research. Active regions are expected to dominate the X-ray images. That is, these images do not have a uniform disk, so solar position from an X-ray image is more challenging and thus they have less accuracy than the visible images.

The following discussion for SAM concerns only the calibrations and processing for the X-ray filter. While the photon-detection technique is known to work in the X-rays for Si detectors, there is no heritage for processing solar X-ray data from LASP's previous satellite instrument programs. The prototype EVE instrument did fly on a sounding-rocket flight in

April 2008, so there are some limited solar data during solar-cycle minimum conditions to validate the SAM processing algorithms prior to the SDO launch.

7.1. Responsivity Algorithm

The responsivity algorithm for SAM is similar to that of MEGS A, B, and P and is given by:

$$R_{\mathrm{SURF}}(i,j,\lambda)=\frac{\sum_{\text{all images}}\frac{N_{\mathrm{photons}}(i,j,t,\lambda,E_{\mathrm{beam}})}{I_{\mathrm{SURF}}(t)}}{F_{\mathrm{SURF}}(\lambda,E_{\mathrm{beam}})\Delta t A_{\mathrm{slit}}\Delta\lambda} \tag{21}$$

Here, N_{photons} is the number of photon events at a wavelength λ within a band $\Delta\lambda$. To improve statistics, events from many images are first normalized by the SURF beam current and then added together. In order to compare the number of photon events measured with the SURF signal, the SURF flux is multiplied by the integration time [Δt]. The slit area [A_{slit}] is the area of the SAM entrance slit (5.31×10^{-10} cm^2). The responsivity [R_{SURF}] consists of the transmission of the Ti – Al – C filter and the quantum throughput [QT] of the CCD. With the CCD QT being almost 100%, the responsivity result is primarily the spectral shape of the filter's transmission.

Each photon event is assumed to be an isolated event. Double or more concurrent and co-located events, which are intended by design to be less than 10% of the time for solar maximum conditions, will effectively be treated incorrectly as a single photon event but with more energy (shorter wavelength). The algorithm to detect a photon event starts by using a dark- and particle hit-corrected MEGS A image, similar to Equation (1). Then the algorithm identifies each pixel within the SAM image area that has a signal of greater than ten DN (≈20 electrons). The surrounding pixels with a signal greater than two DN are added together to get the energy of the photon event in DN. This number is then converted to energy units by multiplying by the gain (G_{SAM}, about two electrons per DN) and the Si band-gap energy (E_{Si}, 3.65 eV per electron-hole pair). If there are two photon events adjacent to each other, the algorithm splits the signal of the intermediate pixel between the two events. Events that are larger than 3×3 pixels are assumed to contain multiple photon events and are not included in the spectrum, although the final spectrum is adjusted to reflect the estimated number of events lost in this manner.

Binning by wavelength does not work well for SAM spectra, as a single DN may represent multiple wavelength bins at longer wavelengths or a single wavelength bin may include several DN at shorter wavelengths. Therefore, the SAM spectrum is accumulated in energy bins, with each bin equivalent to one DN. In order to create an image, the pixel location of photons events is recorded and these photon events are accumulated into an image over several integrations. With a long enough accumulation period (several minutes to an hour or more), one could separate the X-ray images into multiple wavelength bands within the 0.1 to 7 nm range.

Although the sounding-rocket flight demonstrated the effectiveness of this single-photon-event method for the use with solar data, the SURF calibrations could not properly count enough single photon events for this method to provide accurate results. This shortcoming for SURF calibrations is due to the SURF beam's small size, which results in the image being spread over only a couple of pixels (*versus* hundreds of pixels for a solar image). As a result, all of the photon events are concentrated in a few pixels, and multiple events are mistakenly counted as higher-energy single events.

Therefore, in order to properly calibrate SAM, it was necessary to instead use Equations (22 – 24). The measured signal [S_{measured}] in eV second^{-1} is given by:

$$S_{\mathrm{measured}}(t, E_{\mathrm{beam}}) = \sum_{i,j} \frac{C_{\mathrm{image}}(i, j, t, E_{\mathrm{beam}})}{\Delta t} G_{\mathrm{SAM}} E_{\mathrm{Si}} \tag{22}$$

where C_{image} is the number of DN in each pixel, G_{SAM} the gain to convert DN to electrons, and E_{Si} is energy conversion from electrons to eV (3.65 eV for silicon). The predicted signal [$S_{\mathrm{predicted}}$] is:

$$S_{\mathrm{predicted}}(t, E_{\mathrm{beam}}) = \int_0^\infty R_{\mathrm{SAM}}(\lambda) F_{\mathrm{SURF}}(\lambda, E_{\mathrm{beam}}) I_{\mathrm{SURF}}(t) A_{\mathrm{slit}}\, \mathrm{d}\lambda \tag{23}$$

The responsivity of SAM [R_{SAM}] is the transmission model of the Ti – Al – C filter and the CCD detector based on the atomic parameters from Henke, Gullikson, and Davis (1993). If the SAM responsivity is correct, we expect the ratio of S_{measured} to $S_{\mathrm{predicted}}$ to be unity. This ratio represents a scaling factor [f_{SAM}] on how to correct the SAM responsivity.

$$f_{\mathrm{SAM}}(E_{\mathrm{beam}}) = \frac{S_{\mathrm{measured}}(t, E_{\mathrm{beam}})}{S_{\mathrm{predicted}}(t, E_{\mathrm{beam}})} \tag{24}$$

By using multiple beam energies (331, 361, 380, and 408 MeV), the different scaling factors from each beam energy indicate roughly how the responsivity needs to be adjusted in wavelength. The filter thicknesses were adjusted in the transmission model until f_{SAM} was near unity.

The value of the SAM gain factor [G_{SAM}] was determined by using several methods. The primary method involved comparing the solar spectrum created from the SAM data when assuming a gain value of two with the Whole Heliosphere Interval (WHI) solar spectrum generated by SORCE/XPS (Woods *et al.*, 2009). A distinct peak in irradiance occurred in both spectra, but the SAM spectrum had to be shifted to higher energy by a factor of about 1.3 in order for these peaks to occur at the same wavelength. This finding suggests that the SAM gain factor should have a value of 2.67 e$^-$ DN^{-1}. This gain factor, however, only applies to the single-photon-event method used for solar analysis, and does not apply to the signal integration method because it only registers signals greater than two DN. To get the correct gain for SURF calibrations, the signal was extrapolated to determine what the signal would theoretically be if the limit were set at zero DN. Doing this, the signal would be 7% greater, so the gain value used for SURF calibrations, should be 7% less than for the solar calibrations, yielding a value of 2.47 e$^-$ DN^{-1}.

The spectral-binning method takes advantage of the presence of several solar spectrum lines from the MEGS A2 slit that appear next to the SAM image. The wavelengths of these spectral lines are known, so therefore the energies of the photon events in these lines are also known. Therefore, the gain can be directly derived from the ratio of the expected photon signal from a specific wavelength to the measured photon signal. This method generates a gain value of 3.1 $\pm$ 0.44 e$^-$ DN^{-1}. On account of the large uncertainty of this method, we do not use this value for the gain (using instead the 2.47 e$^-$ DN^{-1} from the other method); however, this method does provide limited validation for the other SAM gain values we calculated.

In order to reflect the actual resolution possible with SAM, we smoothed the responsivity model based on the resolution calculated by Equation (25) at each energy before applying the model to the solar data. Again, G_{SAM} is the gain factor, while D_{S} is the dark correction for the CCD image, and F is the Fano factor, which has a value of 0.115 for silicon. hc/λ converts wavelength of the photon into energy, while $1/E_{\mathrm{Si}}$ converts energy to number of electron-hole pairs. We have

$$\Delta E = 8.61\sqrt{\left(G_{\mathrm{SAM}}\sqrt{D_{\mathrm{S}}}\right)^2 + \left(\frac{hc}{\lambda} \times \frac{F}{E_{\mathrm{Si}}}\right)} \tag{25}$$

While the SAM X-ray algorithms are developed, they require significantly more validation. This SAM technique for X-ray spectra and images is at an experimental stage. Much can be learned from a longer time series of solar observations than what is possible over a single sounding rocket flight. The SAM results are not a required part of the EVE results, as the primary source for the 0.1 to 7 nm band is the ESP zeroth-order channel that has the same spectral bandpass as SAM. If the SAM algorithm is found to be properly providing results consistent with the ESP results, then the longer-term degradation for SAM (filter transmission and detector QT) can be tracked with the ESP results. Of particular concern with SAM is the double or more photon events that can contaminate the SAM spectra and images, and how this contamination might evolve with solar activity (cycle minimum to maximum and during flares) and with variation in the number of active regions on the solar disk.

7.2. Uncertainty Algorithm

The uncertainty for the SAM responsivity [R_{SAM}] is derived from Equations (22 – 24) and is as follows:

$$\sigma^2_{R_{\mathrm{SAM}}} = R^2_{\mathrm{SAM}}\left[\left(\frac{\sigma_{C_{\mathrm{image}}}}{C_{\mathrm{image}}}\right)^2 + \left(\frac{\sigma_{\Delta t}}{\Delta t}\right)^2 + \left(\frac{\sigma_{G_{\mathrm{SAM}}}}{G_{\mathrm{SAM}}}\right)^2 + \left(\frac{\sigma_{F_{\mathrm{SURF}}}}{F_{\mathrm{SURF}}}\right)^2 + \left(\frac{\sigma_{I_{\mathrm{SURF}}}}{I_{\mathrm{SURF}}}\right)^2 + \left(\frac{\sigma_{f_{\mathrm{SAM}}}}{f_{\mathrm{SAM}}}\right)^2\right] \tag{26}$$

As SAM counts photons, the uncertainty for C_{image} is estimated as the square root of C_{image}. The uncertainty for the G_{SAM} is about 10%. The relative uncertainties for SURF flux [F_{SURF}] and beam current [I_{SURF}] are both 1.0%. The uncertainty of the scaling factor [σ_f] is 0.084. The combination of these uncertainties yields a relative uncertainty of 16% for SAM responsivity.

7.3. Calibration Results

As mentioned in Section 7.1, the SURF calibrations for SAM did not provide reliable photon events for using Equation (19). Therefore, the SAM responsivity is based on the integrated signal from the SURF calibrations as described by Equations (22 – 24). In addition to the data from the calibrations performed at SURF in August 2007, data from a previous SURF calibration in February 2007 and well as rocket calibrations from October 2007 and January 2009 contributed to the calibration process. The data were analyzed using Equation (22) in order to calculate a scaling factor [f_{SAM}]. The responsivity model was then adjusted by altering the filter thicknesses to achieve values for f_{SAM} close to unity. As shown in Table 6, the adjusted responsivity model generated f_{SAM} values that are within 20% of the desired value. The mean final scaling factor is 0.97 with a standard deviation of 0.084. The consistency of the results between the 2007 and 2009 calibrations indicates that SAM did

Table 6 Scaling factor results for SAM calibration sets. Results are based on the calibrated responsivity model and represent different beam energies. Instances where lower currents had to be used due to lack of data at a higher current are marked with an asterisk (*).

Calibration set	Beam energy (MeV)	Beam current (mA)	Original scaling factor	Final scaling factor
Flight instrument (February 2007)	408	1.6×10^{-4}	0.7602	1.0747
	380	1.8×10^{-4}	0.6934	1.0389
	361	2.0×10^{-4}	0.6185	0.9615
	331*	1.0×10^{-4}	0.4769	0.7819
Flight instrument (August 2007)	408	1.6×10^{-4}	0.7742	1.0945
	380	1.8×10^{-4}	0.6796	1.0181
	361	2.0×10^{-4}	0.6289	0.9777
	331*	1.0×10^{-4}	0.5416	0.8852
Rocket pre-flight (October 2007)	408	2.0×10^{-4}	0.7545	1.0667
	380	2.0×10^{-4}	0.6519	0.9767
	361	2.0×10^{-4}	0.6256	0.9726
	331	2.0×10^{-4}	0.5511	0.9036
Rocket post-flight (January 2009)	408*	2.0×10^{-5}	0.6844	0.9675
	380	2.0×10^{-4}	0.6824	1.0224
	361*	2.0×10^{-5}	0.5556	0.8637
	331	2.0×10^{-4}	0.5616	0.9208

not degrade much, if at all, over two years. The SURF calibration results from the flight and rocket instruments also agree remarkably well with each other. Using Equation (26), the average uncertainty of the SAM responsivity was found to be 7.5% for the main SAM wavelength range (0.1 – 7 nm).

The initial values for the SAM responsivity calculation were C thickness of 40 nm, Al thickness of 100 nm, Ti thickness of 300 nm, SiO thickness of 7 nm, and Si detector thickness of five microns. The scaling factors have a systematic variation of 0.44 for 331 MeV beam to 0.60 for 408 MeV beam. After the filter thicknesses were adjusted so that the scaling factors were close to unity, the thicknesses were 80 nm for C, 200 nm for Al, and 320 nm for Ti. The thickness of the Si detector remained five microns. The calculated responsivity with these parameters is consistent for both the flight and rocket SAM instruments. Figure 16 shows the SAM responsivity as determined from the SURF calibrations. The SAM responsivity, as expected, has the appearance of the Ti – Al – C filter transmission profile. The Ti part of the filter is expected to have a dip in transmission near 2.5 nm and this dip is used as a wavelength calibration for SAM calibrations. Triplett *et al.* (2007) provide additional information about filter calibrations performed prior to the SURF calibrations.

8. Solar EUV Measurements on 14 April 2008

The long-term degradation correction for the flight EVE depends on cross-calibrations with the prototype EVE flown on annual sounding rocket flights. This prototype experiment is calibrated using the same techniques and algorithms described in this paper, and the calibration rocket measurement provides a new reference that can be used to adjust the flight

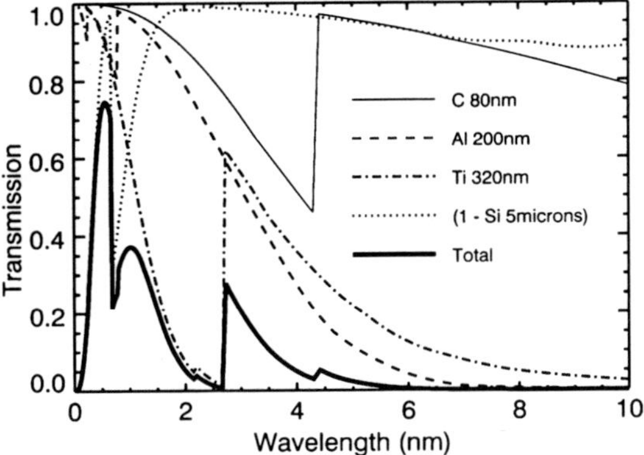

Figure 16 SAM Transmission. This transmission includes the transmission of the foil filter and absorption of the Si CCD (calculated as one minus the Si transmission). These transmissions are calculated using the Henke, Gullikson, and Davis (1993) atomic constants, and the thickness of the filter materials were adjusted to agree with SURF calibration results.

EVE results. Both pre- and post-flight calibrations of the rocket instruments are performed to measure the degradation, if any, of the rocket instruments. This rocket payload for EVE was built and launched to provide the final calibration underflight for the TIMED/SEE instrument on 14 April 2008 from the White Sands Missile Range in New Mexico. The initial results for the solar EUV irradiance measured during this flight were presented by Chamberlin *et al.* (2009), but these results did not include the post-flight calibration where significant improvements were made and applied. The more accurate results from MEGS A and B observations, which include the post-flight SURF calibration, are presented here, along with the first results from MEGS P and SAM. These rocket measurements have been used to update the TIMED/SEE degradation functions (included in SEE Version 10 data products), to provide a reference spectrum for solar-cycle minimum conditions (Chamberlin *et al.*, 2009; Woods *et al.*, 2009), and to improve upon the flight EVE data-processing system.

8.1. MEGS A and B

The calibrated MEGS A1, A2, and B spectra from the sounding-rocket flight are shown in Figure 17, along with the uncertainty of the solar irradiance. These results are very similar to those presented by Chamberlin *et al.* (2009). This latest irradiance result with the prototype EVE includes the post-flight calibrations, which provided improvements especially for the higher-order corrections. No significant degradation was found between the pre- and post-flight calibrations. The main difference between this spectrum and that reported by Chamberlin *et al.* (2009) is in the 35 to 45 nm range where the new results have higher irradiance. This wavelength range is improved now with more accurate, higher-order corrections for MEGS A2 and improved calibrations for MEGS B. The other range of notable difference is 65 to 105 nm, which has slightly higher irradiance than the Chamberlin *et al.* (2009) results.

8.2. MEGS P

MEGS P provides a measurement of the H I Lyman-α emission at 121.6 nm. The April 2008 flight had a few minutes of solar observations, and it was during solar minimum conditions. The data are averaged over the flight at altitudes greater than 250 km so that the contribution from atmospheric absorption is not significant. The MEGS P irradiance [I] is calculated from:

$$I(121.6\ \text{nm}) = \frac{C'(t, \text{filter}) - C_{\text{vis}}(t)}{R_{\text{P}}} f_{\text{degrad}} f_{1\ \text{AU}} \tag{27}$$

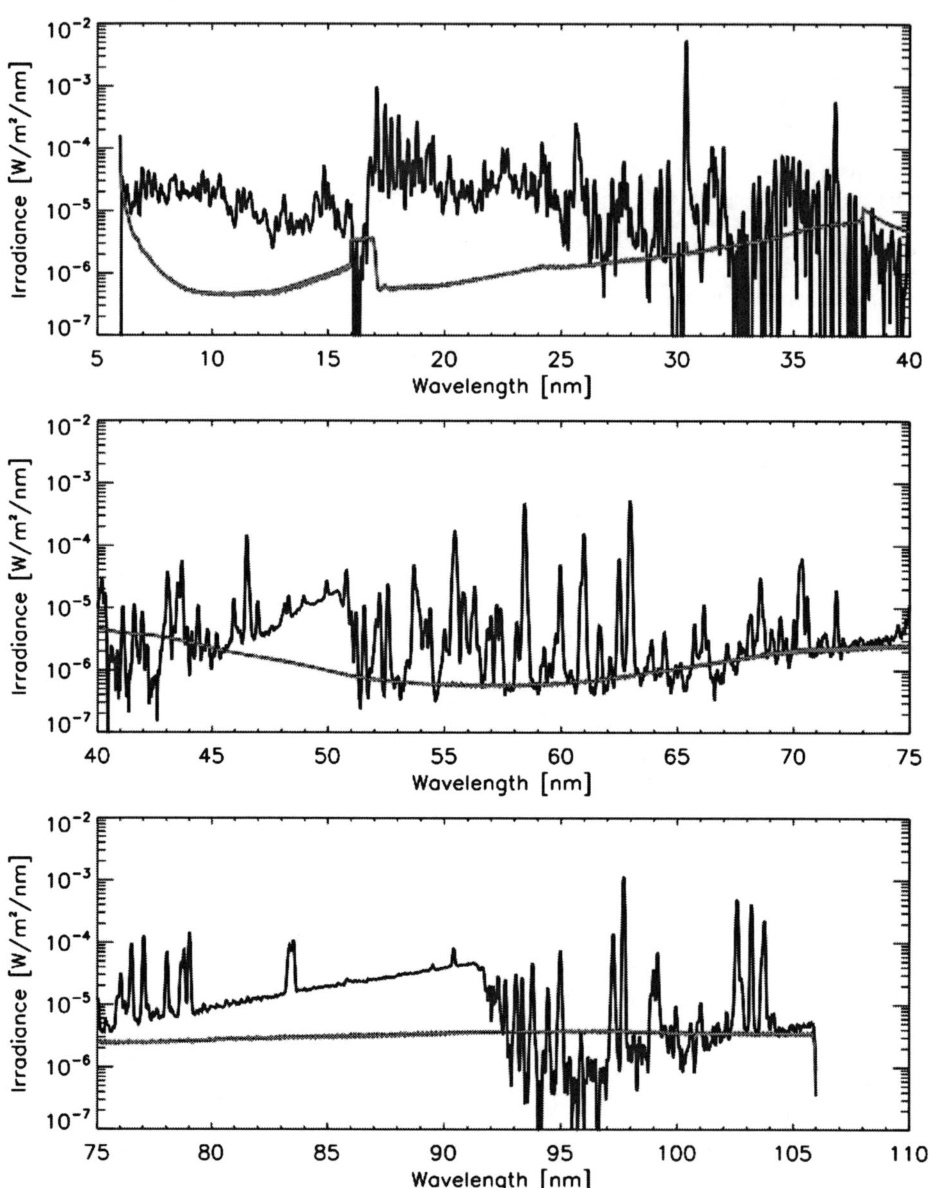

Figure 17 MEGS A1, A2, and B spectra for 14 April 2008. The uncertainty for the solar irradiance is also shown (red line).

where C' [Equation (15)] is the corrected count rate for solar data, C_{vis} is the visible-light correction, R_{P} is the responsivity defined by Equation (17), f_{degrad} is the degradation correction, and $f_{1\text{ AU}}$ is the correction to normalize the irradiance to 1 AU. From SURF calibration analysis and the 2008 flight, the visible-light correction is not necessary ($C_{\text{vis}} = 0.0$), but it will be further validated with in-flight measurements after SDO is launched.

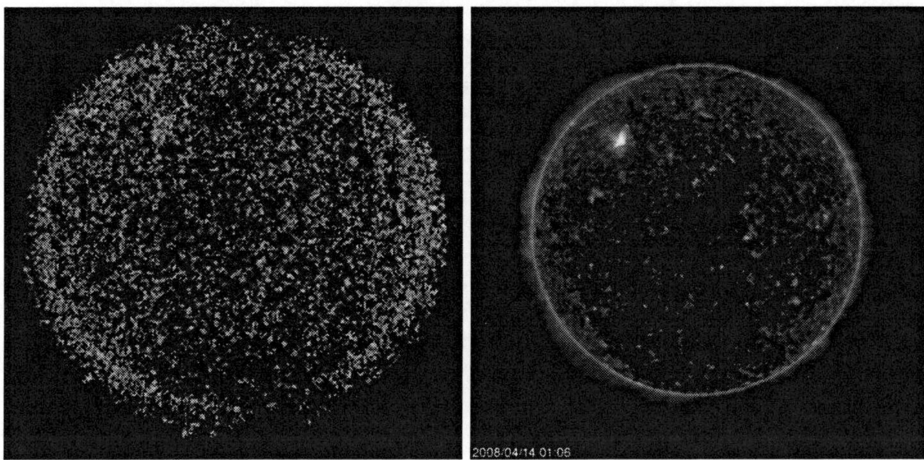

Figure 18 Solar X-ray image from SAM (left) and SOHO/EIT 28.4 nm image (right), courtesy NASA/ESA SOHO/EIT consortium. Both images are representative of the solar corona. The SAM image clearly shows the small active region and coronal emissions above the limb of the Sun.

The MEGS P signal, corrected for dark counts, is 30 DN second^{-1} for this rocket measurement. The irradiance result is 0.0039 $\text{W}\,\text{m}^{-2}$, which is 30% less than the SORCE/SOLSTICE 121 to 122 nm irradiance result of 0.0057 $\text{W}\,\text{m}^{-2}$. This result only uses post-flight MEGS calibrations in January 2009 (pre-flight rocket MEGS-P calibrations were not obtained). It seems unlikely that MEGS P degraded after the rocket flight; therefore, we suspect that the spectral bandpass (profile) for MEGS P is not accurate. The profile is strongly dependent on the Acton Lyman-α filter profile and also the spectral bandpass from the MEGS P first grating; neither is measured directly for MEGS P. Decreasing the $\Delta\lambda_{\text{Sun}}$ and/or increasing the $\Delta\lambda_{\text{SURF}}$ could bring the rocket MEGS P result into better agreement with SOLSTICE. If the flight MEGS P also measures lower irradiance than SOLSTICE, then we will assume the MEGS P spectral bandpass needs to be adjusted to be in agreement with SOLSTICE.

8.3. SAM

Only a few photon events were expected to be recorded during the rocket flight as the flight was during solar minimum. There was one small active region that appeared the day prior to launch and it was located in the northeast quadrant of the observable solar disk. The SAM image is presented in Figure 18, and this image is a compilation of all 17 ten-second images that were obtained during the rocket flight. This image shows the aforementioned active region, as well as the solar corona above the limb of the Sun. Also shown for comparison is the SOHO/EIT image for the day, which has much higher spatial resolution. The SAM image is representative of the coronal bremsstrahlung continuum and emission lines in the 0.1 to 7 nm band, and the EIT 28.4 nm image measures the bright, coronal Fe XV emission line and several other weaker emission lines near 28.4 nm.

As mentioned, the advantage of SAM is that the individual photon events provide the energy, and thus the wavelength, for each photon. Once all of the photon events are processed, a solar spectrum can be produced for the rocket flight, and this spectrum is shown in Figure 19. The WHI spectrum generated by fitting model spectra to XPS broadband measurements is also shown for comparison (Woods *et al.*, 2009).

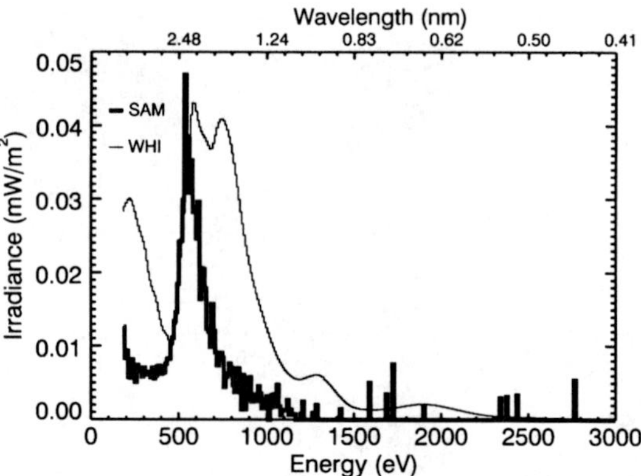

Figure 19 SAM solar spectrum from the April 2008 rocket flight. The SAM spectrum is shown as a thick line. The WHI reference spectrum from XPS is shown for comparison as a thin line.

The solar spectrum generated by SAM is much lower than the WHI reference spectrum. Over the range 177 to 1240 eV (1 to 7 nm), the total SAM irradiance is 5.74×10^{-5} W m^{-2}, while that of WHI is 1.57×10^{-4} W m^{-2}. The ESP irradiance for the 0.1 – 7 nm band is 9.9×10^{-5} W m^{-2}, and the XPS broadband result for 0.1 to 7 nm is 9.3×10^{-5} W m^{-2}. These later values are more accurate as a broadband measurement than the WHI "model" spectrum. Nonetheless, the SAM irradiance is much lower than any of these other results; for example, SAM is 42% lower than the ESP irradiance. This large difference may be related to processing photon events for the SAM solar measurements, but not doing photon event processing for the SAM calibration data. Of particular concern, the solar data might have multiple photon events that are being interpreted as single photon events. Some indication for this concern is that the derived gain for solar measurements indicated 25% differences for the SAM gain value (see Section 7.1). Additional studies are planned to help resolve these differences in the SAM calibrations and solar measurement.

9. Conclusions

These EVE calibration results, along with the first rocket results with the prototype EVE, confirm that the EVE instruments are exceeding design requirements. As discussed in more detail by Woods *et al.* (2010), MEGS is designed to measure the solar EUV irradiance from 5 to 105 nm with 0.1 nm spectral resolution, ten-second temporal cadence, and better than 25% accuracy. The accuracy of the MEGS responsivity ranges from 1% to 10%, with the larger uncertainties occurring at wavelengths where there are low signals during SURF calibrations or associated with larger corrections for grating higher-order contributions. Application of the SURF calibrations to the solar measurement yields irradiance accuracy of 5% to 20%, again wavelength dependent and with larger uncertainties at wavelengths where there are low signals.

These calibrations also highlight some limitations of the EVE instruments. There are a couple of spectral regions where MEGS has low responsivity; consequently, those regions have less accurate calibration and irradiance results. These regions include the 5 to 6 nm range for MEGS A1, the 15 to 17 nm range for MEGS A1 and A2, and 35 to 42 nm range for MEGS A2 and B. From these multiple SURF calibration sets, we have learned how to better select the optimal SURF conditions, such as its beam energy, for each MEGS instrument, thus leading to more accurate responsivity results for the more recent calibrations of the rocket-prototype EVE instruments. These lessons learned will be transferred to the flight EVE with the first EVE underflight rocket calibration, planned for three months after SDO launches. The MEGS P spectral bandpass is the least accurate part of its algorithm, estimated to have 10% uncertainty but perhaps larger. Overlap with SOLSTICE can provide in-flight calibration for MEGS P to improve upon its pre-flight calibrations. The SAM solar-irradiance results have significant differences between its spectrum and the modeled spectrum used in the analysis of XPS broadband data. Some of this difference could be related to the model predictions, but part of this difference might be related to the photon event extraction algorithm. More extended measurements, than just one rocket measurement, are expected to help clarify the differences and lead to improvements for the SAM calibrations and algorithms.

The results from EVE, along with the solar-cycle minimum results from the rocket launch in April 2008, will significantly improve upon the understanding of the solar EUV variability over all time scales from solar-cycle variations over a period of years, to solar-rotation variations that happen on the order of days, as well as solar flares on time scales of seconds to hours. The anticipated EVE irradiance results coupled with the solar-imaging results from SDO's other experiments will advance our understanding of irradiance variations and eruptive events such as flares. These advances are then expected to improve upon Earth's ionospheric and thermospheric models for space-weather operations, to better track satellites, and to manage communication and navigation systems. Furthermore, the EVE observations will provide an important extension of the solar EUV irradiance record currently being made by the SOHO and TIMED satellites.

Acknowledgements The authors would like to express their appreciation to Mitch Furst and the staff at NIST SURF III for all of their help throughout this calibration process, and special thanks go to Mitch Furst for reviewing this paper. We would also like to thank the many dedicated LASP scientists and engineers who designed and built EVE, and the many who also spent many weeks at the NIST calibration facility to obtain these calibration results. We also express special thanks to Vanessa George at LASP for supporting the preparation of this manuscript. We thank the referee for their helpful comments and suggestions. The solar irradiance results from 14 April 2008 can be obtained from the LASP Interactive Solar Irradiance Datacenter (LISIRD, http://lasp.colorado.edu/lisird). The EIT image is from the SOHO/EIT Consortium; SOHO is a joint ESA and NASA program. This research is supported by NASA contract NAS5-02140.

References

Arp, U., Friedman, R., Furst, M.L., Makar, S., Shaw, P.-S.: 2000, *Metrologia* **37**, 357.

Bailey, S.M., Woods, T.N., Eparvier, F.G., Solomon, S.C.: 2006, *Adv. Space Res.* **37**, 209.

Chamberlin, P.C., Woods, T.N., Eparvier, F.G.: 2002, *Opt. Eng.* **45**, 063605.

Chamberlin, P.C., Hock, R.A., Crotser, D.A., Eparvier, F.G., Furst, M., Triplett, M.A., Woodraska, D., Woods, T.N.: 2007, *SPIE Proc.* **6689**, 66890N.

Chamberlin, P.C., Woods, T.N., Crotser, D.A., Eparvier, F.G., Hock, R.A., Woodraska, D.L.: 2009, *Geophys. Res. Lett.* **36**, L05102.

Crotser, D.A., Woods, T.N., Eparvier, F.G., Ucker, G., Kohnert, R., Berthiaume, G., Weitz, D.: 2004, *SPIE Proc.* **5563**, 182.

Crotser, D.A., Woods, T.N., Eparvier, F.G., Triplett, M.A., Woodraska, D.L.: 2007, *SPIE Proc.* **6689**, 66890M.

Didkovsky, L., Judge, D., Wieman, S., Woods, T., Jones, A.: 2010, *Solar Phys.* doi:10.1007/s11207-009-9485-8.

Henke, B.L., Gullikson, E.M., Davis, J.C.: 1993, *At. Data Nucl. Data Tables* **54**, 181.

Hinteregger, H.E., Fukui, K., Gilson, B.R.: 1981, *Geophys. Res. Lett.* **8**, 1147.

Rottman, G.J., Woods, T.N., Sparn, T.P.: 1993, *J. Geophys. Res.* **98**, 10667.

Saloman, E.B.: 1975, *Appl. Opt.* **14**, 1391.

Triplett, M.A., Crotser, D.A., Woods, T.N., Eparvier, F.G., Chamberlin, P.C., Berthiaume, G.D., Weitz, D.M., Vest, R.E.: 2007, *SPIE Proc.* **6689**, 668900.

Westhoff, R.C., Rose, M.K., Gregory, J.A., Berthiaume, G.D., Seely, J.F., Woods, T.N., Ucker, G.: 2007, *SPIE Proc.* **6686**, 668604.

Woods, T.N., Rottman, G.: 2005, *Solar Phys.* **230**, 375.

Woods, T.N., Rottman, G., Vest, R.: 2005, *Solar Phys.* **230**, 345.

Woods, T.N., Tobiska, W.K., Rottman, G.J., Worden, J.R.: 2000, *J. Geophys. Res.* **105**, 27195.

Woods, T.N., Eparvier, F.G., Bailey, S.M., Chamberlin, P.C., Lean, J., Rottman, G.J., Solomon, S.C., Tobiska, W.K., Woodraska, D.L.: 2005, *J. Geophys. Res.* **110**, A01312.

Woods, T.N., Chamberlin, P.C., Harder, J.W., Hock, R.A., Snow, M., Eparvier, F.G., Fontenla, J., McClintock, W.E., Richard, E.C.: 2009, *Geophys. Res. Lett.* **36**, L01101.

Woods, T.N., Chamberlin, P., Eparvier, F.G., Hock, R., Jones, A., Woodraska, D., Judge, D., Didkosky, L., Lean, J., Mariska, J., McMullin, D., Warren, H., Berthiaume, G., Bailey, S., Fuller-Rowell, T., Sojka, J., Tobiska, W.K., Viereck, R.: 2010, *Solar Phys.* doi:10.1007/s11207-009-9487-6.

Solar Phys (2012) 275:179–205
DOI 10.1007/s11207-009-9485-8

EUV SpectroPhotometer (ESP) in *Extreme Ultraviolet Variability Experiment* (EVE): Algorithms and Calibrations

L. Didkovsky · D. Judge · S. Wieman · T. Woods · A. Jones

Received: 2 October 2009 / Accepted: 10 November 2009 / Published online: 16 December 2009

Abstract The *Extreme ultraviolet SpectroPhotometer* (ESP) is one of five channels of the *Extreme ultraviolet Variability Experiment* (EVE) onboard the NASA *Solar Dynamics Observatory* (SDO). The ESP channel design is based on a highly stable diffraction transmission grating and is an advanced version of the *Solar Extreme ultraviolet Monitor* (SEM), which has been successfully observing solar irradiance onboard the *Solar and Heliospheric Observatory* (SOHO) since December 1995. ESP is designed to measure solar Extreme UltraViolet (EUV) irradiance in four first-order bands of the diffraction grating centered around 19 nm, 25 nm, 30 nm, and 36 nm, and in a soft X-ray band from 0.1 to 7.0 nm in the zeroth-order of the grating. Each band's detector system converts the photo-current into a count rate (frequency). The count rates are integrated over 0.25-second increments and transmitted to the EVE Science and Operations Center for data processing. An algorithm for converting the measured count rates into solar irradiance and the ESP calibration parameters are described. The ESP pre-flight calibration was performed at the Synchrotron Ultraviolet Radiation Facility of the National Institute of Standards and Technology. Calibration parameters were used to calculate absolute solar irradiance from the sounding-rocket

The Solar Dynamics Observatory
Guest Editors: W. Dean Pesnell, Phillip C. Chamberlin, and Barbara J. Thompson

L. Didkovsky (✉) · D. Judge · S. Wieman
Space Sciences Center, University of Southern California, Los Angeles, CA, USA
e-mail: leonid@usc.edu

D. Judge
e-mail: judge@usc.edu

S. Wieman
e-mail: wieman@usc.edu

T. Woods · A. Jones
Laboratory for Atmospheric and Space Physics, University of Colorado at Boulder, Boulder, CO, USA

T. Woods
e-mail: tom.woods@lasp.colorado.edu

A. Jones
e-mail: andrew.jones@lasp.colorado.edu

flight measurements on 14 April 2008. These irradiances for the ESP bands closely match the irradiance determined for two other EUV channels flown simultaneously: EVE's *Multiple EUV Grating Spectrograph* (MEGS) and SOHO's *Charge, Element and Isotope Analysis System/Solar EUV Monitor* (CELIAS/SEM).

Keywords Solar extreme ultraviolet irradiance · Instrumentation and data management · Spectrophotometer · Radiometric calibration

1. Introduction

The *Solar Dynamics Observatory* (SDO) is the first NASA Living With a Star mission with its launch planned for February 2010. It will provide accurate measurements of solar atmosphere characteristics with high spatial and temporal resolution at many wavelengths simultaneously. These measurements will help us understand the solar-activity cycle, the dynamics of energy transport from magnetic fields to the solar atmosphere, and the influences of this energy transport on the Earth's atmosphere and the heliosphere. SDO's geosynchronous orbit will allow for continuous observations of the Sun and enable an exceptionally high data rate (over of 100 Megabits per second) through the use of a single dedicated ground station. Dynamic changes of the solar radiation in the extreme ultraviolet (EUV) and X-ray regions of the solar spectrum are efficient drivers of disturbances in the Earth's space-weather environment. The *Extreme ultraviolet Variability Experiment* (EVE) is one of three instrument suites on SDO. EVE provides solar EUV irradiance measurements that are unprecedented in terms of spectral resolution, temporal cadence, accuracy, and precision. Furthermore, the EVE program will incorporate physics-based models of the solar EUV irradiance to advance the understanding of solar dynamics based on short- and long-term activity of the solar magnetic features (Woods *et al.*, 2009).

ESP is one of five channels (Woods *et al.*, 2009) in the EVE suite. It is an advanced version of the SOHO/CELIAS *Solar Extreme ultraviolet Monitor* (SEM) (Hovestadt, Hilchenbach, and Burgi, 1995; Judge, McMullin, and Ogawa, 1998). SEM measures EUV solar irradiance in the zeroth diffraction order (0.1 to 50.0 nm bandpass) and two (plus and minus) first-order diffraction bands (26 to 34 nm bandpasses) centered at the strong He II 30.4 nm spectral line. More than 13 years of EUV measurements have shown that the SEM is a highly accurate and stable EUV spectrometer (Didkovsky *et al.*, 2009) that has suffered only minor degradation, mainly related to deposition of carbon on the SEM aluminum filters.

The ESP design is based on a highly stable diffraction transmitting grating (Schattenburg and Anderson, 1990; Scime, Anderson, and McComas, 1995), very similar to the one used in SEM. ESP has filters, photodiodes, and electronics with characteristics (transmission, sensitivity, shunt resistance, *etc.*) that are susceptible to some change over the course of a long mission. Because of such possible degradation, ESP shall be periodically calibrated throughout the mission. The calibration program includes a pre-flight calibration followed by a number of sounding rocket under-flights with a nominally identical prototype of the flight instrument that is typically calibrated shortly before and shortly after each under-flight. Calibration of the SDO/EVE and EVE/ESP soft X-ray and EUV flight instruments was performed at the National Institute of Standards and Technology (NIST) Synchrotron Ultraviolet Radiation Facility (SURF-III). NIST SURF-III has a number of beamlines (BL) with different specifications and support equipment (*i.e.* monochromators, translation stages, and calibration standards). The beamline used for a given instrument depends on the instrument's characteristics and calibration requirements such as spectral ranges, entrance aperture, alignment, *etc.* Before the flight, ESP was integrated into the EVE package, it was first calibrated on SURF BL-9. BL-9 is equipped with a monochromator capable of scanning through a wide EUV spectral band which allows the instrument efficiency profile for ESP to be measured in increments of 1.0 nm over a spectral window of about 15 to 49 nm. After ESP was mounted onto EVE, the second ESP calibration was performed on SURF BL-2 (Furst, Graves, and Madden, 1993). BL-2 illuminates the instrument with the whole synchrotron irradiance spectrum, which can be shifted in its spectral range by changing the energy of the electrons circulating in the electron storage ring producing the synchrotron radiation. The intensity of the EUV beam is controlled by the current (and therefore number) of circulating electrons. Because the EUV beam consists of a wide spectrum of photon energies, this type of calibration is a radiometric calibration. A major part of this paper is related to the results of the ESP BL-9 and BL-2 calibrations.

The scientific objectives of ESP are given in Section 2. Section 3 presents a short overview of the ESP channel. An algorithm to calculate solar irradiance from count rates measured in each of ESP's photometer bands from measurements conditions (*i.e.* ESP temperature, dark current, band resonance, *etc.*), and from ESP calibration data and orbit parameters is given in Section 4. In Section 5, results from ESP ground tests are shown. Section 6 summarizes ESP pre-flight calibration results. A comparison of ESP solar-irradiance measurements from the sounding-rocket flight of 14 April 2008 with data from other EUV channels (EVE/MEGS sounding rocket instrument and SOHO/SEM) is given in Section 7. Concluding remarks are given in Section 8.

2. ESP Scientific Objectives

The scientific objectives of ESP were developed as a result of analysis of soft X-ray and EUV observations by diode-based photometers with thin-film metal filters, including those on sounding-rocket flights (Ogawa *et al.*, 1990; Judge, McMullin, and Ogawa, 1998), and TIMED/XPS (Woods, Bailey, and Eparvier, 1998). We found several critical limitations with those observations including significant spectral contamination in the filter-based photometers, low duty cycle due to a low-Earth orbit, and relatively low temporal cadence. Some of these limitations have been partially overcome with the SOHO/SEM instrument. More than 13 years of practically uninterrupted SEM observations have resulted in significant improvements in solar modeling with the use of the new $S_{10.7}$ solar index from SEM 26 to 34 nm bands (Tobiska, 2007), in the Earth's atmosphere neutral-density models (Tobiska, Bouwer, and Bowman, 2006; Bowman *et al.*, 2008), and in studying ionospheric responses to solar storms, see *e.g.*, Tsurutani *et al.* (2005). However, some other limitations remain in the SEM observations, such as occasional energetic-particle contamination of the EUV measurements, and a limitation on the data transfer rate, which results in a SEM time cadence of 15 seconds in the SOHO data stream compared to the instrument's native temporal resolution capability of 0.25 seconds. Totally free of these limitations, ESP has the following scientific (and technical) objectives:

i) To provide 40 times better temporal cadence than SOHO/SEM in order to address solar dynamics at the highest EUV cadence to date.

ii) To provide near real-time calibrated solar irradiance with minimal latency for rapid space-weather prediction.

iii) To provide much greater EUV spectral coverage than SOHO/SEM with three times as many data bands permitting atmospheric source region variability studies. ESP's four first-order bands are centered at wavelengths of about 19, 25, 30, and 36 nm. These wavelengths contain strong solar EUV spectral lines traditionally used to study the dynamics of the solar atmosphere, *e.g.* SOHO/EIT has 19.5 nm and 30.4 nm bands, TRACE has a 19.5 nm band, SDO/AIA has 19.3 nm and 30.4 nm bands. These wavelengths, as well as the 0.1 to 7.0 nm bandpass observed with ESP's zeroth-order quad-diode band, are in high demand for space-weather applications, as proxies for changes of the Earth atmospheric neutral density and ionization, and for building more accurate solar EUV models. In addition, the relative activity of the solar E, W, N, and S quadrants on the solar disk and their time dependence will provide data on large-scale relative variability "quadrant to quadrant".

iv) To provide significantly more accurate EUV measurements in order to search for short- and long-term variability and its relationship to solar-produced variability in the Earth's ionosphere. ESP has a more accurate method for compensating for changes in the measured signal related to changes of observing conditions than any other instrument of its class flown to date. This compensation includes a signal correction for changes in detector temperature, for changes of sensitivity due to radiation damage to the electronics, for changes in the amount of, and residual sensitivity to, scattered and stray visible light, and for contamination of the EUV signal due to energetic particles.

All of these factors will significantly improve the accuracy of ESP observations. The scientific objectives of ESP complement those of EVE (Woods *et al.*, 2009) with ESP providing long-term, stable, photometrically accurate, high temporal cadence EUV measurements while the high-resolution spectra in the broad wavelength range will be measured by the EVE/MEGS. ESP calibrated irradiance will be used as reference data for the EVE MEGS-A, MEGS-B, and SAM channels.

In addition to the scientific objectives related to the EUV observations, ESP will detect fast changes of Solar Energetic Particle (SEP) fluxes associated with strong solar-storm events. Geometrically accurate opto-mechanical models of the SEM and ESP were developed (Didkovsky *et al.*, 2007a) to determine the detector's sensitivity to the SEPs by modeling the stopping power for high-energy protons. The ESP stopping-power model shows that a differential signal from the ESP detectors may be used to extract solar proton spectra in an energy range of 38 to 50 MeV with Full Width at Half Maximum (FWHM) of about 6 MeV.

High-cadence and accurate EUV observations provided by the ESP will improve our knowledge of the energy-release spectrum during solar flares, will increase the accuracy of EUV solar models including their predictive capabilities, add to the information about particle initiation and acceleration, and contribute to studies of the ionosphere and to Earth's atmosphere neutral-density models.

3. ESP Overview

SDO/EVE/ESP is an advanced version of the *Solar Extreme ultraviolet Monitor* (SEM) (Judge, McMullin, and Ogawa, 1998) that has been successfully working onboard SOHO as part of the CELIAS experiment since 1995. A comparison of design features for SEM and ESP is shown in Table 1. Each of the ESP features shown in Table 1 that were not incorporated into SEM are enhancements which ultimately improve the accuracy and performance of the channel and, thus, the quality of the data that it contributes to the science of solar EUV variability. For example, the gain-reference mode is an important feature to track changes in each band's gain during the flight lifetime. These changes may be related to both ESP electronics temperature and electronics degradation, *e.g.*, due to the Total Ionizing Dose (TID) radiation damage even though great care was taken to mitigate such change.

Table 1 A comparison of SEM and ESP design features.

Design Feature	SEM	ESP
Number of bands	3	9
Filter wheel	No	Yes, with 5 filters
Dark band	No	Yes
Quad-diode bands	No	Yes
Dark mode for all bands	No	Yes
Gain reference mode	No	Yes

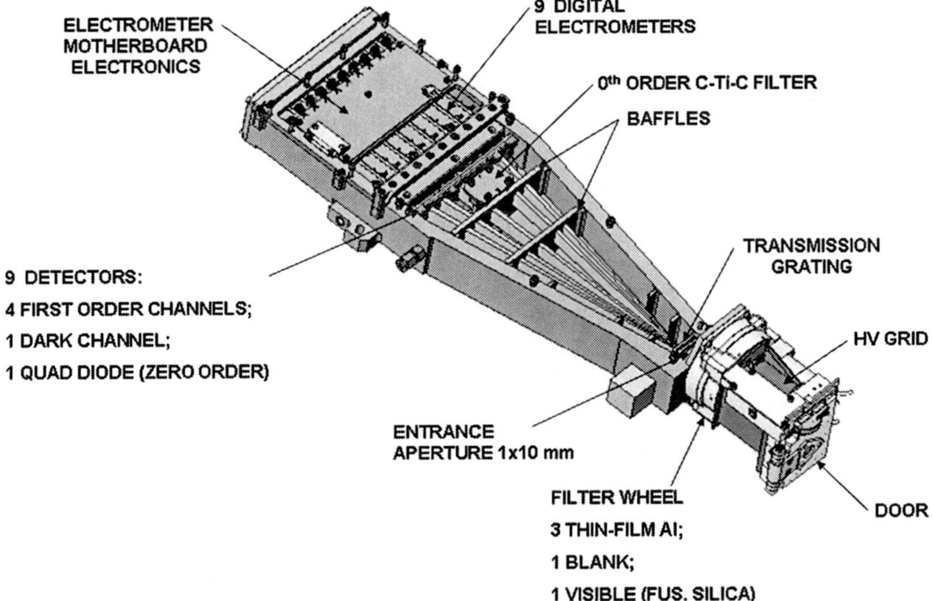

Figure 1 An overview of the ESP configuration (see details in the text).

The ESP instrument configuration is shown in Figure 1. A high-voltage grid in front of ESP, with a potential of 1000 V, eliminates low-energy charged particles that could enter ESP through the entrance slit (1 × 10 mm) and deposit energy in the ESP detectors creating additional output signal beyond that due to EUV photons. The baffles are designed to minimize the amount of visible (scattered and stray) light that reaches the detectors. The limiting apertures on the detectors define the edges of the spectral bands. A filter-wheel mechanism in front of the entrance aperture has three aluminum filters (one primary and two spare), a visible-light (fused-silica) filter, and a dark filter. The thin-film aluminum filter rejects the visible light and transmits the EUV radiation to the detectors. The fused-silica filter is used from time to time to determine the amount of visible light that may reach the EUV detectors and should be subtracted from the EUV-band signals. The dark filter is to measure dark currents on each band directly.

The detector assembly has nine detectors including four first-order bands, four zeroth-order bands in the form of a Quadrant Diode (QD), and a dark band with a diode that is constantly closed off. Nine digital electrometers convert diode currents into count rates using Voltage-to-Frequency Converters (VFC). The electrometer motherboard electronics provide synchronization of operations and the connection of ESP to the EVE electronics.

The first-order bands are centered around 19 nm, 25 nm, 30 nm, and 36 nm. They each have a spectral resolution of about 4 nm at FWHM.

The quad pattern of the zeroth-order bands is centered about the optical axis of ESP. These bands provide information about de-centering ESP from misalignment or pointing errors (see Section 6.4.1). A non-uniformity in the solar-disk irradiance distribution related to the dynamics of active regions, (*e.g.* the growth or decay of such regions or their position changes due to solar rotation) will be detected by the QDs. The signal non-uniformity across the QD bands will be used to determine an angular correction of the first-order band signals, and be used as a real-time solar-flare detector. For normal in-flight operation, solar irradiance to the zero-order band goes through an input aluminum filter installed in the filter wheel and through a free-standing C – Ti – C filter installed between the diffraction grating and the QD detector. These filters constrain the spectral bandpass of the zero-order band, which is from about 0.1 to 7.0 nm at greater than 10% efficiency.

The dispersive element in ESP is a free-standing gold-foil transmission grating with a 200 nm period. Gratings of this type have been used on multiple satellite and sounding-rocket flights and have proven to be reliable for solar EUV irradiance measurements even on long-term space missions. The long-term accuracy of ESP measurements depends on information about possible changes of ESP characteristics such as filter transmissions, detector responsivity, and electronics sensitivity. This information will be provided by regular under-flight calibrations using an ESP sounding-rocket clone instrument. In addition to these under-flight calibrations, a number of methods are in place for monitoring and maintaining ESP calibration onboard. These methods include:

i) in-flight compensation for thermal changes of dark currents, see Equations (3) to (8) in Section 4;
ii) determination and subtraction of signal related to the scattered visible light, see Equations (9), (10) in Section 4;
iii) determination and correction for guiding errors (tilts), see Section 6.4.1;
iv) correction of the measured whole-disk solar EUV irradiance for the location of solar flare, see Section 6.4.1;
v) determination and subtraction of energetic-particle related signals, see Equation (2) in Section 4;
vi) use of spare input aluminum filters in the event of filter damage (*e.g.* pin-hole damage), see Section 6.4.3 and Figure 16;
vii) minimizing TID radiation damage to electronics through the use of an appropriately thick ESP housing and strategically located spot shielding; and
viii) a reference mode for each ESP band to determine and correct possible changes of electronics due to the TID; see Equation (1) in Section 4 and Figure 5.

All of these methods along with sounding rocket under-flights will guarantee highly reliable and accurate EUV measurements by ESP.

4. An Algorithm to Convert ESP Count Rates into Solar Irradiance

EUV solar irradiance is detected by the ESP band detectors as an increase in diode current beyond a base-level dark current that is intrinsic to the diode – electrometer combination and measured even in the absence of photon illumination. These combined (dark + signal) currents are converted by the electrometers into voltages and then into discrete pulses with frequencies that are linear functions of the input signals. The frequencies are measured as counts per a fixed time interval, *e.g.* per 0.25 seconds. They are also referred to as count rates. The count rate of each band is a function of many input parameters: EUV solar intensity and spectral distribution of solar irradiance, dark currents, sensitivity to different wavelengths, distance to the Sun, degradation of ESP, *etc.* An algorithm to convert ESP count rates into solar irradiance has been developed. It combines all of these input parameters via equations described below. Some of the ESP input parameters were determined based on ground tests. Parameters in this group, such as dark currents, reference voltage levels, and thermal changes of these parameters, do not require any input radiation. The parameters of the second group are determined based on the ESP calibration at the NIST SURF-III. These parameters include ESP sensitivity to the angular position of the input source of irradiation, the spectral efficiency of each band, determination of the spectral bandpass profiles, influences from the higher orders of the grating, characteristics of the thin-film filters, *etc.* The parameters of the third group are related to solar observations and will be determined or corrected during the mission. These parameters include a contribution of the visible light and energetic particles to the ESP band signals, a correction of the EUV signal due to changes in the dark currents, and a correction for ESP degradation based on the comparison of the orbit measurements with the sounding-rocket under-flight measurements made simultaneously.

Solar irradiance $[E]$ in units of $\mathrm{W\,m^{-2}}$ for each scientific band $[i]$ is determined as

$$E_i(\lambda, t) = \frac{C_{i,\mathrm{eff}}[1 - \frac{\Delta G_i(T,V,\mathrm{TID})}{\Delta t}]}{A \frac{\int_{\lambda 0-\Delta\lambda}^{\lambda 0+\Delta\lambda} R_i(\lambda,\alpha,\beta) \frac{\lambda}{\hbar c} F_i(\lambda)\mathrm{d}\lambda}{\int_{\lambda 0-\Delta\lambda}^{\lambda 0+\Delta\lambda} F_i(\lambda)\mathrm{d}\lambda} f_{i,\mathrm{degrad}}(t) f_{1\,\mathrm{AU}}(t)} - E_{\mathrm{OS}}, \tag{1}$$

where $C_{i,\mathrm{eff}}$ in the ESP Equation (1) is the band's effective count rate. For the time t_{j}

$$C_{i,\mathrm{eff}}(t_{\mathrm{j}}) = C_{i,\mathrm{meas}}(t_{\mathrm{j}}) - C_{i,\mathrm{ch.dark}}(t_{\mathrm{j}}) - C_{i,\mathrm{particleBG}}(t_{\mathrm{j}}) - \Delta C_{i,\mathrm{vis}}(t_{\mathrm{j}}), \tag{2}$$

which, for each ESP band i, is the result of subtracting the background signal (signal not related to EUV within the designed band-passes) from the total measured signal $C_{i,\mathrm{meas}}$. The background signal is composed of three parts: dark count rate $C_{i,\mathrm{ch.dark}}$, measured when the dark filter is in position, an energetic-particle related signal $C_{i,\mathrm{particleBG}}$, and a difference $\Delta C_{i,\mathrm{vis}}$ between the visible scattered-light count rate when the visible (fused-silica) filter is in place and the dark count rate. We obtain all three of these contributors to the background signal based on ESP design parameters, results of ground tests, and laboratory measurements as described below. All of these contributors will be measured in flight.

One of the nine ESP bands, a dark band C_{dark}, is completely closed at all times to prevent light from reaching it. It is possible to use the C_{dark} measurements (available at any time) as a proxy for dark counts in the science bands, and as a proxy for particle-background signal. The use of the dark-band counts C_{dark} saves time during mission operations by reducing the frequency with which one needs to move the filter wheel back and forth between the dark filter and the primary observing (aluminum) filter. ESP ground tests showed that $C_{i,\mathrm{ch.dark}}$ is sufficiently stable over time periods for which the temperature $[T]$ is stable. We may assume, therefore, that counts, $[C_{i,\mathrm{ch.dark}}(t_{\mathrm{j}})]$ at a given time $[t_{\mathrm{j}}]$ are approximately equal to those $[C_{i,\mathrm{ch.dark}}(t_1)]$, measured at the previous time $[t_1]$, that the dark filter was in place and that the difference between the two is related to the temperature change $[\Delta T_{\mathrm{j}}]$ determined from the ground tests:

$$C_{i,\mathrm{ch.dark}}(t_{\mathrm{j}}, T_{\mathrm{j}}) = C_{i,\mathrm{ch.dark}}(t_1, T_1) + \Delta C_{i,\mathrm{ch.dark}}(t_{\mathrm{j}}, \Delta T_{\mathrm{j}}), \tag{3}$$

where

$$\Delta C_{i,\mathrm{ch.dark}}(t_\mathrm{j}, \Delta T_\mathrm{j}) = C_{i,\mathrm{ch.dark}}(t_\mathrm{j}, T_\mathrm{j}) - C_{i,\mathrm{ch.dark}}(t_1, T_1). \tag{4}$$

Then, the change in a band's dark count rate may be replaced with the dark-band measurements C_{dark} and the band's dark proxy counts $C_{i,\mathrm{chd.proxy}}$,

$$\Delta C_{i,\mathrm{ch.dark}}(t_\mathrm{j}, \Delta T_\mathrm{j}) = \frac{C_{\mathrm{dark}}(t_\mathrm{j}, T_\mathrm{j})}{C_{i,\mathrm{chd.proxy}}(T_\mathrm{j})} - C_{i,\mathrm{ch.dark}}(t_1, T_1). \tag{5}$$

Thus, a band's dark count rate for the time t_j is determined using previous, *e.g.* a couple of days earlier, measurements with the dark filter in place for the time t_1 and the band's dark proxy, where

$$C_{i,\mathrm{chd.proxy}}(T_\mathrm{j}) = \frac{C_{\mathrm{dark}}(T_\mathrm{j})}{C_{i,\mathrm{ch.dark}}(T_\mathrm{j})}, \tag{6}$$

where $C_{\mathrm{dark}}(T_\mathrm{j})$ and $C_{i,\mathrm{ch.dark}}(T_\mathrm{j})$ are determined for a working range of temperatures $[T]$ and tabulated during ground tests. With all the above considerations, the relation for the $C_{i,\mathrm{ch.dark}}(t_\mathrm{j})$ in Equation (2) is

$$C_{i,\mathrm{ch.dark}}(t_\mathrm{j}) = \frac{C_{\mathrm{dark}}(t_\mathrm{j}, T_\mathrm{j})}{C_{i,\mathrm{chd.proxy}}(T_\mathrm{j})}. \tag{7}$$

Another method to determine the $C_{i,\mathrm{ch.dark}}(t_\mathrm{j})$ is to use a change of the dark count rate as a function of the detector temperature $[T_\mathrm{j}]$ determined from the thermal ground tests for each band i

$$C_{i,\mathrm{ch.dark}}(t_\mathrm{j}) = F_{i,\mathrm{ch.dark}}(T_\mathrm{j}). \tag{8}$$

One of these two methods, either the dark-band proxy [Equation (7)] or the function of change of the band's dark count rate [Equation (8)] will be chosen as the most accurate method after engineering measurements on orbit and will be used in Equation (2).

The degree to which each detector is shielded from energetic particles varies slightly depending on its position within the housing (*i.e.* its location with respect to housing walls and other mechanical structures). Because of this positional dependence of shielding thickness and, accordingly, total stopping power for energetic particles, the band responses to energetic-particle flux were studied for both isotropic and concentrated fluxes. This analysis, based on a geometrically accurate three-dimensional (SolidWorks^{TM}) model, showed that the energy deposited in the ESP detectors $C_{i,\mathrm{particleBG}}$ by particles associated with the quiet Sun or with small solar storm events is so small that it is less than the noise. For solar storms associated with extreme solar-flare events, the particle-related band signal may be determined with corresponding position-dependent modeled corrections to the signal extracted from the dark band.

The term $\Delta C_{i,\mathrm{vis}}$ in Equation (2) shows the portion of the band's signal related to scattered visible light. Laboratory measurements, in which a strong visible-light source was placed in front of the ESP entrance aperture (with the aluminum filter removed), showed very little to no visible-light sensitivity in any of the ESP first-order bands. Nevertheless, in case of some change of the conditions of visible-light scattering, the updated $\Delta C_{i,\mathrm{vis}}$ may be found as

$$\Delta C_{i,\mathrm{vis}}(t_\mathrm{j}) = \frac{C_{\mathrm{fus.silica}}(t_\mathrm{j}) - C_{i,\mathrm{ch.dark}}(t_\mathrm{j}) - C_{i,\mathrm{particleBG}}(t_\mathrm{j})}{T_{\mathrm{fus.silica}} + \Delta T_{\mathrm{fus.silica}}} \tag{9}$$

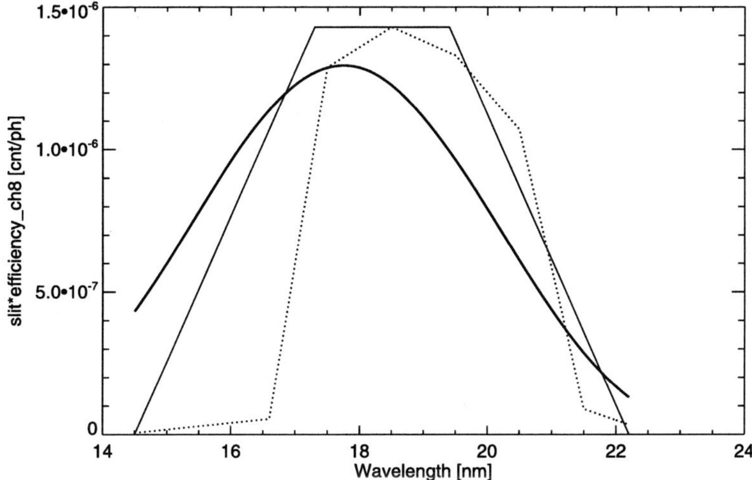

Figure 2 An example of the convolution (thick line) for the Ch8 efficiency profile (dotted line) and exit slit trapezoidal function (thin line).

where $T_{\text{fus.silica}}$ and $\Delta T_{\text{fus.silica}}$ are, respectively, the transmission of the fused-silica filter and its change from pre-flight to flight conditions [Equation (10)]. The amount of incoming visible light reaching the ESP entrance aperture is corrected by the relative distance to the Sun squared.

$$\Delta T_{\text{fus.silica}} = T_{\text{fus.silicaflight}} - T_{\text{fus.silicapreflight}}. \tag{10}$$

When the visible-light signal, measured with the fused-silica filter, is smaller than the sum of the band's dark and particle background signals, then $\Delta C_{i,\text{vis}}$ is set to zero.

Temporal changes of the band's gain $\Delta G(T, V, \text{TID})/\Delta \text{t}$ are caused by the changes of the following variables: temperature $[T_i]$, the reference voltage $[V_i]$ used to calibrate the electrometer's VFC, and the TID, which can affect the parameters of the VFC. If the reference voltage $[V_i]$ is quite stable against TID and its temperature changes are determined during ground tests, in-flight changes of the gain after subtraction of the temperature/voltage changes give us information about the TID influence.

In Equation (1), A is the ESP entrance aperture area, $R(\lambda, \alpha, \beta)$ is the band responsivity to the wavelengths within the range of the band's spectral window, which is also a function of angular alignment, (α and β), and the solar field of view [FOV]. During solar observations, each channel's bandpass is expanded compared to the calibration bandpass. The band responsivity $R(\lambda, \alpha, \beta)$ in Equation (1) is the result of convolving the responsivity profile measured at BL-9 $\xi_i(\lambda)$ and the band's exit slit function $S_\beta(\lambda, \text{FOV})$:

$$R(\lambda, \alpha, \beta) = S_\beta(\lambda, \text{FOV}) * \xi_i(\lambda). \tag{11}$$

Figure 2 shows an example of the convolution for channel eight (Ch8). The relative amplitude of the spectral radiation component $F(\lambda)$ shown in Equation (1) is $F(\lambda)\,\mathrm{d}\lambda$ divided by $\int_{\lambda 0-\Delta\lambda}^{\lambda 0+\Delta\lambda} F_i(\lambda)\,\mathrm{d}\lambda$. $f_{\text{degradation}}(\text{band}, t)$ is the time- and band-dependent degradation factor and $f_{1\ \text{AU}} = 1/r^2$.

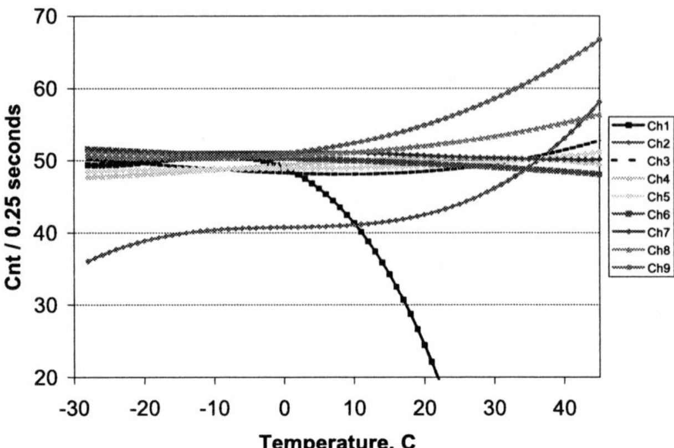

Figure 3 Thermal changes of ESP dark count rates in the temperature range of $-28°\text{C}$ to $34°\text{C}$. The bands are marked with different colors: Ch1 (black), Ch2 (dark green), Ch3 (dotted), Ch4 (light green), Ch5 (cyan), Ch6 (dark blue), Ch7 (violet), Ch8 (pink), and Ch9 (red).

The term E_{OS} is the portion of the band's solar-irradiance measurement $E_i(\lambda, t)$ that is due to shorter wavelengths at higher diffracted orders

$$E_{\text{OS}} = \sum_{n=2,3} R_{\text{OS}} E_n, \tag{12}$$

where R_{OS} is the band's relative sensitivity to higher orders, second and third, $R_{\text{OS}} = R_n / R_1$. E_n is the portion of energy that reaches the band's detector from a higher order.

5. Results from the Ground Tests

Ground tests performed on ESP (as part of the EVE assembly) included thermal-vacuum (TV), electro-magnetic interference (EMI), and electro-magnetic compatibility (EMC) tests. Concurrent EMI and EMC tests were performed and produced some increased (peak) count rates compared to the smooth changes of currents with temperature observed without EMI/EMC. Analysis of thermal changes of both dark count rates and reference count rates described in this section was first applied to the whole set of data, which includes the EMI/EMC peaks. A refined data set, which excludes peaks related to "survival condition" electro-magnetic fields far exceeding those expected under operational conditions, led to similar (within the standard errors) characteristics of the thermal-current changes. This refined set of measurements was used to build the thermal-current curves for each ESP band.

5.1. Thermal Changes of Dark Count Rates

A variety of dark-current changes (dark-count rates) for each of the nine ESP bands are shown in Figure 3. All of the thermal characteristics were approximated with a third-degree polynomial fit to about 5.2×10^6 measurements obtained during TV tests in 2007. The standard-deviation error for these fits are within ± 0.5 corents (0.25 seconds^{-1}) with the mean number of ± 0.38 corents (0.25 seconds^{-1}).

Figure 3 shows that all ESP band dark thermal changes for the temperature range from $-28°\text{C}$ to $34°\text{C}$, except for Ch1, are within a range of count rates of about ± 15 corents (0.25 seconds^{-1}). Ch1 thermal changes are much larger for the positive temperatures. How-

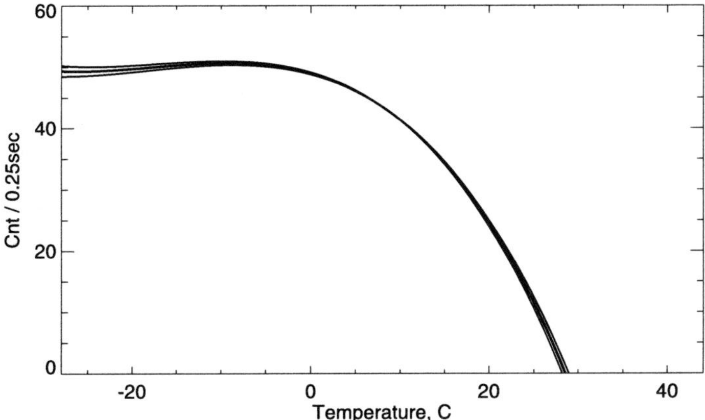

Figure 4 One-σ standard errors for ESP Ch1 signal variations related to the thermal change tests. Each point's measurement uncertainty was assumed to be ± 0.5 corents (0.25 seconds^{-1}).

ever, with the ESP detector planned temperature of about 10°C on orbit, dark-count rate for Ch1 will be suitable for subtraction as long as it remains positive. One-σ standard errors for the Ch1 thermal changes are shown in Figure 4. The thermal changes of dark counts determined during 2007 will be verified and corrected if required during engineering tests on orbit.

5.2. Thermal Changes of Reference Count Rates

The reference mode is designed to determine possible changes of the electronics gain of each VFC. When the reference mode is enabled by software, a stable reference voltage is applied to the outputs of the electrometers and, thus, to the inputs of the VFCs. Even with the thick ESP envelope and additional metal shields at the VFC component locations, the VFC may still be susceptible to some degradation due to the X-ray and particle-radiation environment, which is measured as accumulated TID. Periodic measurements of the count rates in reference mode and comparison of these count rates to the pre-flight values will show these TID related changes. In addition to possible changes of count rates related to TID, reference mode count rates are sensitive to the changes of electronics temperature. Figure 5 shows these thermal changes determined from the 2007 TV tests. The thermal changes of the counts in the reference mode show small linear trends and these changes will be applied to the correction of the electronics gain; see Equation (1), term $\Delta G_i(T, V, \mathrm{TID})$.

6. ESP Pre-flight Calibration

The goal of the pre-flight calibration is to determine all of the channel variables that affect the calculation of solar irradiance [Equation (1)]. Some of these variables may be directly measured during ground tests, others are determined from the SURF calibrations. A spectrum of energetic-particle energies that are sufficient to create a particle background signal on the ESP detectors is determined from the ESP opto-mechanical proton-interaction model.

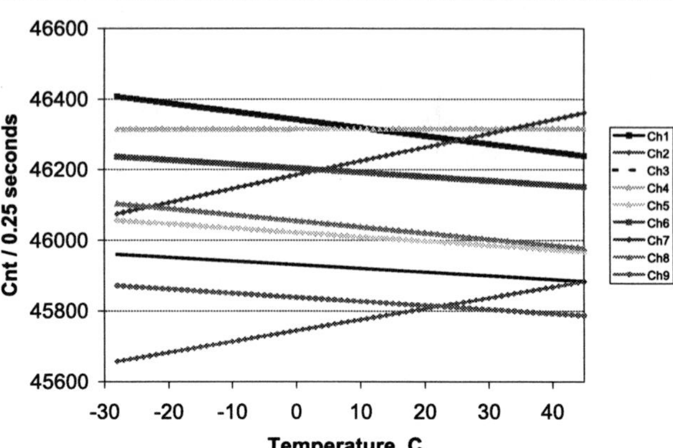

Figure 5 Thermal changes of ESP reference count rates in the temperature range of −28°C to 34°C for ESP bands from Ch1 to Ch9. The bands are marked with different colors: Ch1 (black), Ch2 (dark green), Ch3 (dotted), Ch4 (light green), Ch5 (cyan), Ch6 (dark blue), Ch7 (violet), Ch8 (pink), and Ch9 (red). The vertical axis shows counts per 0.25 seconds.

Table 2 ESP variables: where they are determined (second column) and how accurately (third column). The larger value[1] is for extreme solar-flare events. The larger value[2] is for the zeroth-order bands. The larger number[3] is based on the assumption that there is some detectable amount of visible light. The larger number[4] is for a combination of extreme observing conditions.

ESP parameter and (Equation)	Where it is determined	Relative error [%]
$C_{i,\text{ch.dark}}$ (2) – (9)	G	0.5
C_{dark} (5) – (7)	G, SURF	0.5
$C_{i,\text{chd.proxy}}$ (5) – (7)	G	0.7
$C_{i,\text{particleBG}}$ (9)	Model	0.5 – 5.0[1]
$\Delta C_{i,\text{vis}}$ (2), (9)	G, SURF	0.0 – 5.12[2]
$C_{\text{fus.silica}}$ (9)	G, F	0.0 – 0.5[3]
$T_{\text{fus.silica}}$ (9), (10)	G, F	0.5
$\Delta T_{\text{fus.silica}}$ (9), (10)	G, F	0.5
$C_{i,\text{meas}}$ (2)	G, SURF, F	0.5
$C_{i,\text{eff}}$ (2)	F	0.87 – 7.2[4]
$dG_i(T, V, \text{TID})$ (1)	G, F	0.7
R_{OS} (11)	SURF	4.2
E_{OS} (1), (11)	SURF	5.9
A (1)	G	0.1
$R_i(\lambda, \alpha, \beta)$ (1)	SURF	2.5
$\int R_i(\lambda, \alpha, \beta) F_i(\lambda)\, d\lambda$ (1)	SURF	9.8
$\int F_i(\lambda)\, d\lambda$ (1)	SURF	5.6
$f_{i,\text{degrad}}$ (1)	F, Underflight, SURF	15.9
$f_{1\,\text{AU}}$ (1)	F	0.1
$E_i(\lambda, t)$ (1)	F	21.3[4]

6.1. ESP Variables: Where They Are Measured and How Accurately

Equation (1) for solar irradiance shows all of the ESP variables (parameters) that need to be determined for converting measured count rates into corresponding irradiance. Each of these variables is shown in Table 2 with an indication of the Equations (1) through (11) in which it appears (first column). The second column shows the method of determination, *e.g.* *G* is for ground tests (non-SURF), and the third column shows the relative statistical error of

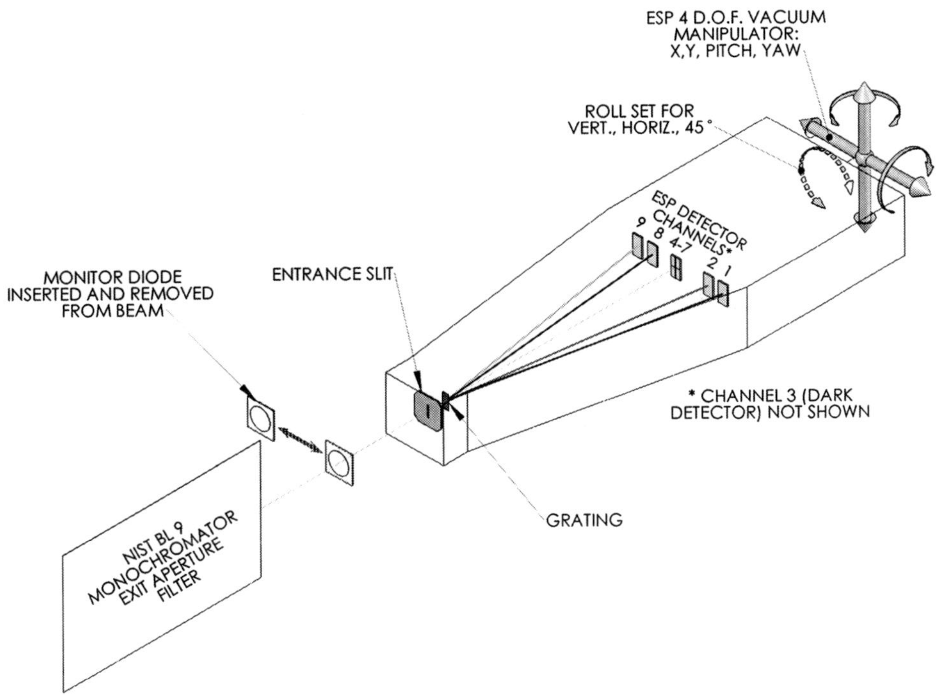

Figure 6 Opto-mechanical layout for ESP BL-9 calibration. The monochromator and BL-9 filters provide 27 test wavelengths over the range from 15.6 to 49.0 nm. A removable monitor diode measures the beam intensity before and after each wavelength's test run. ESP's orientation shown is horizontal. Note, the Al and Ti filters were removed from ESP during the BL-9 calibration.

determination. The largest source of uncertainty (error) is related to long-term degradation trends (see $f_{i,\mathrm{degrad}}$ term in Table 2). This error will be significantly reduced (to about 5%) after three to five sounding rocket under-flights sufficient to understand and model the degradation curve. This reduced degradation error will decrease the uncertainty of determination of solar irradiance from 21.3% shown in Table 2 to about 10%.

6.2. ESP Calibration Overview

ESP calibrations at NIST SURF include initial calibration at the BL-9 where ESP was calibrated before assembling ESP into the SDO/EVE and a radiometric calibration at BL-2 after assembling it to the EVE suite of channels. BL-9 is used primarily for detailed spectral profiles for each first-order band and filter-transmission calibration, and the SURF BL-2 is used primarily for end-to-end radiometric calibration of ESP. On both BL-9 and BL-2, the ESP entrance aperture is illuminated with an essentially non-divergent optical beam.

The BL-9 calibration at SURF consists of measurements of the channel response to a set of narrow wavelengths provided by a monochromator. BL-9 has a small vacuum tank with a translation stage where the instrument, *e.g.* ESP, is installed for testing. The tank provides ESP alignment through linear motions (X and Y) and tilts (Pitch and Yaw). Additionally, to reduce the influence of the beam linear polarization, ESP can be rotated (roll axis) to horizontal, vertical, or 45° orientations. The optical layout for the ESP BL-9 calibration is shown in Figure 6. A wavelength scan allowed for determination of the spectral sensitivity

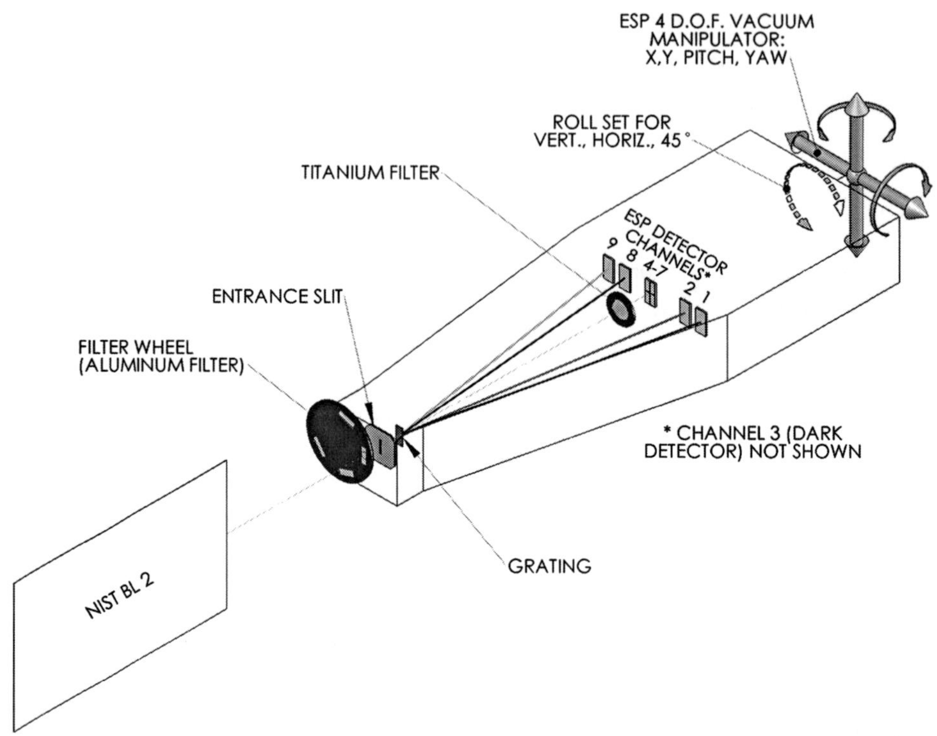

Figure 7 Opto-mechanical layout for ESP BL-2 calibration. ESP is assembled into the EVE but calibrated separately from other EVE channels when the ESP entrance slit and optical axis is aligned to the BL-2 optical beam. ESP's orientation shown is horizontal.

profiles and efficiencies for each of the ESP first-order bands. Due to the small reflectivity of the BL-9 grazing-incidence diffraction grating at short wavelengths, the QD (zeroth-order) bands were not calibrated at BL-9. The BL-9 vacuum tank mechanical dimensions are too small to permit insertion of the whole flight ESP, *e.g.* with the filter wheel, for calibration. To fit the allowed dimensions, the entrance door and the filter wheel with filters were removed from the flight ESP during BL-9 calibration.

For the BL-2 radiometric calibration (Didkovsky *et al.*, 2007b) the EVE suite of channels was installed in a large vacuum tank used for calibration of large NASA instruments. The intensity of BL-2 synchrotron light is accurately known, to about 1%, and is a primary radiometric standard at NIST. The key parameters for the synchrotron source are its beam energy and beam current, and these are provided by SURF who calibrates these parameters on a regular basis. SURF provides beam energy and current, X, Y, Pitch, and Yaw control. These input parameters together with the band's signal are recorded during calibration. The measurements are then used to determine the channel parameters (Table 2, marked as SURF in the second column). The optical layout for ESP BL-2 calibration is shown in Figure 7. The goal of ESP alignment to the beam is to make the ESP optical axis coincident with the beam optical axis. QD bands are usually used for ESP alignment. The alignment consists of a number of scans in two orthogonal directions (X and Y) to find the center of the beam. After finding the center, the Yaw and Pitch tilts are adjusted to have approximately zero X and Y coordinates, calculated from four QDs.

Because synchrotron light is highly polarized, the SURF calibrations usually need to be performed with two orthogonal orientations to account for the band sensitivity to polarization. The ESP grating is sensitive to polarization but photometers detectors are not. ESP efficiency is higher for the orientation in which the grating grooves are aligned with the direction of the synchrotron beam polarization. The gimbal table inside the BL-2 has a mounting ring for the instrument, and this ring can be rotated so the instrument has any angle relative to the synchrotron beam (polarization). Any two orthogonal orientations can be used, such as $-45°$ and $+45°$. However, it has been shown for several grating spectrometers at SURF that a single calibration can be done at 45°, and the same result as averaging horizontal and vertical calibrations is obtained. The ESP pre-flight calibration was done at an orientation of 45°.

6.3. Results from the BL-9 Calibration

ESP was calibrated at BL-9 in two calibration modes, both horizontal [H] and vertical [V] orientations to the SURF beam. These provide a full set of information about sensitivity of ESP to the BL-9 linear polarization. Each of these calibration modes required different mounting of ESP to the tank's vacuum flange and was followed by the alignment of ESP to the beam optical axis. Determination of ESP band sensitivities to the angular position of a source of irradiation in the ESP FOV was provided by corresponding tilts of the ESP's aligned optical axis in the FOV range. The mean $(H + V)/2$ result of calibration was the band's efficiency [ξ] determined as

$$\xi(\lambda)_{0.25\ \text{seconds}} = \frac{(C_{\text{ch}} - C_{\text{dark}})}{F_\lambda}, \tag{13}$$

where F_λ is

$$F_\lambda = \frac{I_{\text{monitor}}}{1.602 \times 10^{-19} \times D_{\text{d.eff}}(\lambda)}, \tag{14}$$

where I_{monitor} is the monitor diode current measured for the whole beam, $D_{\text{d.eff}}$ is the SURF calibrated (against an absolute radiometric standard) wavelength-dependent quantum efficiency of the monitor diode. The size of the beam used for any of H or V orientations was decreased to be smaller than the size of the ESP entrance aperture (1 × 10 mm).

6.3.1. ESP Filter Transmissions

Efficiencies of the first-order bands for the flight ESP were determined in 2006 from the BL-9 calibration with the aluminum filter removed (Figure 6). The correction of these efficiencies for corresponding wavelength-dependent aluminum filter transmission were performed later, in 2008 after measurements at BL-9. ESP aluminum filters have a thickness of 150 ± 5 nm. The result of this measurement is displayed in Figure 8. It shows a typical (small) deviation in the filter transmission related to two sources of uncertainty: the aluminum film thickness and the thickness of the oxide layers of the filters.

The ESP C – Ti – C filter is a composite filter consisting of a 284 nm thick titanium layer between two carbon layers, each 19 nm thick. Figure 9 shows a comparison of the measured transmission of the zeroth-order C – Ti – C filter (with the minimum wavelength limited by the BL-9 design to 3.65 nm) and transmission of a modeled C – Ti – C filter with 284 nm of Ti and 38 nm of C but without a support-mesh structure. This model is based on calculations using the atomic parameters by Henke, Gullikson, and Davis (1993).

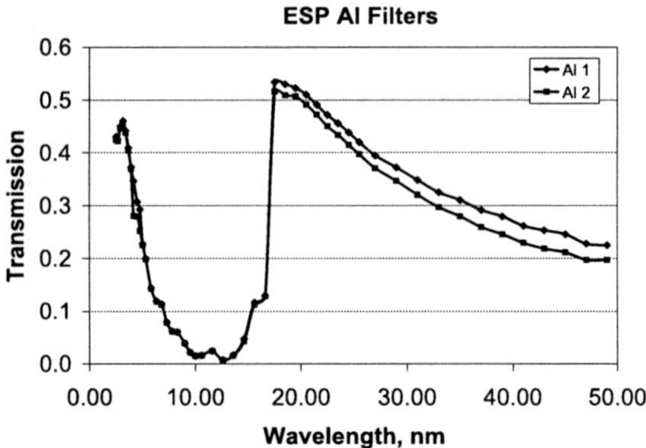

Figure 8 ESP aluminum filters transmission as measured at BL-9. Both filters (Al 1 and Al 2) are from the same batch as was used for the flight ESP filters.

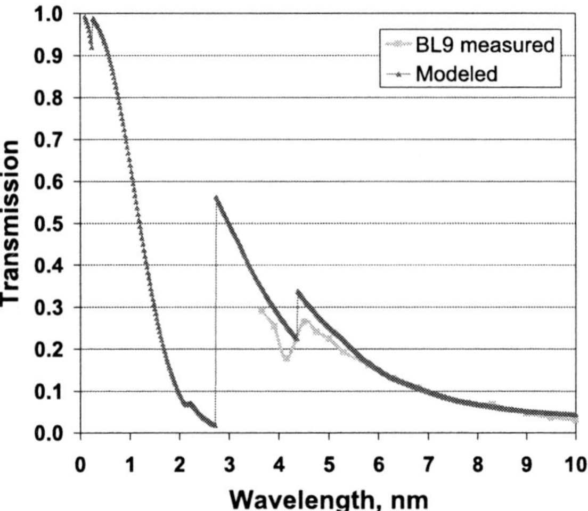

Figure 9 A typical modeled transmission profile (blue line) of a C – Ti – C filter with a thickness of 284 nm in a wavelength range of 0.1 to 10 nm. This modeled transmission profile is compared to that of the ESP titanium filter as measured at BL-9 (green line) for a wavelength range of 3.65 to 49.0 nm. Figure 9 shows transmission for shorter wavelengths, from 10 nm. The ESP Ti filter is from the batch used for the flight ESP filter.

6.3.2. Measured Transmission of the Diffraction Grating

The angle [φ_m] between the perpendicular to the grating and the direction to the transmission maxima is a function of the angle between the ESP optical axis and the direction of the input beam. According to the theory of a diffraction grating, this angle [φ_m] for normal incidence is a function of the wavelength [λ] and the period of the grating [d] (for the flight ESP d = 201 nm):

$$d \times \sin(\varphi_m) = m \times \lambda, \tag{15}$$

where m is the diffraction order, $m = 0, \pm 1, \ldots$. If the input beam does not coincide with the ESP optical axis, *e.g.* is tilted, the output (diffracted) beam is shifted in direction:

$$d \times \left[\sin(\theta) - \sin(\varphi_m)\right] = m \times \lambda, \tag{16}$$

where θ is the angle of the input beam with respect to the grating normal; φ_m is direction to the maxima. Efficiency profiles for the tilted ESP positions will be analyzed in the corresponding Section 6.4.1. Figure 10 shows transmission of the diffraction grating at 30.4 nm in

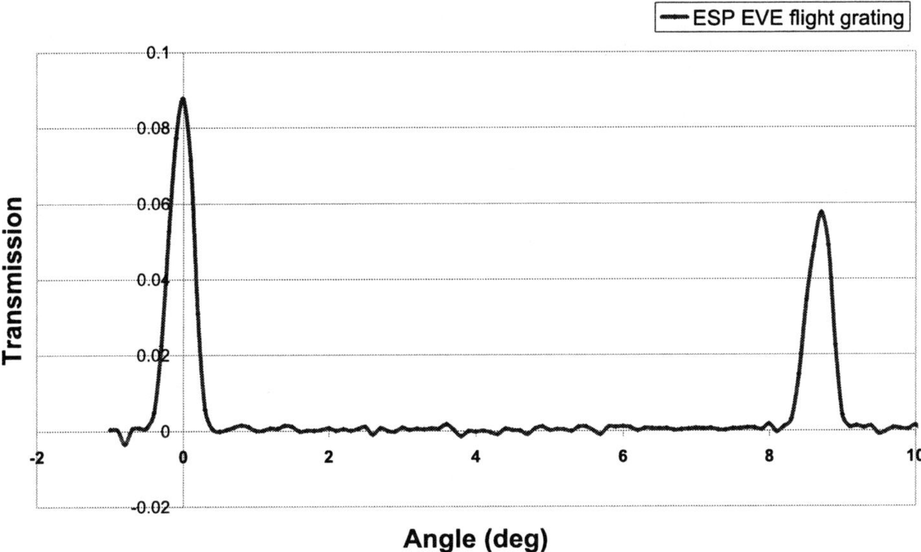

Figure 10 SURF measurements of the transmission of the diffraction grating for Ch9 (30.4 nm) in the zeroth-order (central peak) and the plus first-order.

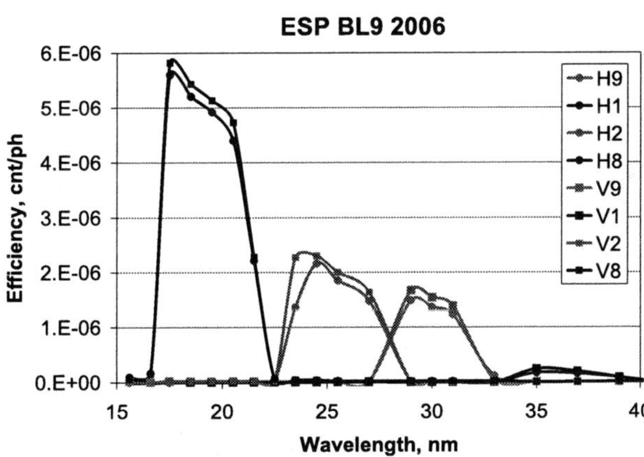

Figure 11 Spectral profiles of ESP first-order band efficiencies. From the shorter wavelengths to the longer wavelengths ESP first-order efficiencies are shown for Ch8 (19 nm, black), Ch2 (25 nm, blue), Ch9 (30 nm, green), Ch1 (36 nm, brown). Top curves (squares) show profiles for vertical (V) orientation, bottom curves (dots) are for horizontal (H) orientation. The larger efficiency for V than for H is due to the beam polarization.

the zeroth- and first-orders for the zero incidence angle beam measured at the SURF reflectometer. The amplitude of the zeroth-order transmission peak is about 9%. The first-order transmission peak at $\varphi_m = 8.7°$ [Equation (15)] is about 6% of the input intensity.

6.3.3. On-axis Efficiencies

The results of the ESP BL-9 calibration for the ESP on-axis position in horizontal and vertical orientations are shown in Figure 11. The QD efficiency spectral profile is shown in Figure 12. The profile is obtained as a combined result of BL-9 measurements (Figure 9), calculations, and BL-2 calibration. Efficiencies and spectral bandpasses for ESP bands are summarized in Table 3.

Figure 12 Spectral profile of ESP QD band efficiency. The efficiency drops about three orders of magnitude between 7 nm and 10 nm. This rapid decrease of the efficiency below 7 nm has a small effect on the calculated zeroth-order irradiance as the QD passband does not extend below 7 nm.

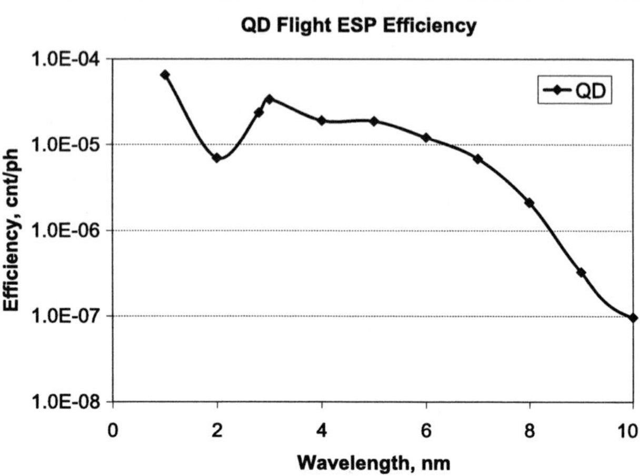

Table 3 Efficiencies and bandpasses of the ESP bands.

ESP band	Maximal (averaged for H and V) efficiency [cnt ph^{-1}]	FWHM edges [nm]	FWHM [nm]	Central wave-length [nm]
Ch1	2.09×10^{-7}	34.0–38.7	4.7	36.35
Ch2	2.23×10^{-6}	23.1–27.6	4.5	25.35
Ch8	5.71×10^{-6}	17.2–20.8	3.6	19.00
Ch9	1.62×10^{-6}	28.0–31.8	3.8	29.90
QD	6.55×10^{-5}	0.1–7.0	6.9	3.55

6.4. Results from the BL-2 Radiometric Pre-flight Calibration

Pre-flight radiometric calibration of ESP-flight (ESPF) was performed on 30 August 2007 at BL-2. ESPF was assembled in EVE and rotated 45°. The goal of that pre-flight radiometric calibration was to determine all characteristics marked as SURF in the second column of the Table 2. The calibration consists of three major parts: alignment to the SURF beam optical axis, tests with ESP tilted positions, tests for ESP optical-axis center point.

The alignment starts with vertical and horizontal scans to determine both the intensity center of the beam and the geometrical center of the beam. In the intensity center of the beam, the ESP is aligned in both yaw and pitch angles (Figure 7) to have differential signals from the zeroth-order QD (bands 4–7) equal to zero. If the differential signal along the dispersion direction of the grating is marked as X_d and in the opposite direction as Y_d, then the equations for these differential signals are:

$$X_\mathrm{d} = \frac{\mathrm{Ch6} + \mathrm{Ch7} - \mathrm{Ch4} - \mathrm{Ch5}}{\mathrm{Ch4} + \mathrm{Ch5} + \mathrm{Ch6} + \mathrm{Ch7}}, \tag{17}$$

$$Y_\mathrm{d} = \frac{\mathrm{Ch5} + \mathrm{Ch6} - \mathrm{Ch4} - \mathrm{Ch7}}{\mathrm{Ch4} + \mathrm{Ch5} + \mathrm{Ch6} + \mathrm{Ch7}}. \tag{18}$$

The tests with ESP tilted positions include a number of FOV maps where ESP responses to the tilts are measured as a matrix of tilts in both α and β directions (in yaw and pitch), and using cruciform scans in these directions.

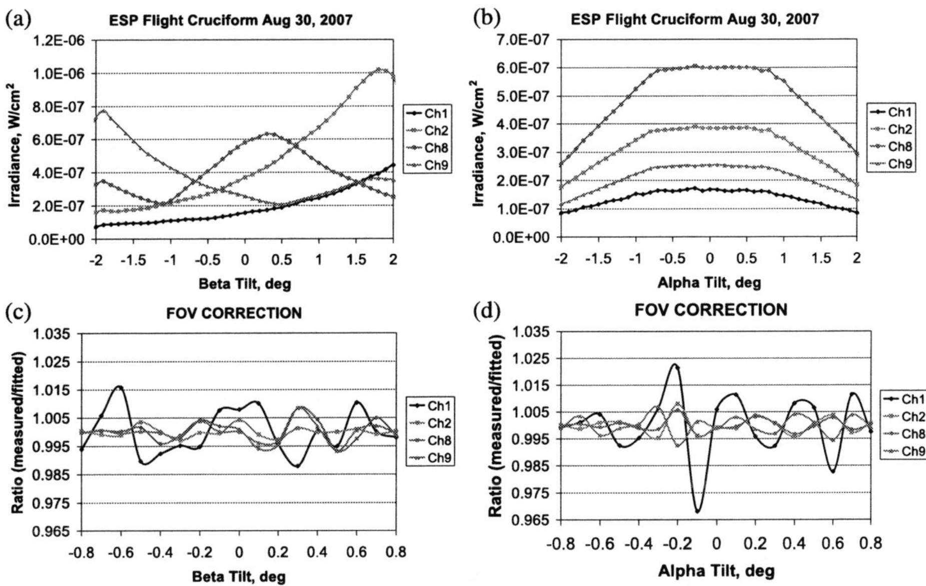

Figure 13 Calculated irradiance [Equation (1)] for the full FOV of ($\pm 2°$) from the BL-2 cruciform in the dispersion direction (a). Same for the perpendicular to the dispersion direction (b). A result of applying correction curves to the cruciform profiles is shown on the bottom panels. The level of the correction is shown as a ratio between the measured values for tilted positions and the fitted values (from the curves) for the dispersion direction (c), and for the perpendicular direction, (d). The correction fit was applied to the FOV of ($\pm 0.8°$), 1.5 times larger than the solar angular size.

6.4.1. ESP Responses for the Tilted Positions

The goals of the ESP off-axis calibration were to determine efficiency changes as a function of angle and QD band responses to the tilts. Some possible misalignment on orbit compared to the on-axis calibration at SURF may occur either as a result of mechanical and/or thermal deformations, or be related to an initial im-perfect co-alignment to the SDO imaging instruments, *e.g.* to AIA. The total misalignment is limited to within ± 3 arcmin. However, ESP may work in a much larger FOV of $\pm 2°$ and BL-2 pre-flight calibration included such $\pm 2°$ tilts as bi-directional cruciforms. The results of calculated BL-2 irradiance [Equation (1)] for β scans (along the dispersion) and α scans (in the perpendicular to dispersion direction) are shown in Figure 13a and b. The plus and minus offsets along the β axis cause wavelength shifts in different directions for plus and minus first-order ESP bands. These shifts increase or decrease calculated irradiance in the vicinity of the optical axis (Figure 13a). In contrast to these offsets along the β axis, the offsets in the perpendicular direction, along the α-axis do not cause wavelength shifts and the changes of calculated irradiance are small for small offsets (Figure 13b). Larger offsets (more than $\pm 0.8°$) cause a decrease of irradiance due to vignetting of the beam by detector masks.

The Figure 13 bottom panels, c and d, show ratios between measured and fitted curves as imperfectly compensated fluctuations with amplitudes of up to $\pm 1.5\%$ (c: dispersion direction), and $\pm 2.7\%$ (d: perpendicular to the dispersion direction).

Another reason to calibrate ESP in tilted positions is to determine the sensitivity of the zeroth-order bands to the tilts. The information about this sensitivity is important for establishing tilts based on the differential signals [Equations (16), (17)]. This zeroth-order

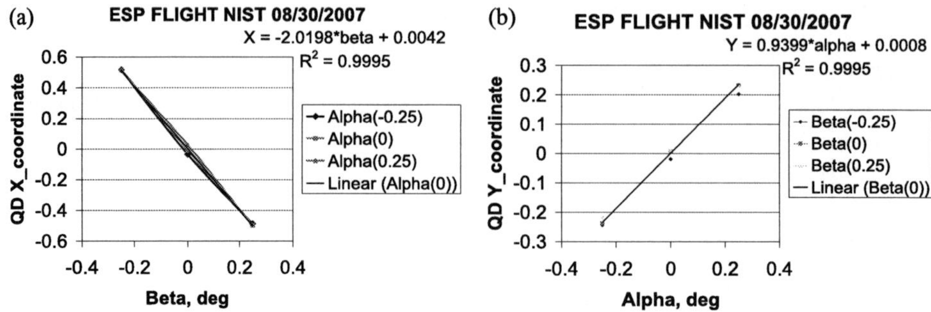

Figure 14 (a) The sensitivity of QD bands to tilts along the dispersion direction. (b) Same for the perpendicular to the dispersion direction. The sensitivity in each direction is a linear function within the tilts of $\pm 0.25°$. The sensitivity in the vertical [α] direction is about two times lower than the sensitivity in the dispersion (β) direction.

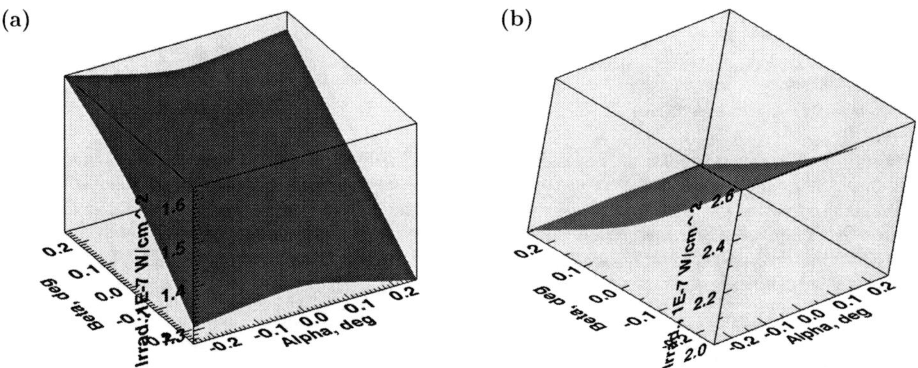

Figure 15 (a) An example of the change of calculated irradiance (in units of 10^{-7} W cm^{-2} in the $\pm 0.25°$ FOV). Note the significantly larger change of the irradiance in the grating dispersion (β) direction than in the α direction related to the shift of the diffraction maximum with the change of θ (15). (b) Same for Ch9. Note that the irradiance change for plus and minus β is opposite to the Ch1 (a) trend. The changes of the irradiance for both (a) and (b) in the α direction are small.

sensitivity will be used on orbit to determine the amount of misalignment and to calculate the position of a solar flare on the disk with near real-time information. The QD sensitivity to the tilts is shown in Figure 14.

Determined sensitivities [Equations (19), (20)] to the tilts within $\pm 0.25°$ are

$$X_{\mathrm{d}} = -2.02 \times \beta + 0.0042, \tag{19}$$

$$Y_{\mathrm{d}} = 0.94 \times \alpha + 0.0008. \tag{20}$$

6.4.2. BL-2 Calibration for ESP FOV Maps

Calibration for the ESP in tilted positions includes two-dimensional, nine-point, FOV maps with eight points in which either α, β, or both are tilted to $\pm 0.25°$ and one point for the center of the FOV. Such maps show the angular change of $R_i(\lambda, \alpha, \beta)$ (1). The results for the plus (Ch1) and minus (Ch9) first diffraction-order bands are shown in Figure 15. ESP

Table 4 A comparison of the BL-2 140 MeV irradiance that entered the ESPF bands (second column) and the measured irradiance (third column). The fourth column shows the spectral bands and the fifth column gives the relative error for this comparison. QD comparison was performed with the irradiance calculated for the beam with energy of 380 MeV.

ESP band	Input irradiance [W cm^{-2}]	Measured by ESPF, irradiance [W cm^{-2}]	Spectral band [nm]	Δ, %
Ch1	1.83×10^{-6}	1.83×10^{-6}	34.3 – 38.5	0.0
Ch2	8.16×10^{-7}	8.25×10^{-7}	23.4 – 28.1	1.1
Ch8	1.69×10^{-7}	1.70×10^{-7}	17.5 – 21.1	0.6
Ch9	1.37×10^{-6}	1.38×10^{-6}	28.0 – 32.7	0.7
QD	1.13×10^{-5}	1.12×10^{-5}	1.0 – 7.0	0.9

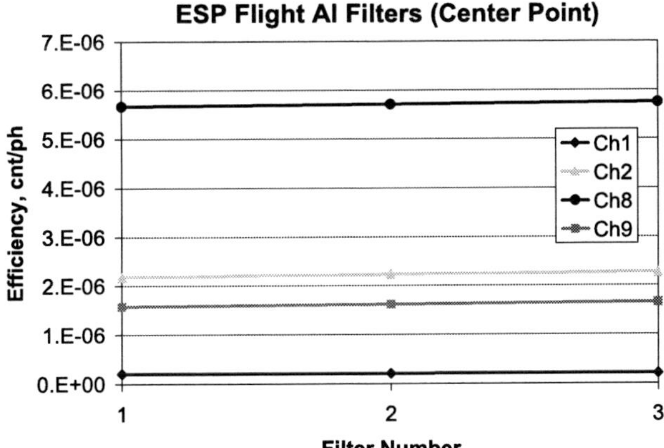

Figure 16 ESP first-order band efficiencies for three aluminum filters installed in the filter wheel of the flight ESP. The measured changes of the efficiencies related to a small difference in transmission for these Al filters are small, about $\pm$ 2.5%.

calibration for the FOV maps with two-dimensional tilts provides additional information to the single α or β cruciform tests (see above) about changes of the measured irradiance in the ESP $\pm 0.25°$ FOV.

6.4.3. BL-2 Calibration for ESP On-axis Position

ESP BL-2 calibration in the on-axis position includes calibration for the set of three aluminum filters in the filter wheel (Figure 1) and calibration with different synchrotron-beam energies, 380 MeV (a primary energy for calibration), 331 MeV, 285 MeV, 229 MeV, 183 MeV, and 140 MeV. These different energies shift the spectral distribution to provide a wide variety of emitted irradiation, from X-ray to visible light, suitable to calibrate the scientific bands and to permit a determination of the amount of higher-order diffraction contamination (order sorting). The results of ESPF pre-flight calibration (30 August 2007) for the beam energy of 140 MeV, which provides the spectral distribution with very low higher-order contamination, is given in Table 4. It shows a comparison of BL-2 irradiance which was entering each ESPF spectral band and the irradiance measured by the ESPF. For the zeroth-order QD band which is free of higher-order contamination, the comparison is given for the beam energy of 380 MeV. All three aluminum filters on the ESP filter wheel show a very similar transmission with deviations less than ±2.5%. Figure 16 shows the variation in efficiencies among these filters (1 through 3).

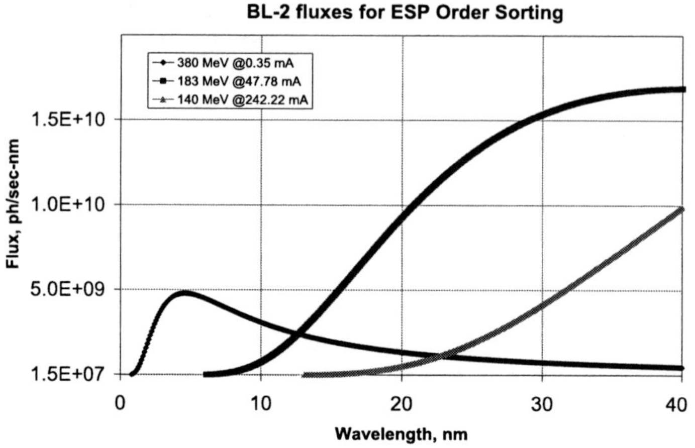

BL-2 fluxes for ESP Order Sorting

Wavelength, nm	0	10	20	30	40

Legend: 380 MeV @0.35 mA; 183 MeV @47.78 mA; 140 MeV @242.22 mA. Y axis: Flux, ph/sec-nm (1.5E+07, 5.0E+09, 1.0E+10, 1.5E+10).

Figure 17 An example of BL-2 fluxes for the beam energies of 380, 183, and 140 MeV, at beam currents of 0.35, 47.8, and 242.2 mA, correspondingly.

6.4.4. ESP Sensitivity to the Higher Orders (BL-2 Order Sorting Test)

The BL-2 radiometric calibration includes a special test called "Order Sorting". This test consists of a number of calibrations at different electron-beam energies, and thus varies the spectral content of the photon beam with which the ESP entrance aperture is illuminated. An example of BL-2 spectra for beam energies of 380 MeV, 331 MeV, and 140 MeV is shown in Figure 17.

Figure 17 shows that a calibration with the beam energy of 140 MeV allows for the detection and measurement of the input irradiance mainly by the first-order bandpasses. For example, the flux for the second-order of Ch9 at about 15 nm is more than two orders of magnitude lower than that of the first-order. These low-energy beams (140 MeV and 183 MeV) allow us to calibrate ESP with a minimal amount of higher-order irradiation. In contrast, the QD calibration use beams with high energy, *e.g.* 380 MeV and 331 MeV. These higher-energy beams provide sufficient signal to the QD zeroth-order channels (0.1 to 7.0 nm) and allow for precise alignment of the ESP to the beam, which is important during the calibration tests. The results of the Order Sorting test are shown in Figure 18.

Figure 18 shows two sets of points (gray and filled with colors for Ch1 (blue diamonds), Ch2 (red triangles), Ch8 (brown circles), and Ch9 (green squares) calculated for the first- through the fourth-order using beam energies of 140 MeV, 183 MeV, and 380 MeV. The efficiencies marked with the gray points were modeled assuming no change of the mean efficiency Eff_i for each order $[i]$ with different beam energies $[E]$ and beam currents $[I_E]$. The mean efficiency was determined as $(1/n_\lambda \sum_{\lambda_1}^{\lambda_2} \mathrm{Eff}(\lambda))$. Thus, effective counts $[C_{\mathrm{eff}}]$ are the mean efficiency multiplied by the integrated flux $\sum_{\lambda_{1i}}^{\lambda_{2i}} \Phi$:

$$C_{\mathrm{eff}} = \sum_{i=1}^{4} \mathrm{Eff}_i \left(\sum_{\lambda_{1i}}^{\lambda_{2i}} \Phi_{\mathrm{E}}(\lambda) \times I_{\mathrm{E}} \times \Delta\lambda \right), \tag{21}$$

where λ_{1i}, λ_{2i} are the edges of the bandpass for the order i.

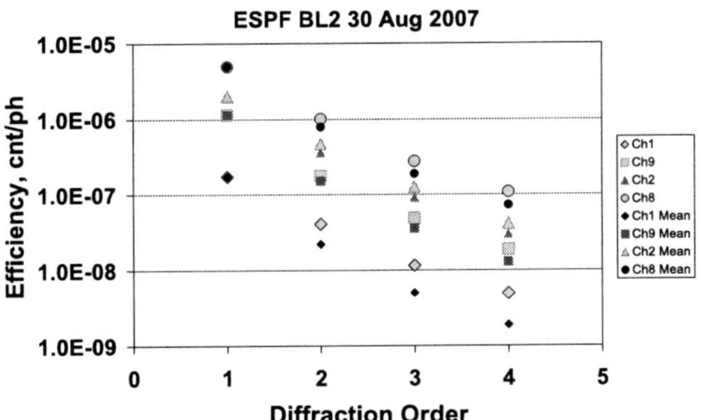

Figure 18 ESP efficiencies for the first-order channels (left side points) and range of the efficiency changes, between gray and colored points for the second, third, and fourth orders. The points were determined using two different modeling approaches.

The efficiencies marked with color-filled points are based on another approach, that allows for a change of the modeled efficiency due to flux shape change inside the channel's bandpass. The change of the spectral flux profile related to the switch from one energy to another (see Figure 17) shifts the mean efficiency for the channel, order, and the beam energy, allowing more accurate modeling of the effective counts measured by the channel. Each higher-order efficiency color point in Figure 18 represents the averaged efficiency for the three beam energies used. The channel's efficiency (one averaged number for all orders) determined as measured effective counts C_{eff} divided by the flux $\Phi(i, E)$ integrated over all order bandpasses $\sum_{i,\lambda} \Phi$ for this approach is compared to the sum of the mean efficiencies for each beam energy $[E]$ as

$$\frac{C_{\text{eff}}(\text{Ch}, E)}{\Phi(i, E)} = \sum_{i=1}^{4} \text{Eff}(\text{Ch}, E, i) \times \frac{\Phi(i, \text{Ch}, E)}{\Phi(i, E)}. \tag{22}$$

The right side of Equation (22) is the sum of efficiencies weighted by the ratio of the fluxes and, thus, sensitive to the shape of the flux profile in each higher-order bandpass.

The higher-order counts for the BL-2 ESP calibration are shown in Figure 19. The ratio between the measured effective counts and the counts extracted in the first-order bandpass is illustrated.

The ratio of 1.0 is within one percent for the beam energy of 140 MeV. It starts to be higher for the beam energy of 183 MeV and reaches the maximum ratio of 1.45 for Ch1 at the beam energy of 380 MeV. These ratios will be significantly lower for solar measurements with narrow spectral lines in contrast to the more continuous profiles characteristic of synchrotron radiation (Figure 17). Analyses for the higher-order contribution during solar measurements with SOHO/SEM showed that for the He II (30.4 nm) solar irradiance for quiet and intermediate activity these contributions are smaller than 10%.

7. A Comparison of ESP Measured Irradiance from the Sounding Rocket Flight with other Measurements

A sounding-rocket flight (NASA Rocket 36.240) with the EVE suite of channels was flown on 14 April 2008. The ESP rocket instrument (ESPR) collected data in both first-order and

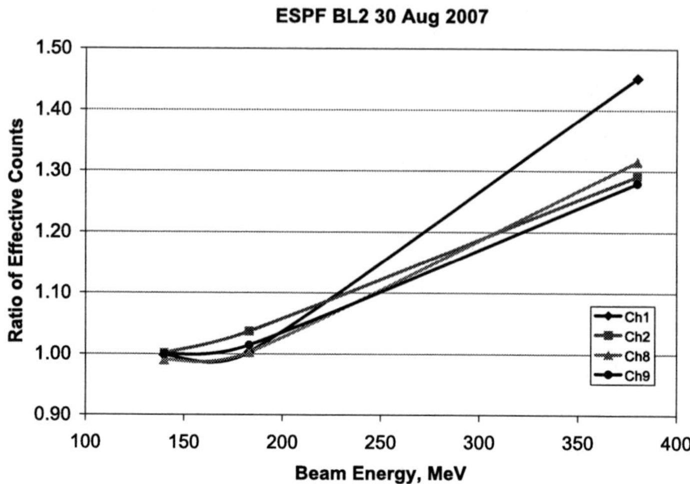

Figure 19 Amount of higher-order irradiance for three BL-2 beam energies, 140, 183, and 380 MeV is shown for the ESP first-order channels as a ratio between the measured effective counts and the counts extracted from the first-order bandpass.

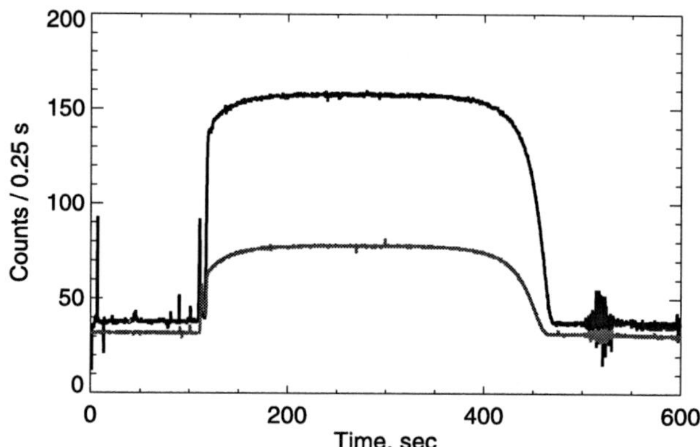

Figure 20 An example of observing profiles (detector counts *vs.* time) for two ESPR bands (Ch8 and Ch9) obtained during EVE sounding-rocket flight 36.240 flown on 14 April 2008. Black line shows Ch8 (19 nm) observing profile, blue line is for Ch9 (30.4 nm) data. The count rate on the plot includes dark counts (about 37.8 cnt (0.25 seconds^{-1}) for Ch8 and 31.9 cnt (0.25 seconds^{-1}) for Ch9). The time starts as the rocket was launched. The apogee point with altitude of about 285 km was achieved 275 seconds into the flight.

zeroth-order bands. ESP operational software was used to convert the count rates measured near apogee into solar irradiance according to Equation (1). The calculated irradiance was corrected for the Earth's atmosphere absorption (which is wavelength, date, and altitude dependent) using the MSIS (Hedin, 1987; Picone, Hedin, and Drob, 2003) atmosphere model. Figure 20 shows an example of the original count rates before subtracting dark currents for two ESPR bands: Ch8 and Ch9. Solar irradiance determined from the sounding-rocket flight ESPR measurements in the first- and zeroth-order bands was compared to the irradiance obtained from the EVE/MEGS spectra integrated over the same wavelengths. The results of

Table 5 A comparison of ESPR and MEGS absolute solar irradiance determined from the 14 April 2008 sounding-rocket flight. MEGS spectral irradiance was integrated over the ESPR wavelengths (fourth column) determined using the convolution as shown in Equation (11) (see also Figure 2 for Ch8 spectral band). The last column shows relative difference between ESP and MEGS irradiance.

ESP band	ESP irradiance [W m^{-2}]	MEGS irradiance [W m^{-2}]	Spectral band [nm]	Δ, %
Ch1	1.28×10^{-4}	1.32×10^{-4}	33.0 – 38.55	3.0
Ch2	N/A	N/A	N/A	N/A
Ch8	4.65×10^{-4}	4.65×10^{-4}	14.5 – 22.2	0.0
Ch9	5.20×10^{-4}	5.28×10^{-4}	26.7 – 33.8	1.5
QD	9.86×10^{-5}	9.77×10^{-5}	0.1 – 7.0	0.9

Table 6 A comparison of BL-2 irradiance that entered the ESPR bands (second column) for the beam energy of 140 MeV with the irradiance measured by the ESPR (third column). The fourth column shows the spectral bands and the fifth column gives relative error of this comparison. QD comparison was performed with the irradiance calculated for the beam with energy of 380 MeV.

ESP band	Input irradiance [W cm^{-2}]	Measured by ESPR irradiance [W cm^{-2}]	Spectral band [nm]	Δ, %
Ch1	2.22×10^{-6}	2.24×10^{-6}	34.2 – 38.5	0.9
Ch2	8.02×10^{-7}	8.12×10^{-7}	23.3 – 27.4	1.2
Ch8	1.79×10^{-7}	1.85×10^{-7}	17.5 – 20.9	3.4
Ch9	1.63×10^{-6}	1.65×10^{-6}	28.0 – 32.7	1.2
QD	1.01×10^{-5}	1.01×10^{-5}	1.0 – 7.0	0.0

this comparison are shown in Table 5. At the time of the sounding-rocket flight ESPR Ch2 was not working due to low shunt resistance of the diode detector. This detector was replaced in December 2008, and currently the ESPR instrument is fully operational and its bands are calibrated. The results of the BL-2 (23 January 2009) radiometric calibration for the updated ESPR are shown in Table 6. Table 6 compares ESPR input and measured irradiance similar to the comparison given in Table 4 for the ESPF. ESP Ch9 (30.4 nm) flux for the 14 April 2008 measurements was also compared to the SOHO/SEM first-order flux (26 to 34 nm). To make this comparison, ESP Ch9 flux was re-calculated for the SEM first-order bandpass and plotted in Figure 21 as a brown circle. Figure 21 shows SEM calibrated absolute first-order flux in the units of photons cm^{-2} seconds^{-1} (blue points) with over-plotted sounding-rocket data from both the SEM-clone instrument (red squares) and the Rare Gas Ionization Cell (RGIC; black triangles). SOHO/SEM data show that a solar minimum occurred by the end of 2008. ESP sounding-rocket flight data are in good agreement with both EVE/MEGS and SOHO/SEM data.

8. Concluding Remarks

As part of the SDO/EVE suite of channels, ESP will provide highly stable and accurate measurements of absolute solar irradiance in five EUV wavelength bands. Its high tempo-

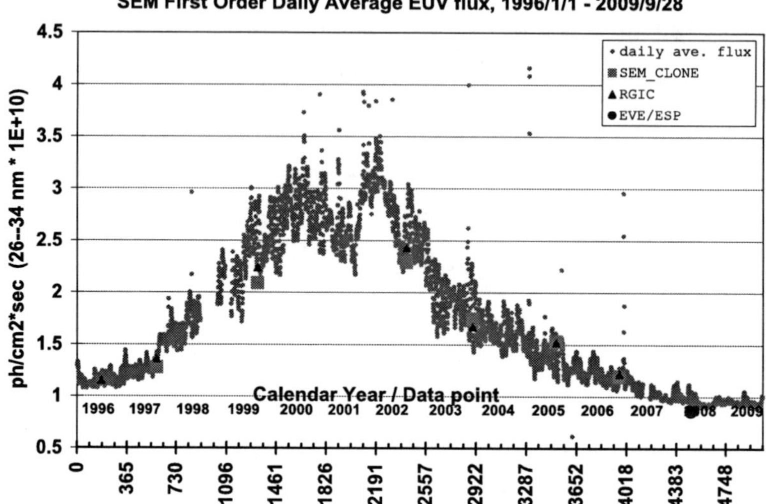

Figure 21 A comparison of SOHO/SEM first-order absolute EUV flux data for 26 to 34 nm (blue diamonds) with the SEM clone sounding-rocket data in the same wavelength bandpass (red squares), RGIC data converted to the SEM bandpass (black triangles) from the same sounding-rocket under-flights, and from EVE/ESP (Ch9 at 28.0 to 31.8 nm) sounding-rocket flight (brown circle). Solar spectral variability in the SOHO/SEM bandpass was calculated based on the SOLERS-22 solar-model composite spectrum (Woods *et al.*, 1998) used for SOHO/SEM calibration since 1996.

ral cadence, low latency, and spectral overlap with the other SDO instruments will provide important details of the dynamics of rapidly changing irradiance related to impulsive phases of solar flares. An algorithm to convert measured count rates and other ESP data into solar irradiance was described. ESP has many significant design improvements over its SOHO/SEM predecessor, including the possibility of measuring on-orbit changes of dark currents, electronics gain changes, contamination degradation, and pinholes of thin-film filters, visible-light scatter, and energetic-particle background. All optical components of ESP (diodes, filters, grating) were also separately tested and calibrated prior to an end-to-end calibration. The ESP was calibrated at SURF BL-9 using quasi-monochromatic wavelengths, and at BL-2 to obtain a radiometric calibration. Efficiency profiles determined during BL-9 calibration were used as reference data for calculation of irradiance in each band during BL-2 calibration. Measurements of solar irradiance from ESPR during the EVE sounding-rocket flight of 14 April 2008 were compared with corresponding EVE/MEGS spectra and SOHO/SEM irradiance measurements. These comparisons show good agreement. Both the ESP flight and rocket instruments are fully calibrated and ready for solar measurements.

Acknowledgements The authors thank Frank Eparvier, Mike Anfinson, Rick Kohnert, Greg Ucker, Don Woodraska, Karen Bryant, Gail Tate, Matt Triplett, and the rest of the LASP EVE Team at the University of Colorado at Boulder for their many contributions during ground tests of the ESP. We also thank Don McMullin of the Space Systems Research Corporation for his discussions and long-term support of this mission and the NIST SURF Team: Rob Vest, Mitch Furst, Alex Farrell, and Charles Tarrio for support of ESP calibration at BL-9 and BL-2 and diffraction grating transmission measurements at the SURF reflectometer facility. This work was supported by the University of Colorado award 153-5979.

References

Bowman, B.R., Tobiska, W.K., Marcos, F.A., Valladares, C.: 2008, The JB2006 empirical thermospheric density model. *J. Atmos. Solar-Terr. Phys.* **70**, 774.

Didkovsky, L.V., Judge, D.L., Jones, A.R., Wieman, S., Harmon, M., Tobiska, W.K.: 2007a, SEP temporal fluctuations related to extreme solar flare events detected by SOHO/CELIAS/SEM, http://www-rcf.usc.edu/~leonid/papers/AIAA-2007-496-973.pdf.

Didkovsky, L.V., Judge, D.L., Wieman, S., Harmon, M., Woods, T., Jones, A., Chamberlin, P., Woodraska, D., Eparvier, F., McMullin, D., Furst, M., Vest, R.: 2007b, SDO EVE ESP radiometric calibration and results. In: *SPIE* **6689**, San Diego, 6689OP1.

Didkovsky, L.V., Judge, D.L., Wieman, S., McMullin, D.: 2009, Minima of solar cycles 22/23 and 23/24 as seen in SOHO/CELIAS/SEM. In: Cranmer, S., Hoeksema, T., Kohl, J. (eds.) *SOHO-23: Understanding a Peculiar Solar Minimum, ASP Conf. Ser.*, submitted.

Furst, M.L., Graves, R.M., Madden, R.P.: 1993, Synchrotron ultraviolet radiation facility (SURF II) radiometric instrumentation calibration facility. *Opt. Eng.* **32**, 2930.

Hedin, A.E.: 1987, MSIS-86 thermospheric model. *J. Geophys. Res.* **92**, 4649.

Henke, B.L., Gullikson, E.M., Davis, J.C.: 1993, X-ray interactions: photoabsorption, scattering, transmission, and reflection at $E = 50-30\,000$ eV, $Z = 1-92$. *At. Data Nucl. Data Tables* **54**(2), 181.

Hovestadt, D., Hilchenbach, M., Bürgi, A., Klecker, B., Laeverenz, P., Scholer, M., Grünwaldt, H., Axford, W.I., Livi, S., Marsch, E., Wilken, B., Winterhoff, H.P., Ipavich, F.M., Bedini, P., Coplan, M.A., Galvin, A.B., Gloeckler, G., Bochsler, P., Balsiger, H., Fischer, J., Geiss, J., Kallenbach, R., Wurz, P., Reiche, K.-U., Gliem, F., Judge, D.L., Ogawa, H.S., Hsieh, K.C., Möbius, E., Lee, M.A., Managadze, G.G., Verigin, M.I., Neugebauer, M.: 1995, CELIAS – charge, element and isotope analysis system for SOHO. *Solar Phys.* **162**, 441.

Judge, D.L., McMullin, D.R., Ogawa, H.S., Hovestadt, D., Klecker, B., Hilchenbach, M., Möbius, E., Canfield, L.R., Vest, R.E., Watts, R., Tarrio, C., Kuehne, M., Wurz, P.: 1998, First solar EUV irradiances from SOHO by the CELIAS/SEM. *Solar Phys.* **177**, 161.

Ogawa, H.S., Canfield, L.R., McMullin, D., Judge, D.L.: 1990, Sounding rocket measurement of the absolute solar EUV flux utilizing a silicon photodiode. *J. Geophys. Res.* **95**, 4291.

Picone, J.M., Hedin, A.E., Drob, D.P., Aikin, A.C.: 2003, NRL-MSISE-00 empirical model of the atmosphere: Statistical comparisons and scientific issues. *J. Geophys. Res.*, doi:10.1029/2002JA009430.

Schattenburg, M.L., Anderson, E.H.: 1990, X-ray/VUV transmission gratings for astrophysical and laboratory applications. *Phys Scr.* **41**, 13.

Scime, E.E., Anderson, E.H., McComas, D.J., Schattenburg, M.L.: 1995, Extreme-ultraviolet polarization and filtering with gold transmission gratings. *Appl. Opt.* **34**(4), 648.

Tobiska, W.K.: 2007, SOLAR2000 v2.30 and SOLARFLARE v1.01: New capabilities for space system operations, http://www.spacewx.com/pdf/SET_AIAA_2007.pdf.

Tobiska, W.K., Bouwer, S.D., Bowman, B.R.: 2006, The development of new solar indices for use in thermospheric density modeling, http://sol.spacenvironment.net/~JB2008/pubs/SET_AIAA_2006_revD.pdf.

Tsurutani, B.T., Judge, D.L., Guarnieri, F.L., Gangopadhyay, P., Jones, A.R., Nuttall, J., Zambon, G.A., Didkovsky, L., Mannucci, A.J., Iijima, B., *et al.*: 2005, The October 28, 2003 extreme EUV solar flare and resultant extreme ionospheric effects: Comparison to other Halloween events and the Bastille Day event. *Geophys. Res. Lett.* **32**, L03S09.

Woods, T.N., Bailey, S., Eparvier, F.G.: 1998, TIMED solar EUV experiment. *SPIE* **3442**, 180.

Woods, T.N., Ogawa, H., Tobiska, K., Farnik, F.: 1998, SOLERS-22 WG-4 and WG-5 report for the 1996 SOLERS-22 workshop. In: Pap, J.M., Fröhlich, C., Ulrich, R.K. (eds.) *Solar Electromagnetic Radiation Study for Solar Cycle 22*, Kluwer, Dordrecht, 511.

Woods, T.N., Chamberlin, P., Eparvier, F.G., Hock, R., Jones, A., Woodraska, D., Judge, D., Didkosky, L., Lean, J., Mariska, J., *et al.*: 2009, Extreme ultraviolet variability experiment (EVE) on the solar dynamics observatory (SDO): Overview of science objectives, instrument design, data products, and model developments. *Solar Phys.*, accepted.

Solar Phys (2012) 275:207–227
DOI 10.1007/s11207-011-9834-2

The *Helioseismic and Magnetic Imager* (HMI) Investigation for the *Solar Dynamics Observatory* (SDO)

P.H. Scherrer · J. Schou · R.I. Bush · A.G. Kosovichev · R.S. Bogart · J.T. Hoeksema · Y. Liu · T.L. Duvall Jr. · J. Zhao · A.M. Title · C.J. Schrijver · T.D. Tarbell · S. Tomczyk

Received: 6 June 2011 / Accepted: 4 August 2011 / Published online: 18 October 2011

Abstract The *Helioseismic and Magnetic Imager* (HMI) instrument and investigation as a part of the NASA *Solar Dynamics Observatory* (SDO) is designed to study convection-zone dynamics and the solar dynamo, the origin and evolution of sunspots, active regions, and complexes of activity, the sources and drivers of solar magnetic activity and disturbances, links between the internal processes and dynamics of the corona and heliosphere, and precursors of solar disturbances for space-weather forecasts. A brief overview of the instrument, investigation objectives, and standard data products is presented.

Keywords *Solar Dynamics Observatory* · Helioseismology · Instrumentation and data management · Magnetic fields, photosphere

1. Overview

The *Helioseismic and Magnetic Imager* (HMI) investigation is part of the NASA *Solar Dynamics Observatory* (SDO) mission, which is the first flight component of the NASA

The solar Dynamics Observatory
Guest Editors: W. Dean Pesnell, Phillip C. Chamberlin, and Barbara J. Thompson

P.H. Scherrer (✉) · J. Schou · R.I. Bush · A.G. Kosovichev · R.S. Bogart · J.T. Hoeksema · Y. Liu · J. Zhao
W.W. Hansen Experimental Physics Laboratory, Stanford University, Stanford, CA 94305-4085, USA
e-mail: pscherrer@solar.stanford.edu
url: http://hmi.stanford.edu

T.L. Duvall Jr.
Laboratory for Astronomy and Solar Physics, NASA Goddard Space Flight Center, Greenbelt, MD 20771, USA

A.M. Title · C.J. Schrijver · T.D. Tarbell
Lockheed Martin Solar and Astrophysics Laboratory, 3251 Hanover St., Palo Alto, CA 94304, USA

S. Tomczyk
High Altitude Observatory, National Center for Atmospheric Research, 3080 Center Green CG-1, Boulder, CO 80301, USA

Living With a Star (LWS) program. HMI is the name of one of the instruments on SDO and is also the name of a science investigation. This article provides a terse overview of both aspects of HMI primarily as an introduction to several more detailed articles. HMI is a joint project of the Stanford University Hansen Experimental Physics Laboratory, the Lockheed Martin Solar and Astrophysics Laboratory, the High Altitude Observatory, and Co-Investigators at 21 additional institutions.

1.1. Top Level Goals

The overarching purpose of SDO is to make progress in the capability to understand and predict solar events that contribute to variability in the Earth's space environment. The solar variability that generates events important for Earth is driven by emergence and evolution of solar magnetic active regions. The primary goal of the HMI investigation is to study the origin of solar variability and to characterize and understand the Sun's interior and the various components of magnetic activity. These top-level objectives are described in the LWS Science Architecture Team Report (Mason and LWS Panel, 2001) and the SDO Science Definition Team Report (Hathaway and SDO SDT Panel, 2001).

The range of science investigations that HMI can contribute to, as understood at the time of the proposal to NASA for the HMI investigation, was described in that proposal (Scherrer and HMI Team, 2002). The expectation is that only some of the scientific research identified in the proposal will be pursued by the core HMI team at Stanford University with many of the studies to be conducted by the Co-Investigator team and by many others who are enticed by the challenging topics enabled by the availability of the HMI data collection. Therefore this article should not be construed as the list of investigations that will be pursued by the HMI team but rather as the list of topics that were examined in some detail to ensure that the observational and pipeline processing capabilities of the HMI project can make significant contributions to their better understanding. A detailed discussion of the initial investigation expectations was provided to NASA as a required document and simultaneously to the community via web availability in November 2004 in the "HMI Science Plan" (Kosovichev and HMI Science Team, 2004). Since this expanded description as well as the original more condensed version in the HMI 2002 proposal have been on the web for several years, the anticipated science investigation topics need not be fully detailed here. While these defining documents anticipated the SDO launch to be in 2006 or 2007 to begin a five-year base mission at a time of low solar activity, the Sun managed to delay the start of the current cycle until the SDO actual launch in early 2010. The earlier documents' description of the science objectives of HMI are still valid and are therefore the primary basis for this article.

The specific scientific objectives of the HMI investigation are to measure and study: convection-zone dynamics and the solar dynamo; the origin and evolution of sunspots, active regions and complexes of activity; the sources and drivers of solar magnetic activity and disturbances; links between the internal processes and dynamics of the corona and heliosphere; and precursors of solar disturbances for space-weather forecasts. These objectives will be described in some more detail below.

1.2. Scope

This article is one of several that describe the HMI investigation, instrument, and processing plans as of roughly the time of the SDO launch. It covers primarily an overview of the planned science investigation with only a very terse mention of the other aspects, which are described in some detail in the associated articles. The detailed description of the HMI instrument is contained in Schou *et al.* (2011) and it should be cited by users of data produced by HMI. Some of the aspects of the instrument and data-analysis pipeline required specific HMI developments and have been detailed in other associated articles. These include detailed discussions of the HMI image quality (Wachter *et al.*, 2011), the HMI filter characteristics (Couvidat *et al.*, 2011), the instrumental polarization calibration (Schou *et al.*, 2010), the vector-field inversion technique (Borrero *et al.*, 2010), and the time–distance helioseismology travel-time measurement methods (Couvidat *et al.*, 2010), and the time–distance helioseismology processing pipeline (Zhao *et al.*, 2011). The expectation is that there will be additional articles describing the actual on-orbit characteristics of data products. These actual data performance articles will not be available until more than a year into the mission. Data-product descriptions and performance notes will be updated and maintained via links to the web at http://hmi.stanford.edu and http://jsoc.stanford.edu for the duration of the SDO mission.

1.3. The *Helioseismic and Magnetic Imager* Instrument

To support these objectives, the HMI instrument has been designed to produce measurements in the form of filtergrams in a set of polarizations and spectral-line positions at a regular cadence for the duration of the mission. The HMI instrument makes measurements of the motion of the solar photosphere to study solar oscillations and measurements of the polarization in a spectral line to study all components of the photospheric magnetic field. The HMI instrument is an enhanced version of the *Michelson Doppler Imager* (MDI) instrument which is part of the *Solar and Heliospheric Observatory* (SOHO: Scherrer *et al.*, 1995). HMI has significant heritage from MDI with modifications to allow higher resolution, higher cadence, and the addition of a second camera to provide additional polarization measurements without interfering with the filtergram cadence needed for helioseismology. HMI provides stabilized one-arcsecond resolution full-disk Doppler velocity, line-of-sight magnetic flux, and continuum proxy images every 45 seconds, and vector magnetic field maps every 90 or 135 seconds depending on the image frame sequence selected. The solar image nearly fills the 4096×4096 pixel CCD camera allowing 0.5 arcsecond per pixel. The HMI instrument was developed at the Lockheed Martin Solar and Astrophysics Laboratory (LMSAL) in collaboration with Stanford University as part of the Stanford Lockheed Institute for Space Research collaboration.

1.4. HMI Data Products

The significant data stream to be provided by HMI must be analyzed and interpreted with advanced tools that permit study of the complex flows and structures deduced from helioseismic inversion together with direct measurements of surface magnetic fields. It is essential to have convenient access to all data products – Dopplergrams, full-disk vector magnetograms, sub-surface flow-field and sound-speed maps, as well as coronal magnetic-field estimates – for any region or event selected for analysis. The other SDO investigation that generates a large dataflow, the *Atmospheric Imaging Assembly* (AIA: Lemen *et al.*, 2011) developed and operated by LMSAL, and HMI have combined operations and data processing capabilities in the form of the HMI/AIA Joint Science Operations Center (JSOC). The JSOC consists of three components: the JSOC Instrument Operations Center (IOC) located at LMSAL, the Science Data Processing facility (SDP) located at Stanford University, and the AIA Visualization Center (AVC) located at LMSAL. The JSOC-SDP provides a data system for archiving the HMI and AIA data and derived data products with convenient access to the data by all interested investigators as well as the general public. JSOC-managed data products are available directly to the user via web tools, with key products also distributed through the Virtual Solar Observatory (VSO). The SDP facility also provides processing capabilities for higher-level HMI data products. Sufficient computing capability is in place to allow the investigation of a subset of the HMI science objectives.

HMI will obtain filtergrams in various positions in the Fe I 617.3 nm spectral line and a set of polarizations at a regular cadence for the duration of the mission. Several higher levels of data products will be produced from the filtergrams. The basic science observables are full-disk Doppler velocity, a continuum brightness proxy, line-of-sight magnetic flux, and vector magnetic field. These are available at full spatial resolution at a 45-second cadence. Additional estimates are generated for the line width and line depth. The basic vector magnetic-field quantity is an array of 24 filtergrams (six wavelengths, four Stokes parameters) averaged over 12 minutes. Additionally, the results of Milne–Eddington inversions and the proper disambiguated field vector will be provided. Derived data products sampled and averaged at various resolutions and cadence and sub-image samples tracked with solar rotation are also available. A selection of these will be made available on a regular basis, and other data products will be made on request. Also of great potential value are derived products such as sub-surface flow maps, far-side activity maps, and coronal and solar-wind models which require longer sequences of observations. A selection of these will also be produced in the processing pipeline in near-real time. Most of the HMI Co-Investigators have specific roles in providing software to enable production of the various standard data products. The processing of HMI data is accomplished in a "pipeline" moving data from the observed filtergrams through multiple steps to basic science products (the "observables") and higher-level derived products such as sub-photospheric flow maps. The basic processing pipeline is shown in Figure 1. The JSOC-SDP is described in more detail on its web page at http://jsoc.stanford.edu.

1.5. Team, Collaborations and Community

The HMI investigation plan identifies a broad range of scientific objectives that can be addressed with HMI observations. HMI provides a unique set of data required for scientific understanding, detailed characterization, and advanced warning of the effects of solar disturbances on global changes, space weather, human exploration and development, and technological systems. HMI also provides correlative data required for accomplishing objectives of the other SDO instruments. A significant research effort must be undertaken to exploit the capabilities of HMI and SDO.

The HMI program itself does not have sufficient resources to pursue all of the science goals possible with HMI data or even all of the goals identified in this article; neither was it planned to. The expectation is that Co-Investigators as well as many other interested investigators will find support as needed to carry out the studies that will in the end make most use of HMI data to achieve the goals of HMI, SDO, and LWS. To help this process, NASA has selected a consortium led by D. Braun at Colorado Research Associates (a division of Northwest Research Associates) to support development of next-generation helioseismology analysis tools. Tools developed by this supporting project will be available in several years. Most HMI analysis is likely to be carried out with support from NASA LWS and Guest Investigator grant programs in the US and from other national science-support agencies in other countries.

To further the goals of nationally funded research it is important that the science teams pursue careful science investigations and tell others about the results – and not just the science community. SDO investigations and HMI in particular have aspects which will be of

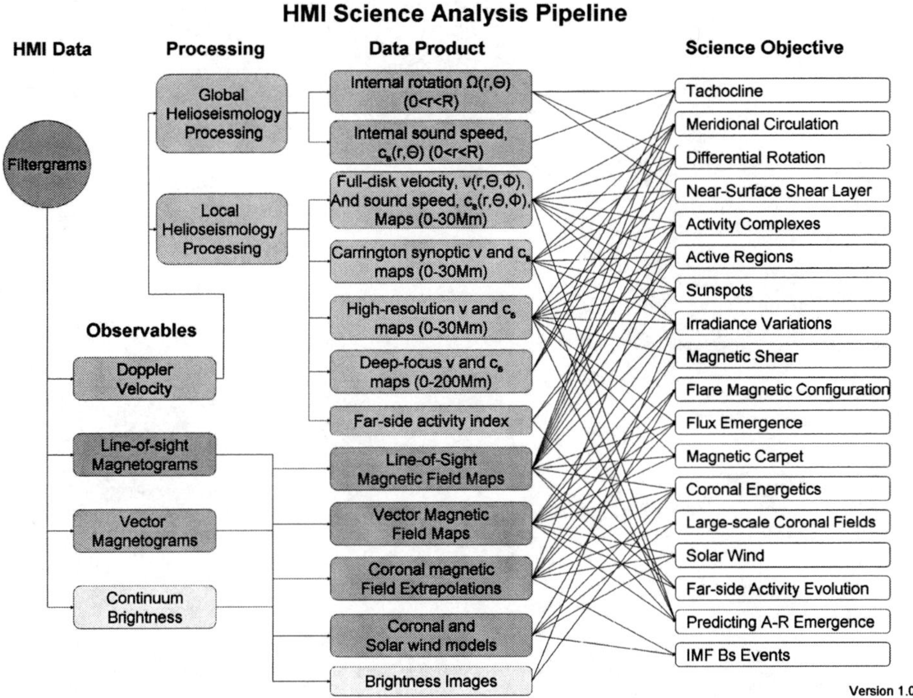

Figure 1 Schematic flow diagram for HMI data-analysis pipeline. The observed filtergrams are used primarily to produce "observables" which are the level-1 data products available to the community. Further higher-level products such as internal rotation and coronal-field extrapolations are produced on a regular cadence and are expected to be the data used for science investigations, shown by the arrows, to achieve the HMI science objectives.

interest to the public at large. Also SDO and HMI will offer excellent opportunities for developing interesting and timely educational material. A highly leveraged, collaborative Education and Public Outreach (E/PO) program is a key part of this investigation (Drobnes *et al.*, 2011). The EPO program is described in components of the http://solar-center.stanford.edu access point.

1.6. History

In January 1996 when the first SOHO/MDI test images became available, Duvall, Kosovichev, and colleagues (Duvall *et al.*, 1993, 1997; Kosovichev, 1996) began an investigation to explore using the newly developed time–distance method of local-area helioseismology with MDI high-resolution data. This method proved to be immediately useful but constrained by the limited field of view of the MDI high-resolution (1.2 arcsec) configuration. The success of the method, limited field of view, and limited telemetry opportunities suggested that some future mission would be needed to fully exploit the opportunity to make detailed probes in the near-surface solar interior. This conclusion was presented at the NASA Roadmap conferences at JHU-APL in April 1996 and JPL in October 1996. In 1998 a proposal was submitted to NASA for a MIDEX mission named "Hale" which included a full-disk high-resolution MDI-like instrument and an EUV imaging instrument. Figure 2, which shows wave speed beneath a sunspot, was made for that proposal with data obtained with

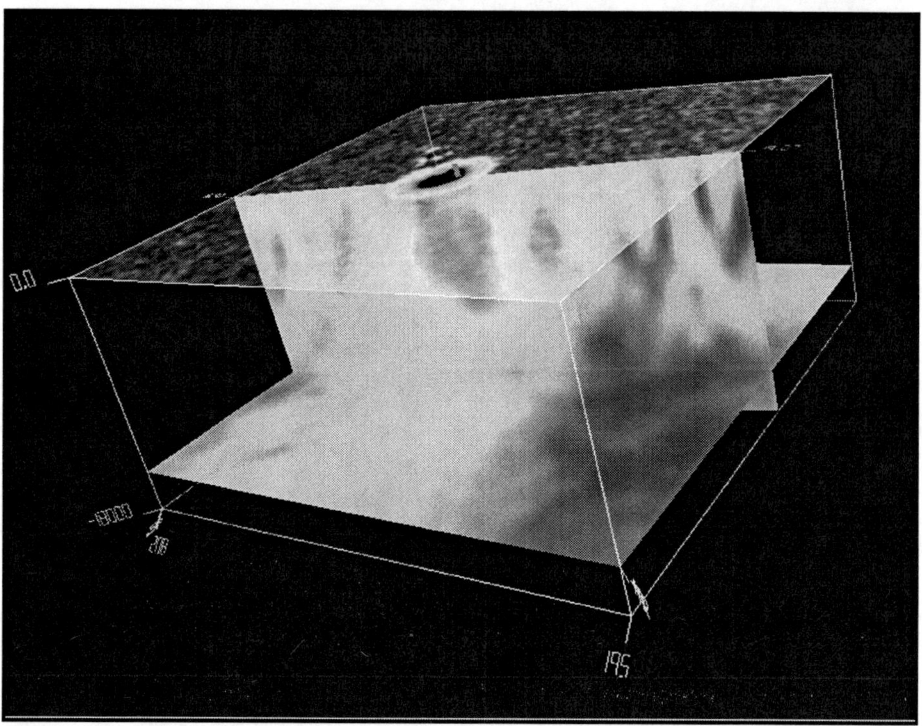

Figure 2 Helioseismic tomography used to map wave speed beneath the sunspot of NOAA AR 8243 on 18 June 1998 computed with the method developed by Duvall *et al.* (1997). The top surface is the MDI continuum-intensity proxy and interior slices are wave speed with blue tones for wave speed slower than average for the box and red colors for higher speed. The box is about 195 Mm square and 18 Mm deep. Analysis method was described by Kosovichev, Duvall, and Scherrer (2000) and the figure is from a proposal to the NASA 1998 MIDEX opportunity – Hale: Exploring Solar Activity.

SOHO/MDI in June 1998. This was the first sunspot that traversed the MDI high-resolution field while high-rate telemetry was available and was in the week before the 1998 SOHO mission disruption. The MIDEX proposal was not selected but the concept was included in the Sun–Earth Connections Roadmap of 2000 as a mission named SONAR which contained the essence of the eventual SDO. Over the next two years these concepts helped to form the SDO mission concept as part of the then new Living With a Star program. Broader community input into the SDO Definition Study helped to define a "strawman" HMI instrument which led to the full HMI proposal for the SDO mission in April 2002. About fourteen years elapsed between the first suggestion of a mission with HMI capabilities and the beginning of the science phase of SDO in May 2010.

2. HMI Science Goals

The HMI investigation encompasses three primary LWS objectives: first, to determine how and why the Sun varies; second, to improve our understanding of how the Sun drives global change and space weather; and third, to determine to what extent predictions of space weather and global change can be made and to prototype predictive techniques.

2.1. Science Overview

The Sun is a magnetic star. High-speed solar wind, the structure of heliospheric current sheets, coronal holes, coronal mass ejections, flares, and variable components of irradiance are all linked to the variability of magnetic fields that pervade the solar interior and atmosphere. Many of these events can have profound impacts on our technological society, so understanding them is a key objective for LWS. A central question is the origin of solar magnetic fields. Most striking is that the Sun exhibits 22-year cycles of global magnetic activity involving magnetic active regions with very well-defined polarity rules producing global-scale magnetic patterns. Coexisting with these large-scale magnetic structures and concentrated active regions are ephemeral active regions and other compact and intense flux structures (Zwaan, 1987) that emerge randomly over much of the solar surface forming a "magnetic carpet" (Title and Schrijver, 1998).

The extension of these changing fields, at all scales, into the solar atmosphere creates coronal activity, which is the source of space-weather variability. The HMI scientific investigation addresses the fundamental problems of solar variability with studies in all interlinked temporal and spatial domains, including global scale, active regions, small scale, and coronal connections. One of the prime objectives of the Living With a Star program is to understand how well predictions of evolving space-weather variability can be made. The HMI investigation will examine these questions in parallel with the fundamental science questions of how the Sun varies and how that variability drives global change and space weather.

The tools that will be used in the HMI program include: helioseismology to map and probe the solar convection zone where a magnetic dynamo likely generates this diverse range of activity; measurements of the photospheric magnetic field, which results from the internal processes and drives the processes in the atmosphere; and brightness measurements, which can reveal the relationship between magnetic and convective processes and solar irradiance variability.

Helioseismology, which uses solar oscillations to probe flows and structures in the solar interior, is providing remarkable new perspectives about the complex interactions between highly turbulent convection, rotation, and magnetism. It has revealed a region of intense rotational shear at the base of the convection zone (Schou *et al.*, 1998), called the tachocline, which is the likely seat of the global dynamo. Convective flows also have a crucial role in advecting and shearing the magnetic fields, twisting the emerging flux tubes and displacing the photospheric footpoints of magnetic structures present in the corona. Flows at all spatial scales influence the evolution of the magnetic fields, including how the fields generated near the base of the convection zone rise and emerge at the solar surface, and how the magnetic fields already present at the surface are advected and redistributed. Both of these mechanisms contribute to the establishment of magnetic-field configurations that may become unstable and lead to eruptions that affect the near-Earth environment (*e.g.*, van Driel-Gesztelyi and Culhane, 2009).

Local-area helioseismology methods have begun to reveal the great complexity of rapidly evolving 3D magnetic structures and flows in the sub-surface shear layer in which the sunspots and active regions are embedded. Most of these techniques were developed by members of the extended HMI team for analysis of MDI and GONG observations. As useful as they are, the limitations of MDI telemetry and the limited field of view at high resolution and limited GONG resolution have prevented the full exploitation of the methods to answer important questions about the origins of solar variability. By using these techniques on continuous, full-disk, high-resolution observations, HMI will enable detailed probing of dynamics and magnetism within the near-surface shear layer, and provide sensitive measures of variations in the tachocline.

Just as existing helioseismology experiments have shown that new techniques can lead to new understanding; methods to measure the full vector magnetic field have been developed and shown the potential for significantly enhanced understanding of magnetic evolution and connections. What existing and planned ground-based programs cannot do, and what *Hinode* cannot do, is to observe the full-disk vector magnetic field continuously at a cadence sufficient to follow the development of activity. The HMI vector magnetic-field measurement capability, in combination with the other SDO instruments and other programs (*e.g.*, STEREO, *Hinode*, and SOLIS) will provide data important for connecting solar variability in the solar interior to variability in the solar atmosphere, and to the propagation of solar variability in the heliosphere.

HMI brightness observations will provide information about the areal distribution of magnetic and convective contributions to irradiance variations, and also about variations of the solar radius and shape.

2.2. Scientific Goals

The broad goals described above will be addressed in a coordinated investigation of a number of parallel studies. These segments of the HMI investigation listed in Section 1.1 are described in some more detail here.

The goals represent long-standing problems that can be addressed by a number of more immediate tasks. The description of these tasks reflects our current level of understanding and will obviously evolve in the course of the investigation. Some of the currently most-important tasks are described below. The five broad objectives above are themselves described as sets of component objectives. Figure 1 shows a conceptual flow diagram connecting HMI planned data products with these component objectives.

2.2.1. Convection-Zone Dynamics and the Solar Dynamo

Fluid motions inside the Sun are believed to generate solar magnetic fields. Complex interactions between turbulent convection, rotation, large-scale flows, and magnetic field produce regular patterns of solar activity changing quasi-periodically with the solar cycle. How are variations in the solar cycle related to the internal flows and surface magnetic field? How is the differential rotation produced? What is the structure of the meridional flow and how does it vary? What roles do the zonal-flow pattern and the variations of the rotation rate in the tachocline play in the solar dynamo? These issues are usually studied only in zonal averages by global helioseismology, but the Sun is longitudinally structured. Local helioseismology has revealed the presence of large-scale flows within the near-surface layers of the solar convection zone (*e.g.*, Howe, 2008). These flows possess intricate patterns that change from one day to the next, accompanied by more gradually evolving patterns such as banded zonal flows and meridional circulation cells. These flow structures have been described as Solar Sub-surface Weather (SSW). Successive maps of these weather-like flow structures suggest that solar magnetism strongly modulates flow speeds and directions. Active regions tend to emerge in latitudes with stronger shear. The connections between SSW and active-region development are presently unknown.

2.2.1.1. Structure and Dynamics of the Tachocline Observation of the deep roots of solar activity in the tachocline is of primary importance for understanding the long-term variability of the Sun. HMI will use global and local helioseismic techniques to observe and investigate the large-scale character of the convection zone and tachocline. Tests with MDI observations and simulated data have shown that variations at the tachocline can be measured with local-helioseismology methods (Zhao *et al.*, 2009). This and related helioseismology techniques will be used to study deep structures.

2.2.1.2. Variations in Differential Rotation Differential rotation is a crucial component of the solar cycle and is believed to generate the global-scale toroidal magnetic field seen in active regions. HMI will extend this useful data product with better near-surface resolution. Topics include solar differential rotation, relations between variations of rotation and magnetic fields, longitudinal variation of zonal flows ("torsional oscillations"), relations between the torsional pattern and active regions, sub-surface shear and its variations with solar activity, and the origin of the "extended" solar cycle (*e.g.*, Howe *et al.*, 2011).

2.2.1.3. Evolution of Meridional Circulation Precise knowledge of meridional circulation in the convection zone is crucial for understanding the long-term variability of the Sun. Helioseismology has found evidence for variation of the internal poleward flow during the solar cycle (*e.g.*, Komm *et al.*, 2011). To understand the global dynamics we must follow the evolution of the flow. HMI will enable measurement of the meridional circulation to significantly higher latitudes than has been possible with MDI and will generate continuous data for detailed, 3D maps of the evolving patterns of meridional circulation providing information about how flows transport and interact with magnetic fields throughout the solar cycle.

2.2.1.4. Dynamics in the Near-Surface Shear Layer Helioseismology has revealed that significant changes in solar structure over the solar cycle occur in the near-surface shear layer. However, the physics of these variations and their role in irradiance variations are still unknown. HMI will characterize the properties of this shear layer, the interaction between surface magnetism and evolving flow patterns, and the changes in structure and dynamics as the solar cycle advances. It will assess the statistical properties of convective turbulence over the solar cycle, including the kinetic helicity and its relation to magnetic helicity – two intrinsic characteristics of dynamo action.

2.2.1.5. Origin and Evolution of Sunspots, Active Regions, and Complexes of Activity Observations show that magnetic flux on the Sun does not appear randomly. Once an active region emerges, there is a high probability that additional eruptions of flux will occur nearby (activity nests). How is magnetic flux created, concentrated, and transported to the solar surface where it emerges in the form of evolving active regions? To what extent are the appearances of active regions predictable? What roles do local flows play in their evolution? HMI will address these questions by providing tracked sub-surface sound-speed and flow maps for individual active regions and complexes under the visible surface of the Sun combined with surface magnetograms. Suggestions have been made that flux emerging in active regions originates in the tachocline. Flux is somehow ejected from the depths in the form of loops that rise through the convection zone and emerge through the surface. Phenomenological flux-transport models show that the observed photospheric distribution of the flux does not require a long-term connection to flux below the surface. Rather, field motions are described by the observed poleward flows, differential rotation, and surface diffusion acting on emerged flux of active regions. Does the active-region magnetic flux really disconnect from the deeper flux ropes after emergence? Are emerging flux regions really originate at the tachocline?

2.2.1.6. Formation and Deep Structure of Magnetic Complexes of Activity HMI will allow exploration of the nature of long-lived complexes of solar activity, the principal sources of solar disturbances. "Activity nests" have been a puzzle for many decades. They may continue from one cycle to the next, and may be related to variations of solar activity on the scale of one to two years and short-term "impulses" of activity. HMI will probe beneath these features to 0.7 R_{Sun} the bottom of the convection zone, to search for correlated flow or thermal structures.

2.2.1.7. Active Region Source and Evolution By using acoustic tomography we can image sound-speed perturbations that accompany magnetic-flux emergence and disconnection that may occur. Vector magnetograms can give evidence on whether flux leaves the surface pre-dominantly as "bubbles", or whether it is principally the outcome of local annihilation of fields of opposing polarity. With a combination of helioseismic probing and vector-field measurements, HMI will provide new insight into active-region flux emergence and removal.

2.2.1.8. Magnetic Flux Concentration in Sunspots Formation of sunspots is one of the long-standing questions of solar physics. Observations from MDI have revealed complicated flow patterns beneath sunspots and indicated that the highly concentrated magnetic flux in spots is accompanied by converging mass flows in the upper 3 – 4 Mm beneath the surface (Figure 3j). The evolution of these flows is not presently known. Detailed maps of subsurface flows in deeper layers, below 4 Mm, combined with surface fields and brightness for up to nine days during disk passage will allow investigation of the relations between flow dynamics and flux concentration in spots.

2.2.1.9. Sources and Mechanisms of Solar Irradiance Variation Magnetic features – sunspots, active regions, and network – that alter the temperature and composition of the solar atmosphere are primary sources of irradiance variability. How exactly do these features cause the irradiance variations? HMI together with SDO/AIA and SDO/*EUV Variability Experiment* (EVE: Woods *et al.*, 2010) will study physical processes that govern these variations. The relation between interior processes, properties of magnetic field regions, and irradiance variations, particularly the UV and EUV components that have a direct and significant effect on Earth's atmosphere, will be studied.

2.2.1.10. Sources and Drivers of Solar Activity and Disturbances It is commonly believed that the principal driver of solar disturbances is stressed magnetic field. The stresses are released in the solar corona producing flares and coronal mass ejections (CME). The source of these stresses is believed to be in the solar interior. Flares usually occur in areas where the magnetic configuration is complex, with strong shears, high gradients, long and curved neutral lines, recent flux emergence, *etc*. This implies that the trigger mechanisms of flares are controlled by critical properties of magnetic field that lead eventually to MHD instabilities. But what kinds of instability actually govern, and under what conditions they are triggered, is unknown. With only some theoretical ideas and models, there is no certainty of how magnetic field is stressed or twisted inside the Sun or just what the triggering process is.

2.2.1.11. Origin and Dynamics of Magnetic Sheared structures and α-type Sunspots Some spots contain two umbrae of opposite magnetic polarity within a common penumbra and are the source of powerful flares and CMEs. Such α-type sunspot regions are thought to inject magnetic flux into the solar atmosphere in a highly twisted state. It is important to determine what processes beneath the surface lead to development of these spots and allow them to become flare and CME productive. This investigation will be carried out by analysis of evolving internal mass flows and magnetic-field topology of such spots.

2.2.1.12. Magnetic Configuration and Mechanisms of Solar Flares Vector magnetic-field measurements can be used to infer field topology and vertical electric current, both of which are believed to be essential to understand the flare process. Observations are required that can continuously track changes in magnetic field and electric current with sufficient spatial resolution to reveal changes of field strength and topology before and after flares. HMI will provide these unique measurements of the vector magnetic field over the whole solar disk with reasonable accuracy and at high cadence.

2.2.1.13. Emergence of Magnetic Flux and Solar Transient Events Emergence of magnetic flux is closely related to solar transient events. MDI, GONG, and BBSO data show that there can be impulsive yet long-lived changes to the fields associated with eruptive events. The emergence of magnetic flux within active regions is often associated with flares. Emerging magnetic-flux regions near filaments lead to eruption of filaments. CMEs are also found to accompany emerging flux regions. Further, emergence of isolated active regions can proceed without any eruptive events. This suggests that magnetic flux emerging into the atmosphere interacts with pre-existing fields leading to loss of magnetic-field equilibrium. Observations of electric-current and magnetic topology differences between newly emerging and pre-existing fields will likely lead to the understanding of why emerging flux causes solar transient events. Vector polarimetry provided by HMI will enable these quantitative studies.

2.2.1.14. Evolution of Small-Scale Structures and Magnetic Carpet The quiet Sun is covered with small regions of mixed polarity, sometimes termed "magnetic carpet", contributing to solar activity on short timescales. As these elements emerge through the photosphere they interact with each other and with larger magnetic structures. They may provide triggers for eruptive events, and their constant interactions may be a source of coronal heating. They may also contribute to irradiance variations in the form of enhanced network emission. While HMI will certainly not see all of this flux, it will allow global-scale observations of the small-scale element distribution, their interactions, and the resulting transformation of the large-scale field.

2.2.2. Links Between the Internal Processes and Dynamics of the Corona and Heliosphere

Intrinsic connectivity between multi-scale magnetic-field patterns increases coronal structure complexity leads to variability. For example, CMEs apparently interact with the global-scale magnetic field, but many CMEs, especially fast CMEs, are associated with flares, which are believed to be local phenomena. Model-based reconstruction of 3D magnetic structure is one way to estimate the field from observations. More realistic MHD coronal models based on HMI high-cadence vector-field maps as boundary conditions will greatly enhance our understanding of how the corona responds to evolving, non-potential active regions.

2.2.2.1. Complexity and Energetics of the Solar Corona Observations from SOHO and TRACE have shown a variety of complex structures and eruptive events in the solar corona. However, categorizing complex structures has not revealed the underlying physics of the corona and coronal events. Two mechanisms have been proposed to generate stressed magnetic fields: photospheric shear motions and the emergence of already sheared magnetic fields; and both may, in fact, be at work on the Sun. But which plays the dominant role and how the energy injection is related to eruptive events are unknown. Magnetic helicity is an important characteristic of magnetic complexity and its conservation intrinsically links the generation, evolution, and reconnections of the magnetic field. HMI vector magnetic-field and velocity data (from both helioseismology and vector magnetic-field time-series-based velocity-inversion techniques) will allow better estimations of injections of energy and helicity into active regions. Observations from SDO/AIA and available coronagraphs (SOHO/LASCO and STEREO) will show the subsequent response and propagation of complexity into the corona and heliosphere, relating the build-up of helicity and energy with energetic coronal events such as CMEs.

2.2.2.2. Large-Scale Coronal Field Estimates Models computed from line-of-sight photospheric magnetic maps have been used to reproduce coronal forms that show multi-scale closed-field structures as well as the source of open field that starts from coronal holes but spreads to fill interplanetary space. Modeled coronal field demonstrates two types of closed field region: helmet streamers that form the heliospheric current sheet and a region sandwiched between the like-polarity open field regions (Zhao and Webb, 2003). There is evidence that most CMEs are associated with helmet streamers and with newly opened flux. HMI will provide uniform magnetic-field coverage at a high cadence, and together with simultaneous AIA and STEREO coronal images will enable the development of coronal magnetic-field models and study of the relationship between pre-existing patterns, newly opening fields, long-distance connectivity, and CMEs.

2.2.2.3. Coronal Magnetic Structure and Solar Wind MHD simulation and current-free coronal-field modeling based on magnetograms are two ways to study magnetic-field structuring of solar-wind properties. These methods have proven effective and promising, showing potential in applications of real-time space-weather forecasting. It has been demonstrated that modeling of the solar wind can be improved with increased cadence of the input magnetic data. By providing full-disk vector-field data at high cadence, HMI will enable these models to describe the distribution of the solar wind, coronal holes and open field regions, and how magnetic fields in active regions connect with interplanetary magnetic-field lines.

2.2.3. Precursors of Solar Disturbances for Space-Weather Forecasts

The number, strength, and timing of the strongest eruptive events are unpredictable at present. We are far from answering simple questions such as "will the next cycle be larger than the current one?" "When will the next large eruption occur?" Or even "when will there be several successive quiet days?" As we learn more about the fundamental processes through studies of internal motions, magnetic-flux transport and evolution, relations between active regions, UV irradiance, and solar-shape variations we will be vigilant for opportunities to develop prediction tools. Nevertheless, there are several near-term practical possibilities to improve the situation with HMI observations.

2.2.3.1. Far-Side Imaging and Activity Index A procedure for solar far-side imaging was developed using data from MDI and has led to the routine mapping of the Sun's far side (Braun and Lindsey, 2001). Acoustic travel-time perturbations are correlated with strong magnetic fields, providing a view of active regions well before they become visible as they rotate onto the disk at the east limb. Synoptic images, which are now able to cover the entire far hemisphere of the Sun, will provide the ability to forecast the appearance on the visible disk of large active regions up to two weeks in advance and allow the detection of regions which emerge just a few days before rotating into view. HMI's full coverage to the limb will allow lower-noise far-side estimates.

2.2.3.2. Predicting Emergence of Active Regions by Helioseismic Imaging Rising magnetic-flux tubes in the solar convection zone may produce detectable seismic signatures which would provide warning of their impending emergence. Helioseismic images of the base of the convection zone will employ a similar range of p modes as those used to construct images of the far side. A goal is to detect and monitor seismic signatures of persistent or recurring solar activity near the tachocline. Success here could lead to long-term forecasts of solar activity.

2.2.3.3. Determination of Magnetic Cloud B_S Events Potentially valuable information for geomagnetic forecasts – predictions of magnetic cloud B_S (southward field) events – can be obtained from the vector-field measurements. Long intervals of large southward interplanetary magnetic field and high solar-wind speed are believed to be the primary cause of intense geomagnetic disturbances, with the B_S component being the more important quantity. It has been shown that magnetic-field orientation in "clouds" remains basically unchanged while propagating from the solar surface to Earth's orbit. This provides a plausible chain of related phenomena that should allow prediction of the geo-effectiveness of CMEs directed toward Earth to be made from solar observations (see Siscoe and Schwenn, 2006, for a review). Estimates of embedded B_S should be significantly improved by incorporating frequently updated vector magnetic-field maps into coronal-field projections with the potential addition of coronagraphic observations from SOHO, SDO/AIA, and STEREO.

3. Theoretical Support and Modeling

The maximum scientific benefit from HMI can be obtained only with a specific theoretical program that includes numerical simulations of wave excitation and propagation in magnetized plasma, magneto-convection, local and global dynamo processes, magnetic-flux emergence, magnetic structures and MHD processes in the solar corona. Theoretical models for inversion of helioseismic and magnetic data are also extremely important for HMI data analyses. There are some key analysis capabilities that are not yet well understood. These include the interpretation of phases of acoustic waves measured in the presence of magnetic fields; the extraction of robust vector-field inferences with limited wavelength coverage; the problem of inferring coronal magnetic-field topology and properties (such as magnetic free energy) from the photospheric measurements, using force-free extrapolations, MHD models or other approaches; and understanding the helioseismic signals that might be detectable from sub-surface convection and fields. All of these areas are important to achieving the science goals of HMI. The inversion algorithms needs to be sufficiently fast to provide the scientific data products that are required for studying precursors of solar disturbances, such as full-disk maps of sub-surface flows, far-side images, and vector magnetograms, in almost real time. As a part of this project we will operate a 3D radiation MHD code that is being developed by HMI Co-Is at the NASA Ames Research Center. The members of the HMI research team will actively participate in the coordinated SDO Guest Investigator and LWS Theory Modeling and Data-Analysis Programs. It is clear that recent and ongoing developments in simulations and modeling will play an increasingly important role in progress for most HMI science goals.

4. Data Products

The HMI project has set the goal of producing a regular set of data products at a higher level of processing than has been the norm for most solar missions. In addition to the normal "Level-1 Science Data," which we interpret to be physical quantities such as Dopplergrams, magnetograms, continuum proxy, *etc.* which are already one stage beyond the observed filtergrams, we expect to provide standardized higher-level products such as sub-photospheric flow maps and coronal magnetic-field extrapolations. Such higher-level products have in the past been produced as byproducts of scientific analysis rather than a base product of the mission. These higher-level products will be based on the best available algorithms at the time that they are first put into production. The plan was to have the science team port the best existing code into the HMI processing pipeline late in the pre-launch phase to be ready early in the flight phase of the mission. This scenario counted on support to develop the techniques coming from outside the mission funding. In some cases this did not happen, so some of the planned products may not be available at the start. These data products are expected to evolve during the mission as the techniques of analysis continue to be developed.

There are some significant issues to be resolved before some of the products, *e.g.* robust inferences of sub-surface flows in active regions, can be provided.

Figure 3 shows samples of analyses that can be made using the planned standard data products.

The scientific data products can be divided into five main areas: global and local helioseismology, line-of-sight and vector magnetography, and continuum-intensity studies. These five primary scientific analyses cover all main HMI objectives, and have the following characteristics.

4.1. Global Helioseismology

The traditional normal-mode method will produce large-scale axisymmetrical distributions of sound speed, density, adiabatic exponent, and flow velocities through the whole solar interior from the energy-generating core to the near-surface convective boundary layer. These diagnostics will be based on frequencies and frequency splitting of modes of angular degree up to 1000, obtained for several-day intervals as needed and up to $\ell = 300$ for each 72-day interval. These will be used to produce a regular sequence of internal rotation and sound-speed inversions to allow observation from the tachocline to the near-surface shear layer. The techniques used here have been adapted from those used for MDI (Larson and Schou, 2011).

4.2. Local-Area Helioseismology

Local-area helioseismology methods such as time–distance techniques, ring-diagram analysis, and acoustic holography represent powerful tools for investigating physical processes

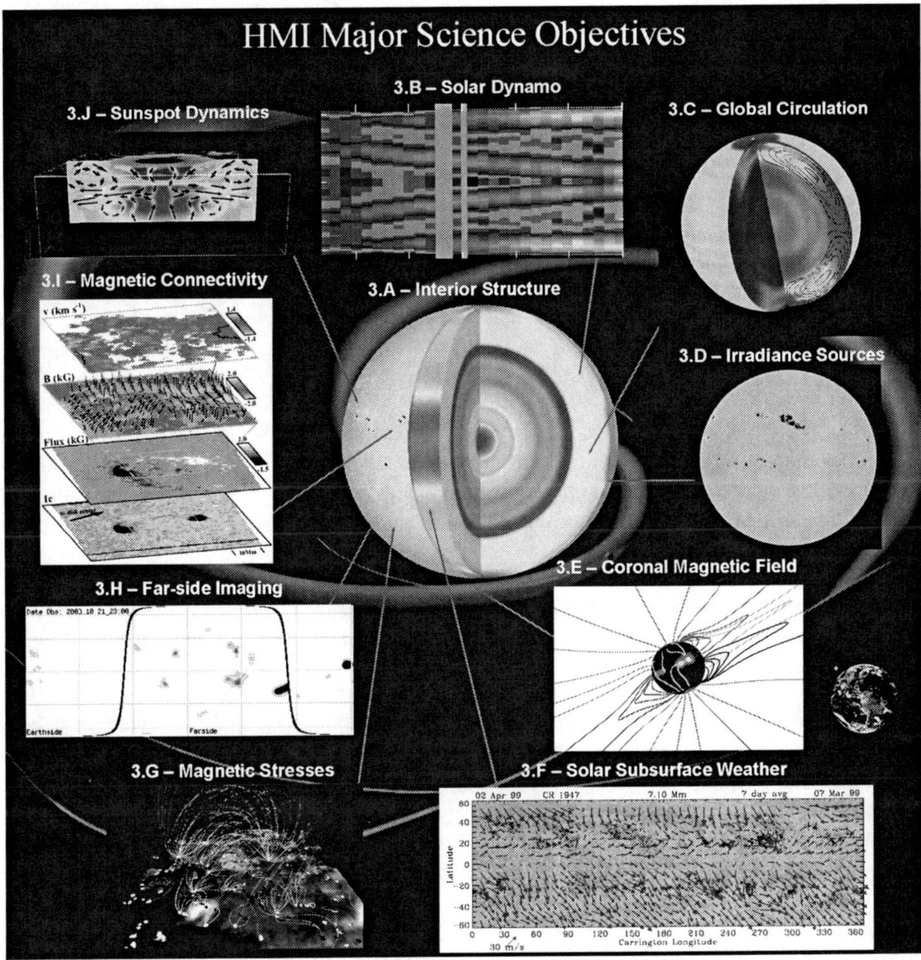

Figure 3 Samples of images that can be created using the planned HMI standard data products. The background is an artist's conception of the interplanetary current sheet as implied by extrapolated coronal fields (Wilcox, Hoeksema, and Scherrer, 1980). The overlays show: (A) Sound-speed variations relative to a standard solar model (Kosovichev, 2003); (B) Solar-cycle variations in the sub-photospheric rotation rate (Schou, private communication 2002); (C) Solar meridional circulation and differential rotation (http://soi.stanford.edu/press/ssu8-97/); (D) Sunspots and plage contribute to solar irradiance variation (SOHO/MDI); (E) MHD model of the magnetic structure of the corona (Linker and Mikic, 1997); (F) Synoptic map of the sub-surface flows at a depth of 7 Mm (Haber *et al.*, 2002); (G) EIT image and magnetic–field lines computed from the photospheric field (Title, 1997); (H) Active regions on the far side of the Sun detected with helioseismology (SOHO/MDI); (I) Vector field image showing the magnetic connectivity in sunspots (Lites, Skumanich, and Martinez Pillet, 1998); (J) Sound-speed variations and flows in an emerging active region (Zhao, Kosovichev, and Duvall, 2001). This figure is from Scherrer and HMI Team (2002) with far-side image updated.

inside the Sun. These methods based on measuring local properties of acoustic and surface-gravity waves, such as travel times, frequency and phase shifts will provide images of internal structures and flows on various spatial and temporal scales and depth resolution. The targeted high-level regular data products include the following.

4.2.0.4. Full-Disk Velocity and Sound-Speed Maps Maps of the upper convection zone (covering the top 20 Mm) obtained every eight hours with the time–distance methods on a Carrington grid of 120 by 120 degrees with a computational resolution of 0.12 degrees. These products are described by Zhao *et al.* (2011).

4.2.0.5. Synoptic Maps of Mass Flows and Sound-Speed Perturbations Synoptic maps are made for the upper convection zone for each Carrington rotation from averages of full-disk time–distance maps. These are made from day-wide North–South strips centered on the central meridian averaged from the eight-hour products described above.

4.2.0.6. Synoptic Maps of Horizontal Flows in Upper Convection Zone Maps are also made for each Carrington rotation with a 5-, 15-, and 30-degree resolution from ring-diagram analyses. These products are described by Bogart *et al.* (2011a, 2011b).

4.2.0.7. Higher-Resolution Maps Zoomed on Particular Active Regions Maps of sunspots and other targets, obtained with four- to eight-hour resolution for up to nine days continuously, with the time–distance method can be created on request as described in Zhao *et al.* (2011). Specific active regions or other regions of interest can also be processed through the "rings" pipeline upon request.

4.2.0.8. Deep-Focus Maps Maps spanning the whole convection zone depth, 0 – 200 Mm, with 10 – 15 degree resolution are planned but not yet developed to a data-product level of confidence.

4.2.0.9. Far-Side Images of the Sound-Speed Perturbations Maps associated with large active regions every 12 hours are computed using the GONG far-side pipeline.

In addition to this standard set of techniques, new approaches such as waveform tomography of small-scale structures will be deployed as they become developed.

The techniques for producing maps described above have been tested over several years with data from MDI and GONG and some with early HMI data. We are reasonably confident in inferences made in regions with little or no magnetic activity. However in regions of strong fields there are a number of issues that are either known to, or likely to, cause errors in inferences of physical parameters such as sound speed and flow velocities. Most of these issues are described by Kosovichev (2011) and references therein. Issues of phase variation with height as sampled are discussed by Rajaguru *et al.* (2010). Robust inferences of conditions in the vicinity of strong fields using local-area helioseismology can be achieved when these issues are properly accounted for in the analysis procedures.

4.3. Magnetography

The traditional line-of-sight component of the magnetic field is produced every 45 seconds as a co-product with the Doppler-velocity observations used for helioseismology. A 12-minute average of the longitudinal field is also computed from the Stokes vector obtained as described in the next section. This observable has proven to be very useful in tracking magnetic-field evolution at all temporal and spatial scales. HMI will provide complete coverage of magnetic processes in the visible photosphere. Magnetograms will be assembled into synoptic and synchronic (similar to traditional Carrington maps but corrected for differential rotation) maps and extracted HMI active-region patches will be tracked across the Sun. Several products will be computed with various cadence and resolution for use as input to coronal magnetic-field and solar-wind models, activity forecasting, and correlative studies.

4.4. Vector Magnetic Field

The vector field is one of the most-often absent important physical observables of the active solar atmosphere. HMI will produce several standard data series of vector fields. A 12-minute cadence full-disk analysis will be computed routinely and continuously to provide boundary conditions for large-scale coronal modeling. HMI, with the help of a robust inversion technique, will also provide AR-tracked and full-disk on-request vector magnetic field, and thermodynamic parameters of the photospheric plasma, with reasonable error estimates. These data will be used to quantitatively measure the free energy of the magnetic field, magnetic stresses, and helicity, providing important input to many prime science objectives and tasks of HMI and other SDO investigations. The HMI data products include the raw Stokes vector at six wavelengths across the line obtained each $90-135$ seconds and ordinarily averaged into a 12-minute product. Milne–Eddington inversions are routinely computed from these as described by Borrero *et al.* (2010). A further product is computed as the final vector field by applying a field direction disambiguation technique, since the Stokes parameters cannot distinguish the polarity of the component of the field normal to the line-of-sight (Metcalf, 1994; Metcalf *et al.*, 2006; Leka *et al.*, 2009). HMI provides a new challenge and opportunity for vector-field studies: observations are available for the full disk at all times. Thus while including all active regions with strong fields where the vector magnetic-field methods are expected to yield reasonably accurate measurements, the observations also include large areas of isolated fields where spatial averages may be useful but are hard to achieve in a robust manner. The optimization code for inference of non-linear force-free coronal magnetic fields is based on work by Wiegelmann (2004) and Wiegelmann, Inhester, and Sakurai (2006). We will also produce inferred velocities of the magnetic plasma, helicity, and Poynting fluxes based on tracking techniques (Welsch *et al.*, 2007; Schuck, 2008).The magnetic data products are described in more detail at http://jsoc.stanford.edu/jsocwiki/MagneticField.

4.5. Continuum Intensity

The observations of an estimate of intensity in the continuum near the HMI spectral line will give a very useful measure of spot, faculae area, and other sources of irradiance. This will be important for studying the relationship between the MHD processes in the interior and lower atmosphere and irradiance variations. The continuum data will be also used for limb shape analysis, and for public information and education purposes. As side products of the Doppler and Continuum products HMI will produce line-depth and line-width estimates for the Fe I 617.3 nm line.

4.6. Real-Time Products

Data for observation planning for collaborative observing programs and for prediction of space-weather events are also produced as standard products. HMI Active Region Patches (HARPs) will be identified and tracked each 12 minutes. Line-of-sight magnetic field, continuum images, and other data products are available for space-weather investigations and forecasts in near-real time. Vector-field quantities can be computed with a slightly longer lag. Time series of computed space-weather quantities, such as total flux and neutral-line length, will be computed for each HARP on a rapid cadence. The near-real-time products are available with lags from the time of observation of less than 30 minutes. On a regular basis (daily or more often) we will provide full-disk maps of sound-speed distribution and mass flows in the upper convection zone, coronal magnetic field, predicted solar-wind velocity, and far-side activity-index maps.

5. Summary

The HMI instrument is now on orbit as part of SDO and is producing data that appear to be of sufficient quality to address the goals described here. The JSOC data center is distributing data as needed. The data are available to the broad community and we hope it is productive to use HMI observations and data products to advance the broad goals of SDO.

Acknowledgements The HMI project is supported by NASA contract NAS5-02139. Efforts to develop science analysis code for the pipeline have been supported by NASA grants NNG05GM85G, NNG06GE40G, NNX07AP61G, NNX09AB10G, NNX09AG81G, and NNX10AC55G. We wish to thank Elizabeth Citrin the NASA SDO Program Manager, Barbara Thompson and Dean Pesnell the SDO project scientists, Madhulika Guhathakurta the NASA LWS program scientist, Arthur Poland, George Withbroe, and Richard Fisher at NASA Headquarters for both formative and continuing support.

The authors thank the entire HMI team for help in developing the investigation, instrument, and JSOC data system. In particular the partnership with the team at LMSAL was essential for the project to proceed. A project such as HMI is clearly a team effort with contributions from many. In addition to the authors, at least the following people contributed to the development of the HMI investigation and instrument (institutions as of launch): *Hansen Experimental Physics Laboratory, Stanford University*: Jim Aloise, Art Amezcua, John Beck (Co-I), Kelly Beck, Sudeepto Chakraborty, Millie Chethik, Keh-Cheng Chu, Carl Cimilluca, Sebastien Couvidat, Nancy Christensen, Romeo Durscher, Thomas Hartlep, Keiji Hayashi, Tim Huynh, Stathis Ilonidis, Kevin Kempter, Irina Kitiashvili, Rasmus Larsen, Tim Larson, Leyan Lo, Haruko Makatani, Rakesh Nigam, Konstantin Parchevsky, Bala Poduval, Brian Roberts, Kim Ross, Deborah Scherrer, Jeneen Sommers, Jennifer Spencer, Margie Stehle, Xudong Sun, Hao Thai, Karen Tian, Richard Wachter, Jeff Wade, XuePu Zhao (Co-I), and HEPL Staff, *Lockheed Martin Solar and Astrophysics Laboratory*: Dave Akin, Brett Allard, Ron Baraze, Mitch Baziuk, Tom Berger, E. Bogle, Bob Caravalho, Brock Carpenter, C. Cheung, Cathy Chou, Roger Chevalier, K. Chulick, Tom Cruz, Jerry Drake, Dexter Duncan, Jay Dusenbury, Janet Embrich, Chris Edwards, Cliff Evans, Peter Feher, Barbara Fischer, Charles Fischer, Samuel Freeland, Frank Friedlander, Glen Gradwohl, Henry Hancock, Gary Heyman, Bob Honeycutt, Elizabeth Hui, Bruce Imai, Jerry Janecka, Ramona Jimenez, Dwana Kacensky, Peter Kacensky, Claude Kam, Noah Katz, Karen Kao, Dave Kirkpatrick, Gary Kushner, Michael Levay, Russ Lindgren, Gary Linford, Andrea Lynch, Dnyanesh Mathur, Ed McFeaters, John Miles, Keith Mitchell, Sarah Mitchell, Ruth Mix, Margaret Morgan, Rose Navarro, Tom Nichols, Tracey Niles, Jackie Pokorny, Roger Rehse, Rick Rairden, John-Paul Riley, Lomita Rubio, David Schiff, Isela Scott, Ralph Sequin, Cheryl Seeley, Lawrence Shing, Araya Silpikul, Larry Springer, Richard Shine, Bob Stern, Louis Tavarez, Edgar Thomas, Ryan Timmons, Darrel Torgerson, Angel Vargas, Shan Varaitch, Dale Wolf, C. Jacob Wolfson, Carl Yanari, Ross Yamamoto, Kent Zickuhr, *High Altitude Observatory*: Juan Borrero Santiago, Gregory L. Card, Anthony Darnell, Rebecca C. Elliott, David Elmore, Jonathan Graham, Bruce Lites, Arturo Lopez Ariste, Matthias Rempel, Hector Socas-Navarro, *Jet Propulsion Laboratory*: Michael Turmon, JILA: Benjamin Brown, Gwen Dickenson, Nicholas Fetherstone, Deborah Haber, Bradley Hindman, Swati Routh, Regner Trampedach, Juri Toomre (Co-I), *NASA Ames Research Center*: Nagi Mansour (Co-I), Alan Wray (Co-I), *New Jersey Institute of Technology/Big Bear Solar Observatory*: Phil Goode (Co-I), Vasyl Yurchyshyn, *North West Research Associates/Colorado Research Associates*: Graham Barnes, Aaron Birch, Doug C. Braun (Co-I), Ashley Crouch, K.D. Leka (Co-I), Charles A. Lindsey (Co-I), Tom Metcalf (Co-I, deceased), Orion Poplawski, Martin Woodard, *National Solar Observatory*: Walter Allen, Olga Burtseva, Irene González Hernández, Frank Hill (Co-I), Rachel Howe (Co-I), Rudi Komm, Igor Suarez Sola, Sushanta Tripathy, Kiran Jain, Shakur Kholikov, *Predictive Science Inc.*: Michael Choy, Jon Linker (Co-I), Zoran Mickic, Pete Riley, Timofey Titov, Janvier Wijaya, *Smithsonian Astrophysical Observatory*: Alisdair Davey, Sylvain Korzennik (Co-I), *University of California Los Angeles*: Roger Ulrich (Co-I), *University of Hawaii*: Marcelo Emilio, Jeffrey R. Kuhn (Co-I), Isabelle Scholl, *University of Maryland College Park*: Judit Pap (Co-I), *University of Southern California*:Shawn Irish, Johann Reiter, Edward J. Rhodes Jr. (Co-I), Anthony Spinella, *Yale University*: Charles Baldner, Sarbani Basu (Co-I), *Aarhus University*: Jørgen Christensen-Dalsgaard (Co-I), *Cambridge University*: Douglas Gough (Co-I), *European Space Agency*: Bernhard Fleck (Co-I), *Indian Institute of Astrophysics*:Dipankar Banerjee, S. Paul Rajaguru, Siraj Hasan, *James Cook University*: Aimee Norton, *Max Planck Institute for Solar System Research*: Raymond Burston, Laurent Gizon, Yacine Saidi, Sami Solanki (Co-I), *Mullard Space Science Laboratory*: Elizabeth Auden, Len Culhane (Co-I), *National Astronomical Observatory of Japan*: Kaori Nagashima, Takashi Sekii (Co-I), *Rutherford Appleton Laboratory*: Richard Harrison (Co-I), Sarah Dunkin, Matthew Calpp, Nick Waltham, *Sheffield University*: Michael Thompson (Co-I), *University of Tokyo*: Hiromoto Shibahashi (Co-I), *e2v*: Gary Auker, Rob Wilson, Paul Jerram, *LightMachinery* : John Hunter, Ian Miller, Jeff Wimperis, *Andover Corporation & Zygo Corporation*: John Cotton, *H. Magnetics*: Ralph Horber.

References

Bogart, R.S., Baldner, C., Basu, S., Haber, D.A., Rabello-Soares, M.C.: 2011a, HMI ring diagram analysis I. The processing pipeline. *J. Phys. C* **271**(1), 012008. doi:10.1088/1742-6596/271/1/012008.

Bogart, R.S., Baldner, C., Basu, S., Haber, D.A., Rabello-Soares, M.C.: 2011b, HMI ring diagram analysis II. Data products. *J. Phys. C* **271**(1), 012009. doi:10.1088/1742-6596/271/1/012009.

Borrero, J.M., Tomczyk, S., Kubo, M., Socas-Navarro, H., Schou, J., Couvidat, S., Bogart, R.: 2010, VFISV: Very fast inversion of the stokes vector for the helioseismic and magnetic imager. *Solar Phys.* doi:10.1007/s11207-010-9515-6.

Braun, D.C., Lindsey, C.: 2001, Seismic imaging of the far hemisphere of the Sun. *Astrophys. J. Lett.* **560**, L189 – L192. doi:10.1086/324323.

Couvidat, S., Zhao, J., Birch, A.C., Kosovichev, A.G., Duvall, T.L., Parchevsky, K., Scherrer, P.H.: 2010, Implementation and comparison of acoustic travel-time measurement procedures for the solar dynamics observatory/helioseismic and magnetic imager time–distance helioseismology pipeline. *Solar Phys.* doi:10.1007/s11207-010-9652-y.

Couvidat, S., Schou, J., Shine, R.A., Bush, R.I., Miles, J.W., Scherrer, P.H., Rairden, R.L.: 2011, Wavelength dependence of the helioseismic and magnetic imager (HMI) instrument onboard the Solar Dynamics Observatory (SDO). *Solar Phys.* doi:10.1007/s11207-011-9723-8.

Drobnes, E., Littleton, A., Pesnell, W.D., Buhr, S., Beck, K., Durscher, R., Hill, S., McCaffrey, M., McKenzie, E., Scherrer, D., Wolt, A.: 2011, The SDO education and outreach (E/PO) program: Changing perceptions one program as a time. *Solar Phys.* accepted.

Duvall, T.L. Jr., Jefferies, S.M., Harvey, J.W., Pomerantz, M.A.: 1993, Time–distance helioseismology. *Nature* **362**, 430 – 432. doi:10.1038/362430a0.

Duvall, T.L. Jr., Kosovichev, A.G., Scherrer, P.H., Bogart, R.S., Bush, R.I., de Forest, C., Hoeksema, J.T., Schou, J., Saba, J.L.R., Tarbell, T.D., Title, A.M., Wolfson, C.J., Milford, P.N.: 1997, Time–distance helioseismology with the MDI instrument: initial results. *Solar Phys.* **170**, 63 – 73.

Haber, D.A., Hindman, B.W., Toomre, J., Bogart, R.S., Larsen, R.M., Hill, F.: 2002, Evolving submerged meridional circulation cells within the upper convection zone revealed by ring-diagram analysis. *Astrophys. J.* **570**, 855 – 864. doi:10.1086/339631.

Hathaway, D., SDO SDT Panel: 2001, Solar Dynamics Observatory report of the science definition team. Technical report. Accessed 13 May 2011. http://www.nswp.gov/sdo/sdo_sdt_report.pdf.

Howe, R.: 2008, Helioseismology and the solar cycle. *Adv. Space Res.* **41**, 846 – 854. doi:10.1016/j.asr.2006.12.033.

Howe, R., Hill, F., Komm, R., Christensen-Dalsgaard, J., Larson, T.P., Schou, J., Thompson, M.J., Ulrich, R.: 2011, The torsional oscillation and the new solar cycle. *J. Phys. C* **271**(1), 012074. doi:10.1088/1742-6596/271/1/012074.

Komm, R., Howe, R., Hill, F., González Hernández, I., Haber, D.: 2011, Solar-cycle variation of zonal and meridional flow. *J. Phys. C* **271**(1), 012077. doi:10.1088/1742-6596/271/1/012077.

Kosovichev, A.G.: 1996, Tomographic imaging of the Sun's interior. *Astrophys. J. Lett.* **461**, L55 – L57. doi:10.1086/309989.

Kosovichev, A.G.: 2003, What helioseismology teaches us about the Sun. In: Wilson, A. (ed.) *Solar Variability as an Input to the Earth's Environment* **535**, ESA, Noordwijk, 795 – 806.

Kosovichev, A.G.: 2011, Local helioseismology of sunspots: Current status and perspectives. *Solar Phys.*, submitted.

Kosovichev, A.G., HMI Science Team: 2004, HMI Science Plan. Accessed 13 May 2011. http://hmi.stanford.edu/doc/HMI-S014.pdf.

Kosovichev, A.G., Duvall, T.L. Jr., Scherrer, P.H.: 2000, Time–distance inversion methods and results (Invited review). *Solar Phys.* **192**, 159 – 176.

Larson, T., Schou, J.: 2011, HMI global helioseismology data analysis pipeline. *J. Phys. C* **271**(1), 012062. doi:10.1088/1742-6596/271/1/012062.

Leka, K.D., Barnes, G., Crouch, A.D., Metcalf, T.R., Gary, G.A., Jing, J., Liu, Y.: 2009, Resolving the 180° ambiguity in solar vector magnetic field data: Evaluating the effects of noise, spatial resolution, and method assumptions. *Solar Phys.* **260**, 83 – 108. doi:10.1007/s11207-009-9440-8.

Lemen, J.R., Title, A.M., Akin, D.J., Boerner, P.F., Chou, C., Drake, J.F., Duncan, D.W., Edwards, C.G., Friedlaender, F.M., Heyman, G.F., Hurlburt, N.E., Katz, N.L., Kushner, G.D., Levay, M., Lindgren, R.W., Mathur, D.P., McFeaters, E.L., Mitchell, S., Rehse, R.A., Schrijver, C.J., Springer, L.A., Stern, R.A., Tarbell, T.D., Wuelser, J.-P., Wolfson, C.J., Yanari, C., Bookbinder, J.A., Cheimets, P.N., Caldwell, D., Deluca, E.E., Gates, R., Golub, L., Park, S., Podgorski, W.A., Bush, R.I., Scherrer, P.H., Gummin, M.A., Smith, P., Auker, G., Jerram, P., Pool, P., Soufli, R., Windt, D.L., Beardsley, S., Clapp, M., Lang, J., Waltham, N.: 2011, The atmospheric imaging assembly (AIA) on the Solar Dynamics Observatory (SDO). *Solar Phys.* doi:10.1007/s11207-011-9776-8.

Linker, J.A., Mikic, Z.: 1997, Extending coronal models to Earth orbit. In: Crooker, N., Joselyn, J., Feynman, J. (eds.) *Coronal Mass Ejections: Causes and Consequences, Geophys. Monogr. Ser.* **99**, AGU, Washington.

Lites, B.W., Skumanich, A., Martinez Pillet, V.: 1998, Vector magnetic fields of emerging solar flux. I. Properties at the site of emergence. *Astron. Astrophys.* **333**, 1053 – 1068.

Mason, G., LWS Panel: 2001, Living with a star science architecture team report to SECAS. Technical report. Accessed 13 May 2011. http://www.nswp.gov/lwsgeospace/SECAS/LWSSAT_SECASreport_30Aug01.pdf.

Metcalf, T.R.: 1994, Resolving the 180-degree ambiguity in vector magnetic field measurements: The 'minimum' energy solution. *Solar Phys.* **155**, 235 – 242. doi:10.1007/BF00680593.

Metcalf, T.R., Leka, K.D., Barnes, G., Lites, B.W., Georgoulis, M.K., Pevtsov, A.A., Balasubramaniam, K.S., Gary, G.A., Jing, J., Li, J., Liu, Y., Wang, H.N., Abramenko, V., Yurchyshyn, V., Moon, Y.-J.: 2006, An overview of existing algorithms for resolving the 180°ambiguity in vector magnetic fields: Quantitative tests with synthetic data. *Solar Phys.* **237**, 267 – 296. doi:10.1007/s11207-006-0170-x.

Rajaguru, S.P., Wachter, R., Sankarasubramanian, K., Couvidat, S.: 2010, Local helioseismic and spectroscopic analyses of interactions between acoustic waves and a sunspot. *Astrophys. J. Lett.* **721**, L86 – L91. doi:10.1088/2041-8205/721/2/L86.

Scherrer, P.H., HMI Team: 2002, Helioseismic and magnetic imager for Solar Dynamics Observatory. Accessed 13 May 2011. http://hmi.stanford.edu/doc/HMI-S001.pdf.

Scherrer, P.H., Bogart, R.S., Bush, R.I., Hoeksema, J.T., Kosovichev, A.G., Schou, J., Rosenberg, W., Springer, L., Tarbell, T.D., Title, A., Wolfson, C.J., Zayer, I., MDI Engineering Team: 1995, The solar oscillations investigation – Michelson Doppler Imager. *Solar Phys.* **162**, 129 – 188. doi:10.1007/BF00733429.

Schou, J., Antia, H.M., Basu, S., Bogart, R.S., Bush, R.I., Chitre, S.M., Christensen-Dalsgaard, J., di Mauro, M.P., Dziembowski, W.A., Eff-Darwich, A., Gough, D.O., Haber, D.A., Hoeksema, J.T., Howe, R., Korzennik, S.G., Kosovichev, A.G., Larsen, R.M., Pijpers, F.P., Scherrer, P.H., Sekii, T., Tarbell, T.D., Title, A.M., Thompson, M.J., Toomre, J.: 1998, Helioseismic studies of differential rotation in the solar envelope by the solar oscillations investigation using the Michelson Doppler Imager. *Astrophys. J.* **505**, 390 – 417. doi:10.1086/306146.

Schou, J., Borrero, J.M., Norton, A.A., Tomczyk, S., Elmore, D., Card, G.L.: 2010, Polarization calibration of the helioseismic and magnetic imager (HMI) onboard the Solar Dynamics Observatory (SDO). *Solar Phys.* doi:10.1007/s11207-010-9639-8.

Schou, J., Scherrer, P.H., Bush, R.I., Wachter, R., Couvidat, S., Rabello-Soares, M.C., Bogart, R.S., Hoeksema, H.T., Liu, T., Duvall, J.T.L., Akin, D.J., Allard, B.A., Miles, J.W., Rairden, R., Shine, R.A., Tarbell, T.D., Title, A.M., Wolfson, C.J., Elmore, D.F., Norton, A.A., Tomczyk, S.: 2011, The helioseismic and magnetic imager instrument design and ground calibration. *Solar Phys.* doi:10.1007/s11207-011-9842-2.

Schuck, P.W.: 2008, Tracking vector magnetograms with the magnetic induction equation. *Astrophys. J.* **683**, 1134 – 1152. doi:10.1086/589434.

Siscoe, G., Schwenn, R.: 2006, CME disturbance forecasting. *Space Sci. Rev.* **123**, 453 – 470. doi:10.1007/s11214-006-9024-y.

Title, A.M.: 1997, Solar mystery near solution with data from SOHO spacecraft, the Sun's newly-discovered magnetic carpet may explain coronal heating. Accessed 13 May 2011. http://soi.stanford.edu/press/ssu11-97/.

Title, A.M., Schrijver, C.J.: 1998, The Sun's magnetic carpet. In: Donahue, R.A., Bookbinder, J.A. (eds.) *Cool Stars, Stellar Systems, and the Sun* **154**, Astron. Soc. Pac., San Francisco, 345.

van Driel-Gesztelyi, L., Culhane, J.L.: 2009, Magnetic flux emergence, activity, eruptions and magnetic clouds: Following magnetic field from the Sun to the heliosphere. *Space Sci. Rev.* **144**, 351 – 381. doi:10.1007/s11214-008-9461-x.

Wachter, R., Schou, J., Rabello-Soares, M.C., Miles, J.W., Duvall, T.L., Bush, R.I.: 2011, Image quality of the helioseismic and magnetic imager (HMI) onboard the solar dynamics observatory (SDO). *Solar Phys.* doi:10.1007/s11207-011-9709-6.

Welsch, B.T., Abbett, W.P., De Rosa, M.L., Fisher, G.H., Georgoulis, M.K., Kusano, K., Longcope, D.W., Ravindra, B., Schuck, P.W.: 2007, Tests and comparisons of velocity-inversion techniques. *Astrophys. J.* **670**, 1434 – 1452. doi:10.1086/522422.

Wiegelmann, T.: 2004, Optimization code with weighting function for the reconstruction of coronal magnetic fields. *Solar Phys.* **219**, 87 – 108. doi:10.1023/B:SOLA.0000021799.39465.36.

Wiegelmann, T., Inhester, B., Sakurai, T.: 2006, Preprocessing of vector magnetograph data for a nonlinear force-free magnetic field reconstruction. *Solar Phys.* **233**, 215 – 232. doi:10.1007/s11207-006-2092-z.

Wilcox, J.M., Hoeksema, J.T., Scherrer, P.H.: 1980, Origin of the warped heliospheric current sheet. *Science* **209**, 603 – 605. doi:10.1126/science.209.4456.603.

Woods, T.N., Eparvier, F.G., Hock, R., Jones, A.R., Woodraska, D., Judge, D., Didkovsky, L., Lean, J., Mariska, J., Warren, H., McMullin, D., Chamberlin, P., Berthiaume, G., Bailey, S., Fuller-Rowell, T., Sojka, J., Tobiska, W.K., Viereck, R.: 2010, Extreme ultraviolet variability experiment (EVE) on the Solar Dynamics Observatory (SDO): Overview of science objectives, instrument design, data products, and model developments. *Solar Phys.* doi:10.1007/s11207-009-9487-6.

Zhao, J., Kosovichev, A.G., Duvall, T.L. Jr.: 2001, Investigation of mass flows beneath a sunspot by time–distance helioseismology. *Astrophys. J.* **557**, 384 – 388. doi:10.1086/321491.

Zhao, J., Hartlep, T., Kosovichev, A.G., Mansour, N.N.: 2009, Imaging the solar tachocline by time–distance helioseismology. *Astrophys. J.* **702**, 1150 – 1156. doi:10.1088/0004-637X/702/2/1150.

Zhao, J., Couvidat, S., Bogart, R.S., Parchevsky, K.V., Birch, A.C., Duvall, T.L., Beck, J.G., Kosovichev, A.G., Scherrer, P.H.: 2011, Time–distance helioseismology data-analysis pipeline for helioseismic and magnetic imager onboard Solar Dynamics Observatory (SDO/HMI) and its initial results. *Solar Phys.* doi:10.1007/s11207-011-9757-y.

Zhao, X.P., Webb, D.F.: 2003, Source regions and storm effectiveness of frontside full halo coronal mass ejections. *J. Geophys. Res.* **108**, 1234. doi:10.1029/2002JA009606.

Zwaan, C.: 1987, Elements and patterns in the solar magnetic field. *Annu. Rev. Astron. Astrophys.* **25**, 83 – 111. doi:10.1146/annurev.aa.25.090187.000503.

Solar Phys (2012) 275:229–259
DOI 10.1007/s11207-011-9842-2

Design and Ground Calibration of the *Helioseismic and Magnetic Imager* (HMI) Instrument on the *Solar Dynamics Observatory* (SDO)

J. Schou · P.H. Scherrer · R.I. Bush · R. Wachter · S. Couvidat · M.C. Rabello-Soares · R.S. Bogart · J.T. Hoeksema · Y. Liu · T.L. Duvall Jr. · D.J. Akin · B.A. Allard · J.W. Miles · R. Rairden · R.A. Shine · T.D. Tarbell · A.M. Title · C.J. Wolfson · D.F. Elmore · A.A. Norton · S. Tomczyk

Received: 7 June 2011 / Accepted: 9 August 2011 / Published online: 4 October 2011

Abstract The *Helioseismic and Magnetic Imager* (HMI) investigation (*Solar Phys.* doi:10.1007/s11207-011-9834-2, 2011) will study the solar interior using helioseismic techniques as well as the magnetic field near the solar surface. The HMI instrument is part of the *Solar Dynamics Observatory* (SDO) that was launched on 11 February 2010. The instrument is designed to measure the Doppler shift, intensity, and vector magnetic field at the solar pho-

The Solar Dynamics Observatory
Guest Editors: W. Dean Pesnell, Phillip C. Chamberlin, and Barbara J. Thompson

Electronic supplementary material The online version of this article (doi:10.1007/s11207-011-9842-2) contains supplementary material.

J. Schou (✉) · P.H. Scherrer · R.I. Bush · R. Wachter · S. Couvidat · M.C. Rabello-Soares · R.S. Bogart · J.T. Hoeksema · Y. Liu
W.W. Hansen Experimental Physics Laboratory, Stanford University, Stanford, CA 94305-4085, USA
e-mail: schou@sun.stanford.edu

T.L. Duvall Jr.
Laboratory for Astronomy and Solar Physics, NASA Goddard Space Flight Center, Greenbelt, MD 20771, USA

D.J. Akin · B.A. Allard · J.W. Miles · R. Rairden · R.A. Shine · T.D. Tarbell · A.M. Title · C.J. Wolfson
Lockheed Martin Solar and Astrophysics Laboratory, 3251 Hanover St., Palo Alto, CA 94304, USA

J.W. Miles
USRA NASA Ames Research Center, Mail Stop 211-3, Moffett Field, CA 94035, USA

D.F. Elmore · A.A. Norton · S. Tomczyk
High Altitude Observatory, National Center for Atmospheric Research, 3080 Center Green CG-1, Boulder, CO 80301, USA

Present address:
D.F. Elmore
National Solar Observatory/Sacramento Peak, 3010 Coronal Loop, Sunspot, NM 88349, USA

Present address:
A.A. Norton
James Cook University, School of Engineering and Physical Sciences, Townsville, QLD, 4810, Australia

tosphere using the 6173 Å Fe I absorption line. The instrument consists of a front-window filter, a telescope, a set of waveplates for polarimetry, an image-stabilization system, a blocking filter, a five-stage Lyot filter with one tunable element, two wide-field tunable Michelson interferometers, a pair of 4096^2 pixel cameras with independent shutters, and associated electronics. Each camera takes a full-disk image roughly every 3.75 seconds giving an overall cadence of 45 seconds for the Doppler, intensity, and line-of-sight magnetic-field measurements and a slower cadence for the full vector magnetic field. This article describes the design of the HMI instrument and provides an overview of the pre-launch calibration efforts. Overviews of the investigation, details of the calibrations, data handling, and the science analysis are provided in accompanying articles.

Keywords Solar Dynamics Observatory · Helioseismology, observations · Instrumentation and data management · Magnetic fields, photosphere

1. Introduction

The *Helioseismic and Magnetic Imager* (HMI) instrument was built as part of the HMI investigation (Scherrer *et al.*, 2011) and is designed to measure the Doppler shift, line-of-sight magnetic field, intensity, and vector magnetic field at the solar photosphere using the 6173 Å Fe I absorption line (Norton *et al.*, 2006).

To a significant extent the HMI instrument design is based on the highly successful *Michelson Doppler Imager* (MDI) instrument (Scherrer *et al.*, 1995), but it does have a few significant improvements. Among these improvements are two cameras instead of one, better spatial resolution and temporal coverage, the capability to observe the full Stokes vector, full-disk filtergram data that are all downlinked without significant processing, and a significant level of redundancy. The minimal onboard processing ensures that optimal processing can be done, and if necessary redone, on the ground.

Like MDI, the HMI instrument is what has become known as a filtergraph. In other words it takes a series of images at different wavelengths and polarizations that are combined to derive the physical parameters. This is in contrast with spectrographs that take spectra while scanning across the image.

Below we start with a description of the various components of the instrument (Section 2). This is followed by an overview of the calibration in Section 3 and conclusions in Section 4. Rather than describing all the instrument details and calibrations in one article, a number of other articles describe the main components of the calibration efforts. Wachter *et al.* (2011) describe the image quality and CCD performance, Couvidat *et al.* (2011) the filter performance, and Schou *et al.* (2010) the polarization properties. In all cases we have restricted ourselves to ground-based results, with updates to be made and published after analysis of on-orbit HMI data.

The HMI instrument was built by the Lockheed Martin Solar and Astrophysics Laboratory (LMSAL). Many of the components used in HMI are essentially identical to those used for the *Atmospheric Imaging Assembly* (AIA) instrument also built at LMSAL, for which more details can be found in Lemen *et al.* (2011).

2. Instrument Description

The HMI instrument consists of three main parts: an optics package, an electronics box, and a harness to connect the two.

The HMI Optics Package (HOP) layout is described in Section 2.1. We present details of the imaging optics in the HOP in Section 2.2, polarization selectors in Section 2.3, filters in Section 2.4, the Image-Stabilization System (ISS) in Section 2.5, the various mechanisms in Section 2.6, the thermal control in Section 2.7, and the CCDs and cameras in Section 2.9. The contents of the HMI Electronics Box are described in Sections 2.8 and 2.10, while the flight software is described in Section 2.11. The harness is not described in detail. Finally, Section 2.12 contains a comparison of MDI and HMI.

For reference, the masses are 47 kg for the optics package, including Multi-Layer Insulation (MLI), 19 kg for the electronics package, and 7 kg for the harness, for a total of 73 kg. The HMI instrument uses about 95 W of power during normal operations, of which about 15 W are consumed by each camera.

2.1. HMI Optics Package

The HOP houses the optical components of the instrument and protects them from the environment. The bottom of the HOP is an optical bench consisting of two face sheets and a honeycomb core. The side panels and a number of internal partitioning walls are mounted on this bench and in turn support the lid of the box. The box is packaged in MLI to provide thermal stability and to minimize the amount of power needed to keep the HOP near the design temperature of around 20 C. A number of heaters are attached to the bottom and top of the box. Their placement and operation is described in Section 2.7.1. The combination of the insulation system and the multiple heaters means that it is possible to keep the HOP at an adjustable and stable temperature.

Two CCDs and attached radiators are mounted on top of the box along with two Camera Electronics Boxes (CEBs), also with radiators. The camera components are described in Section 2.9.

The HOP mounts to the spacecraft using four fiberglass legs, two of which are also used to point the HOP, as described in Section 2.6.4.

Finally, a connector panel on the side of the box (near the shutters) connects to the harness. Vents ensure that the pressure differential between the inside and outside of the HOP stays low during launch and provide an escape path for any volatile contaminants in the box.

To provide an overview, a schematic of the optical path of HMI is shown in Figure 1 and a photograph of the inside of the HOP in Figure 2. Sunlight travels through the instrument from the front window at the upper right to the cameras at the lower right of the schematic. The majority of the optical components are mounted to the optical bench. The components along the optical path are described in more detail in the following subsections. A raytrace is shown in Figure 3. For a prescription of the optical components see Appendix B.1. In order to avoid radiation-induced darkening, all glass elements used are made of radiation-resistant glass, except for one element in the front window and parts of the Michelsons.

The front window is a 50 Å bandpass filter that reflects most of the incident sunlight. It is followed by a 14-cm diameter refracting telescope consisting of a primary and secondary lens.

Two focus/calibration mechanisms, three polarization-selection mechanisms, and the ISS tip–tilt mirror are located between the telescope and the polarizing beamsplitter feeding the tunable filter. The filter is mounted in a precisely temperature-controlled oven, containing the following elements:

- A telecentric lens
- A blocking filter
- A Lyot filter with a single tunable element

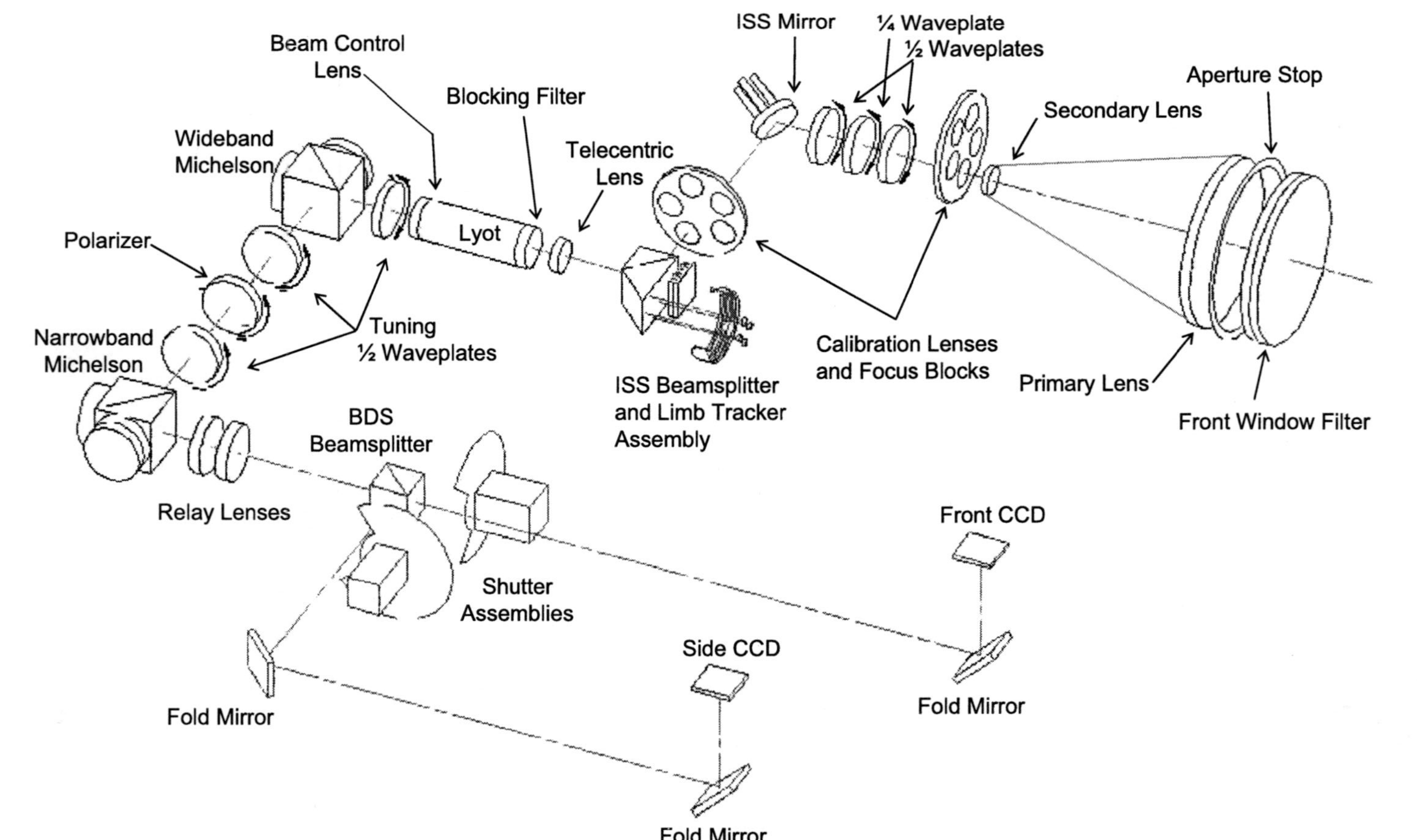

Figure 1 Schematic of the HMI optical layout with various elements annotated. The calibration LED behind the BDS beamsplitter is not shown. See Figure 3 for scale (Couvidat *et al.*, 2011).

Figure 2 Photograph of the inside of the optics package. Light enters through the telescope at the upper left, passes through the chain of filters near the bottom, and is detected by the CCD cameras at the upper right. The box is roughly 88 cm long by 55 cm wide by 16 cm high. The various elements are identified in Figure 1.

- A beam control lens
- Two tunable Michelson interferometers
- Reimaging optics

Following the oven is the Beam Distribution System (BDS) beam splitter that feeds two identical shutters each mounted at a pupil image, and the CCD camera assemblies at the focal plane.

In a number of places 1/4 waveplates are included to reject unwanted reflections. These are not always shown or discussed further, but are included in the Zemax® prescription in Appendix B.1. In total the light traverses 80 or 84 elements, depending on the path though the Michelsons (and counting multiple traverses of the same element) and is reflected seven or nine times.

There are two mechanisms external to the optics package: a front door, which protects the front window during launch, and an alignment mechanism used to adjust the optics-package pointing.

The following subsections describe each subsystem in more detail.

2.2. Imaging Optics

The main telescope is a two-element refracting telescope with a 140 mm clear aperture, giving an image with a nominal diffraction limit (λ/D) of 0.91 arcsec. The primary lens is convex aspherical on the front surface and convex spherical on the back surface, while

Table 1 Key numbers for the imaging optics. EFL is the Effective Focal Length. Image Size is the diameter of the solar image for a solar radius of 960 arcsec, corresponding, roughly, to the mean over the year. Numbers for the primary lens only include the front window and primary lens. Numbers for the telescope include only the front window, primary, and secondary lenses. Path lengths are from the front surface of the front window.

Subsystem	EFL	Focal Ratio	Image Size	Angular Magn.	Path Length
Primary Lens	586 mm	f/4.18	5.45 mm	1.00	615 mm
Telescope	2468 mm	f/17.65	23.0 mm	5.43	895 mm
Complete	4953 mm	f/35.42	46.1 mm	8.24	2218 mm

the secondary (Barlow) lens is bi-concave. The primary and secondary lenses are connected with a low coefficient-of-thermal-expansion metering tube made of a carbon fiber composite to maintain focus. Key parameters of the optical system are given in Table 1. All surfaces other than the front of the primary are spherical.

The focal ratio (in vacuum) through the Lyot is f/18.3 and through the Michelsons f/21.7. The primary to secondary magnification is thus 2.01 of which a factor of 1.63 is accomplished by the reimaging lenses.

The telephoto ratios of the telescope and the entire system are 2.72 and 2.23, respectively, measuring from the front face of the front filter.

The HMI calibration and focus-adjustment system consists of two calibration–focus wheels (see Section 2.6.2), each of which contain an empty position, three positions with optical flats of varying thickness, and a lens in the fifth position. By making the (identical) steps in optical thickness in the one wheel four times larger than the steps in the other wheel, it is possible to select one of 16 uniformly spaced focus steps of roughly 1 mm at the focal plane, corresponding to 2/3 of a depth of focus. Note that the heaters on the front-window mounting ring (see Section 2.7.1) can also be used to fine tune the focus by inducing a center-to-edge temperature gradient in the front window, thereby causing it to act as a weak lens. Besides allowing best focus to be set on orbit, this capability also provides a highly repeatable means for measuring the instrument focus and assessing image quality through phase-diversity analysis.

In calibration mode, a lens in the fifth position of each wheel images the entrance pupil onto the focal plane to provide uniformly integrated sunlight. This is illustrated in Figure 4. Calibration-mode images are used to provide Doppler calibrations, monitor the instrument transmission and assess variations in the flatfield. The calibration-mode image has a diameter of 50.1 mm, large enough to cover the largest solar image seen.

The light is folded by the ISS mirror and then split by a polarizing beamsplitter to send the s-component light to the filters while passing the orthogonal light onto the limb sensor. The limb sensors receive the full 50 Å bandwidth of the front window, while light for the rest of the instrument continues through an 8 Å blocking filter located just inside the oven.

A telecentric lens at the entrance of the filter oven produces a telecentric beam for the Lyot filter. This ensures that the angular distribution of light passing through the Lyot filter is identical for each image point, minimizing the variation of the central wavelength across the image as well as contrast loss. However, different parts of the solar image pass through different parts of the filters.

At the exit of the Lyot filter a beam-control lens minimizes the clear-aperture requirements for the Michelsons by making the extreme rays parallel. This means that the beam is not telecentric through the Michelson interferometers.

Doing so causes a center-to-edge variation of the center wavelength and a degradation of the contrast. However, this degradation, as well as the consequent increase in the noise,

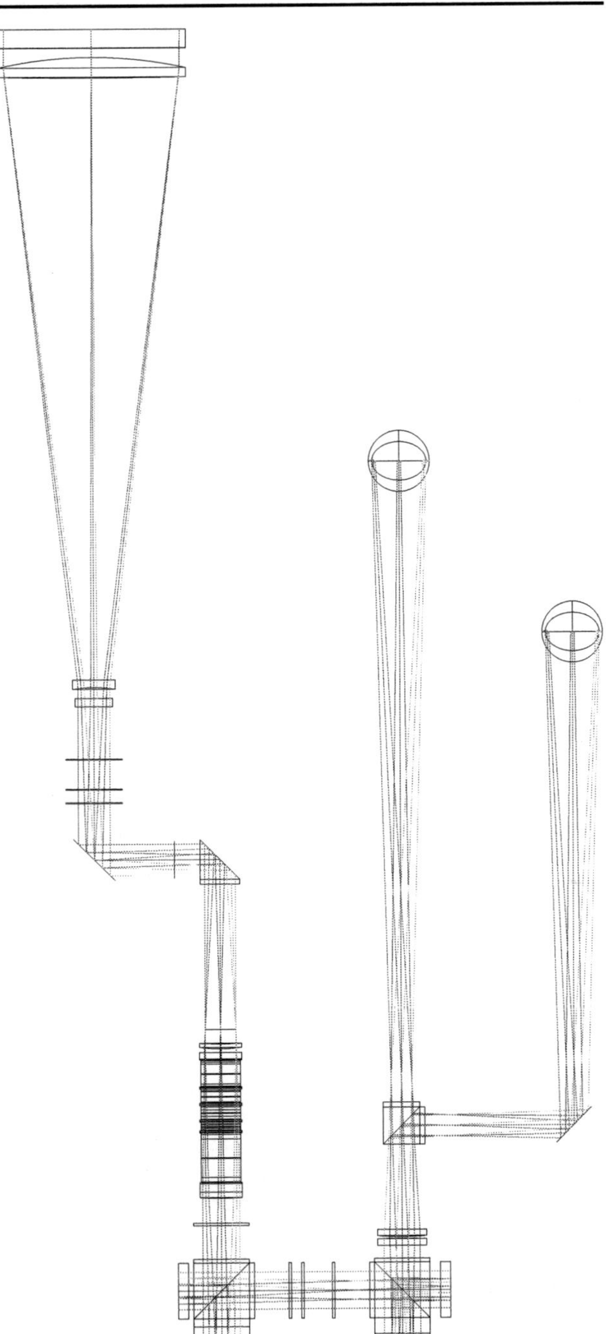

Figure 3 Raytrace of HMI, excluding the path to the limb sensor. The CCD detectors are above the plane of the plot. Different colors indicate different places on the solar image, at center and limb on each side. See Figure 1 for identification of the elements. The vertical distance from the front of the front window to the back of the second Michelson is 106 cm.

is insignificant both in an absolute sense and relative to that introduced by imperfections in the manufacture of the Michelsons. As the actual center-wavelength variation and contrasts are measured (see Section 3.3), the various causes are largely of academic interest.

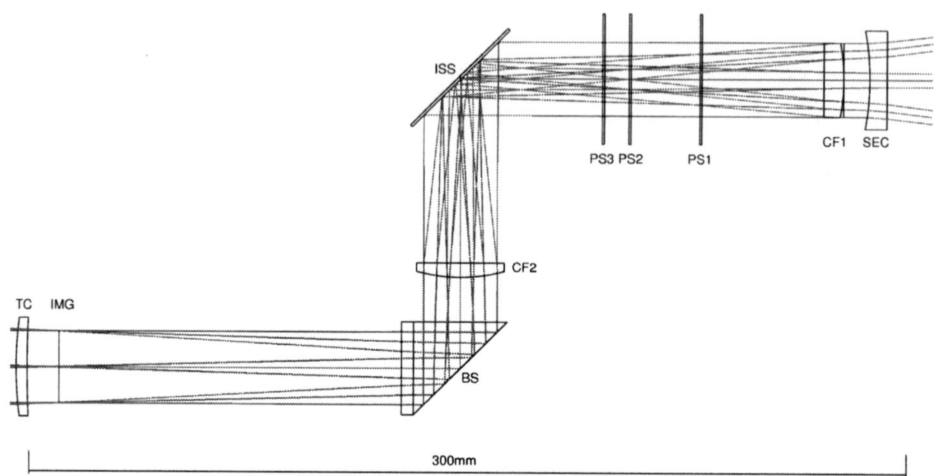

Figure 4 Raytrace with obsmode rays shown in blue and calmode rays shown in red (see Section 2.2). In the order the light travels (from upper right to lower left) the elements shown are: Secondary lens (SEC), first calibration lens (CF1), three polarization selectors (PS1, PS2, and PS3), ISS fold mirror (ISS), second calibration lens (CF2), polarizing beamsplitter (BS), a line showing the primary focus (IMG), and telecentric lens (TC) (Schou *et al.*, 2010).

At the exit of the oven, a pair of lenses reimages the primary focus onto the detectors. A beamsplitter evenly divides the light between the two camera paths with folding mirrors used to provide convenient placement of the cameras on top of the optics package. In order to prevent non-uniform movement of the shutter blades causing exposure gradients across the image, the shutters are placed at a pupil image. This has the minor disadvantage of briefly apodizing the aperture causing a slight broadening of the PSF. (Note that the shutter blade takes roughly 10 ms to cross the beam, which is small, but not negligible, compared to a nominal exposure time of around 150 ms.) The alternative of using a focal-plane shutter was rejected as it is likely to cause uncorrectable spatial non-uniformities in the exposures.

The unused port of the beamsplitter has a light trap. For the purpose of having a light source to test the cameras when there is no external light source, an LED was placed at this port. There is no need to use this LED on orbit after the front door has been opened.

2.3. Polarization Selectors

The polarization selectors rotate quartz waveplates to convert the desired incoming polarization into the fixed linear polarization selected by the polarizing beamsplitter. Each retarder is mounted in a Hollow Core Motor (HCM) that allows it to be rotated to any of 240 uniformly spaced angles (see Section 2.6.1). The nominal retardances of the waveplates are 10.5, 10.25, and 10.5 waves, in the order encountered. This design has a three for two redundancy, meaning that any one of the three waveplates can be left in any position, while preserving the ability to measure all polarization states (subject to the finite number of positions). High-order waveplates were chosen for mechanical convenience, but this choice does have implications for the polarization performance, as discussed by Schou *et al.* (2010). In particular, it leads to artifacts in calmode where the angles of incidence are substantially larger than in observation mode (obsmode), and thus means that the calibration mode (calmode) results cannot be applied to obsmode without corrections.

2.4. Filters

The HMI filter system consists of the front window, a blocker filter, a Lyot filter with a single tunable element, and two tunable Michelson interferometers. Both the Lyot filter and the Michelsons have temperature-compensating designs, and all of the filters, except the front window, are mounted in an oven stable to better than $\pm 0.01\ \mathrm{C\,hour}^{-1}$, as discussed in Section 2.7.2. The filter system enables narrow-band filtergrams to be made across the Fe I 6173 Å line by co-tuning one Lyot tunable element and both Michelson interferometers. The combined filter bandpass is 76 mÅ FWHM (nominally at center tuning) with a tunable range of 690 mÅ. Details of the filters are given by Couvidat *et al.* (2011) and summarized below in the order encountered by the light.

2.4.1. *Front Window*

The front window, the main purpose of which is to limit the heat input to the instrument, is a 50 Å bandpass filter made by Andover Corporation. In the order in which the light travels it consists of 6 mm of Schott BK7G18 radiation-resistant clear glass, 3 mm of Schott GG495 green glass, a bandpass coating, an infrared rejection coating, and another 6 mm of BK7G18 glass. The BK7G18 glass, among other things, serves as a radiation shield for the non-radiation-resistant GG495 glass. The GG495 glass serves two important purposes: One is to block the blue and ultraviolet light from reaching the bandpass filter, which can be damaged by exposure to those wavelengths. Another is to cause the window to absorb a small amount of light, thereby heating it up and limiting the center-to-edge temperature gradient.

A particular concern for the front window is that it experiences a very large temperature perturbation (tens of C) when going in and out of eclipses, causing the focus to change dramatically due to the induced center-to-edge temperature gradient (and to some extent due to the temperature change of the primary lens, which is radiatively coupled to the window). The eclipses are up to 72 minutes in length and so in order to limit the additional loss of data while recovering, a limit on the recovery time of 60 minutes was imposed. Models of the front window indicate an e-folding time of order ten to fifteen minutes for any temperature perturbations, which should be adequate to meet the requirement. Models indicate that raising the temperature of the window mounting ring (using the operation heaters described in Section 2.7.1) during the eclipses should help with the recovery.

Despite the use of radiation-resistant glass, a certain increase in the absorptivity, as seen with MDI, is likely to occur, which may in turn change the focus of the instrument. This focus change may be compensated by changing the focus position being used and/or fine tuned by changing the front-window heater setpoints.

Due to manufacturing problems there was a significant wavefront error in all of the front windows produced and as a result a compensating error was polished into the surface of the selected front window. This is further discussed by Wachter *et al.* (2011).

Some of the optical elements, including the front window, cause fringes in obsmode and/or calmode, as discussed by Couvidat *et al.* (2011).

2.4.2. *Blocking Filter*

The blocking filter is a three-period, all-dielectric, interference filter, also made by Andover Corporation, with a FWHM bandpass of 8 Å. The blocker rejects the unwanted orders of the Lyot and Michelson filters and limits the heat input into the oven.

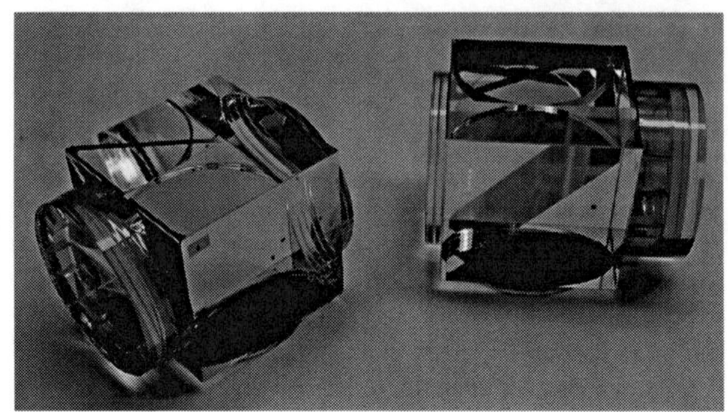

Figure 5 A pair of Michelson interferometers. The protective yellow tape on the vacuum legs was removed before launch. The cubes are 45 mm on each side. Photo courtesy of LightMachinery Inc.

2.4.3. Lyot Filter

The wide-field, temperature-compensated Lyot filter uses the same basic design as the MDI (Scherrer *et al.*, 1995) filter with the addition of a fifth tuned element in order to accommodate the wide range in radial velocity due to the geosynchronous orbit of SDO. The five-element (named E1 through E5 in order of increasing FWHM) Lyot filter has a 1:2:4:8:16 design (actual order is given in Section 2.4.5). This is unlike the MDI filter where some of the elements were doubled in order to reduce unwanted sidelobes in the untuned part. The final (narrowest) filter element of the Lyot filter is tuned by a rotating half-wave plate mounted in a HCM identical to those used for the polarization selectors. The Lyot tuning waveplate is not redundant. The FWHM of the untuned part is 612 mÅ, nominally. The Lyot components are held in optical contact by optical grease to avoid stressing the elements with changing temperature, and are keyed to hold the elements in proper relative alignment.

Each Lyot element contains nine pieces: A fused-silica spacer, a polarizer, a piece of potassium dihydrogen phosphate (KDP) or ammonium dihydrogen phosphate (ADP), a piece of calcite, a half waveplate, a piece of calcite, a piece of KDP or ADP, a quarter waveplate, and a fused-silica spacer. In addition the first element (E2) has a quarter waveplate between the first spacer and polarizer. By pairing KDP (for E1, E2, and E3) or ADP (for E4 and E5) elements with the calcite elements, the temperature sensitivity in the calcite is compensated by an opposite change in the KDP or ADP.

Details of the construction of the individual elements can be found in the supplementary material described in Appendix B.1. Physical diameters of the Calcite, ADP, and KDP elements are 32 mm with a clear aperture of 30 mm.

Due to a drawing error, the clear aperture at the exit of the Lyot holder is only marginally larger than the solar image. This results in some vignetting if the image is not accurately centered with respect to the obstruction. The position of the obstruction is expected to be very stable and can be determined from the regular flatfield observations taken every few months. For details see Wachter *et al.* (2011).

2.4.4. Michelson Interferometers

The final filters are two wide-field, tunable, solid Michelson interferometers with clear apertures of 32 mm and nominal free spectral ranges of 345 mÅ and 172 mÅ (172 mÅ and 86 mÅ FWHM pass bands, respectively, both made by LightMachinery). The design incorporates a polarizing beamsplitter with a vacuum leg and a solid glass leg. The vacuum leg is maintained with temperature-compensating standoffs made of calcium fluoride. The Michelsons are pictured in Figure 5.

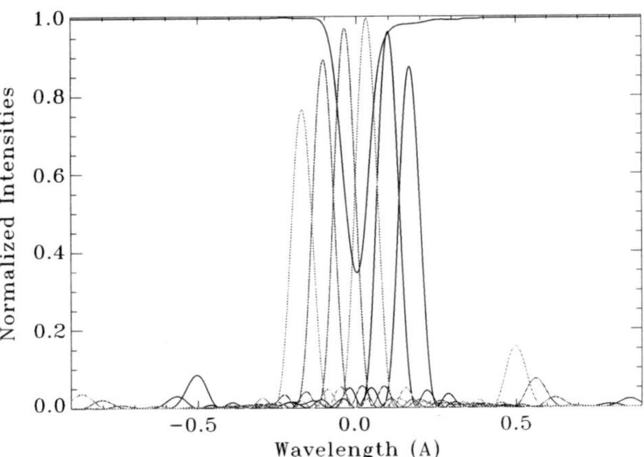

Figure 6 Example of HMI tuning-position profiles obtained from the wavelength-dependence calibration procedure. Six tuning positions are shown here with respect to the Fe I solar line at disk center and at rest. The line profile was provided by R.K. Ulrich and obtained at the Mount Wilson Observatory. Details of the calibration from which the profiles are derived are given in Section 3.3.

Tuning of the two Michelsons is accomplished by rotating a combination of a half-wave plate, a polarizer, and a half-wave plate, each of which is mounted in a HCM identical to those used for the polarization selectors. Unlike the Lyot tuner, this design has a full three-for-two redundancy, meaning that any one of the three mechanisms can fail in any position without impacting the ability to tune either Michelson. Ordinarily only the waveplates are used for tuning.

In order to ensure the best spatial uniformity of the tuning, regular non-radiation-resistant BK7 glass was used for the Michelson cubes. Given the substantial amount of glass between the edge of the beam and the amount of additional shielding around the Michelsons, no significant amount of radiation-induced darkening is expected.

2.4.5. *Filter Summary*

Figure 6 shows a model of the transmission profiles of the resulting filter. Details of the actual performance are given by Couvidat *et al.* (2011).

As with the MDI Michelsons, some amount of wavelength drift may be expected with time. Any such drift can be compensated by changing the HCM positions corresponding to each desired tuning position.

In summary the nominal filter FWHMs are, in the order in which the light travels:

- Front window: 50 Å
- Blocking filter: 8 Å
- Lyot element E2: 0.69 Å
- Lyot element E3: 1.38 Å
- Lyot element E4: 2.76 Å
- Lyot element E5: 5.52 Å
- Lyot element E1: 344 mÅ (tuned)
- Wide Band (WB) Michelson: 172 mÅ (tuned)
- Narrow Band (NB) Michelson: 86 mÅ (tuned)
- Untuned part: 612 mÅ
- Complete filter: 76 mÅ (at center tuning position)

Using the HCMs with half waveplates for tuning and having 240 steps means that one step is 1/60 of the Free Spectral Range (FSR) or 1/30 of the Full Width at Half Maximum (FWHM) of the corresponding tuned element.

2.5. Image-Stabilization System

The HMI Image-Stabilization System is a closed-loop system with a tip–tilt mirror to remove jitter measured at a primary image within HMI.

The ISS uses the light rejected by the polarization-selection beamsplitter to image the Sun at the guiding-image focal plane, located at the opposite side of the ISS beamsplitter from the Lyot filter (see Figure 1). Four detectors are positioned on the north, south, east, and west limbs of the solar image, each consisting of a redundant photodiode pair. The electronic limb-sensor photodiode preamplifier has two gains, ground-test mode and Sun mode, and selectable prime or redundant photodiodes. In order to avoid the introduction of noise between the limb sensors and the electronics package the (redundant) limb tracker pre-amp board is mounted inside the optics package.

The mirror uses low-voltage piezoelectric transducer (PZT) actuators to remove errors in the observed limb position, derived as the difference between the intensity reaching opposite photo diodes. The mirror design has a first resonance of ≈ 800 Hz, which is much higher than the structural modes of the HMI optics package and the expected jitter frequencies. This enables a simple analog control system.

The range of the tilt mirror is approximately ± 18 by ± 15 arcsecs in image space. A larger range for coarse pointing is provided by adjustable instrument legs described in Section 2.6.4.

The servo gains and other parameters are adjustable by ground commands. In particular, offsets can be added to the X- and Y-axis error signals to change the nominal pointing while maintaining lock. Individual PZT actuator offsets can be specified to set the nominal position of the mirror anywhere in its range during open-loop operation or during calibrations.

The error and mirror signals are continually sampled, and downlinked at a cadence of four to eight seconds to monitor jitter and drift. For calibration purposes, these signals can be downlinked at a rate of up to 512 Hz allowing for a more detailed measurement of the performance.

2.6. Mechanisms

The instrument contains a number of different types of mechanisms, each of which is described below. In most cases the details of their operation are of little consequence to the scientific operation of the instrument, in other cases they have important implications. The mechanisms are controlled by Field Programmable Gate Arrays (FPGAs) commanded by the flight software.

2.6.1. Hollow Core Motors

The HCMs are used to rotate the polarization selectors and wavelength tuning optics. Each motor has an encoder with 240 positions. To move from one position to another the motor is first commanded to move in the desired direction. Once the encoder indicates that a particular position has been reached, the FPGA waits a programmable time after which the motor windings are shorted out, causing the motor to brake and stop some distance away. By selecting an appropriate encoder position and delay time (selectable in units of 1 μs), the motors can be made to stop at any desired position (there is very little detent in the motors). However, in order to be able to control the stopping positions more accurately, only positions at the places where the encoder changes are used. This is further discussed in Section 2.11.5. A HCM is shown in Figure 7.

Figure 7 A Hollow Core Motor. The size of the housing is 88.9 mm by 101.6 mm by 24.0 mm.

The 360° spin time of a HCM is around one second. By going in the shortest direction it is thus possible to change between arbitrary positions in less than about 500 ms.

Since the HCMs are planned to move as often as once every 1.875 seconds and thus will experience about 80 million moves in five years, an extensive life-test program was performed in which 375 million moves were performed without problems, leading to a high confidence that they will perform as desired.

The standard deviation of the stopping positions depends significantly on the motor being tested and the starting and stopping positions, but is generally of order tens of arcsecs of rotation of the motors.

When the adjustment algorithm discussed in Section 2.11.5 is used, the accuracy of the absolute stopping positions is set by the accuracy of the encoder, which is specified to be 0.15° (0.1 steps) or better. Separate measurements with a precision encoder show that the accuracy is indeed better than this.

2.6.2. *Focus Wheels*

The focus wheels are very similar to the HCMs, except that the optics are placed around the rotation axis in a wheel, rather than on axis. While they can, in principle, be positioned at any one of 180 positions, only the five positions for which the optics are centered are used. The focus wheels also take about one second to rotate, and since the longest distance necessary to move is 0.4 revolutions, the maximum move time is 400 ms. A photograph of a filterwheel is shown in Figure 8.

As relatively few moves are needed and since similar mechanisms have been flown on LMSAL instruments before, no formal life test was performed.

2.6.3. *Shutters*

The shutters are very similar to those on MDI and the *Solar X-ray Imager* for GOES (Lemen *et al.*, 2004) and provide exposure measurements with a resolution of 1 μs. The combination of knowing the actual exposure times and downlinking all of the images means that it is possible to correct for any exposure variations. Thus even variations more than an order of magnitude larger than those seen on MDI after about 150 million operations (currently the errors are of order a part in 10 000 of the exposure time) on orbit will cause no detrimental effects. A shutter is shown in Figure 9.

Figure 8 Focus wheel. In this case the one closest to the telescope. Clockwise from the top are the thinnest focus block, the thickest focus block, the blank position, the medium focus block, and the calibration lens. The size of the housing is 109.98 mm by 121.03 mm by 19.05 mm.

Figure 9 Shutter mechanism and blade. For scale, the shutter blade has a diameter of 84.07 mm.

In normal operation, the shutters can be commanded to expose the CCDs from about 35 ms to about 16.7 s in increments of 1 ms. The expected exposure time on orbit is around 150 ms.

Since the shutters will move roughly once per 3.75 seconds, 40 million exposures are expected in five years. A life test to 85 million exposures was performed, again without problems. At the end of the life test, the typical standard deviation in the exposure time was roughly 10 μs, meeting the required performance.

2.6.4. Alignment Mechanism

The HMI alignment mechanism has two legs, each of which can adjust the optics package pointing in steps of 0.32 arcsec at a rate of 16 steps per second. The directions of image motion caused by the two legs are roughly orthogonal and correspond roughly to diagonal motions in the CCD image. The legs do not have built-in encoders, rather they rely on a hardware center indicator and a software step counter. There are also end-of-range indicators, which if tripped will cause the FPGA to stop moving the leg. These indicators are at positions in the range 1667 to 1850 steps from the center position. The flight software also limits the motion to ± 1500 steps or about ± 480 arcsecs in order to not rely on the end-of-range indicator.

The alignment legs will have to adjust the pointing by a few steps, perhaps every couple of weeks, to keep the solar image centered in the ISS dynamic range. Otherwise they are not planned to be used after launch. Given this, no formal life test was performed.

2.6.5. Front Door

The front door has two identical actuators, either of which can open the door regardless of the position of the other, thus providing redundancy. To close the door, both mechanisms must be in the closed position, but since it is only planned to operate (open) the door once after launch (the purpose of the door is to protect against damage and contamination during integration, test, and launch), there is no need for this to be redundant. The door mechanism has switches that indicate when either of the actuators is in the closed position, and switches that indicate when the door is fully opened (with either actuator).

2.7. Thermal Control

2.7.1. Heaters

Three different types of heaters are used on the instrument, each of which is redundant.

The survival heaters keep the instrument warm enough to avoid damage when the instrument is not powered, or in case of instrument failure. These heaters are powered directly by the spacecraft (bypassing the instrument electronics) and are controlled by separate built-in thermostats.

Operational heaters are used to maintain the instrument at the desired operating temperature. There are seven such heaters on the instrument: Two on the bottom of the HOP, two on top, one on the telescope tube, and two on the window mounting ring. Each of these zones is controlled by flight software using temperature measurements from thermistors distributed around the optics package, as further discussed in Section 2.11.4.

The placement of the various thermistors and the operational and survival heaters is included as supplementary material. See Appendix B.2.

The CCDs have their own radiators and are ordinarily kept below -50 C (without any thermal control) to minimize dark current and radiation damage. Decontamination heaters are used during the commissioning phase to warm up the CCDs in order to prevent them from accumulating condensable materials outgassing from the rest of the instrument after launch. These are mounted on the back of the CCDs and are thermostatically controlled. They can be individually enabled and disabled by command. If needed, they can also be used later in the mission to drive off any condensed material or to anneal the CCDs.

2.7.2. Filter Oven

To further stabilize the performance of the most temperature-sensitive elements of the filter system, *i.e.* the Lyot filter and Michelsons, are placed in a filter oven.

The oven can be set to eight uniformly spaced temperatures between 28 C and 35 C and is designed to maintain a temperature stable to 0.01 C hour^{-1}, with the best estimate of the actual performance being more than an order of magnitude better. To achieve this performance, the oven uses an analog proportional heater control. The oven-control electronics and heaters are redundant. In order to stabilize the temperature, the oven pre-amp and controller boards are mounted on the oven itself, thereby making them temperature controlled.

The oven includes a multi-stage thermal isolation system to ensure that the time scale of any residual filter-temperature variations is very long compared to that of the solar p-mode oscillations.

The telecentric and relay lenses are the entrance and exit windows of the oven.

2.8. HMI Electronics Box

The power supplies, control electronics, and data-processing electronics are housed in a separate HMI Electronics Box (HEB), which is mounted inside the spacecraft bus. The HEB contains a total of 17 items:

- Two Power Converter Subsystems
- Two PCI to Local Bus Bridge and 1553 Interface Boards
- Two RAD6000 Processor Boards
- Four combined Mechanism and Heater Controller Boards (Sections 2.11.4 and 2.11.5)
- One Housekeeping Data Acquisition Board (Section 2.11.3)
- Two Camera Interface (CIF) Boards (Section 2.10.1)
- Two combined Data Compression and High Rate Interface (DCHRI) Boards (Section 2.10.2)
- One ISS Limb Tracker Board (Section 2.5)
- One ISS PZT Driver Board (Section 2.5)

The main control is performed by a redundant pair of BAE Systems RAD6000 CPUs, either of which can control the instrument, as selected when power is applied. The CPU communicates with the spacecraft over a MIL-STD-1553 bus through the Bridge Boards which are also used to connect to the other boards in the HEB.

The power supplies are redundant and are both connected to each of two spacecraft power buses.

The Bridge Boards each contain a very stable temperature-compensated clock, with a finely adjustable frequency, used for time-stamping the images. By monitoring the clock-to-spacecraft and spacecraft-to-ground time differences and adjusting the frequency, it is possible to both keep an accurate time (better than 100 ms) and to avoid jumps in time, as would be seen if one relied on either the spacecraft time or needed to reset the HMI clock to make adjustments.

As described in Sections 2.5, 2.7.2, and 2.9, the HOP and cameras also contain a number of circuit boards.

A simplified block diagram is shown in Figure 10.

2.9. CCDs and Camera Electronics

The CCDs and camera electronics are very similar to those used for AIA (Lemen *et al.*, 2011), except that the HMI CCDs are front-side illuminated, while the AIA CCDs are back-side illuminated.

HMI contains two identical CCD detectors, with a 4096^2 pixel format, made by E2V. The CCDs have 12 μm pixels and a full-well capacity of close to 200 000 electrons. The CCD assemblies are mounted to the optics package using fiberglass insulator tubes in order to minimize the amount of heat transferred to them.

The CCDs are cooled passively using radiators mounted on top of the CCD assemblies, pointed perpendicular to the direction to the Sun. The expected CCD temperatures are below − 50 C, resulting in a negligible dark current.

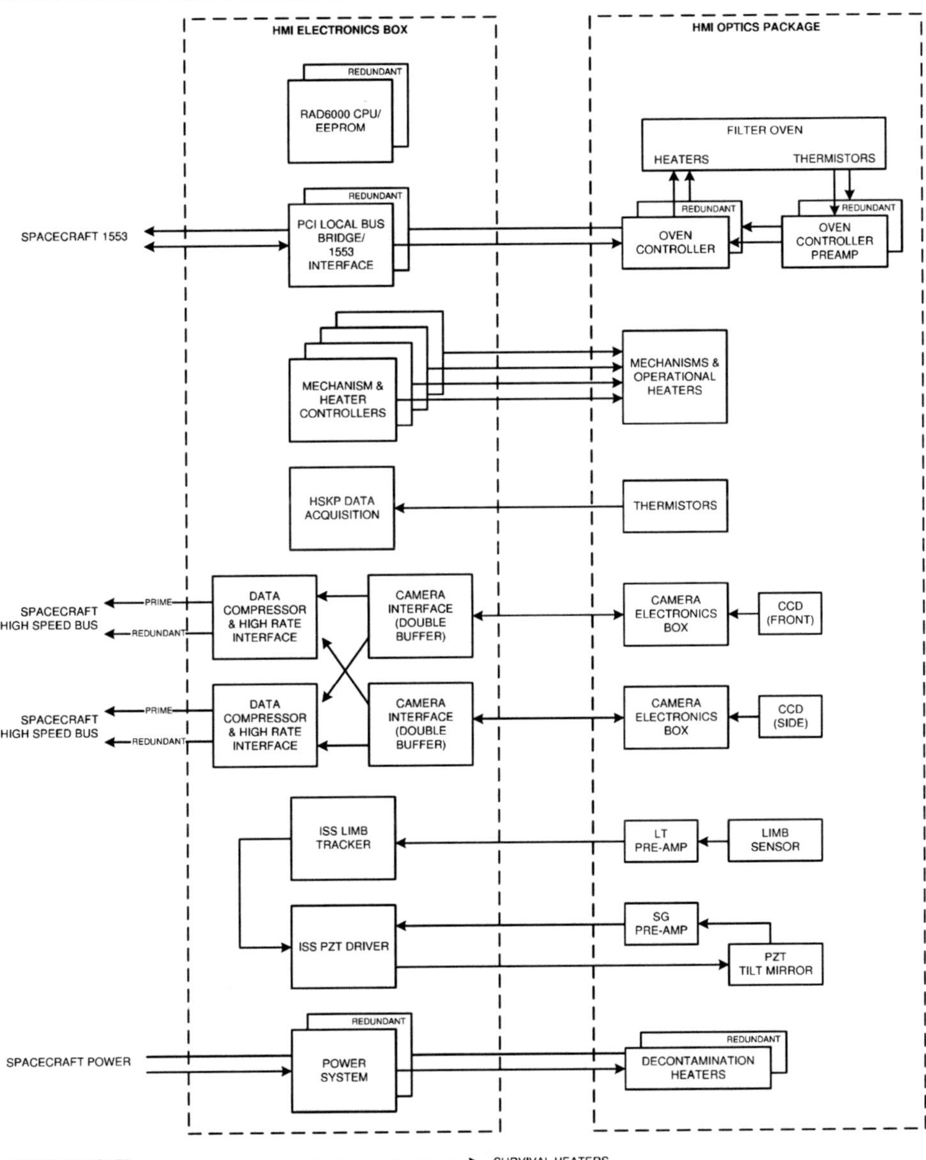

Figure 10 A simplified electronics block diagram of HMI. For simplicity the power and command–telemetry connections between the different components in the HEB have not been shown.

The cameras, made by Rutherford Appleton Laboratory (RAL), have a readout rate of 2 Mpixels s^{-1} through each of four quadrant readout ports. Using all four amplifiers, a CCD can be read out with 14 bits per pixel in about 2.3 seconds. The actual performance is discussed in Wachter *et al.* (2011). In case of failures, it is also possible to read out using any two adjacent amplifiers or any single one, with correspondingly increased readout times, thus providing for a somewhat graceful degradation.

In typical operation a programmable number of CCD rows (generally corresponding to well over a full image) are first shifted rapidly to eliminate any built-up charge, the shifting is stopped while the shutter is opened, and finally the CCD is read out. Generally the CCD is overscanned by one or two columns and rows in order to track performance.

2.10. Data Processing Electronics

The data from the CCDs are downlinked without any processing other than hardware compression and encoding. The different components of the electronics are described below.

2.10.1. Camera Interface Board

Each camera is connected to one CIF board using a spacewire (IEEE 1355) cable. Each CIF contains two image buffers, allowing one to be filled with new data while the other is being read out.

2.10.2. Data Compression and High Rate Interface

Next, the data are sent from the CIF boards to the combined Data Compression and High Rate Interface (DCHRI) boards. Normally each DCHRI board receives data from one CIF; however, the boards are cross-strapped allowing the data from both CIFs (and thus both cameras) to go to one DCHRI, thus providing significant redundancy.

The DCHRI boards do the following processing of the images (in order): crop (*e.g.* to not transmit data far above the solar limb), select which bits are to be used (*e.g.* to not transmit photon noise), put the resulting values through a lookup table (in order to better match the truncation to the photon noise at different intensities), and perform a lossless compression. The parameters of each of these steps can be set via loadable tables and can be different for each image in a framelist (see Section 2.11.6).

Each DCHRI connects to two separate redundant interfaces on the spacecraft data bus using spacewire links, thereby providing further redundancy. If these redundancies are used only a very small reduction in data rate results. The data rate allocated to HMI is 55 Mbit s^{-1}, which may be adjusted by transmitting new tables to the spacecraft, with each DCHRI able to deliver in excess of the currently allocated rate.

2.11. Flight Software

The flight software (FSW) consists of several different modules. Some, *e.g.* those related to loading other modules, have little direct effect on the scientifically important aspects of the instrument. Others, such as the sequencer, are important for understanding the instrument capabilities. The following sections describe the scientifically important parts of the FSW.

2.11.1. Overall Control

The core of the FSW is a kernel running under a VxWorks real-time operating system. It is read from ROM at startup and is responsible for loading other modules and allocating CPU time for their execution.

Ordinarily the instrument is booted and loads the various modules from EEPROM into RAM. While not planned for routine use, it is possible to reload these parts of the software to the EEPROM. It is also possible to load new FSW directly to the RAM for testing or in case the EEPROM becomes non-functional.

In addition to being able to load the flight software itself, it is possible to load tables to memory for use by other modules, as described below.

2.11.2. Commanding Interface

The FSW has the capability to receive and execute a large number of commands to load software and tables, control various subsystems such as mechanisms and heaters, configure various onboard control systems, initiate downloading of diagnostics data, and so forth.

It is also possible to execute simple onboard scripts. Typical examples of such scripts include ones to change the instrument heater settings during eclipses when the Earth blocks the Sun or ones to run different observing sequences during the eclipses or transits of the Moon or planets.

2.11.3. Housekeeping

The instrument provides a large variety of data in the housekeeping data stream. These include such items as temperatures, voltages, currents, limb-sensor voltages, states of subsystems, mechanism positions, command counts, and so forth. These are generally downlinked every four or eight seconds.

In addition, a copy of certain housekeeping values relevant to each image are also downlinked as a so-called Image Status Packet (ISP) with the image data. Currently a maximum of nine ISPs are allocated per 16 seconds, so although the cameras can run somewhat faster, this effectively sets the maximum frame rate at one image per 1.778 seconds.

The total allocated housekeeping data rate is 2353 bits s^{-1}. Of this 1714 bits s^{-1} are currently allocated to the regular housekeeping and 612 bits s^{-1} for the ISPs.

It is also possible to more intensively downlink selected hardware values at rates from 128 Hz to 512 Hz for about 3.5 minutes. This is typically used for sampling the ISS, but may also be used for such items as motor currents. These diagnostic data are downlinked over a separate channel, shared between various spacecraft subsystems, at a rate of 10 kbits s^{-1}.

Finally, the instrument has a number of sensors connected directly to the spacecraft for health and safety monitoring when the instrument is not powered on.

2.11.4. Thermal Control

Each of the seven operational heaters are controlled by the FSW. In addition to being left constantly on or off, the FSW can run them at specified duty cycles using a pulse-width modulation scheme in which the heaters are on for N time increments out of M, with $2 \leq M \leq 100$, $0 \leq N \leq M$, and the time increments are fixed at 1/8 second.

The FSW also provides basic thermostatic temperature control of the instrument. The scheme uses three temperatures: deadband low, target, and deadband high, and three duty cycles: cold, deadband, and off. The modes use the same M, but different N. At any given time, the heater zone is in one of three states: Rising state, in which the cold duty cycle is applied; Maintaining state, in which the deadband duty cycle is applied; and Dropping state, in which the duty cycle is 0. The heater zone changes from Rising to Maintaining state when the target temperature is reached, and from Maintaining to Rising state if the temperature drops below the deadband low. If the temperature rises above the deadband high, the state is changed to Dropping. The state changes from Dropping to Maintaining if the temperature drops below the target.

The consequence of this control algorithm is that one has to either have a large deadband and accept that the temperature drifts with the environment, typically with a 24-hour period, or have a tight deadband and accept frequent, but smaller, changes in the temperatures as the duty cycles change.

2.11.5. Mechanism Control

The mechanisms are also controlled by the FSW.

The front-door and alignment mechanisms are simple stepper motors and as such require little control. For the alignment mechanism, the FSW can seek for the zero crossing and, since there is no encoder, keep a count of steps taken. As mentioned earlier, the FSW limits the allowed positions to $\pm$ 1500 steps from the center position.

The focus-mechanism and shutter interfaces are simple and require little active control (Sections 2.6.2 and 2.6.3).

The HCMs are substantially more complicated. While they can be operated by simple commanding to move clockwise or counterclockwise to a given target, two other features are implemented in the FSW. First, in order to minimize the time to perform a move, the FSW can be set to automatically choose the shorter direction. Second, since the stopping positions are inherently sensitive to such parameters as voltage and temperature, which are themselves somewhat variable, an optimization algorithm has been implemented. As discussed in Section 2.6.1, the HCMs use an adjustable delay to fine tune the stopping position. An encoder tells whether the motion overshot or undershot the position. The optimization algorithm uses this information to adjust the delay times up or down slightly to always end up near the desired position, regardless of temperature or voltage variations. Since the optimal delay also depends on move distance and to a lesser extent on the starting position, a separate delay is kept for each motor, start position, and stop position combination. A side effect of the use of this algorithm is that a given move has to be performed a number of times before the stopping position is accurate and it is thus desirable to perform changes in the set of tuning positions used during times when no sensitive data are being taken (*e.g.* during eclipses). Another side effect is that, since the delays are given as integer μs (corresponding, roughly, to a part in 10^6 of a rotation), the algorithm may introduce a small but systematic pattern in the stopping position. (Note that there was an error in this algorithm early in the ground testing and that data taken then need special correction. However, data taken 1 June 2007 and later should be free of this error.)

2.11.6. Sequencer

The part of the FSW most directly affecting the operation of the instrument from a scientific point of view is probably the sequencer. The sequencer is responsible for taking the images at the correct times and with the correct settings, as well as for executing calibrations at scheduled times.

The sequencer has two main levels of control: The top level determines which set of observations, described in standard framelists (see below), to execute at any given time, while the lower level schedules the individual frames.

At the top level, the sequencer uses a prioritized list to determine which framelist to execute next. By carefully selecting these priorities and the desired execution times for the different framelists, it is possible to interrupt the regular observables framelist to run a calibration sequence or to run different sequences depending on the time of day, for example.

The main control of the images taken is done by using so-called framelists. These contain, among other things, a list of relative times to take the individual filtergrams, as well as the wavelength, polarization, focus, exposure time, camera, and compression settings. At any given time only one framelist is executing. An example of a framelist and more information about what they contain are given in Appendix A.

Once the start time of a framelist is known, the exact time for each of the images can be calculated from the start time and the relative offset specified in the framelist. From these times, the time of execution of each instrument activity required to make an image can in turn be determined. Below is a list of the four main events scheduled during the taking of an image. The relative times of these are set by ground-commandable registers:

- Mechanism Move: This is the time at which the polarization selectors, wavelength selectors, and focus wheels start moving.
- Clear: This is the time at which the camera starts clearing the CCD and thereby initiates the image taking. See Section 2.9 for details.
- Shutter Open: This is the time at which the shutter opens.
- Readout: This is the time at which the readout of the image from the cameras to the CIF is initiated.

The events for the two cameras are independent, however a warning is issued if incompatible events overlap, *e.g.* if a HCM or focus wheel is moving between the start of the shutter move and the start of the readout.

2.12. HMI–MDI Comparison

To aid users of MDI in transitioning to use HMI, Table 2 compares key properties of the two instruments.

3. Calibration

Since HMI operates at a visible wavelength (unlike *e.g.* AIA) it was possible to perform a wide variety of end-to-end ground calibrations. We next describe the basic setups used for the calibration followed by a summary of the results obtained. Calibration details are described in separate articles for image quality and CCD performance (Wachter *et al.*, 2011), for filter performance (Couvidat *et al.*, 2011), and for polarization properties (Schou *et al.*, 2010).

3.1. Calibration Setups

As described in Wachter *et al.* (2011) most ground calibrations were performed in one of two configurations. The most common setup uses a stimulus telescope that is a reverse of the HMI telescope. The stimulus telescope consists of primary and secondary lenses manufactured to the same specifications as the flight optics. As a result there is significant field curvature in the stimulus telescope.

The stimulus telescope can be fed with either white light from a stabilized lamp or with laser light. The lamp light is fed into a fiber bundle that is then imaged onto the pupil using a condenser lens. This setup scrambles the light and ensures a uniform illumination of both the image and pupil. At the focus of the stimulus telescope it is possible to mount a number of different targets, most of which are made with a metal film deposited on a glass slide. The stimulus telescope focus can be found in auto-collimation using a large optical flat.

The same basic setup is used with one of two laser light sources. A tunable dye laser is fed through an optical fiber from another room and typically illuminates a rotating diffuser. A tunable solid-state laser was used for some tests. The solid-state laser mounts directly on the stimulus telescope and uses a simple diffuser. Unfortunately the solid-state laser is not very bright, which limited its use for calibrations.

The other main setup feeds sunlight to HMI. A heliostat on the roof tracks the Sun and directs sunlight through a tube to a fold mirror in front of the instrument.

Table 2 Comparison of MDI and HMI. For MDI: FD refers to full-disk mode, HR to high-resolution mode, and structure to the continuous data. Note that some numbers for HMI (*e.g.* the cadences and filtergram positions) are nominal and may be changed as operational experience is gained.

Property	MDI	HMI
Target line	Ni I 6768 Å	Fe I 6173 Å
Aperture	12.5 cm	14.0 cm
Optical resolution (λ/D)	1.17 arcsec	0.91 arcsec
Pixel size	1.98 arcsec (FD) 0.61 arcsec (HR)	0.505 arcsec
CCD	One 1024 × 1024 21 μm	Two 4096 × 4096 12 μm
Front window FWHM	50 Å	50 Å
Blocking filter FWHM	8 Å	8 Å
Lyot design	1:1:2:2:4:8	1:2:4:8:16
Untuned FWHM	465 mÅ	612 mÅ
Tunable element FWHMs	94 mÅ Michelson 189 mÅ Michelson	86 mÅ Michelson 172 mÅ Michelson 344 mÅ Lyot
Final filter FWHM	85 mÅ	76 mÅ
Spectral resolution (λ/FWHM/1000)	80	81
Polarization	$I \pm Q$ and $I \pm V$	All
Filtergram cadence	3 seconds	3.75 seconds per camera 1.875 overall
Filtergram positions	5 at 75 mÅ spacing	6 at 69 mÅ spacing
Nominal observables cadence	60 seconds	45 seconds
Data rate	160 kbit s^{-1} (FD and HR) 5 kbit s^{-1} (Structure)	55 Mbit s^{-1}

The stimulus telescope, as used for image-quality measurements, as well as the overall calibration setup are described in more detail by Wachter *et al.* (2011). Uses for filter characterization are described by Couvidat *et al.* (2011), as is the use of the heliostat.

Tests were performed in two different environments: In one case the instrument resides in a vacuum tank with an optical-quality window. In the vacuum tank the CCDs could be cooled to dramatically reduce the dark current and thus photon noise. However the optical and polarization characteristics of the window are not known reliably. A sophisticated temperature control system was constructed to perform thermal tests on the instrument.

In the other case the instrument is operated in a clean tent. This is more convenient, but has disadvantages of its own. At room temperature there is a significant and often variable amount of dark current, requiring compensation in the analysis. The presence of air in the instrument changes the focus dramatically. To compensate for this, an air-to-vacuum corrector consisting of a plano–concave and plano–convex lens pair with identical radius of curvature and spaced by an adjustable amount is placed between the stimulus telescope and the instrument. Alternatively, the targets in the stimulus telescope can be moved a very specific distance from the position determined by the auto-collimation. Finally, the tuning of the Michelsons changes by an amount that depends on air pressure, temperature, and humidity.

Since the stimulus telescope is in either case in air, the presence of air currents causes small image distortions and degradation of the image quality.

Standard calibration sequences were developed for the Comprehensive Performance Tests (CPTs) which were performed regularly before launch to monitor the performance of the instrument during integration and testing. The instrument orientation changed 90° in roll around the line-of-sight direction, with respect to the stimulus telescope as well to the direction of gravity, when it was moved from LMSAL in Palo Alto, where the instrument was assembled and tested, to NASA Goddard Space Flight Center (GSFC), where the integration to the spacecraft was performed. The orientation at Astrotech Science Operations in Florida, where the spacecraft was prepared for launch and some of the tests preformed, was the same as at GSFC.

Finally, it should be noted that some of the calibrations have been deferred to on-orbit for practical reasons.

3.2. Image Quality

Details of the image-quality calibration setup, procedures, and results are given by Wachter *et al.* (2011). Some of the main results are given below. The Strehl ratio is measured to be 0.74 ± 0.03 or better. The uncertainty is due to difficulties with compensating for the test setup, and is thus expected to be better on orbit. Since a wavefront error in the front window was polished out, thereby substantially improving the Modulation Transfer Function (MTF) (Wachter *et al.*, 2011), no measurable changes of the MTF have been observed during the CPTs and other optical-performance tests.

The optical distortion has been modeled with Zernike polynomials and is as large as two pixels at the edge of the field of view. Residuals are of the order of a few hundredths of a pixel, and are dominated by irregularities in the CCD lattice.

Residual image motions of up to half a pixel are induced by the rotation of the HCMs. These motions have been consistently observed, and are corrected for in the observables calculations.

The performance of the ISS is shown in Figure 11. As can be seen, the attenuation is significant over a wide frequency range.

Small measured variations in the best focus across the field of view can be expressed as a field curvature (0.4 mm focal-plane variation from center to edge) and a field tilt (up to 0.5 mm focal-plane variation from edge to edge). Given that the HMI focus steps are 1 mm, corresponding to less than a depth of focus, this additional variation does not degrade the image quality significantly. The best focus has been set near the middle of the available focus range (focus position 9 in a range from 1 through 16), with a bias toward a higher focus position, because adjustment of the heaters at the edge of the front window can only bring the best focus to a lower position.

The two HMI cameras are aligned (with a residual shift of less than ten pixels), and the shift and the rotation (0.08°) have been monitored. During the calibration period, the cameras have shifted slightly (by a couple of pixels) relative to each other, possibly due to changing mechanical stresses in the instrument. Small shifts are not a concern because they can be constantly monitored by tracking the solar limb or correlating the images.

The flat field shows the expected structure originating in the design of the CCD, contamination in the optics, and large-scale optical variations. The internal vignetting (see Section 2.4.3) results in a sharp drop off in the flat field at the outer edge of the field of view. The large-scale flat field could be determined with an accuracy of 0.2%. Drifts observed in the small-scale flat field in vacuum are believed to be caused by outgassing material condensing

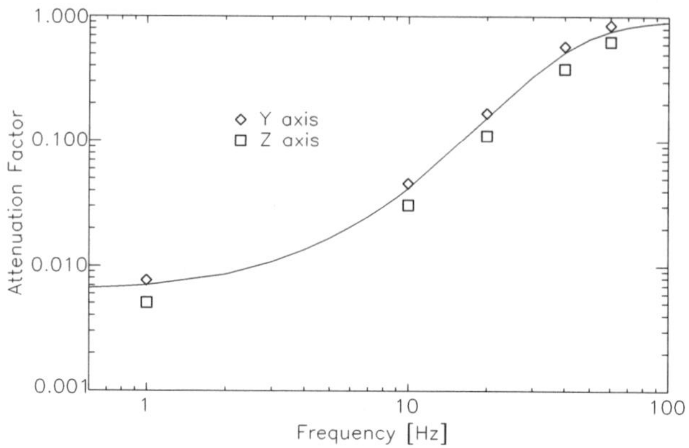

Figure 11 Example of the ratio of the observed signal on the limb sensors (sampled at a high rate, as described in Section 2.5) with and without the ISS turned on for a variety of frequencies. Also shown is a model of the response. Both the measured data and the model depend on the instrument gain settings.

on the cold CCDs. Given the lengthy outgassing period planned during on-orbit commissioning, during which the CCDs are kept warmer than the rest of the instrument, and the ability to drive any condensed material off by warming the CCD, the amount of material to condense is expected to be small in space. Small-scale flat-field monitoring will be performed to keep the flat field up to date at the aforementioned level of accuracy (Wachter and Schou, 2009).

The CCD full well is around 12 000 DN, which with an inverse gain of roughly $16e^{-}\ \mathrm{DN}^{-1}$ corresponds to roughly 200 ke^{-}. The camera non-linearity is better than about 1% up to the point where the system saturates. Since the exposure time can be controlled accurately over a large range, it is expected that the maximum exposure level near disk center will be close to full well, but the determination of the exact target has been deferred to commissioning.

The dark current is negligible when the CCDs are operated at their design temperature. The read noise is negligible compared to the photon noise.

3.3. Wavelength Dependence

The instrument also underwent a significant calibration effort in the area of filter performance. This is described in detail by Couvidat *et al.* (2011) and summarized below.

Observable sequences taken by HMI to measure the Doppler velocity and complete magnetic-field vector at the solar surface consist of full-disk filtergrams observed at six wavelengths spanning the range ± 172.5 mÅ around the target wavelength of 6173 Å. In order to accurately derive the Doppler velocities and vector magnetic fields from these six wavelengths and various polarization states, the transmission profiles of all the components of the HMI optical and filter system described in Section 2.4 need to be known: this is the purpose of the wavelength-dependence calibration.

The five stages of the Lyot filter and the two Michelson interferometers have transmission profiles, which can be fully characterized, in a first approximation, by only three parameters: the FSR, the relative phases, and the contrasts (ranging from 0 to 1). Both the phase and contrast vary across the element aperture and therefore are given as 2D maps. The FSRs of the elements are, to a very good approximation, equal to twice their FWHM. To determine these parameters we took cotune and detune sequences with various light sources. A detune sequence is a series of 27 or 31 HMI images for which the three wavelength selectors are tuned so that the peaks of maximum transmittance of the tunable elements do not necessarily coincide. For a cotune sequence the tunable elements are tuned together.

From a detune sequence with laser light, the phase and contrast maps of the three tunable elements can be determined. Such maps can be obtained both in obsmode and calmode. Detune sequences taken at different laser wavelengths give us access to the FSRs of the tunable elements, and also the phase and contrast maps of the non-tunable elements (Lyot elements E2 to E5). The FSRs of the non-tunable elements and the transmission profiles of the front window and blocking filters have to be measured separately while these parts are outside the instrument. Detune sequences in sunlight also allow a determination of the phases of the tunable elements. In sunlight, the contrasts of these elements cannot be assessed and the phases can only be derived in calmode. We fit for the phases of the tunable elements as well as the parameters defining the Fe I solar line (approximated by a Gaussian profile for convenience). Numerous detune sequences taken on the ground showed the high quality of the HMI filter elements. The phase ranges of the three tunable elements are relatively small (less than 20°), and their contrasts are very high (>0.95 on average), especially compared to the Michelson interferometers in the MDI instrument. Their FSRs are well within specifications. For reasons that are not fully understood, the average phases of the tunable elements are slightly different in obsmode and calmode (especially for E1). This is partly due to the non-negligible angular dependence of these elements. Similarly, for the narrow-band Michelson there seems to be a small difference between phase maps obtained from the front and side cameras. The overall mean transmission profile of the non-tunable elements is centered around $+16$ mÅ from the target wavelength, with a FWHM of ≈ 624.5 mÅ for the main peak. Two small main sidelobes are located at about ± 1 Å.

The best estimates of the FWHM of the various elements are:

- Front window: 48.5 Å
- Blocking filter: 8.43 Å
- Lyot element E2: 0.704 Å
- Lyot element E3: 1.390 Å
- Lyot element E4: 2.84 Å
- Lyot element E5: 5.68 Å
- Lyot element E1: 346.75 mÅ $\pm$ 2.3 mÅ (tuned)
- Wide Band (WB) Michelson: 170.95 mÅ $\pm$ 0.8 mÅ (tuned)
- Narrow Band (NB) Michelson: 85.45 mÅ $\pm$ 0.45 mÅ (tuned)
- Untuned part: 624.5 mÅ
- Complete filter: 76.15 mÅ (at center tuning position)

In some cases uncertainties are given, for others no good estimates are available, although the Lyot element uncertainties are likely less than 1%.

The blocking filter and front window are composed of multiple glass blocks that unfortunately act as weak Fabry–Pérot interferometers because of partial reflections at the glass interfaces. This produces a fringe pattern visible in the filtergrams and in the phase and contrast maps of the filter elements. The fringe pattern can be characterized by taking detune sequences in white light (from a lamp) in calmode and obsmode. The transmission profile of the non-tunable part of the HMI optical filter can be expanded into a Fourier series. The first seven Fourier coefficients are accessible through this analysis. The Fourier-coefficient maps at these different spatial frequencies show the different fringe patterns and allow us to pinpoint their origin. An issue is that the phases of these patterns depend on the temperature of the front window and blocking filter. Therefore these fringes will not be stable for some time (with an e-folding time similar to the roughly ten to fifteen minutes for the focus stabilization) after each daily eclipse during the Spring and Fall eclipse seasons (each of which lasts roughly 21 days).

The detailed wavelength profiles of the blocking filter and front window also means that each detune- or cotune-sequence position does not receive exactly the same amount of incident light: the integral value of the transmission profiles varies, resulting in an I-ripple. Detune sequences in white light allow us to measure the overall I-ripple of HMI. Fine-tuning sequences in which each element is tuned separately allow us to measure the individual I-ripples of the tunable elements. These individual I-ripples are $< 1\%$ (relative peak-to-peak variation of transmitted intensity), while the overall I-ripple is slightly below 2%. The I-ripple was not taken into account in the analysis.

Transmission profiles, fringe patterns, and I-ripples may depend on the angle of incidence of light rays, and on the temperature of the optical-filter elements. Hence the need to characterize these angular and temperature dependencies. The angular dependence of the tunable elements is obtained in calmode from detune sequences taken by moving a small field stop to various positions in the image plane. The same can be done in obsmode by moving an aperture stop across the front window. The Lyot element E1 is the tunable element with the largest angular dependence, followed by the NB and the WB Michelsons. This was expected, as the angular dependence of a wide-field Lyot element varies as θ^2 (where θ is the angle of incidence) while it varies as θ^4 for a wide-field Michelson interferometer. E1 also seems to show a significant amount of azimuthal dependence. The temperature dependence of the tunable elements is satisfactory. Indeed, the wavelength drift is only about 15 mÅ C^{-1} at 30 C. Fortunately the I-ripple does not vary with temperature or with a change of focus blocks.

Finally, the overall throughput (ratio of transmitted over input intensity) of HMI was measured at $\approx 1.35\%$, which is more than adequate for the 45-second cadence of the observable sequence.

An example of the resulting six tuning position transmission profiles is shown in Figure 6, with respect to the Fe I solar line profile at rest and at disk center (R.K. Ulrich, private communications).

After launch the orbital velocity of SDO may be used to improve the instrument calibration on orbit. The Sun–SDO radial velocity is known with an exquisite precision that allows us, in theory, to reach a much better accuracy on the FSRs of the tunable elements than from ground tests, and to reduce some uncertainties in the filter transmission profiles. However, due to the limited range of this radial velocity ($\pm 3.5\ \mathrm{km\,s^{-1}}$ from the orbit), it is not clear to what extent the calibration can be improved. This will be further investigated after launch.

As an illustration of the end-to-end performance, Figure 12 shows the expected photon noise in the Dopplergrams.

3.4. Polarization

The final major calibration area is polarization. The details of this effort are described in Schou *et al.* (2010).

For the polarization tests, a Polarization Calibration Unit (PCU), described by Schou *et al.* (2010), is inserted into the beam between the stimulus telescope and HMI. The PCU allows for a polarizer and an approximately quarter-wave plate to be inserted in the beam, each of which can be rotated to arbitrary angles.

By taking a carefully crafted set of images with various settings of the PCU and of the HMI polarization selectors, it is possible to derive almost all of the relevant properties of the stimulus telescope, PCU, front window, and polarization selectors. The undetermined parameters (polarization-angle zero point and one of the polarization-selector rotation angles) were determined from independent measurements.

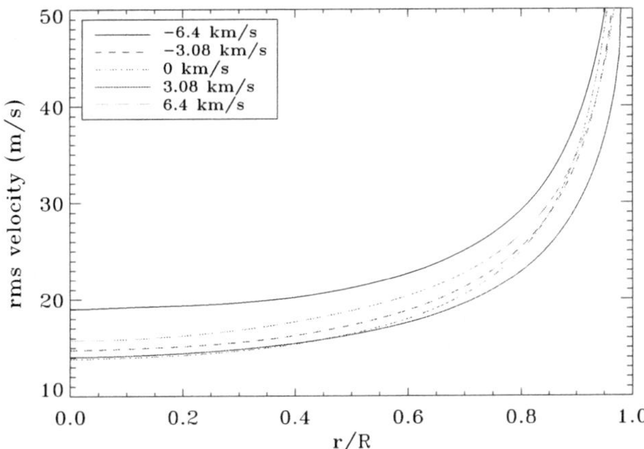

Figure 12 Expected Doppler noise from photon statistics calculated from the results of the wavelength calibration and averaged over azimuth. Results are shown for a variety of line-of-sight velocities. The variation with azimuth is small relative to the variation with velocity.

From these measurements it is in turn possible to determine a model of the response of the instrument to a given input Stokes vector. The main results are that the polarimetric model indicates that the crosstalk, after calibration, between Q, U, and V is less than 1% across the FOV and that the leakage from Q and U to V without calibration is below 5%, both meeting the requirements.

Some of the terms in the calibration, in particular the I to Q, U, and V leakages (telescope polarization) are difficult to determine using ground-based measurements, but are straightforward to determine on orbit, and the determination of those terms have thus been deferred. It may be noted that apart from these terms it is generally difficult to do a proper polarization calibration on orbit as there are no well-defined polarization sources available.

Several of the parameters describing the polarimetric model have a significant temperature dependence, but it has also proven possible to model this with sufficient accuracy to meet specifications.

4. Conclusion

The HMI instrument provides full-disk solar imaging at one arcsecond resolution with unprecedented temporal coverage and resolution of both line-of-sight observables and full Stokes parameters.

The instrument was designed based on the highly successful MDI instrument, incorporating improvements in several areas of the optics and electronics. Among the improvements the substantial redundancy means that the instrument will be able to continue to provide high-quality data, even in case of the malfunction of significant components.

The instrument has gone through a substantial calibration effort to be able to provide reliable inferences about the Sun. We have been able to characterize all relevant aspects of the instrument with a very good accuracy.

Given the above we are confident that the instrument will provide results of unprecedented quality for a long period of time.

Updated documentation about the instrument will be available from the project website http://hmi.stanford.edu and the observables from the instrument will be available from http://jsoc.stanford.edu shortly after the observations are taken.

Acknowledgements A large number of people in the Stanford Solar Physics group provided invaluable help during the design and test of the instrument, including: Jim Aloise, Art Amezcua, John Beck, Keh-Cheng Chu, Carl Cimilluca, Romeo Durscher, Sarah Gregory, Tim Larson, Kim Ross, Jeneen Sommers, and Hao Thai. We would also like to thank Robert Byer, Carsten Langrock, Joe Schaar, and Karel Urbanek at Stanford's Ginzton Laboratory for help with designing and assembling the tunable solid-state laser. At LMSAL a large number of engineers and other personnel helped with design and test, including Ron Baraze, Jerry Drake, Dexter Duncan, Jay Dusenbury, Chris Edwards, Barbara Fischer, Glen Gradwohl, Gary Heyman, Noah Katz, Dwana Kacensky, Dave Kirkpatrick, Gary Kushner, Russ Lindgren, Gary Linford, Dnyanesh Mathur, Edward McFeaters, Keith Mitchell, Rose Navarro, Tom Nichols, Roger Rehse, J.-P. Riley, Larry Springer, Bob Stern, Louie Tavarez, Edgar Thomas, Darrel Torgerson, Ross Yamamoto, Carl Yanari, and Kent Zickhur. We also thank Tom Anderson, Lisa Bartusek, Michael Bay, Liz Citrin, Peter Gonzales, Juli Lander, Eliane Larduinat, Wendy Morgenstern, Dean Pesnell, Chad Salo, Mike Scott, and Barbara Thompson at GSFC, Juan Manuel Borrero, Greg Card, Rebecca Centeno, Tony Darnell, and Bruce Lites at HAO, KD Leka at NWRA/CORA, Matthew Clapp, Sarah Dunkin, and Nick Waltham at RAL, Gary Auker and Rob Wilson at E2V, John Hunter, Ian Miller, and Jeff Wimperis at LightMachinery, and Gerard Gleeson at ASO. Finally we would like to thank Jack Harvey for useful comments and advice. This work was supported by NASA contract NAS5-02139 to Stanford University. The HMI data used are courtesy of NASA/SDO and the HMI science team.

Appendix A: Sequencer and Framelist Examples

Table 3 shows an example of a framelist. In order, the columns are the following:

- FID: The Frame ID is simply downlinked with the image and has no effect on the sequencer execution. In the present case it is used to indicate the position in the line and the polarization state. It can also be used to indicate frames intended for calibration only.
- RelTime: This is the time (in ms) of the frame relative to the start time of the sequence.
- Img: The type of readout (1, 2, or 4 port), cropping and compression used.
- PL: The index of the desired polarization setting.
- WL: The index of the wavelength setting desired.
- CF: The focus position used. 1 – 16 indicate normal images. 17 calmode.
- Exp: The index of the exposure time desired.
- Obspath: The camera used. IMAGE indicates that an exposure is taken. DARK indicates that a dark frame is taken.

Note that most of the settings use an index into another table. That table then details the settings of the individual mechanisms. In many cases a default value is specified. The corresponding values are kept separately in registers in the FSW. This allows for changing parameters such as the focus position and exposure time without remaking framelists.

Similarly PL positions 258 and 259 correspond to LCP and RCP while 250 through 253 are four positions allowing for the determination of I, Q, U, and V.

In the example shown in Table 3, WL positions 465, 467, 469, 471, 473, and 475 correspond, in order, to I5, I4, I3, I2, I1, and I0, which in turn correspond to increasing wavelength.

As can be seen, the framelist loops twice through the wavelengths in a particular non-sequential order I3, I4, I0, I5, I1, I2. Combined with choosing the starting point such that the center wavelengths (I2 and I3) are centered on the target times (0 seconds and 45 seconds), this minimizes the errors in the inferred Doppler velocities.

Table 3 Framelist taking LOS data on the front and full IQUV on the side camera. See text for details.

FID	RelTime	Img	PL	WL	CF	Exp	ObsPath
10098	0	DEFAULT	258	469	DEFAULT	DEFAULT	FRONT2_IMAGE
10090	1875	DEFAULT	250	469	DEFAULT	DEFAULT	SIDE1_IMAGE
10099	3750	DEFAULT	259	469	DEFAULT	DEFAULT	FRONT2_IMAGE
10091	5625	DEFAULT	251	469	DEFAULT	DEFAULT	SIDE1_IMAGE
10079	7500	DEFAULT	259	467	DEFAULT	DEFAULT	FRONT2_IMAGE
10070	9375	DEFAULT	250	467	DEFAULT	DEFAULT	SIDE1_IMAGE
10078	11250	DEFAULT	258	467	DEFAULT	DEFAULT	FRONT2_IMAGE
10071	13125	DEFAULT	251	467	DEFAULT	DEFAULT	SIDE1_IMAGE
10158	15000	DEFAULT	258	475	DEFAULT	DEFAULT	FRONT2_IMAGE
10150	16875	DEFAULT	250	475	DEFAULT	DEFAULT	SIDE1_IMAGE
10159	18750	DEFAULT	259	475	DEFAULT	DEFAULT	FRONT2_IMAGE
10151	20625	DEFAULT	251	475	DEFAULT	DEFAULT	SIDE1_IMAGE
10059	22500	DEFAULT	259	465	DEFAULT	DEFAULT	FRONT2_IMAGE
10050	24375	DEFAULT	250	465	DEFAULT	DEFAULT	SIDE1_IMAGE
10058	26250	DEFAULT	258	465	DEFAULT	DEFAULT	FRONT2_IMAGE
10051	28125	DEFAULT	251	465	DEFAULT	DEFAULT	SIDE1_IMAGE
10138	30000	DEFAULT	258	473	DEFAULT	DEFAULT	FRONT2_IMAGE
10130	31875	DEFAULT	250	473	DEFAULT	DEFAULT	SIDE1_IMAGE
10139	33750	DEFAULT	259	473	DEFAULT	DEFAULT	FRONT2_IMAGE
10131	35625	DEFAULT	251	473	DEFAULT	DEFAULT	SIDE1_IMAGE
10119	37500	DEFAULT	259	471	DEFAULT	DEFAULT	FRONT2_IMAGE
10110	39375	DEFAULT	250	471	DEFAULT	DEFAULT	SIDE1_IMAGE
10118	41250	DEFAULT	258	471	DEFAULT	DEFAULT	FRONT2_IMAGE
10111	43125	DEFAULT	251	471	DEFAULT	DEFAULT	SIDE1_IMAGE
10098	45000	DEFAULT	258	469	DEFAULT	DEFAULT	FRONT2_IMAGE
10092	46875	DEFAULT	252	469	DEFAULT	DEFAULT	SIDE1_IMAGE
10099	48750	DEFAULT	259	469	DEFAULT	DEFAULT	FRONT2_IMAGE
10093	50625	DEFAULT	253	469	DEFAULT	DEFAULT	SIDE1_IMAGE
10079	52500	DEFAULT	259	467	DEFAULT	DEFAULT	FRONT2_IMAGE
10072	54375	DEFAULT	252	467	DEFAULT	DEFAULT	SIDE1_IMAGE
10078	56250	DEFAULT	258	467	DEFAULT	DEFAULT	FRONT2_IMAGE
10073	58125	DEFAULT	253	467	DEFAULT	DEFAULT	SIDE1_IMAGE
10158	60000	DEFAULT	258	475	DEFAULT	DEFAULT	FRONT2_IMAGE
10152	61875	DEFAULT	252	475	DEFAULT	DEFAULT	SIDE1_IMAGE
10159	63750	DEFAULT	259	475	DEFAULT	DEFAULT	FRONT2_IMAGE
10153	65625	DEFAULT	253	475	DEFAULT	DEFAULT	SIDE1_IMAGE
10059	67500	DEFAULT	259	465	DEFAULT	DEFAULT	FRONT2_IMAGE
10052	69375	DEFAULT	252	465	DEFAULT	DEFAULT	SIDE1_IMAGE
10058	71250	DEFAULT	258	465	DEFAULT	DEFAULT	FRONT2_IMAGE
10053	73125	DEFAULT	253	465	DEFAULT	DEFAULT	SIDE1_IMAGE
10138	75000	DEFAULT	258	473	DEFAULT	DEFAULT	FRONT2_IMAGE
10132	76875	DEFAULT	252	473	DEFAULT	DEFAULT	SIDE1_IMAGE
10139	78750	DEFAULT	259	473	DEFAULT	DEFAULT	FRONT2_IMAGE
10133	80625	DEFAULT	253	473	DEFAULT	DEFAULT	SIDE1_IMAGE
10119	82500	DEFAULT	259	471	DEFAULT	DEFAULT	FRONT2_IMAGE
10112	84375	DEFAULT	252	471	DEFAULT	DEFAULT	SIDE1_IMAGE
10118	86250	DEFAULT	258	471	DEFAULT	DEFAULT	FRONT2_IMAGE
10113	88125	DEFAULT	253	471	DEFAULT	DEFAULT	SIDE1_IMAGE

In both halves of the framelist, the front camera does LCP and RCP at each wavelength, thereby allowing for the Doppler and LOS field to be obtained. The side camera, on the other hand, does two of the four polarizations in the first half and the other two in the second half, thereby allowing for a 90-second cadence using data from that camera only.

Many other framelists with various tradeoffs are, of course, possible. Apart from ones with different details in the regular observing sequences, they include ones taking detunes, focus sweeps, linearity data, and so forth.

Appendix B: Electronic Supplementary Material

B.1. Optical Prescription

For reference two Zemax optical prescriptions are included as supplementary material. Both files are for the instrument as designed and do not include any known or suspected aberrations. Polarization has not been included.

The first model (HMI_FOLD_2005-05-19.ZMX and the corresponding text prescription HMI_FOLD_2005-05-19.txt (supplementary text)) has configurations for the various focus settings and calmode. For clarity it does not include the final fold mirrors below the CCDs.

The second model (HMI_FOLD_2005-05-19_paths.ZMX and the corresponding text prescription HMI_FOLD_2005-05-19_paths.txt (supplementary text)) is for best focus only, but includes the alternate paths through the Michelsons, the path to the second camera, and the final fold mirrors.

None of the models include the path to the limb sensor.

B.2. Heater Zones and Thermistor Locations

For reference, the placement of the various thermistors is shown in the file thermistors.pdf (supplementary pdf). The heaters zones and heaters are included as HEATERS.PDF (supplementary pdf).

References

Couvidat, S., Schou, J., Shine, R.A., Bush, R.I., Miles, J.W., Scherrer, P.H., Rairden, R.L.: 2011, Wavelength dependence of the Helioseismic and Magnetic Imager (HMI) instrument onboard the Solar Dynamics Observatory (SDO). *Solar Phys.* doi:10.1007/s11207-011-9723-8.

Lemen, J.R., Duncan, D.W., Edwards, C.G., Friedlaender, F.M., Jurcevich, B.K., Morrison, M.D., Springer, L.A., Stern, R.A., Wuelser, J., Bruner, M.E., Catura, R.C.: 2004, The solar X-ray imager for GOES. In: Fineschi, S., Gummin, M.A. (eds.) *SPIE* **5171**, 65 – 76. doi:10.1117/12.507566.

Lemen, J.R., Title, A.M., Akin, D.J., Boerner, P.F., Chou, C., Drake, J.F., Duncan, D.W., Edwards, C.G., Friedlaender, F.M., Heyman, G.F., Hurlburt, N.E., Katz, N.L., Kushner, G.D., Levay, M., Lindgren, R.W., Mathur, D.P., McFeaters, E.L., Mitchell, S., Rehse, R.A., Schrijver, C.J., Springer, L.A., Stern, R.A., Tarbell, T.D., Wuelser, J.P., Wolfson, C.J., Yanari, C., Bookbinder, J.A., Cheimets, P.N., Caldwell, D., Deluca, E.E., Gates, R., Golub, L., Park, S., Podgorski, W.A., Bush, R.I., Scherrer, P.H., Gummin, M.A., Smith, P., Auker, G., Jerram, P., Pool, P., Soufli, R., Windt, D.L., Beardsley, S., Clapp, M., Lang, J., Waltham, N.: 2011, The Atmospheric Imaging Assembly (AIA) on the Solar Dynamics Observatory (SDO). *Solar Phys.* doi:10.1007/s11207-011-9776-8.

Norton, A.A., Graham, J.P., Ulrich, R.K., Schou, J., Tomczyk, S., Liu, Y., Lites, B.W., López Ariste, A., Bush, R.I., Socas-Navarro, H., Scherrer, P.H.: 2006, Spectral line selection for HMI: A comparison of Fe I 6173 Å and Ni I 6768 Å. *Solar Phys.* **239**, 69 – 91. doi:10.1007/s11207-006-0279-y.

Scherrer, P.H., Bogart, R.S., Bush, R.I., Hoeksema, J.T., Kosovichev, A.G., Schou, J., Rosenberg, W., Springer, L., Tarbell, T.D., Title, A., Wolfson, C.J., Zayer, I., MDI Engineering Team: 1995, The solar oscillations investigation – Michelson Doppler Imager. *Solar Phys.* **162**, 129 – 188. doi:10.1007/BF00733429.

Scherrer, P.H., Schou, J., Bush, R.I., Kosovichev, A.G., Bogart, R.S., Hoeksema, J.T., Liu, Y., Duvall, T.L., Title, A.M., Tomczyk, S., HMI Science Team: 2011, The helioseismic and magnetic imager investigation. *Solar Phys.* doi:10.1007/s11207-011-9834-2.

Schou, J., Borrero, J.M., Norton, A.A., Tomczyk, S., Elmore, D., Card, G.L.: 2010, Polarization calibration of the Helioseismic and Magnetic Imager (HMI) onboard the Solar Dynamics Observatory (SDO). *Solar Phys.* doi:10.1007/s11207-010-9639-8.

Wachter, R., Schou, J.: 2009, Inferring small-scale flatfields from solar rotation. *Solar Phys.* **258**, 331 – 341. doi:10.1007/s11207-009-9406-x.

Wachter, R., Schou, J., Rabello-Soares, M.C., Miles, J.W., Duvall, T.L., Bush, R.I.: 2011, Image quality of the Helioseismic and Magnetic Imager (HMI) onboard the Solar Dynamics Observatory (SDO). *Solar Phys.* doi:10.1007/s11207-011-9709-6.

Solar Phys (2012) 275:261–284
DOI 10.1007/s11207-011-9709-6

THE SOLAR DYNAMICS OBSERVATORY

Image Quality of the *Helioseismic and Magnetic Imager* (HMI) Onboard the *Solar Dynamics Observatory* (SDO)

R. Wachter · J. Schou · M.C. Rabello-Soares · J.W. Miles · T.L. Duvall Jr. · R.I. Bush

Received: 16 March 2010 / Accepted: 4 January 2011 / Published online: 10 February 2011

Abstract We describe the imaging quality of the *Helioseismic and Magnetic Imager* (HMI) onboard the *Solar Dynamics Observatory* (SDO) as measured during the ground calibration of the instrument. We describe the calibration techniques and report our results for the final configuration of HMI. We present the distortion, modulation transfer function, stray light, image shifts introduced by moving parts of the instrument, best focus, field curvature, and the relative alignment of the two cameras. We investigate the gain and linearity of the cameras, and present the measured flat field.

Keywords Helioseismology, observations · Instrumental effects · Solar Dynamics Observatory

1. Introduction

The *Helioseismic and Magnetic Imager* (HMI) is a 14-cm aperture imaging solar instrument with an effective focal ratio of 37.4. It is built to operate in essentially monochromatic light at 6173 Å. The spectral filters are realized by the front window, a wide-band blocker filter, a tunable five-element Lyot filter, and two Michelson interferometers. The images are stabilized against spacecraft jitter by an Image Stabilization System (ISS) and recorded by

The Solar Dynamics Observatory
Guest Editors: W. Dean Pesnell, Phillip C. Chamberlin, and Barbara J. Thompson

R. Wachter (✉) · J. Schou · M.C. Rabello-Soares · R.I. Bush
Stanford University, Stanford, CA 94305, USA
e-mail: richard@sun.stanford.edu

J.W. Miles
Lockheed Martin Advanced Technology Center, Palo Alto, CA 94304, USA

J.W. Miles
SOFIA-USRA, NASA Ames Research Center MS N211-3, Moffett Field, CA 94035, USA

T.L. Duvall Jr.
Solar Physics Laboratory, NASA/Goddard Space Flight Center, Greenbelt, MD 20771, USA

two cameras, both equipped with a 4096^2 pixel CCD. Two legs at the back end of the HMI instrument allow a tilt of the optical axis to correct for long-term alignment drift of HMI *versus* the *Solar Dynamics Observatory* (SDO) satellite. Focus adjustment is achieved by placing a pair of fused-silica plane-parallel plates into the optical path. Two focus wheels with five openings, each with three plates of varying thickness, an open position, and a lens, which is used to image the aperture for calibration purposes, provide 16 discrete focus positions. For a detailed description of the optical design of the instrument, see Schou *et al.* (2011).

The optical calibration of the instrument aims at optimizing and characterizing the instrument in a way that allows us to calculate observables from the raw data (filtergrams) which are free from avoidable instrumental artifacts. The observables, the most important being magnetic fields and Doppler velocities, are fed into a data processing pipeline that resolves the surface and subsurface structure of the Sun (Couvidat *et al.*, 2010).

The instrumental effects that can be corrected through image processing must be known as accurately as possible, whereas the instrumental influences that cannot be corrected must be sufficiently small to meet the observable accuracy requirements. The instrumental artifacts that can be corrected in the observable calibration are the flat-field (detector and optical), the distortion, the relative alignment of the two cameras, and the residual image motion caused by movable parts in the instrument. Not easily correctable is a less-than-perfect modulation transfer function (MTF), caused by a variety of optical aberrations and leaking charges on the CCD, scattered light, and field curvature.

The calibration efforts therefore focus on:

i) bringing the point spread function close to the optimum set by the diffraction limit and measuring the remaining optical aberrations. The stray light, which contributes to the tail of the point spread function, is measured separately.
ii) measuring the distortion as accurately as possible.
iii) setting the best focus in the middle of the available focal range for both cameras.
iv) minimizing and characterizing field curvature and tilt.
v) determining the misalignment of the cameras as accurately as possible.
vi) minimizing the image motion introduced by the moving parts of the instrument, and obtaining consistent knowledge of the remaining movements.
vii) measuring the nonlinearity of the camera response to light exposure, and verifying the saturation level.
viii) measuring the gain of the detector amplifiers.
ix) measuring the flat field on all scales, and monitoring its change in different environments.

We note that the above-mentioned characterizations can be repeated after launch, although certain tests must be performed in a different way. In some cases, more accurate calibration statistics are expected from the space measurements and will be made available. A detailed presentation of our results from the ground measurement and its later comparison with the in-flight measurements will show both the temporal drifts of the instrument and any effects of the launch, which will be instructional for future calibrations of imaging space instruments.

2. Calibration Setup

2.1. Instrument and Stimulus Telescope

All image-quality calibration measurements are performed with a stimulus telescope that provides a test beam with the correct Lagrange invariant. The target is mounted in the focal

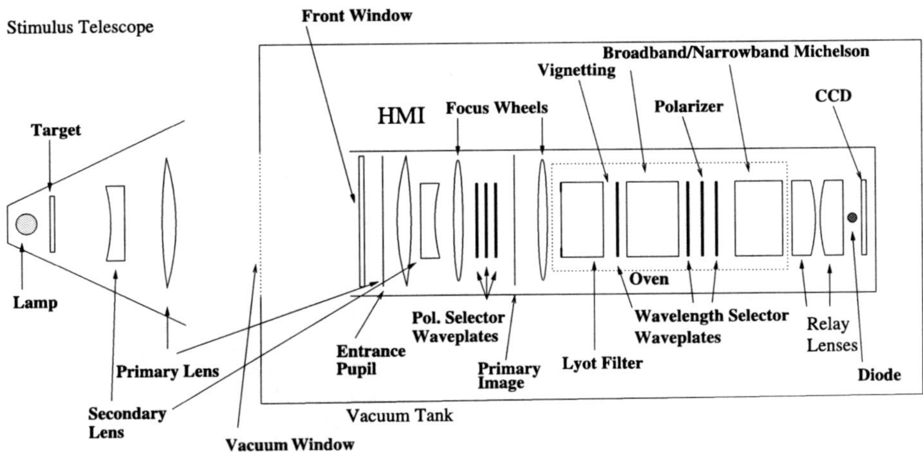

Figure 1 A schematic drawing of the calibration measurement setup showing only the elements most relevant for the image quality. Note also that the setup is not drawn to scale.

plane of the stimulus telescope with a lamp providing a source of white light. The optical axes of the stimulus telescope and the instrument are aligned (except for measurements where a controlled misalignment is required).

The system consisting of the stimulus telescope and the instrument was subject to thermal and vibrational perturbations in its environment. While the instrument has thermal control through various heaters and sensors at the instrument (with the spectral filters placed in the thermally controlled oven), the stimulus telescope has no such control. Instrument and stimulus telescope are stabilized against vibrational perturbations by use of a floating table. Blurring from turbulent air currents in the optical path and temperature variations are the major sources of noise in image-quality measurements.

Most of the time the instrument was subject to the same environment as the stimulus telescope ("in-air measurements"). However, the instrument was occasionally placed in a vacuum tank in order to test it at temperatures that resemble flight conditions. In particular, the dark current in the CCD detectors is drastically reduced, because the CCDs operate at much lower temperatures. In this setup, the collimated beam between the stimulus telescope and the instrument transits a glass window in the wall of the vacuum tank. This window is an additional uncontrolled element in the optical path of the system, introducing a temperature-dependent amount of optical power (and possible other aberrations) into the system.

Figure 1 shows a simplified layout drawing of the calibration setup including the optical elements of the instrument that are most important for the image quality. For a detailed optical prescription of the instrument, see Schou *et al.* (2011).

2.2. Test Facilities

The tests were carried out at the Lockheed Martin Advanced Technology Center (LMATC) in Palo Alto, CA before November 2007, at Goddard Space Flight Center (GSFC), Greenbelt, MD between November 2007 and July 2009, and at the Astrotech Facilities (ASO) in Titusville, FL until launch. The results from each series of measurements reflect the different thermal and vibrational perturbations of the environment. Also, the orientation of the stimulus telescope with respect to the instrument at LMATC was 90° different from the ori-

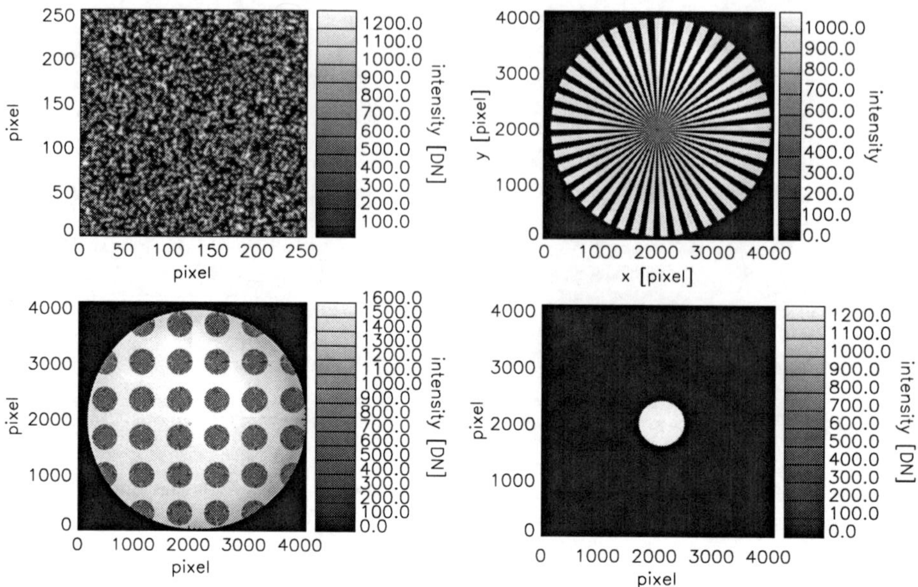

Figure 2 Focused instrumental images of different targets: the random dot target (upper left), the star target (upper right), the star array target (lower left), and the 5-mm field stop (lower right).

entation at the other two facilities. The instrument was put into the vacuum tank three times, but the instrument could be optically stimulated only for the first two times.

2.3. Targets

The most common stimulus telescope targets used for imaging are shown in Figure 2. They are:

i) The random dot target: A distribution of randomly placed dots of different diameters. The target provides a strong signal on all spatial scales. It is nonrepetitive at small scales, so that displacements can be measured uniquely and accurately. This is important for camera alignment and distortion measurements. The target is isotropic and has a translationally invariant power spectrum, which means that the relative MTF can be measured with the random dot target at any position and in any spatial direction, without any artifacts arising as a consequence of target properties. Best focus and the field curvature and tilt have been measured with the random-dot target.

ii) The star target: This target consists of a star with triangular rays of light and shadow. A star target is created to perform absolute measurements of the MTF. The response at each single spatial frequency is inferred from the different radial distances from the center. However, the different frequencies are measured at different field positions; the highest frequencies are measured at the center of the field, while the lowest frequencies are measured toward the edge of the field.

iii) The star-array target: Similar to the star target, but with multiple, smaller stars covering the optical field. While the very low spatial frequencies are not available, the star-array target examines the variation of the MTF across the field of view.

iv) The field stop target: A round field stop, which limits the field of view. We use field stops with diameter range from 5.0 mm to 27.4 mm, of which the largest is very close to the edge of the field of view. Large field stops provide guidance about the alignment of the stimulus telescope and the instrument, while they introduce no signal for the HMI nominal field of view.

2.4. Light Sources

For most of the tests, we used an intensity-stabilized lamp, which feeds an optical-fiber bundle. The exit of the optical-fiber bundle is projected onto the aperture of the stimulus telescope.

Another light source is a Light Emitting Diode (LED), which sits behind the beam splitter in front of the cameras. The light is not transmitted through large parts of the instrument's optics. In particular, it is not spectrally filtered, and the LED wavelength is different from the HMI target wavelength. With the LED, images can be taken even if it is technically not possible to feed light into the instrument. In this article, only the gain and linearity tests use data obtained with the LED.

3. MTF

3.1. Direct Measurement

The square-wave MTF can be directly measured by using a star target (see upper right panel of Figure 2). The response of the system to a single spatial frequency is derived from the image of the azimuthal square wave that corresponds to a given distance from the star center. The star is remapped in polar coordinates, and a sine wave is fitted for each radial distance.

For the incoherent imaging system under consideration, the amplitude $[A]$ at the spatial frequency λ is related to the MTF by

$$A(\lambda) = \mathrm{MTF}(\lambda)\frac{2}{N}\left|\int_0^N \mathrm{d}x \exp(2\pi \mathrm{i}\lambda x) m(x)\right|, \tag{1}$$

where $m(x)$ is a cut through the remapped, normalized target image at a single frequency, and N is the dimension of the CCD in units of pixels. Representing an azimuthal cut through the star, the target object is given by

$$m(x) = 2\Theta\bigl(\sin(2\pi x\lambda)\bigr), \tag{2}$$

where $\Theta(x)$ is the Heaviside function.

The continuous integral is a sum of integrals over single pixels:

$$A(\lambda) = \mathrm{MTF}(\lambda)\frac{4}{N}\left|\sum_{n=0}^{N/s-1}\sum_{m=ns}^{(n+\frac{1}{2})s-1}\int_{-\frac{1}{2}}^{\frac{1}{2}} \mathrm{d}p \exp 2\pi \mathrm{i}(m+p)\lambda\right|, \tag{3}$$

where N is the number of pixels, $s = 1/\lambda$ is the wavelength, and p is the lateral length of a pixel. The pixel integration can be performed explicitly, which leads to

$$A(\lambda) = \mathrm{MTF}(\lambda)\frac{4\sin(\pi\lambda)}{N\pi\lambda}\left|\sum_{n=0}^{N/s-1}\frac{2}{1-\exp 2\pi \mathrm{i}\lambda}\right|. \tag{4}$$

For $\lambda \ll 1$, we obtain

$$A(\lambda) = \mathrm{MTF}(\lambda)\frac{4\sin(\pi\lambda)}{\pi^2\lambda}, \tag{5}$$

and hence

$$\mathrm{MTF}(\lambda) = A(\lambda)\frac{\pi^2\lambda}{4\sin(\pi\lambda)}. \tag{6}$$

Although $\lambda \ll 1$ is not valid close to the Nyquist frequency, simulations show that expression (6) is valid to within 0.01 up to 80% of the Nyquist frequency where the MTF is measured.

The star-array target (see Figure 2) provides information about the large-scale spatial variation of the MTF.

The MTF can also be derived from the images of the random-dot target by dividing out the power spectrum of the target:

$$\mathrm{MTF}(\lambda) = \frac{\int \mathrm{d}\phi\, p(\lambda,\phi)}{\int \mathrm{d}\phi\, t(\lambda,\phi)}, \tag{7}$$

where $p(\lambda,\phi)$ is the power spectrum of the image in polar coordinates, and $t(\lambda,\phi)$ is the power spectrum of the target in polar coordinates. The azimuthal average provides enough averaging at all relevant frequencies to make both numerator and denominator representative of their relative expectation values. Figure 3 shows estimates of the MTF derived from the star target and the random dot target.

Using the setup consisting of stimulus telescope and instrument, the wavefront error due to the stimulus telescope and instrument cannot be disentangled. To obtain the MTF of the instrument itself, the interferometric measurement of the aberrations in the stimulus telescope is subtracted from the measurement of the aberrations of the combined system (see Section 3.3).

We observed fluctuations in the measured MTF on small time scales, leading to a variation in the Strehl ratio of several percent. The magnitude of the fluctuations depends on the particular environment in which the measurements have been obtained. Both jitter and air currents are responsible for the random blurring on small time scales, while thermal degradations take effect on time scales of several minutes or longer. We conclude that air currents, rather than jitter, are the major source of noise, because the ISS removes most of the environmental jitter effectively. To take out the residual jitter, we placed the instrument and the stimulus telescope on a floating table, without noticeable improvements.

The test facilities at GSFC did not provide a stable room temperature, which caused the aberrations to vary with time, and, at the same time, degrade the MTF. In a thermally stable environment, the measured Strehl ratio is 0.74 ± 0.03 (see Figure 4). We define the Strehl ratio here as the peak intensity of the combined system of instrument and stimulus telescope divided by the peak intensity for the system as designed. Taking the point spread function as given by the diffraction limit of the instrument as a reference, the respective Strehl ratio of the full system (instrument + stimulus telescope) as designed is 0.980, whereas the Strehl ratio of the instrument as designed alone is 0.996. These numbers represent the height of the peak of the point spread function relative to the point spread function that represents the diffraction limit given by the aperture and the wavelength.

Strehl ratios above 0.80 have been measured. As we have no indications that the stimulus telescope is canceling out large parts of the wavefront errors of the instrument, we expect

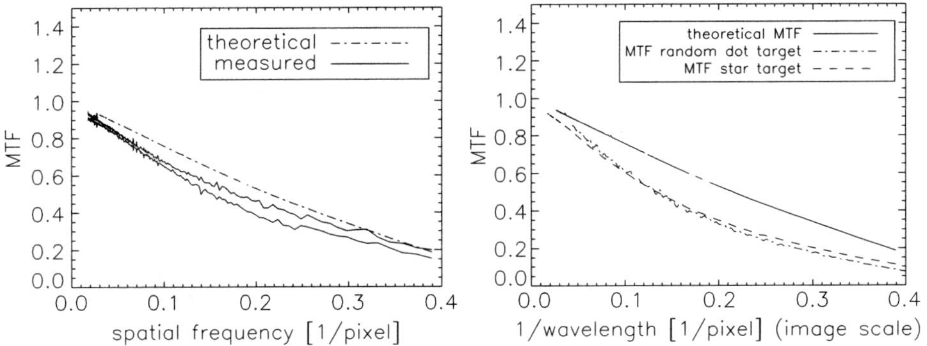

Figure 3 Left panel: The MTF measured with the star target. The theoretical curve represents the system of stimulus telescope and the instrument together. The two solid curves represent the MTF in the two perpendicular directions with the maximum and the minimum amplitude. Right panel: The azimuthally averaged MTF measured with the random dot target and the star target. The MTF can be inferred from the image of the random dot target by dividing out the known power spectrum of the target. Note that the left and right panels represent the MTF measured at different times. While we have not found any systematic changes, we observed random changes due to environmental perturbations.

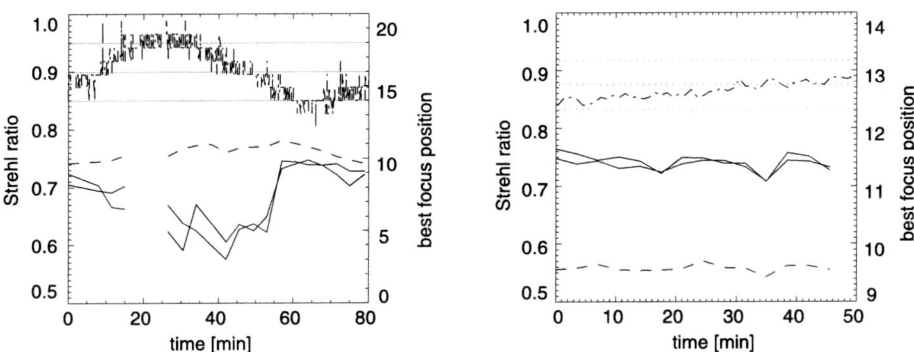

Figure 4 Left panel: Strehl ratio at best focus measured from a series of focus sweeps on a random-dot target in the test facilities of Goddard Space Flight Center in September 2008. The dashed curve gives the variation of the best focus. The upper curve shows the room temperature, with the three dotted lines representing 20.7, 20.8, and 20.9°C. Right panel: The same quantities as in the left panel, measured in the test facilities of Astrotech in August 2009. The horizontal lines represent 23.6, 23.7, and 23.8°C.

the Strehl ratio of the instrument to normally be at the upper end of all the measured values. Thermal perturbations and air currents introduce wavefront perturbations that generally degrade the MTF.

Our initial MTF measurements revealed a severe degradation of the MTF caused by the front window. We suspect that internal tensions caused by the layered front window design introduced large wavefront errors at the entrance pupil. As a remedy, the surface of the front window was polished to compensate for the internal wavefront error. Additionally, the known field-independent astigmatism of the instrument's telescope is now compensated by the front window surface. The procedure resulted in a dramatic improvement of the MTF. Apart from the MTF variations caused by temperature gradients, we have not seen a degradation of the MTF over time.

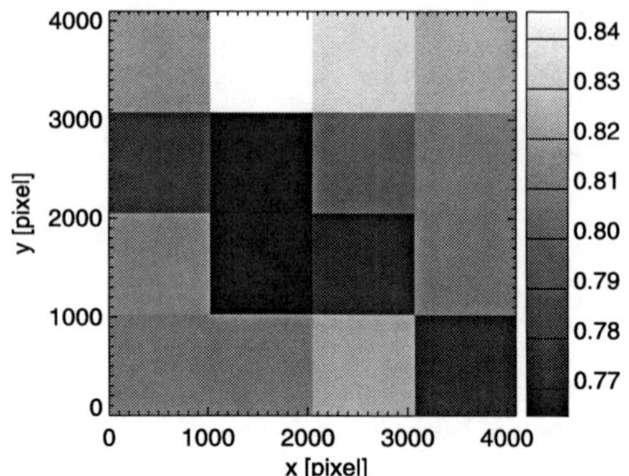

Figure 5 Strehl ratio as a function of field position measured with the star-array target.

Figure 3 provides an estimate for the MTF of the combined system, comparing it to the function expected for a perfect instrument and stimulus telescope. This MTF represents a circular area of radius 400 pixels (about 200 arcsec) at the center of the image. As we have a clearly measurable amount of astigmatism in the system, we show the MTF in the two directions with maximum and minimum blur. For all other directions, the MTF will be in between these two lines.

3.2. Field Variations of MTF

For the variation of the MTF across the field, we need to assume that the target and the turbulent disturbances are homogeneous across the field. The star-array target shows an MTF that has a higher Strehl ratio off a diagonal from the upper left to the lower right corner of the CCD. Figure 5 shows the Strehl ratio measured for the central 16 stars (see Figure 2) of the star array target.

3.3. Phase Diversity Measurements

In order to further characterize the optical imperfections, we also performed a phase diversity analysis. To this end, focus sweeps of the random-dot target were used. As the deconvolved images are of no interest and as we were not successful in applying a traditional phase diversity algorithm, a alternative approach was taken.

To see how this method works, note that the cross spectrum of two images taken at different focus positions is the power spectrum of the true signal times the OTF at one position times the complex conjugate of the OTF at the other position. Like power spectra, but unlike Fourier transforms, cross spectra for multiple realizations may be averaged together to yield an estimate of the limit cross spectrum.

In the present case we averaged cross spectra from 16 128 pixel by 128 pixel patches near the center of the image. This was done for six pairs of focus position, using the best focus and the six adjacent focus positions. A model of the cross spectra was then fitted to the observed spectra while marginalizing over the power spectrum of the target.

This method was then tested with artificial data based on a ray trace and found to accurately reproduce the assumed aberrations.

The phase diversity analysis reveals several interesting results. The consistently most significant term is a spherical aberration of order -0.12 wave. The interferometry of the stimulus telescope indicates that it contributes about -0.02 wave and that the instrument thus has -0.10 wave, substantially more than the as-designed value of -0.01 wave.

The other terms are somewhat smaller. Unfortunately, it has proved difficult to correct them for the effects of the stimulus telescope, which was only measured on one occasion. The values we obtained from measurements at different test facilities were not consistent in all cases. A possible reason for this is that the stimulus telescope is not stable in time, because the mounts are not stress free.

We have also attempted to determine the Strehl values for the combined system from the phase diversity numbers. We found that while the numbers generally agree in magnitude, their variation does not agree with the variation estimated directly from the images. The cause of this is unclear, but it may be related to nonuniformities in the stimulus-telescope illumination.

4. Scattered Light

The instrumental scattered light can lead to important systematic effects, most importantly in the umbrae of sunspots where the scattered light may be a significant fraction of the intensity (Bray and Loughhead, 1979). Properly characterized, scattered light can be deconvolved, as was done by Jefferies and Duvall (1991) and Toner, Jefferies, and Duvall (1997).

To characterize the scattered light in the HMI instrument, a field stop of radius 400 pixels is imaged onto the CCD through the optical system. Ideally, this would appear in intensity as a disk convolved with the diffration limit MTF. In Figure 6, the azimuthally averaged intensity is shown out to about 200 pixels from the edge, where it falls to 10^{-3} of the disk intensity. Overplotted on the data is the result of a model calculation.

We would like to develop the point spread function (PSF) of this instrument to include the scattering that leads to the extended tail in Figure 6. A simple empirical model of the PSF, with parameters describing a narrow core and an extended tail of the PSF, is given by Pierce and Slaughter (1977):

$$A(r) = (1-\epsilon)\mathrm{e}^{-(r/w)^2} + \frac{\epsilon}{1+(r/W)^{\kappa}}, \tag{8}$$

where r is the radial distance, ϵ determines the relative sizes of the two parts, w is the 1/e width of the central Gaussian core, W is the half-width at half-maximum width of the extended tail, and κ determines how fast the extended tail drops off. To determine a set of parameters that fits the observations, a disk of radius 400 pixels (which is a model for the field stop) is convolved with the model PSF. The parameters $\epsilon = 0.1$, $\kappa = 3.0$, $w = 1.8$, and

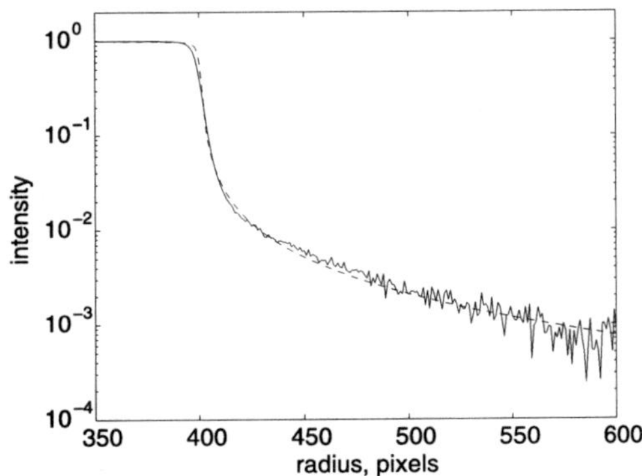

Figure 6 The azimuthal average intensity *versus* radial distance from the center of the field stop.

$W = 3.0$ provide the best fit. The result is plotted together with the data in Figure 6. While the overall fit is good, the small deviation near the edge can be attributed to the simplified representation of the core function as a Gaussian. However, here we are mainly concerned with the extended tail, as we believe the stray light to result from high-angle scatter in the optical elements of the instrument.

Using this function, the scattered light can be removed by deconvolution in regions where it is important, such as sunspot umbrae.

5. Distortion

5.1. Distortion Measurement

To measure the distortion, the random-dot target was mounted in the stimulus telescope. Images were taken at various alignment leg positions. Small areas of the images (256 × 256 pixels), corresponding to the same location on the target, but different positions in the optical field, were cross-correlated. The estimated shift between them corresponds to the difference in the leg position of the images plus the difference in the amount of distortion of the regions. This method of measuring distortion has the advantage that it does not depend on knowledge of the absolute properties of the stimulus telescope and the target. The distortion as a function of field position is expanded in Zernike polynomials, defined as

$$Z_n^m(\rho,\phi) = N_n^m R_n^m(\rho)\cos(m\phi) \quad \text{for } m \geq 0, \tag{9}$$

$$Z_n^m(\rho,\phi) = -N_n^m R_n^m(\rho)\sin(m\phi) \quad \text{for } m < 0, \tag{10}$$

where ϕ is the azimuthal angle and the ρ is the normalized radial distance ($0 \leq \rho \leq 1$). N_n^m is the normalization factor. R_n^m is the radial polynomial where $n - m$ must be even.

We used Zernike polynomials up to the 23rd order; *i.e.* we fitted the coefficients $a(n,m)$ in the expression

$$\mathbf{D}^{\mathrm{f}}(\rho,\phi) = \sum_{n=2}^{23} \sum_{m=-n}^{n} \mathbf{a}(n,m) Z_n^m \tag{11}$$

to the observed data. No significant improvement was found by including higher-order Zernike polynomials.

Zernike polynomials were chosen because they are orthonormal over a unit disk. In our model, the terms with $n = \{0, 1\}$ are omitted, forcing the value at the CCD center to zero, as well as the terms describing the image scale, the roll, and the ellipticity. Those terms are unconstrained by the offset method to determine the distortion. Our model was fitted to the observed shifts using a nonlinear least-square fit. The Zernike coefficients are estimated together with the errors in the nominal leg positions. This way, the obtained distortion does not depend on an exact knowledge of the displacement caused by offsetting the alignment legs. Several sets of images were taken from June 2007 to August 2009. We fitted the distortion for each of 29 sets.

5.2. Distortion Results

To get a best estimate of the distortion, the fitted distortion at a given focus position and camera was averaged over all image sets. This best estimate of the distortion is shown in

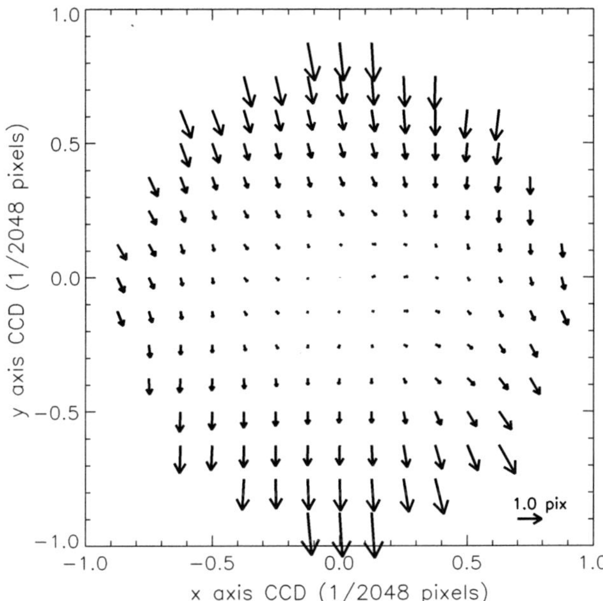

Figure 7 Estimated distortion near best focus (focus position nine) for the front camera.

Figure 7 for the front camera near nominal best focus (focus nine). The distortion is smaller than two pixels at any field position.

We measure the magnitude of the residuals by

$$R_j = \sqrt{\sum_{\mathbf{r}} \left|\mathbf{D}_j(\mathbf{r}) - \mathbf{D}^{\mathrm{f}}_j(\mathbf{r})\right|^2 / (2N_{\mathbf{r}})}, \tag{12}$$

where $\mathbf{D}_j$ is the distortion vector measured at a particular set, and $\mathbf{D}^{\mathrm{f}}_j$ is the best fit for each of these measurements. We average over all $N_{\mathbf{r}}$ points in the spatial grid. This number represents the error of individual measurements. The mean over R_j and its standard deviation are 0.043 ± 0.005 pixel for the front camera and 0.044 ± 0.005 pixel for the side camera near best focus. This number increases away from best focus, because shifts obtained from cross correlations are less accurate for blurred images.

We compare this number with the deviation from the mean of all measurements:

$$S_j = \sqrt{\sum_{\mathbf{r}} \left|\mathbf{D}^{\mathrm{f}}_j(\mathbf{r}) - \overline{\mathbf{D}^{\mathrm{f}}(\mathbf{r})}\right|^2 / (2N_{\mathbf{r}})}. \tag{13}$$

The mean and the standard deviation of S_n are 0.057 ± 0.027 pixel for the front camera and 0.050 ± 0.026 pixel for the side camera. This number increases to 0.10 pixel when the image is out of focus. This means that the changes seen over time are not significant beyond the two σ level. A closer look at individual sets suggests that the variation increases slightly when the temperature of the environment is unstable. This indicates that, within the accuracy of our measurements, the distortion is only weakly sensitive to the different thermal environments in which the measurements have been taken, and has remained constant throughout the integration of the instrument into the satellite.

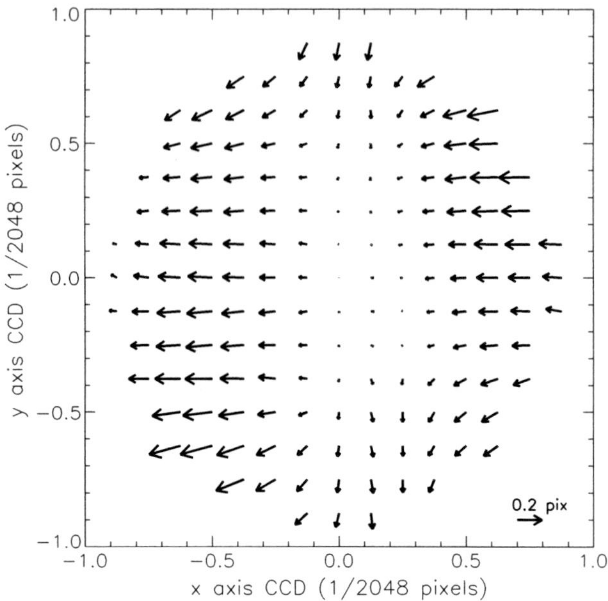

Figure 8 Difference between the distortion of the front camera and the side camera for focus position nine. The difference has been obtained as an average from all available sets of measurements.

5.3. Variation of the Distortion with Camera

The mean difference between the distortion in the two cameras was calculated by averaging the difference between the fitted distortion for each camera over all sets at a given focus position. Figure 8 shows the mean difference between the two cameras at focus position nine (which is the nominal best focus of the instrument). The difference is everywhere smaller than 0.28 pixel. The differences in the distortion in the horizontal direction are statistically significant ($\approx 10\sigma$) and larger than in the vertical direction, which are barely significant ($\approx 2\sigma$), where σ is a standard deviation derived from the difference between the two cameras for different data sets. The additional distortion is caused by an element behind the beam splitter that separates the optical path of the two cameras, or by the CCD itself. We suspect the folding mirror to be responsible for the observed difference in the cameras.

5.4. On Orbit Rolls and Offsets

The image scale, ellipticity, and roll angle will be measured in space by looking at the size and shape of the solar disk, and by determining the relative position coordinates of the solar and lunar disk during lunar transits.

The elliptical distortion can be distinguished from anisotropies in the spectral solar image by rolling the spacecraft around the spacecraft – Sun axis and measuring the changing shape of the solar limb.

By a combination of rolls and offsets, it is also possible to determine further terms in the distortion by tracking the limb and supergranulation, as demonstrated with the Michelson Doppler Imager (MDI) (Korzennik, Rabello-Soares, and Schou, 2004). The plan is to do such calibrations on a regular but infrequent basis.

6. Focus and Camera Alignment

To determine the best focus position, we take an image of the random-dot target at all available focus positions ("focus sweep") and look for the maximum spatial power at intermediate frequencies, where the power depends most strongly on focus (between 13% and 28% of the pixel Nyquist frequency). Fitting a quadratic function to the average power as a function of focus position allows us to determine the best focus in fractions of focus steps. The random blur introduced by air currents limits the accuracy of single measurements of the best focus to 0.1 focus step.

As the setup of the stimulus telescope tries to achieve a nearly collimated beam entering the instrument, focusing on the target is equivalent to focusing on the Sun. An air-to-vacuum corrector in front of the instrument corrects for the fact that there is air in the instrument. We aim at a best focus in the middle of the available focus from position one through sixteen. The best focus of the two cameras should not differ by more than 0.5 focus step. Table 1 (right columns) gives the focus of both cameras for different measurements. Focus position nine is just above the middle of the available focus range, and is the nominal best focus of the instrument. We tried to bring the best focus of the instrument slightly above the middle of the focus range, because the heaters at the front window are able to lower the best focus position, but not to raise it.

The measured focus of the instrument shows some variation. We know that a temperature gradient across the front window introduces refracting power. While the heaters at the rim holding the front window control the temperature at the edge, the center of the window is largely dominated by the air temperature. Therefore, a good control of the front window gradient is difficult to achieve in the absence of a strictly controlled room temperature. This is reflected in the fluctuating focus values. However, the focus difference between the front and side cameras are well within the tolerance limits, with a marginally higher focus (≈ 0.2 focus step) for the front camera.

Note that the best focus cannot be reliably measured when the instrument is in the vacuum tank, as the window of the vacuum tank introduced unknown refracting power to the system.

Some of the proposed observables frame lists require us to combine the raw images of both cameras. To do so, the coalignment of the cameras must be known accurately. The alignment of the cameras can be determined from simultaneous focus sweeps on both cameras by using cross-correlation techniques. We derived a relative lateral shift and a relative rotation for the images of the two cameras, and found the lateral shift to slightly drift over

Table 1 Relative lateral shift between the two cameras (x, y), the relative rotation (rot) in degrees, and the best focus position (range: 1 – 16) for the front and side cameras for different measurements during the calibration campaign.

Date	x	y	rot	Focus front	Focus side
18-Feb-2008	−6.6	5.5	0.082	10.9	10.9
14-Feb-2008	−6.4	5.7	0.081	9.7	9.4
30-Jan-2008	−6.2	5.3	0.082	8.2	8.2
03-Nov-2007	−6.1	4.3	0.079	10.4	10.3
02-Nov-2007	−6.2	4.3	0.080	9.9	9.7
28-Oct-2007	−6.7	3.5	0.080	8.8	8.7
14-Oct-2007	−4.5	4.5	0.082	9.1	9.1

time scales of several months. The relative shifts and rotations are given in Table 1. The observed long-term changes are most likely caused by the changing mechanical stresses on the instrument or the changing thermal environment. We are able to measure the shifts very accurately in space and to monitor them continuously.

7. Field Curvature

Field-curvature measurements use the same data sets as distortion measurements. Image offsets, using the alignment legs to vary the pointing, allow us to distinguish between the properties of the stimulus telescope and the instrument. The images in the front camera and the side camera were recorded alternately. Using the random-dot target, we measure the best focus as a function of field position, and fit the model:

$$f(x,y) = a_{\mathrm{H}}x + b_{\mathrm{H}}y + c_{\mathrm{H}}\left(x^2+y^2\right) + a_{\mathrm{S}}(x-x_{\mathrm{L}}) + b_{\mathrm{S}}(y-y_{\mathrm{L}}) + c_{\mathrm{S}}\left((x-x_{\mathrm{L}})^2+(y-y_{\mathrm{L}})^2\right) + d. \tag{14}$$

Here, x and y are coordinates of the HMI image and x_{L} and y_{L} are the image shifts introduced by offsetting the optical axis of the instrument using the alignment legs. x_{L} and y_{L} can be measured by cross-correlating the images of the random dot target. a_{H} and b_{H} measure the horizontal and vertical component of the HMI field tilt, and c_{H} measures the (quadratic) field curvature. a_{S}, b_{S}, and c_{S} measure the respective quantities for the stimulus telescope. Using all 13 or 25 leg positions, the coefficients $a_{\mathrm{H,S}}, b_{\mathrm{H,S}}, c_{\mathrm{H,S}}, d$ are overdetermined and can be obtained from a least-square fit. Note that the constant focus term [d] does not separate between the stimulus telescope and the instrument.

We found that while there is consistency in the inferred quadratic term, the linear-gradient term fluctuates wildly.

Field-curvature measurements at GSFC were performed in an unstable temperature environment, which leads to a changing of the focus with time. Figure 9 shows a measurement of the best focus together with a record of the outside temperature. It indicates that the focus position follows the temperature gradient with a time lag.

As the model described by Equation (14) assumes a focus independent of time, temporal variation of the focus led to spurious field tilts.

In order to understand the fluctuations of the measurements, we performed a series of 12 field curvature sequences with identical setup. Figure 10 shows the strong anti-correlation between the gradients measured for the stimulus telescope and the instrument.

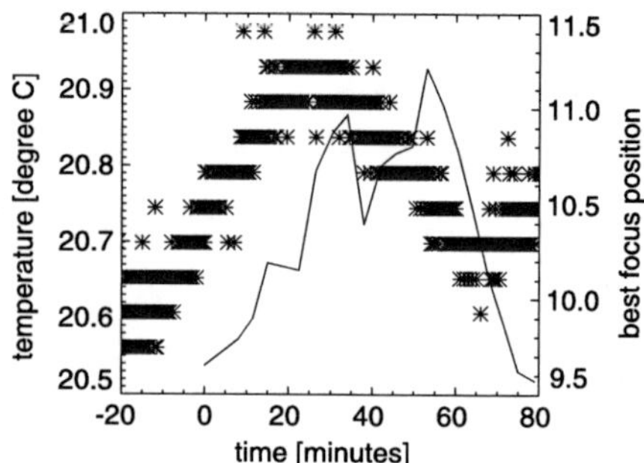

Figure 9 The stars represent the measured air temperature during a continuous focus measurement, and the line shows the measured focus. The focus follows the temperature with a time lag of 18 minutes.

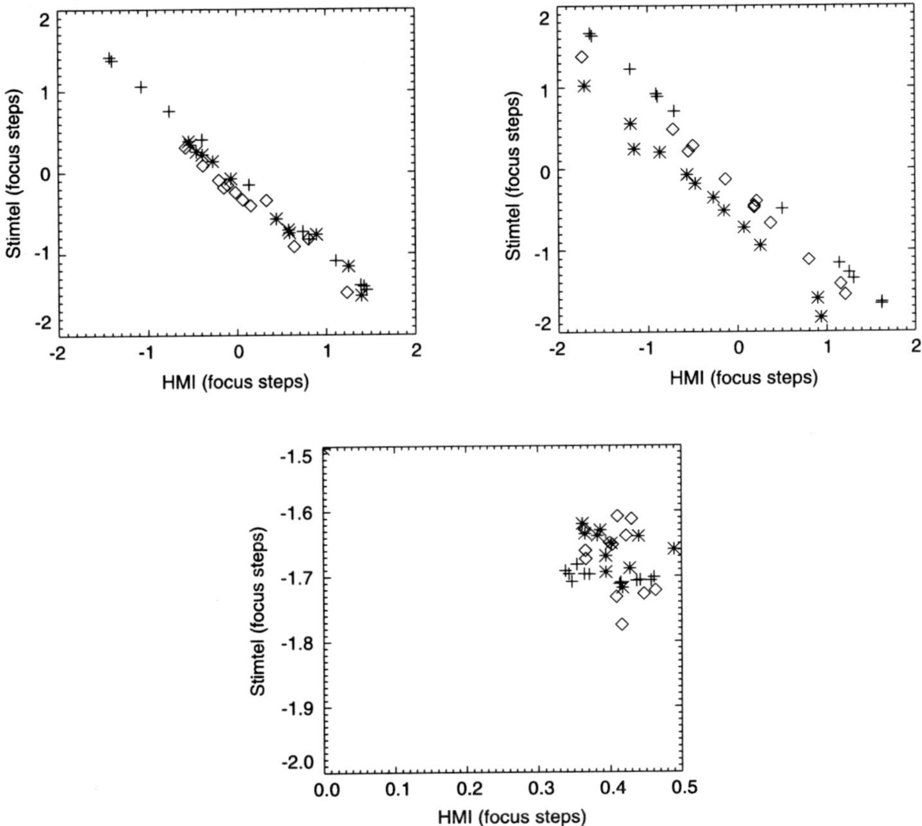

Figure 10 Horizontal (upper left) and vertical (upper right) field tilt as measured for the front camera (diamonds) and the side camera (stars), together with the result of the simulated measurements (crosses). The simulations enable us to explain the variations as an artifact caused by an unstable outside temperature. The lower panel shows the quadratic term, which has no strong sensitivity to the temperature drift.

We simulated a sinusoidal focus drift with a period of 2.2 hours, and an amplitude of 0.75 focus steps, and a random phase. These parameters represent observed focus drifts when we measured the focus continuously, showing that the major properties can be reproduced. For the field-tilt parameters, we see the characteristic, nonphysical anti-correlation of stimulus telescope parameters and instrument parameters. Still, the field-curvature parameters are not biased if the measurements are started at a random time, and therefore their average value provides an estimate for the actual tilt.

We observed that the coefficients for the front and side cameras are strongly correlated. This suggests that jitter or air currents, which act on the short time lag between the cameras and would destroy the correlation, play a minor role in the observed variation in the field tilt parameters. Table 2 gives the mean value and standard deviation of the 12 consecutive measurements.

Figure 11 shows all of the field-curvature measurements that we obtained at the different test facilities. While the temperature fluctuations at GSFC caused large fluctuations of the measured values, the environments at LMATC and ASO were more stable, which resulted in more consistent values.

Table 2 Field tilt and curvature for the front and side cameras. The numbers represent the focus change in units of nominal steps from edge to edge for the focus gradient, and the change from center to edge for the quadratic focus term. The numbers have been derived from a series of twelve field-curvature measurements in an unstable temperature environment at GSFC.

	Grad x	Grad y	Quadratic
Front cam. June 2008	0.01 ± 0.15	-0.13 ± 0.24	0.40 ± 0.01
Side cam. June 2008	0.09 ± 0.17	-0.45 ± 0.20	0.41 ± 0.01

We conclude that the side camera has a vertical gradient in the best focus position with the top of the field being half a focus step lower than the bottom. The gradient observed for the front camera is not significantly different from zero. This suggests that the plane of the side camera CCD slightly deviates from vertical to the optical axis. The field curvature increased initially, but later settled at a value of 0.4 focus step from center to edge.

8. Image Motion

There are several moving parts in the instrument that can potentially lead to image offsets. While they can be corrected before the observables are calculated, they must be known precisely, because any error will lead to crosstalk between the spatial domain and the wavelength domain. Both *a priori* knowledge of shifts introduced by particular waveplates and Sun-center coordinates obtained by fitting the solar limb in individual filtergrams provide information about the instrumental image motion.

There are six rotating waveplates and one rotating polarizer in the instrument. Three waveplates are used for the polarization selection, and three of the waveplates and the rotating polarizer are used to tune the Michelson interferometers and the tunable Lyot filter.

In the following, we call them POL1 ... POL3 (polarization selectors) and WL1 ... WL4 (wavelength selectors), in the order in which the light is passing through them (see Figure 1). All waveplates have been separately rotated by 360° in steps of 30°. We determine the relative image shifts by cross-correlating each image with a reference image. Figure 12 shows an example of the detected image motions.

The most striking feature is the vertical motion with an amplitude of about half a pixel which is produced by the two wavelength-selector wave plates WL2 and WL4, *i.e.*, the waveplates tuning the two Michelson interferometers. We do not fully understand the reason why this vertical motion does not result in a phase-shifted horizontal motion, as would be suggested by a geometric distortion or a mounting error of the waveplates. However, the shift has been consistently seen on the ground, and it will be measured in-flight.

We note that while we detected a solid image shift with the rotating waveplates, the distortion introduced by the waveplates is generally only a few hundredths of a pixel. Occasionally observed shifts up to 0.1 pixel are not reproducible, and therefore are most likely thermally introduced distortions.

Other moving parts of the instrument are the two focus wheels. Because some glass blocks are not perfectly mounted or show a wedge, they shift the image by several pixels. Figure 13 shows the offset as a function of the position of the focus wheels. Unlike the polarization and wavelength-selector waveplates, the focus wheels do not move during regular observations. We note that focus-sweep observations, which will be performed regularly in space to determine the best focus position and the MTF, must take into consideration the finite reaction time of the ISS to the image shifts introduced by the moving focus wheels.

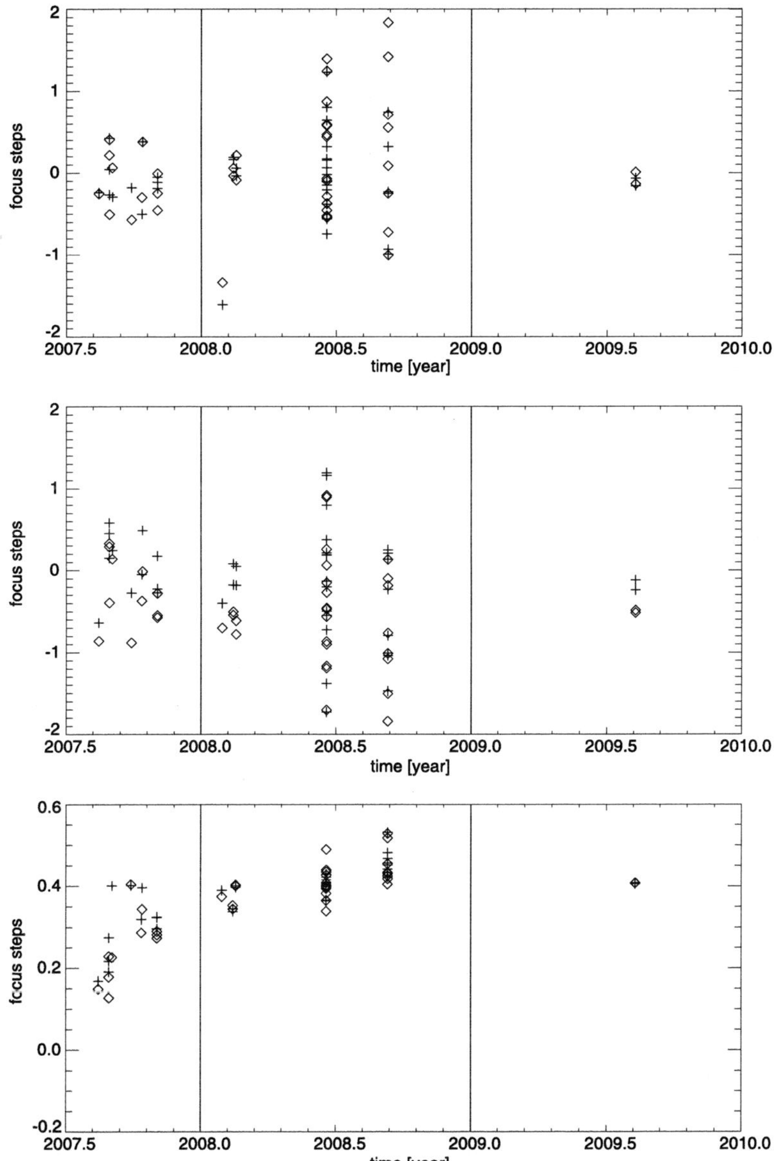

Figure 11 Measured value of the field tilt in horizontal (upper panel) and vertical (center panel) direction, and the quadratic field curvature (lower panel). The numbers represent the focus change in units of nominal focus steps from edge to edge for the focus gradient (left to right and bottom to top), and the change from center to edge for the quadratic focus term. The crosses show the data for the front camera, and the diamonds show the data for the side camera. The horizontal axis represents the time of the measurement. The left section represents the field values obtained at LMATC, the center section represents the values obtained at GSFC, and the right section represents the values obtained at ASO. At GSFC, temperature changes during the field-curvature sequence (which takes about one hour) introduce large fluctuations in the measured parameters, and only averages of a large number of measurements are meaningful. The values measured at LMATC and ASO suggest a field tilt of one-half focus step in the vertical direction for the side camera. The field curvature, which can be measured even in unstable temperature environments, has initially increased, but has stabilized later on.

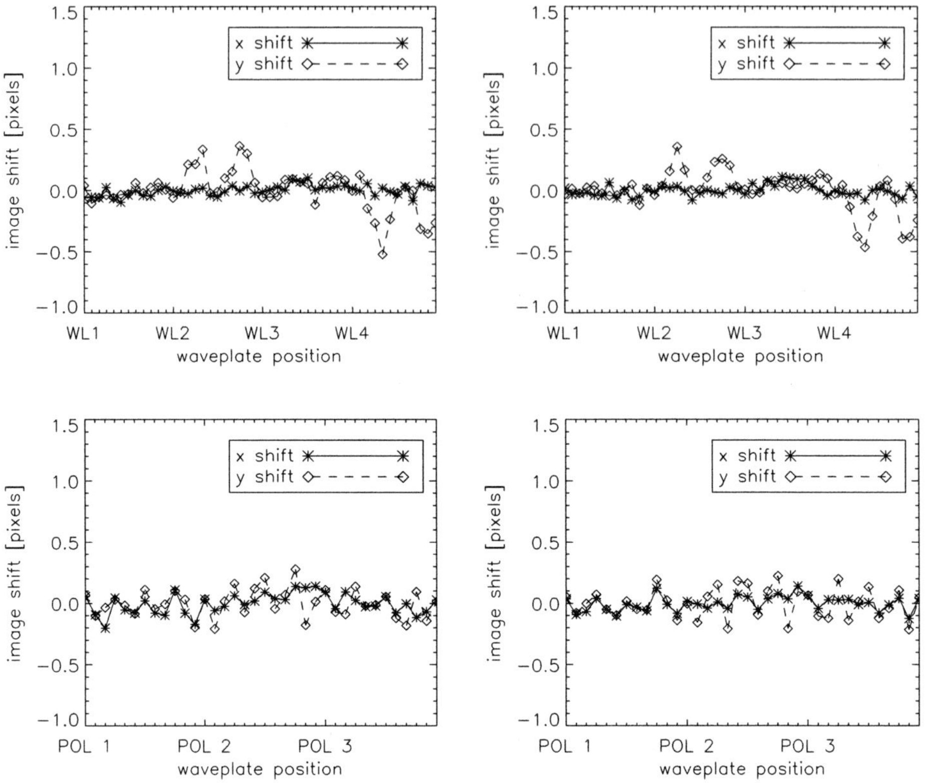

Figure 12 Image motion introduced by moving waveplates (upper panels) and polarizers (lower panels). The left panels show the front camera; the right panels show the side camera. The rotation of the indicated waveplate starts at the major tick marks, with a 30° increment for each minor tick mark.

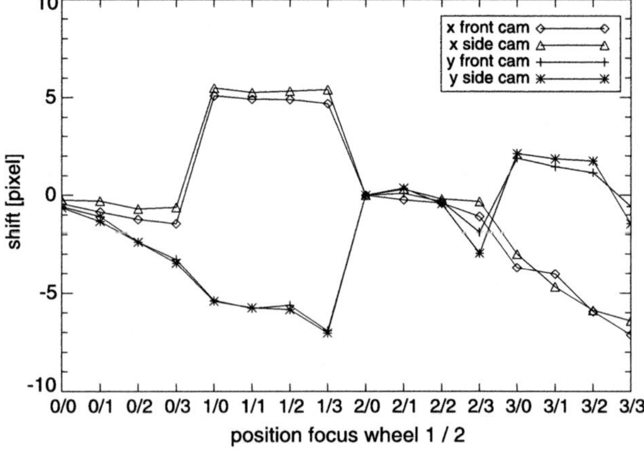

Figure 13 The image shifts introduced by the focus blocks. The labels on the horizontal axis show the thickness of the first and second glass plates in the optical path. The thickness is a multiple of 3.4 mm for the first focus wheel and a multiple of 0.85 mm for the second focus wheel.

9. Flat Field

The flat field has been determined by the method presented by Kuhn, Lin, and Loranz (1991) and Toussaint, Harvey, and Toussaint (2003). Similar to the field curvature and distortion measurements, one can distinguish between the instrumental flat field and the inhomogeneous illumination of the stimulus telescope by tilting their optical axes with respect to each other. We used a 27.4-mm diameter field-stop target, which cuts off only the extreme edge of the field of view. It is large enough to not limit the field of view substantially, but it still gives guidance about the relative displacement of instrument and stimulus telescope. We note that, because the illumination varies only on large scales, the flat field does not depend critically on an extremely accurate knowledge of the image offsets.

The resulting flat field (see Figure 14) shows the different gains of the four quadrants of the CCD, as well as four slabs per quadrant, which are a result of the manufacturing process. This is not a property of the CCD surfaces, but of their respective amplifiers. The small-scale

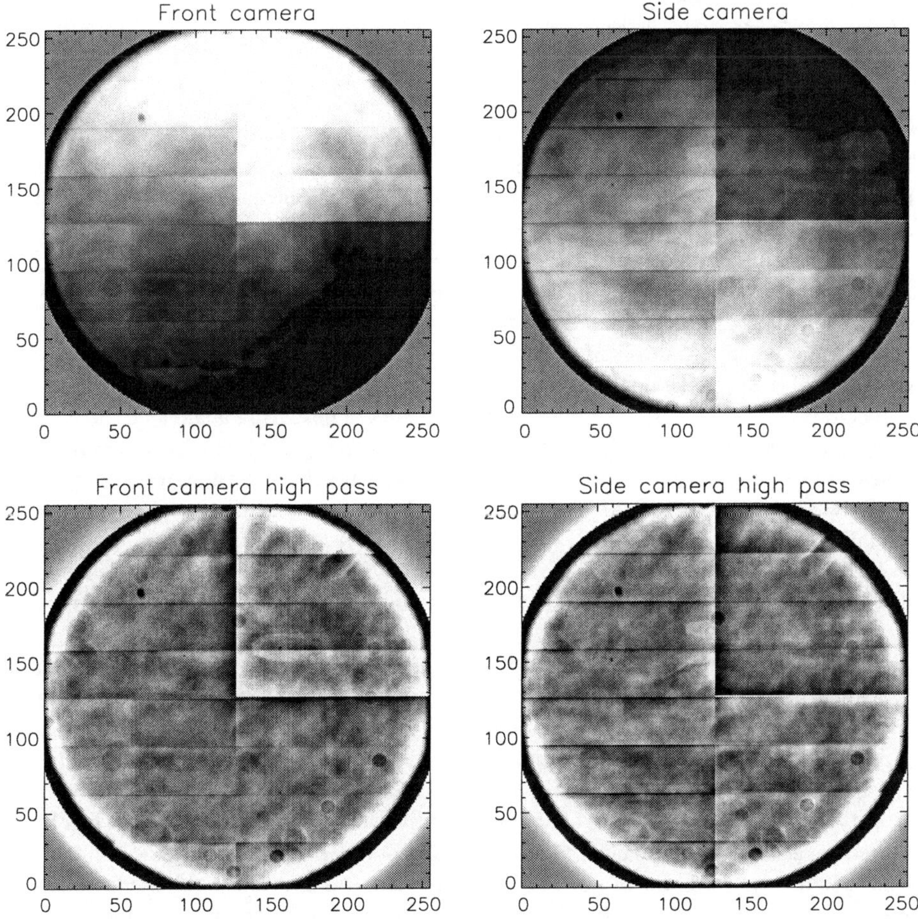

Figure 14 The panels show the flat field of the front and side cameras (left and right columns), with the high-pass-filtered version in the lower panels. In order to show the structure more clearly, the gray scale is saturated to $\pm 5\%$ for the flat field and $\pm 2\%$ for the high-pass-filtered version.

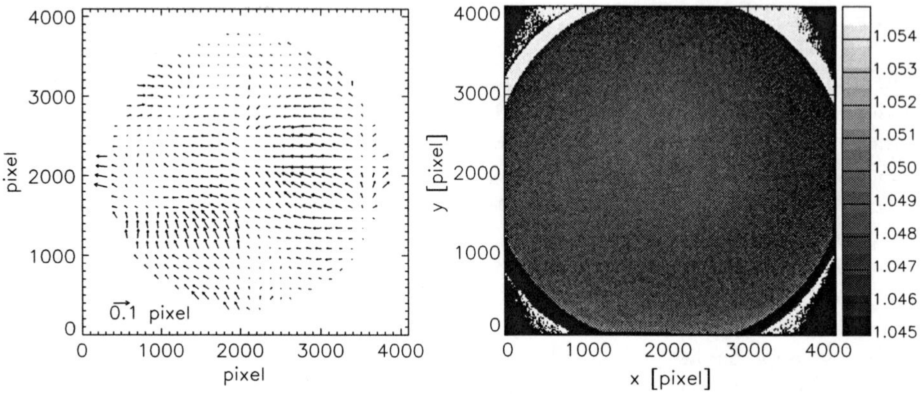

Figure 15 Left panel: Residual shifts between the images after removing a lateral shift, a rotation, and a fourth-order polynomial for the mutual distortion. The spatial structure is dominated by the 512 × 2048 pixel slabs of the CCD. Right panel: Averaged residuals from the offpoint flat field measurements. The residuals give an estimate of the error in the large-scale variation of the flat field. The average error inside the vignetting radius is 0.2%.

features of the flat field can be made visible by applying a high-pass filter to the flat field. The circular structures show out-of-focus dust specks that are part of the optical flat field. The optical flat field is largely identical for both cameras, whereas the CCD flat fields are independent.

The flat field falls off sharply near the edge of the field of view. This is a result of vignetting, caused by the opening at the back end of the Lyot filter being too small, which slightly obstructs the clear aperture of the instrument. The unvignetted part of the optical field is large enough to always fully contain a properly centered solar image. The observables should therefore be unaffected by the vignetting.

There is a larger dust speck on the side CCD camera obscuring an area of about twelve pixels (and several smaller ones on both cameras), reducing the intensity of several pixels below 50% of the normal value. The signal in these pixels is too low to be simply corrected by applying the flat field. The affected pixels need to be treated as missing pixels and spatially interpolated.

We also discovered a stable horizontal structure with a period of roughly 43 pixels. The structure shows different strengths for both CCDs and for the different quadrants, but no apparent temporal variation.

As a measure of the large-scale flat-field quality, we look at the magnitude of the residuals. Figure 15 (right panel) shows the average residuals from the offset images. The images have been rebinned to reduce the photon noise. The residuals are probably caused by slight variations in the aperture illumination, which are ignored in the flat-field derivation. The average residual inside the vignetting radius is 0.2%.

We noticed condensation on the CCD, which produced geometric patterns when the instrument was in the vacuum tank. Figure 16 shows how these patterns became stronger with time as long as the instrument was in the vacuum tank. Images taken with the internal LED at GSFC showed similar patterns. We expect the outgassing of the instrument at the beginning of the mission to be so thorough that condensation does not become a problem during normal operations. If contaminants should build up on the CCD, heaters can be switched on to clean up the CCD.

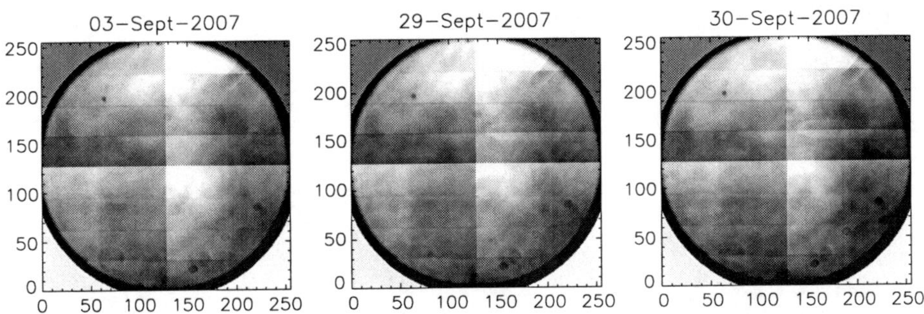

Figure 16 The three panels show the observed flat field in vacuum. The finger-like horizontal structure in the upper right quadrant becomes stronger with time. The gray scale is saturated with a range from 0.95 to 1.05 for better visibility of the structure. Also, the quadrants have been corrected for the different quadrant gains.

There are more stringent requirements on the small-scale flat field, because it is critical to the subpixel spatial interpolation of the filtergrams. The accuracy must be better than 0.1% at any given time. Besides the above-mentioned algorithm, small-scale flat fields can be obtained from any low-contrast image. The condensation patterns shown in Figure 16 lead to small-scale flat field changes over the entire CCD. Although the rate of the flat-field changes depends strongly on the particular thermal environment of the instrument, an order of magnitude estimate of the change may be derived from the fact that we saw a change of 0.2% in the area-average small-scale flat field within a period of five weeks in which the instrument was kept in the vacuum tank. A mechanism to monitor the small-scale flat field from regular filtergram observations at a temporal cadence of one day has been implemented to deal with any changes in the flat field (Wachter and Schou, 2009). This way, we expect to achieve a flat field knowledge of better than 0.1% at all times.

10. Gain and Linearity

10.1. Camera Gain

The gain of the amplifiers of the four quadrants of the CCD is determined by varying the exposure time and determining the slope of the variance *versus* the signal. A number of exposures from 0 to 16 seconds were taken with the LED while the instrument was in the vacuum tank. Figure 17 shows the average signal *versus* an estimate of the variance. The variance is estimated from a pixel-by-pixel difference of two identical images. We see a clear linear dependence before saturation is reached at around 12 000 DN. The variance drops after saturation, because the pixel stops collecting charge, and electric charge is leaking to neighboring pixels. Figure 18 shows the inverse gain as a function of CCD position for the front and the side camera. As expected, the gain is different for each quadrant. The variation within the quadrant is probably a residual effect from the very inhomogeneous illumination of the CCD, which results in a different weight across the intensity range. The average gain for each quadrant is given in Table 3.

10.2. Linearity

Another important aspect of the CCD and camera performance is the linearity of the amplifiers. The linearity can be derived from the same data set as the gain. The center four pixels

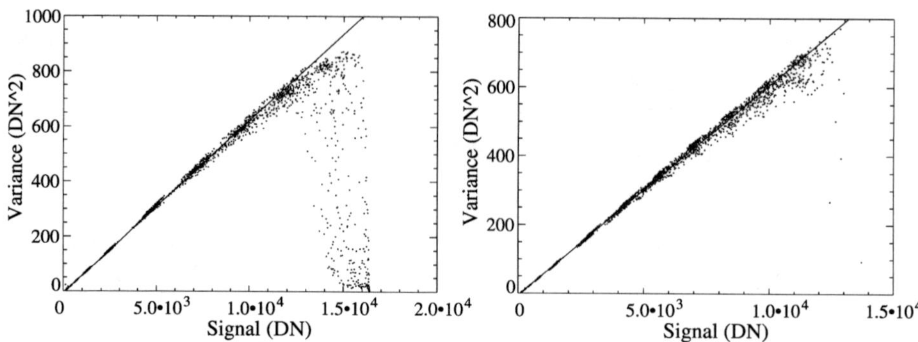

Figure 17 The variance as a function of signal intensity for front (left) and side (right) cameras. The slope of this curve in the linear range is the gain of the camera amplifiers.

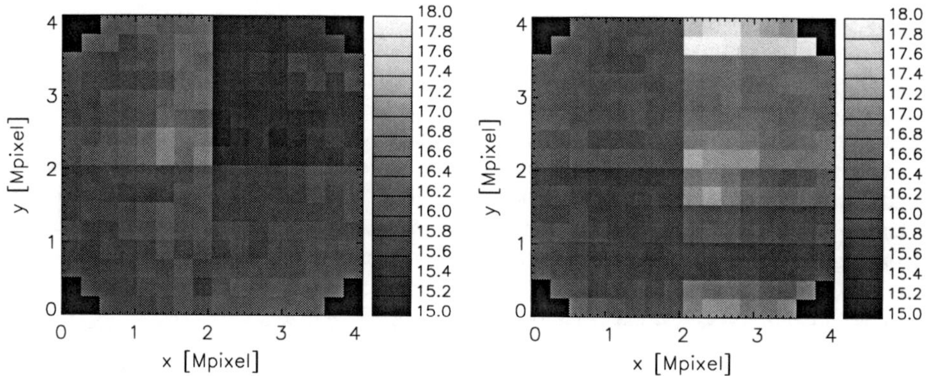

Figure 18 Inverse gain as a function of field position. The gray scale ranges from 15 through 18. The main signal comes from the four different amplifiers; the rest is due to inhomogeneities in the CCD.

Table 3 The inverse gain for the front and side cameras obtained from a variation in exposure.

	Front camera	Side camera
Lower left	15.91	15.83
Lower right	15.91	16.27
Upper left	16.27	16.10
Upper right	15.45	16.92

in a 64×64 rebinned imaged were fitted to a linear function of the exposure time (over the linear range) and the residuals plotted *versus* the intensity.

As can be seen from Figure 19, the intensities saturate at around 12 000 DN for both cameras. At intensities below that value, the nonlinearity is of the order 1%. The nonlinearity is generally very repeatable, but does show a weak temperature dependence.

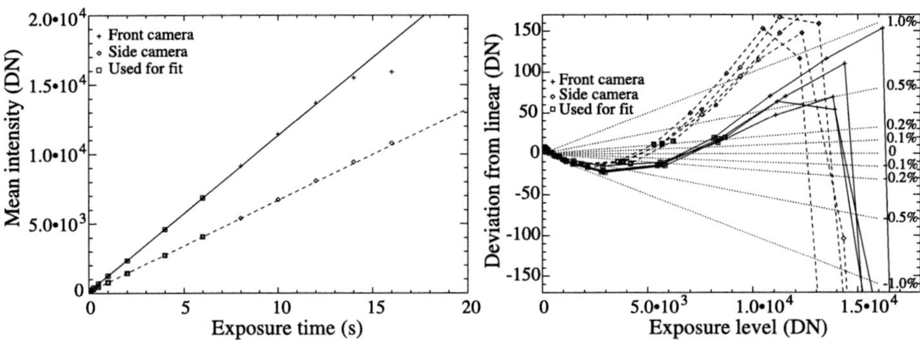

Figure 19 Left panel: Intensity as a function of exposure time for the front and side cameras. Only exposure times far away from the saturation level are used for the linear fit. Right panel: Residuals from a linear fit of the CCD intensity to the exposure time *versus* intensity. The four curves for each camera show the results for the four pixels closest to the center of the CCD in 64×64 rebinned images (*i.e.*, one per quadrant). All fits are done over the same set of exposure times.

11. Combining the Cameras

Depending on the choice of frame list (Couvidat *et al.*, 2011), it may be required to combine filtergrams from both cameras to produce a single observable. To avoid systematic errors from the procedures, an accurate knowledge and a high degree of stability for the shifts between the two cameras is required. Moreover, we need to know the flat fields of both cameras and their MTF.

The alignment of the cameras and their relative distortion can be measured by cross-correlating images. We measured the relative distortion and alignment with the random-dot target and fitted a fourth-order polynomial to fit the relative distortion. Doing so, we obtained a residual distortion with a magnitude of a few hundredths of a pixel, which reflects the structure of the CCD (see Figure 15, left panel). The obvious cause for this pattern is small inhomogeneities in the lattice of the CCD pixels. This structure is stable and can be modeled if necessary. Residual errors have a random, time-varying structure and are probably introduced by thermal fluctuations. They are of the order of a few hundredths of a pixel.

While the accuracy of the large-scale flat field that we obtain from the offpointing method for each individual camera is not sufficient to combine the cameras, we can very accurately determine the relative flat field from any image that is nearly simultaneously recorded in both cameras. The ratio of those images readily provides the relative flat field.

As shown in Section 3, we could not reliably measure any significant difference between the best focus MTF of both cameras on the ground. However, we know that the best focus of the cameras differs by 0.2 focus step, which leads to small differences in the MTF. As the main cause of uncertainty is the air currents, we may be able to find differences in the relative MTF and correct for them once we get data from space.

12. Conclusion

The *Helioseismic and Magnetic Imager* is a high-precision, near-monochromatic optical imager. The on-ground optical calibration shows that HMI meets all performance requirements and guarantees our ability to produce observables of sufficient quality to reach the HMI science goals. It also gives a clear guideline for the in-orbit calibration.

HMI is expected to achieve a Strehl ratio of 0.8 or better. The distortion of up to two pixels at the edge of the field of view has been determined with an accuracy of 0.05 pixel. The large-scale flat field can be determined with an accuracy of 0.2%, and for the small-scale flat field we confirmed that knowing the flat field in-flight at any time with an accuracy of 0.1% or better is realistic. The instrument focus varies from center to edge by 0.4 mm, and from edge to edge by less than 0.5 mm. The best focus is in the middle of the adjustable focus range, and the offset and rotation of the two cameras have been determined to fractions of a pixel.

Modeling helped us to understand the behavior of the instrument in a sometimes unstable temperature environment and also helped us ensure that the instrument will meet the specifications when operated in space. Continuous monitoring of the instrument's performance during a time period of more than two years has made us confident that HMI will provide data of outstanding quality during its mission.

Acknowledgements This work has been supported by the NASA grant NAS5-02139 (HMI). Many people were involved in the calibration efforts for the instrument. We thank staff members at LMATC, GSFC, and ASO for their support. In particular, we thank Dave Kirkpatrick, Darrel Torgerson, Bob Stern, and Brett Allard. We also thank Phil Scherrer, Ted Tarbell, and Alan Title for their input.

References

Bray, R.J., Loughhead, R.E.: 1979, *Sunspots*, Dover, New York.

Couvidat, S., Zhao, J., Birch, A.C., Kosovichev, A.G., Duvall, T.L., Parchevsky, K., Scherrer, P.H.: 2010, Implementation and comparison of acoustic travel-time measurement procedures for the solar dynamics observatory/helioseismic and magnetic imager time-distance helioseismology pipeline. *Solar Phys.* doi:10.1007/s11207-010-9652-y.

Couvidat, S., Schou, J., Shine, R.A., Scherrer, P.H., Bush, R.I.: 2011, Wavelength dependence of the helioseismic and magnetic imager instrument. *Solar Phys.*, submitted.

Jefferies, S.M., Duvall, T.L. Jr.: 1991, A simple method for correcting spatially resolved solar intensity oscillation observations for variations in scattered light. *Solar Phys.* **132**, 215 – 222. doi:10.1007/BF00152283.

Korzennik, S.G., Rabello-Soares, M.C., Schou, J.: 2004, On the determination of Michelson Doppler imager high-degree mode frequencies. *Astrophys. J.* **602**, 481 – 516. doi:10.1086/381021.

Kuhn, J.R., Lin, H., Loranz, D.: 1991, Gain calibrating nonuniform image-array data using only the image data. *Publ. Astron. Soc. Pac.* **103**, 1097 – 1108.

Pierce, A.K., Slaughter, C.D.: 1977, Solar limb darkening. I – At wavelengths of 3033-7297. *Solar Phys.* **51**, 25 – 41. doi:10.1007/BF00240442.

Schou, J., Scherrer, P.H., Bush, R.I., Wachter, R., Couvidat, S., Rabello-Soares, M.C., Liu, Y., Hoeksema, J.T., Bogart, R.S., Duvall, J.T.L., Miles, J.W., Title, A.M., Shine, R.A., Tarbell, T.D., Allard, B.A., Wolfson, C.J., Tomczyk, S., Norton, A.A., Elmore, D.F., Borrero, J.M.: 2011, The helioseismic and magnetic imager instrument design and calibration. *Solar Phys.*, in preparation.

Toner, C.G., Jefferies, S.M., Duvall, T.L. Jr.: 1997, Restoration of long-exposure full-disk solar intensity images. *Astrophys. J.* **478**, 817 – 827. doi:10.1086/303836.

Toussaint, R.M., Harvey, J.W., Toussaint, D.: 2003, Improved convergence for CCD gain calibration using simultaneous-overrelaxation techniques. *Astron. J.* **126**, 1112 – 1118. doi:10.1086/376846.

Wachter, R., Schou, J.: 2009, Inferring small-scale flatfields from solar rotation. *Solar Phys.* **258**, 331 – 341. doi:10.1007/s11207-009-9406-x.

Solar Phys (2012) 275:285–325
DOI 10.1007/s11207-011-9723-8

Wavelength Dependence of the *Helioseismic and Magnetic Imager* (HMI) Instrument onboard the *Solar Dynamics Observatory* (SDO)

Sébastien Couvidat · Jesper Schou · Richard A. Shine · Rock I. Bush · John W. Miles · Philip H. Scherrer · Richard L. Rairden

Received: 30 December 2009 / Accepted: 8 February 2011 / Published online: 4 March 2011

Abstract The *Helioseismic and Magnetic Imager* (HMI) instrument will produce Doppler-velocity and vector-magnetic-field maps of the solar surface, whose accuracy is dependent on a thorough knowledge of the transmission profiles of the components of the HMI optical-filter system. Here we present a series of wavelength-dependence calibration tests, performed on the instrument from 2005 onwards, to obtain these profiles. We obtained the transmittances as a function of wavelength for the tunable and non-tunable filter elements, as well as the variation of these transmittances with temperature and the angle of incidence of rays of light. We also established the presence of fringe patterns produced by interferences inside the blocking filter and the front window, as well as a change in transmitted intensity with the tuning position. This thorough characterization of the HMI-filter system confirmed the very high quality of the instrument, and showed that its properties are well within the required specifications to produce superior data with high spatial and temporal resolution.

Keywords Sun: helioseismology · Instrument: SDO/HMI

The Solar Dynamics Observatory
Guest Editors: W. Dean Pesnell, Phillip C. Chamberlin, and Barbara J. Thompson

S. Couvidat (✉) · J. Schou · R.I. Bush · P.H. Scherrer
W.W. Hansen Experimental Physics Laboratory, Stanford University, 491 S. Service Road, Stanford, CA 94305-4085, USA
e-mail: couvidat@stanford.edu

R.A. Shine · R.L. Rairden
Lockheed Martin Solar and Astrophysics Laboratory, Bldg. 252, Org. ADBS, 3251 Hanover Street, Palo Alto, CA 94304, USA

J.W. Miles
USRA/SOFIA, NASA Ames Research Center, Moffett Field, CA 94035, USA

1. Introduction

The *Helioseismic and Magnetic Imager* instrument (HMI: Schou *et al.*, 2011) onboard the *Solar Dynamics Observatory* satellite (SDO) makes measurements, in the Fe I absorption line centered at the in-air wavelength of 6173.3433 Å (hereafter the target wavelength; see *e.g.* Dravins, Lindegren, and Nordlund, 1981; Norton *et al.*, 2006), of the motion of the solar photosphere to study solar oscillations and of the polarization to study all three components of the photospheric magnetic field. HMI samples the neutral iron line at (nominally) six positions (hereafter sampling positions) symmetrical around the line center at rest. Each sampling-position transmission profile must be as close to a δ function as possible: to this end, HMI is built around an optical-filter system composed of a front window, a blocking filter, a five-stage Lyot filter, and two Michelson interferometers. This system produces a narrow bandpass transmission profile, or transmittance, with a nominal Full Width at Half Maximum [FWHM] of 76 mÅ and a nominal FWHM of 612 mÅ for the non-tunable (fixed) part. The knowledge of the transmittance of each element constituting the optical-filter system is required to access the characteristics of the HMI sampling positions (especially their transmission profile as a function of wavelength), and thus to derive accurate Doppler velocities and vector magnetic fields from the corresponding filtergrams. The term filtergram refers to a picture of the Sun obtained through a specific sampling position, *i.e.* at a specific wavelength, and polarization. From December 2005 to August 2009, the assembled instrument has been subjected to many tests, whose purpose was twofold: first, to detect any defect before launch; second, to calibrate the instrument.

In this article we present the calibration of the wavelength dependence of HMI: this is the process by which we characterize the components of the HMI optical-filter system in terms of their transmittances and how these transmittances are affected by temperature, angle of incidence of the rays of light, change of focus blocks, and so on. In Section 2 we remind the reader of the theoretical transmittances of the Lyot elements, of the Michelson interferometers, and of the origin of interference fringes produced by the front window and blocking filter on the HMI filtergrams. In Section 3 we present the main calibration results: wavelength, angular, and thermal dependences. In Section 4 we present other calibration results, *e.g.* thermal stability and instrument throughput, and also preliminary plans for on-orbit calibration. We conclude in Section 5.

2. Theoretical Wavelength Dependence

The core of the HMI instrument is its optical-filter system, composed of a front window, a blocking filter, a five-stage Lyot filter, and two Michelson interferometers. The blocking filter, Lyot filter, and Michelson interferometers are enclosed in a temperature-stabilized oven. The requirement for the oven stability is $\pm 0.01^{\circ}\mathrm{C\,h^{-1}}$, but the actual performance is better. The nominal operating oven temperature is 30°C. In this section we derive the theoretical transmittances of the filter elements. The layout of HMI is presented on Figure 1. HMI can operate in two main modes: OBSMODE and CALMODE. In CALMODE, the instrument aperture is imaged instead of the Sun, and therefore the front window is in focus. CALMODE is used only for calibration purposes, while OBSMODE is used during normal operations.

2.1. Lyot Filter

The HMI Lyot filter is a five-element, wide-field (hereafter WF), tunable birefringent filter. These elements (stages) are referred to as E2, E3, E4, E5, and E1 (in order of appearance from the blocking filter to the wide-band Michelson interferometer). The narrowest element E1 (in terms of the FWHM of its transmittance) is made tunable by the addition of a rotating

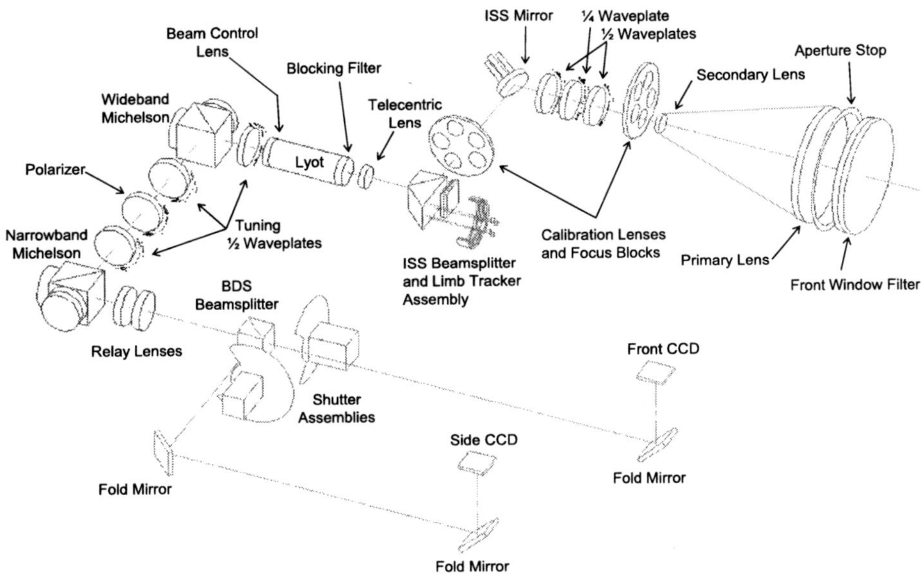

Figure 1 Layout of the HMI instrument. The optical-filter system is drawn in red.

half-wave plate at the exit. The other wave plates are fixed. A telecentric lens at the entrance of the filter oven produces a collimated beam for the Lyot filter: the angular distribution of light is identical for each image point. In order to minimize the temperature sensitivity of the filter, we used the same basic design employed for the SOHO/MDI instrument (Scherrer *et al.*, 1995). Therefore, each Lyot element is made of two birefringent crystals: calcite and either ADP (Ammonium Dihydrogen Phosphate) or KDP (Potassium Dihydrogen Phosphate). KDP is used for the thinner elements E4 and E5 because the necessary ADP thicknesses are too small to be easily fabricated. ADP/KDP crystals are used in combination with calcite to cancel out the wavelength change with temperature of the calcite crystals.

2.1.1. Time Delay

A birefringent crystal decomposes a ray of light into two rays propagating at different velocities: the ordinary and extraordinary rays. Uniaxial birefringent crystals (crystals with a single optical axis) such as calcite, ADP, and KDP are therefore characterized by two refractive indices: n_e for the extraordinary axis (parallel to the optical axis) and n_o for the ordinary axis (perpendicular to the optical axis). The birefringence of such a crystal is defined as $\Delta n = n_e - n_o$. Some optical and physical properties of calcite, ADP, and KDP are listed in Table 1. Some physical properties of the five temperature-compensated WF-Lyot elements are listed in Table 2, while Figure 2 shows an exploded view of the element E1. Note that, except for the extra ADP (or KDP) elements, this is the usual design of a WF-Lyot element as discussed in detail by Title and Rosenberg (1979).

The on-axis time delay of a WF element that is not temperature-compensated is the same as the time delay of a simple Lyot element with the same total thickness of birefringent material, but the angular dependence is greatly reduced. The time delay [Δt] between the ordinary and extraordinary rays is

Table 1 Properties of optical crystals used in HMI at wavelength $\lambda = 6173$ Å and temperature $T_e = 25$°C. The values of the refractive indices n_e and n_o are derived from Sellmeier's equations, with the coefficients taken from the United Crystals and AD Photonics websites (www.unitedcrystals.com/KDPProp.html and www.adphotonics.com/products/calcite_specifications.pdf). The change in birefringence $[\Delta n]$ with wavelength is $\alpha = \partial \Delta n / \partial \lambda$. The thermal expansion coefficients $\alpha_{\shortparallel} = (1/d)\partial d/\partial T_e$, where d is the length of the element, are taken from the Quantumtech website (quantumtech.com/PDF/702.PDF).

Crystal	n_o	Δn	α (Å^{-1})	$\frac{\partial \Delta n}{\partial T}$ (°C^{-1})	$\alpha_{\shortparallel}$ (°C^{-1})
calcite	1.65656	−0.1707	3.2592×10^{-6}	9.98×10^{-6}	5.68×10^{-6}
ADP	1.52311	−0.0396	1.2679×10^{-6}	5.23×10^{-5}	4.20×10^{-6}
KDP	1.50780	−0.0394	1.2530×10^{-6}	1.43×10^{-5}	44.0×10^{-6}

Table 2 Nominal lengths of the Lyot stages of HMI (mm). (A) refers to ADP, (K) to KDP. The FSRs are nominal (FSR_n) full spectral ranges in mÅ at 6173 Å. $\partial\lambda/\partial T_e$ is the measured wavelength shift with temperature, in mÅ °C^{-1}, around 30°C. Δn_1 and Δn_2 are the birefringences, and d_1 and d_2 are the crystal lengths. The index 1 refers to the calcite crystal, while 2 refers to either ADP or KDP.

Element	calcite	ADP/KDP	$\lvert d_1\Delta n_1 - d_2\Delta n_2\rvert$	FSR_n	$\partial\lambda/\partial T_e$
E1	30.620	7.056 (A)	4.89561	690	−3.06
E2	15.310	3.528 (A)	2.44781	1380	−1.36
E3	7.655	1.764 (A)	1.22390	2758	−4.10
E4	4.662	4.500 (K)	0.61451	5516	6.87
E5	2.330	2.250 (K)	0.30725	11032	10.04

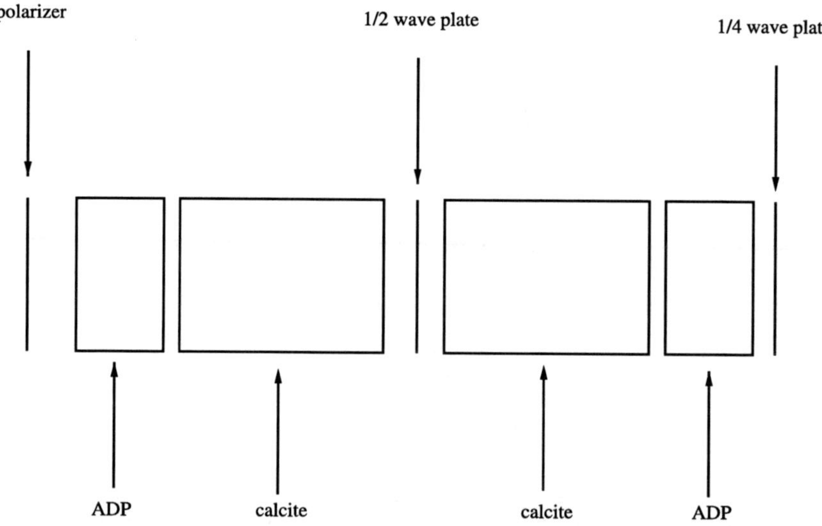

Figure 2 Exploded view of the Lyot element E1. Light enters through the polarizer and exits through the quarter-wave plate. The tuning half-wave plate is not shown, but it is located behind the quarter-wave plate.

$$\Delta t = \frac{D}{2c}\left\{n_e\left[1-\left(\frac{a^2}{n_o^2}+\frac{b^2}{n_e^2}\right)\right]^{1/2} - n_o\left[1-\left(\frac{a^2+b^2}{n_o^2}\right)\right]^{1/2}\right\}$$
$$+ \frac{D}{2c}\left\{n_e\left[1-\left(\frac{a'^2}{n_o^2}+\frac{b'^2}{n_e^2}\right)\right]^{1/2} - n_o\left[1-\left(\frac{a'^2+b'^2}{n_o^2}\right)\right]^{1/2}\right\}, \tag{1}$$

where

$$a = \sin(\theta)\cos(\phi), \qquad b = \sin(\theta)\sin(\phi) \tag{2}$$

and

$$a' = \sin(\theta)\cos(\phi+\pi/2), \qquad b' = \sin(\theta)\sin(\phi+\pi/2). \tag{3}$$

Here θ is the angle between the ray and the optical axis (angle of incidence), while ϕ is the azimuthal angle. Note that n_e and n_o are functions of wavelength and temperature. c is the speed of light. Finally, D is the total thickness of both calcite components separated by the half-wave plate (again, we ignore the ADP or KDP components for the present) and the two major terms are for these two calcite components. The ϕ terms cancel, and on expanding out to second order in θ, the angular dependence near the axis (*i.e.* for small θ) reduces to

$$\Delta t \approx \frac{D}{c}(n_e - n_o)\left[1 - \frac{\theta^2}{4n_o^2}\left(\frac{n_e - n_o}{n_e}\right)\right]. \tag{4}$$

This results in a relative wavelength shift in the off-axis directions of

$$\frac{\delta\lambda}{\lambda} \approx -\frac{\theta^2}{4n_o^2}\left(\frac{n_e - n_o}{n_e}\right) \tag{5}$$

which should be compared with the simple Lyot (not WF) results of

$$\left.\frac{\delta\lambda}{\lambda}\right|_{\phi=0} \approx -\frac{\theta^2}{2n_o^2}, \qquad \left.\frac{\delta\lambda}{\lambda}\right|_{\phi=\frac{\pi}{2}} \approx +\frac{\theta^2}{2n_o n_e} \tag{6}$$

at the two indicated azimuthal angles (Title and Rosenberg, 1979). For the calcite components, the improvement of a WF design in angular dependence is about a factor of 11 in either of these directions, with the additional advantage of no azimuthal dependence.

Along the axis ($\theta = 0$) we have

$$\Delta t \approx \frac{D(n_e - n_o)}{c} = \frac{D\Delta n}{c}. \tag{7}$$

Taking into account the presence of ADP or KDP blocks paired with calcite in the HMI Lyot stages, the total theoretical delay for a normal ray becomes

$$\Delta t \approx \frac{d_1\Delta n_1 - d_2\Delta n_2}{c}, \tag{8}$$

where d_1 and d_2 are the total lengths of, respectively, calcite and ADP (or KDP), while Δn_1 and Δn_2 are their birefringences.

2.1.2. Wavelength Dependence

At $\theta = 0$, the retardance $\Gamma(\lambda, \theta)$ for a calcite WF element is

$$\Gamma(\lambda, 0) = \frac{2\pi c \Delta t}{\lambda} \approx \frac{2\pi D \Delta n}{\lambda}. \tag{9}$$

The corresponding transmittance $T(\lambda, \theta)$ is (Evans, 1949)

$$T(\lambda, 0) = \cos^2\left(\frac{\Gamma(\lambda, 0)}{2}\right) = \frac{1 + \cos(\Gamma(\lambda, 0))}{2} \tag{10}$$

which can be approximated as

$$T(\lambda, 0) \approx \frac{1 + \cos(2\pi\lambda / \text{FSR})}{2}, \tag{11}$$

where FSR is the free spectral range (defined and calculated in the next section).

The on-axis transmittance modeled by Equation (11) is for the ideal case of a perfect Lyot element. Actual Lyot stages in HMI have defects, *e.g.* misalignments and/or imperfect retardances of waveplates or polarizer, and to account for some of these defects their transmittance at $\theta = 0$ is modeled as

$$T(\lambda, 0) = \frac{1 + B\cos(2\pi\lambda / \text{FSR} + \Phi)}{2}, \tag{12}$$

where B is the contrast of the element, and Φ is its relative phase. A contrast smaller than 1 and a gradient in the relative phase across the aperture indicate a defect. For the tunable element E1, the potential presence of an I-ripple (variation in the transmitted intensity as a function of the tuning position; see Section 2.4) is ignored in this model. Characterizing the Lyot elements' wavelength dependence entails determining the values of $B(x, y)$ and $\Phi(x, y)$ at every location (x, y) on the HMI CCD(s) and for the specific distribution of angles of incidence inside HMI, as well as determining the FSRs of the different elements.

2.1.3. Free Spectral Range

The Free Spectral Range [FSR] of a filter element is defined as the wavelength separation between the peaks of maximum transmittance. From Equation (9) it is clear that for a constant Δn this would approximately be

$$\text{FSR} \approx \left|\frac{2\pi\lambda}{\Gamma(\lambda)}\right| \approx \left|\frac{\lambda^2}{D\Delta n}\right|, \tag{13}$$

the more exact formula from Evans (1949) is

$$\text{FSR} \approx \left|\frac{\lambda^2}{D\Delta n}\left(1 - \frac{\lambda}{\Delta n}\frac{\partial \Delta n}{\partial \lambda}\right)^{-1}\right|. \tag{14}$$

As discussed by Title (1974), the extra term makes about a 10% difference for visible wavelengths.

2.1.4. Thermal Compensation

The transmission peaks of a calcite Lyot element shift with temperature due to a combination of thermal expansion and a change in Δn. This shift is independent of the element thickness and Title *et al.* (1976) give the following formula based on experimental data:

$$\frac{\partial \lambda}{\partial T_e} = \left(1 + 5.88 \times 10^{-8} \lambda T_e\right) \times \left(1.711 \times 10^{-3} - 3.126 \times 10^{2} \lambda^{-1} + 6.86 \times 10^{-5} \lambda - 7.55 \times 10^{-10} \lambda^{2}\right), \quad (15)$$

where T_e is the temperature in °C and λ is in Å. For the Fe I line at the target wavelength and at the planned operating temperature of 30°C, the wavelength shift is 0.35 Å °C^{-1}.

This temperature dependence can be greatly reduced by adding retarding crystals to each side of the WF element in such a way as to cancel out the calcite wavelength shift $\partial\lambda/\partial T_e$. ADP and KDP crystals have a smaller birefringence but a larger value of $\partial\lambda/\partial T_e$. Unfortunately, the derivatives are of the same sign, meaning that we have to add the ADP/KDP components in subtraction (*i.e.*, with their fast axis rotated by 90° relative to the calcite's fast axis). Therefore, the two calcite components in a WF Lyot element including ADP (or KDP) components are thicker than the calcite components of a Lyot element with no extra crystals but the same FSR.

The condition for thermal compensation at a specific wavelength λ is

$$\partial(d_1 \Delta n_1 - d_2 \Delta n_2)/\partial T_e = 0 \quad (16)$$

which implies that we can make $\partial\lambda/\partial T_e$ vanish at the target wavelength and $T_e = 30$°C for any element thickness with the proper choice of d_1/d_2. We determined the optimum values of $R_{ADP} = d_1/d_{ADP}$ and $R_{KDP} = d_1/d_{KDP}$ experimentally by fabricating sets of ADP and KDP crystals with thicknesses that gave us our estimated R_{ADP} and R_{KDP}, plus thicknesses of about 5% higher and lower. Using a spectrograph, the transmission profile of each calcite – ADP and calcite – KDP pair (enclosed in an oven) was measured from 25 to 35°C. Then, $\partial\lambda/\partial T_e$ was fit by a second-order polynomial and the slope at 30°C was computed. The results are shown in Figure 3. The final ratios R_{ADP} and R_{KDP} for a zero wavelength shift were then determined by fitting a straight line to the three values in each set. The results are $R_{ADP} = 4.339$ and $R_{KDP} = 1.036$, and the calcite thicknesses must be increased by factors of, respectively, 1.056 and 1.286, compared with the basic WF design with no ADP/KDP crystals. Note that even the 5% error cases show a $\partial\lambda/\partial T_e$ value much lower than the 0.35 Å °C^{-1} value for calcite alone. The actual measured values of $\partial\lambda/\partial T_e$ for each of the five elements before assembly are shown in Table 2.

Fortunately, the additional ADP/KDP crystals do not greatly affect the angular performance of the WF design but do degrade it slightly, mostly because of the increased thickness of the calcite components compared to a calcite-only design with the same FSR. With the addition of ADP/KDP crystals, Equation (5) becomes

$$\frac{\delta\lambda}{\lambda} \approx -\frac{\theta^2}{4n_{o_1}^2}\left(\frac{n_{e_1} - n_{o_1}}{n_{e_1}}\right)\left[\frac{d_1}{D} - \frac{d_2}{D}\frac{n_{o_1}^2}{n_{o_2}^2}\frac{\Delta n_2^2}{\Delta n_1^2}\frac{n_{e_1}}{n_{e_2}}\right], \quad (17)$$

where we have factored out the equivalent calcite result and the terms in the square brackets represent the effect of the temperature-compensating components. D is the total thickness of the calcite components for the corresponding calcite-only design. n_{o_1} and n_{e_1} are the

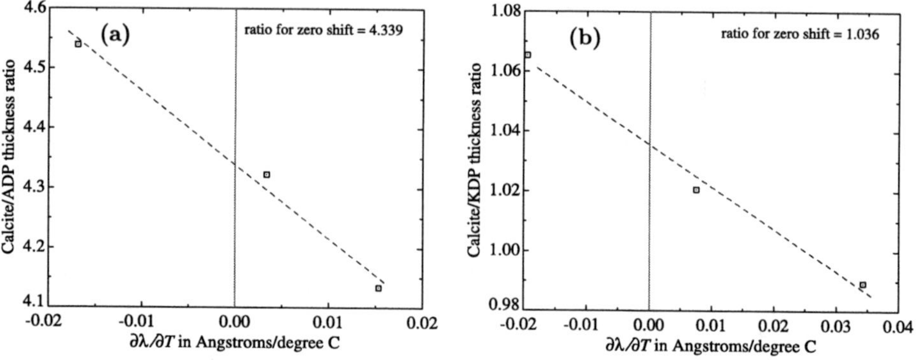

Figure 3 Thickness ratios for the crystals used in the Lyot filter as a function of the temperature dependence of the transmission profile. (a) The squares are the measured $\partial\lambda/\partial T_e$ values for a calcite crystal (9.26 mm thick) and three ADP crystals with different thicknesses intended to span the $\partial\lambda/\partial T_e = 0$ condition. The dashed line is a linear fit and the intersection with the vertical line at zero gives the optimal ratio d_1/d_2 (see text). (b) Same for the KDP case using a calcite crystal 2.77 mm thick.

refractive indices for calcite, while n_{o_2} and n_{e_2} are the refractive indices for ADP/KDP. For the calcite + ADP design, $\delta\lambda/\lambda$ is increased by a factor 1.041 compared to calcite-only, and for the calcite + KDP design it is increased by 1.206. The d_1/D term (the necessary increase in calcite thickness) is the dominant factor.

2.2. Michelson Interferometers

The HMI Michelson interferometers are located after the beam-control lens placed at the end of the Lyot filter (see Figure 1). They are outside the telecentric beam, meaning the chief rays are not parallel to the optical axis anymore. A consequence is that the angular distribution of light is not identical at each image point: the rays reaching a specific point on the CCD are distributed in a cone-shaped beam inside the optical filter, and different image points in the Michelsons will have different diameters and distributions of angles of incidence. Similarly to the Lyot elements, the two Michelson interferometers are WF. Compared to a standard Michelson interferometer, a glass block (here made of N-BK7 optical glass) is added to one of the legs, and the Michelson is made tunable through the addition of a half-wave plate placed outside the interferometer and not through movable mirrors. Moreover, an entrance polarizer, a polarizing beamsplitter, an output quarter-wave plate, and two mirror quarter-wave plates are also added to the usual Michelson design (see an exploded view on Figure 4).

2.2.1. *Time Delay*

The optical-path difference $[\Delta(\theta)]$ as a function of angle of incidence $[\theta]$ through a Michelson interferometer is (Gault, Johnston, and Kendall, 1985)

$$\Delta(\theta) = 2(n_1 d_1 - n_2 d_2) - \left(\frac{d_1}{n_1} - \frac{d_2}{n_2}\right)\sin^2\theta - \left(\frac{d_1}{n_1^3} - \frac{d_2}{n_2^3}\right)\frac{\sin^4\theta}{4} - \left(\frac{d_1}{n_1^5} - \frac{d_2}{n_2^5}\right)\frac{\sin^6\theta}{8}, \quad (18)$$

where n_1 and n_2 are the refractive indices of, respectively, the solid leg (note that n_1 depends on λ and T_e) and the vacuum leg ($n_2 = 1$ in vacuum), and d_1 and d_2 are the corresponding

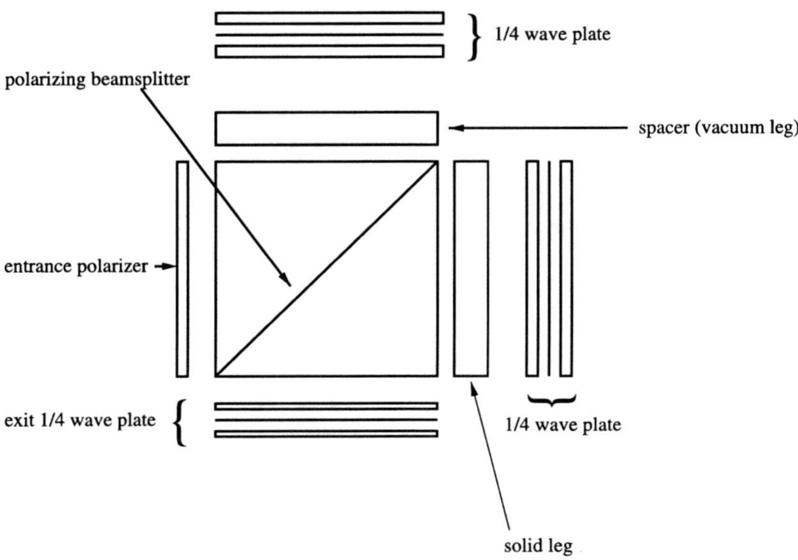

Figure 4 Exploded view of the wide-band HMI Michelson interferometer. The narrow-band Michelson is built the same way, but the light enters through the "exit" quarter-wave plate and exits through the "entrance" polarizer. The tuning half-wave plate is not shown.

leg lengths. The WF condition is $d_1/n_1 = d_2/n_2$, because, in this case, the $\sin^2\theta$ term is canceled and

$$\Delta(\theta) = 2(n_1 d_1 - n_2 d_2) - \left(\frac{d_1}{n_1^3} - \frac{d_2}{n_2^3}\right)\frac{\sin^4\theta}{4} - \left(\frac{d_1}{n_1^5} - \frac{d_2}{n_2^5}\right)\frac{\sin^6\theta}{8}. \tag{19}$$

The time delay $[\Delta t]$ is related to the optical-path difference by $\Delta t = \Delta(\theta)/c$.

2.2.2. *Wavelength Dependence*

From the optical-path difference we derive the retardance $[\Gamma(\lambda,\theta) = 2\pi\,\Delta(\theta)/\lambda]$ and the transmittance $[T(\lambda,\theta)]$ (Title and Ramsey, 1980):

$$T(\lambda,\theta) = \cos^2\left(\pi\frac{\Delta(\theta)}{\lambda} + 2\phi\right) = \frac{1 + \cos\left(2\pi\frac{\Delta(\theta)}{\lambda} + 4\phi\right)}{2}, \tag{20}$$

where ϕ is the tuning angle, *i.e.* the angle of the fast axis of the tuning half-wave plate to the fast axis of the output quarter-wave plate. We added the tuning angle in the equation of the transmittance of the Michelsons, and not in the equation of the transmittance of the Lyot elements (Equations (11) and (12)), because both Michelsons are tunable while only the Lyot element E1 is. Note that Equation (20) differs from Equation (11) of Title and Ramsey (1980) because in HMI the tuning is done by half-wave plates and not polarizers (the polarization axis is rotated twice as fast with a half-wave plate).

A non-zero incidence angle produces a theoretical wavelength change of only about 0.3 mÅ for a collimated beam at $\theta = 1.75°$ (half angle for the HMI field of view) and at target wavelength, if the WF compensation is perfectly achieved. However, the WF condition does not hold at wavelengths different from the target one, because the condition

$(1/n_1)\partial n_1/\partial\lambda = (1/n_2)\partial n_2/\partial\lambda$ is not valid for all λ since the vacuum leg has a constant refractive index.

Following Section 2.1.2, the transmittance of a tunable Michelson at $\theta = 0$ can be expressed as

$$T(\lambda, 0) \approx \frac{1 + \cos(2\pi\lambda/\,\mathrm{FSR} + 4\phi)}{2}, \tag{21}$$

where the FSR is calculated in the next section. However the actual Michelson interferometers in HMI have imperfections (*e.g.* transmission and reflection coefficients smaller than one for the beam splitter, potentially incorrect retardances for the wave plates, misalignment of these wave plates, and so on), also the wavelength dependence of their transmittance at $\theta = 0$ is modeled as follows:

$$T(\lambda, 0) = \frac{1 + B\cos(2\pi\lambda/\,\mathrm{FSR} + \Phi + 4\phi)}{2}, \tag{22}$$

where, as with the Lyot elements, B is the contrast and Φ is the relative phase, and both vary across the aperture. As previously mentioned, this simple transmittance model ignores any potential I-ripple (see Section 2.4).

In HMI, the peak transmittances of the two Michelson interferometers are expected to coincide for $\phi_2 = 2\phi_1$, where the index 1 refers to the Wide-Band (WB) Michelson and 2 refers to the Narrow-Band (NB) Michelson. This relation stems from $\Delta_2 = 2\Delta_1$, which is so because the nominal FSR of the WB Michelson (344 mÅ) is twice that of the NB Michelson (172 mÅ).

The following relation gives the change $[\delta\lambda]$ in central wavelength of the transmission profile as a function of the change $[\delta\phi]$ in the angle of the tuning half-wave plate:

$$\delta\lambda = \frac{\mathrm{FSR}}{\pi} 2\delta\phi. \tag{23}$$

The *modus operandi* to cotune the instrument is that the NB Michelson will be tuned by $\delta\phi = 36°$ steps (*i.e.* 24 Hollow-Core Motor [HCM] steps: one HCM step is 1.5°). The WB Michelson will be tuned by $\delta\phi = 18°$ (or 12 HCM steps). Finally, E1 will be tuned by $\delta\phi = 9°$ (or 6 HCM steps). Therefore, the tuning positions, *i.e.* sampling positions, will nominally be spaced by 68.8 mÅ.

2.2.3. *Free Spectral Range*

For a normal ray ($\theta = 0$) the FSR of a Michelson interferometer can be approximated as

$$\mathrm{FSR} \approx \frac{\lambda^2}{\left|\Delta(\theta) - \lambda\frac{d\Delta(\theta)}{d\lambda}\right|}. \tag{24}$$

We can further simplify this equation by assuming that $\Delta(\theta)$ does not depend on λ, and we obtain

$$\mathrm{FSR} \approx \frac{\lambda^2}{|\Delta(\theta)|}. \tag{25}$$

The FWHM, as for the Lyot elements, is equal to

$$\mathrm{FWHM} \approx \frac{\mathrm{FSR}}{2}. \tag{26}$$

2.2.4. *Thermal Compensation*

The Michelson interferometers are also thermally compensated: the vacuum leg is maintained with temperature-compensating copper standoffs. The "Stonehenge" design of the copper spacers reduces stress. This temperature compensation produces, to lowest order, a theoretical zero sensitivity to temperature variation.

2.3. Front Window and Blocking Filter

The blocking filter is an interference filter made of two glass substrates and three cavities acting like Fabry – Pérot interferometers and located in between the substrates. Its nominal FWHM is 8 Å, and it is centered close to the target wavelength (the center depends on the distribution of angles of incidence and of the temperature). The front window is made of three blocks of glass: a 6-mm thick Schott BK7 glass block, a 3-mm thick colored Schott GG495 glass block, a coating, and another 6-mm thick Schott BK7 glass block. Unwanted interferences arise inside the front window and blocking filter: they are engendered in the glass blocks, which, because of partial reflections at their interfaces, also act like weak Fabry – Pérot interferometers. These interferences produce a fringe pattern visible on HMI filtergrams and which needs to be characterized. These fringes contaminate the phase and contrast maps of the other filter elements as determined by the calibration procedures, and they move with temperature. This will be a problem after the regular eclipses that HMI will be subjected to due to the geosynchronous orbit of SDO: after such events, the spatially averaged temperature and the temperature gradient on the front window will need a certain time to stabilize.

2.3.1. *Time Delay*

In a Fabry – Pérot interferometer placed in vacuum, the phase difference $[\Delta\Phi]$ between two consecutive reflections in a glass block of thickness d is

$$\Delta\Phi(\lambda,\theta) = \frac{2\pi}{\lambda} 2nd\cos(\theta), \tag{27}$$

where θ is the angle of incidence of the rays measured inside the cavity and n is the refractive index of the glass between the two reflecting surfaces. The time delay $[\Delta t]$ is the optical-path difference $[\Delta(\theta)]$, with $\Delta(\theta) = 2nd\cos(\theta)$, divided by the speed of light.

2.3.2. *Wavelength and Angular Dependences*

The transmittance $[T(\lambda,\theta)]$ of a Fabry – Pérot cavity where the two surfaces have a reflectance R is given by

$$T(\lambda,\theta) = \frac{(1-R)^2}{1+R^2-2R\cos(\Delta\Phi(\lambda,\theta))}, \tag{28}$$

Neglecting the change in n with wavelength and using Equation (27), the Fabry – Pérot cavity placed in vacuum has the following angular dependence, expressed in terms of wavelength change $\delta\lambda$:

$$\delta\lambda \approx \lambda\left(\sqrt{1-\left(\frac{1}{n}\right)^2 \sin^2(\theta')} - 1\right), \tag{29}$$

where θ' is the angle of incidence measured outside the cavity (the Snell – Descartes law states that $n\sin(\theta) = \sin(\theta')$).

The blocking filter operates as a Fabry – Pérot cavity that is crossed by a beam of rays of light covering a range of θ' angles. Also, following the recommendation of the blocking filter manufacturer, the Andover Corporation, the overall filter angular dependence can be approximated as (see www.andovercorp.com):

$$\delta\lambda \approx \frac{\lambda}{2}\left(\sqrt{1 - \left(\frac{n_e}{n}\right)^2 \sin^2(\theta')} - 1\right), \tag{30}$$

where θ' is now the maximum angle of incidence, λ is the central wavelength of the transmission profile at $\theta' = 0$, n is the effective refractive index of the filter cavity, n_e is the refractive index of the external medium, and Andover Corporation calculated $n_e/n = 1/2.05$. Compared to Equation (29), the factor $1/2$ is used to approximate the impact of a beam *versus* a single ray, while the ratio n_e/n expresses the fact that the external medium may have a refractive index different from 1. This equation is a good approximation for a beam of light with a maximum incidence angle $\theta' < 20°$, which is exactly the kind of beams crossing the blocking filter in HMI.

2.3.3. *Free Spectral Range*

From Equation (27), the FSR of a Fabry – Pérot cavity is, in a first approximation (neglecting the change in n with wavelength):

$$\mathrm{FSR} \approx \frac{\lambda^2}{2nd\cos(\theta) + \lambda}. \tag{31}$$

There is an implicit temperature dependence, as, for instance, the thickness of the glass block varies with T_e.

2.3.4. *Thermal Compensation*

There is no specific thermal-compensation mechanism in the front window and blocking filter.

2.4. I-ripple

When taking a sequence of filtergrams in white light (*e.g.* with a lamp as source) we observe a variation in the transmitted intensity measured on the HMI CCDs from one sequence position to another. This variation is called an I-ripple. A small amount of I-ripple, expressed as the relative peak-to-peak difference of spatially averaged intensities measured on the CCDs, is inevitable even for a perfect instrument: the integral values of the transmission profiles of the sequence positions are different, mainly because of the presence of the blocking filter and front window (which limit the wavelength range). In the case of a perfect instrument with a blocking filter centered at the target wavelength and with a Gaussian transmission profile, the amount of I-ripple expected is of the order of a couple of tenths of a percent, at most.

Unfortunately, imperfections in the instrument increase this I-ripple. Possible causes of an I-ripple for a Lyot element include: error in the retardance of the half-wave plate; misalignment of the half-wave plate (which is difficult to detect and is the most likely cause of an I-ripple produced by E1); misalignment of both the entrance polarizer and the exit quarter-wave plate, misalignment of the entrance polarizer and retardation error in the tuning half-wave plate, and so on. Possible causes of an I-ripple for a Michelson interferometer include: a misalignment of both the entrance polarizer and the exit quarter-wave plate, an imperfection in the polarizing beamsplitter (*e.g.* polarization leaks), and a misalignment of the exit quarter-wave plate, a misalignment of the entrance polarizer and a retardation error in the tuning half-wave plate, and so on.

Jones calculus applied to a tunable Lyot element (A.M. Title, 2009, private communication) shows that its transmittance $T(\lambda, 0)$ in the presence of an imperfect half-wave plate can be modeled, in a second-order approximation, as

$$T(\lambda, 0) = K_0 \frac{1 + \cos(2\pi\lambda/\mathrm{FSR} + 4\phi)}{2} + \frac{1}{2}\left(K_1 \cos(2\phi) + K_2 \sin(2\phi)\right)^2. \tag{32}$$

where the constants K_1 and K_2 are related to, respectively, the retardance error and the misalignment (tilt) of the half-wave plate, while $K_0 = 1 - K_1^2/2 - K_2^2/2$. The transmitted intensity $[I]$ at a given position of a white-light sequence in which only the Lyot element is tuned can be approximated as

$$\frac{I}{\bar{I}} = K_0 + \left(K_1 \cos(2\phi) + K_2 \sin(2\phi)\right)^2, \tag{33}$$

where $\bar{I}$ is the transmitted intensity averaged over a period of the tuning angles ϕ. To account for other imperfections in the tunable Lyot element, we could generalize its transmittance as

$$T(\lambda, 0) = K_0 \frac{1 + \tilde{B} \cos(2\pi\lambda/\mathrm{FSR} + 4\phi + \Phi)}{2} + \frac{1}{2}\left(K_1 \cos(2\phi) + K_2 \sin(2\phi)\right)^2, \tag{34}$$

where Φ is the relative phase, and $\tilde{B}$ is related to the contrast B of the element. Computing the Jones matrices characterizing defects, other than an imperfect half-wave plate, in the Lyot element E1 and in the Michelson interferometers shows that Equation (33) is a very good approximation for I-ripples produced in most cases studied, but Equation (34) is valid only for a Lyot element with an imperfect half-wave plate and is a more-or-less good approximation in all other cases.

3. Measured Wavelength Dependence

3.1. Calibration Hardware

Until November 2007, calibration tests were performed in a cleanroom at the Lockheed Martin Solar and Astrophysics Laboratory (LMSAL) facilities, in Palo Alto (California), then at the NASA Goddard Space Flight Center (Maryland), and finally after July 2009 at Astrotech Space Operations (Florida). Different light sources were used to feed HMI: the Sun through a heliostat, a dye laser (massive tunable laser using organic dye), a tunable infrared laser operated around 1234 nm (frequency-doubled using a periodically poled lithium-niobate crystal, and assembled at Stanford University), and a lamp to provide white light (with a flat spectrum in the narrow wavelength range of interest). A stimulus telescope (the reversed HMI telescope) was used with the laser and lamp sources, so that the rays of light entering HMI have angles and trajectories similar to the ones HMI will encounter once SDO is launched. A diffuser was added when using the laser sources. Several field and aperture stops were also used. The field stop most often utilized was a solar-size stop (24.6 mm in diameter): placed at the focal plane of the stimulus telescope, it roughly provides the same incidence-angle distribution as the Sun. For the angular-dependence test, a small field stop (5 mm in diameter) was used. Calibration tests were performed both in air and in vacuum. For the tests performed in air, an air-to-vacuum corrector was sometimes added to the system and the stimulus telescope was despaced. In order to keep the CCDs cold and thereby reduce the dark current, a vacuum chamber was needed for some tests (cooling down in air is not possible due to condensation). Among other impacts, the phases [Φ] of the Michelson interferometers differ in air and vacuum. Before the instrument was fully assembled, isolated elements of the optical-filter system were also tested separately at LMSAL using additional devices, *e.g.* spectrographs. Table 3 summarizes the main calibration tests performed on the assembled instrument, and their configuration.

3.2. Detune and Cotune Sequences

Characterizing the wavelength dependence of all of the filter elements requires the taking of a great many filtergrams. These filtergrams have to be ordered in special sequences selected for the amount of information they provide. Two kinds of sequences were frequently used: detune and cotune sequences. A detune sequence is a series of filtergrams taken with the Lyot tunable element E1, the NB Michelson, and the WB Michelson out of phase: the tuning phases applied to these elements are such that their peak transmittances do not necessarily coincide. Conversely, a cotune sequence is a sequence in which the three tunable elements are in phase and their peak transmittances coincide. Cotune sequences are used for the normal HMI operations to produce the observables (Doppler velocity, line-of-sight [l.o.s.] magnetic field, full vector magnetic field, and continuum intensity) while detune sequences are used only for calibration purpose. The detune sequence most often applied throughout the HMI calibration phase is a 27-position sequence in which the tuning phases (the 4ϕ of Equation (22)) are either 0, 120, or 240° (the total number of variations with repetition of three elements at three positions is 27, hence the length of the sequence). This choice of phases separated by 120° stems from the fact that the sum of the cosines of a set of equally spaced fractions of 360° is zero, but for a given tunable element, three tuning angles (*i.e.* a separation of 120°) is the minimum number needed to derive the three unknowns (contrast and phase of this tunable element, and average value of the rest of the transmission profile over the wavelength range).

When the light source emits in a very narrow wavelength range, like the dye laser, this detune sequence allows for an analytic determination of the phases and contrasts of the Michelsons and E1. No fitting procedure is required. This is indeed the only method that we used to determine the contrasts $B(x,y)$ of the tunable elements, at the same time as their phases $\Phi(x,y)$. When using the Sun as the light source, it becomes necessary to fit for the solar Fe I line profile (here conveniently modeled as a Gaussian) concurrently with the phases of the tunable elements. The central wavelength of the solar line is derived from an ephemeris returning the Sun–Earth velocity. With sunlight, it is not possible to obtain the contrasts of the elements, because they are partly degenerate with the ratio of the line depth over the continuum intensity. Finally, when using the lamp as light source, the detune sequence allows for a characterization (in terms of Fourier coefficients) of the interference fringes produced by the front window and blocking filter.

To check whether or not the tuning polarizer in between the two Michelson interferometers is working properly, we added four positions to the 27-position detune sequence (see

Table 3 Summary of main ground-test configurations (tests performed on assembled instrument only). OBSMODE is denoted by O, CALMODE by C.

Test	Mode	Light source	Hardware used
wavelength dependence of tunable elements	O/C	dye laser with diffuser	stimulus telescope 24.6-mm field stop in air or vacuum
wavelength dependence of tunable elements	C	heliostat	in air or vacuum
wavelength dependence of non-tunable elements	O/C	dye laser with diffuser	stimulus telescope 24.6-mm field stop wave and power meters in air or vacuum
interference fringes	O/C	lamp	stimulus telescope 24.6-mm field stop
I-ripples total and individual	O/C	lamp	stimulus telescope 24.6-mm field stop
temperature dependence of tunable elements	O/C	dye laser with diffuser	stimulus telescope 24.6-mm field stop in vacuum
thermal stability of tunable elements	O/C	dye laser with diffuser	stimulus telescope 24.6-mm field stop in vacuum
angular dependence of tunable elements	O	dye laser with diffuser	stimulus telescope aperture stop (PCU) 24.6-mm field stop in air or vacuum
angular dependence of tunable elements	C	dye laser with diffuser	stimulus telescope 5-mm field stop in air or vacuum
throughput	C	heliostat	power meter in air or vacuum
tuning polarizer check	O/C	dye laser with diffuser	stimulus telescope 24.6-mm field stop in air or vacuum
artifact check	NA	lamp	stimulus telescope 24.6-mm field stop in air or vacuum

Section 4.4). These four positions correspond to four different settings of the HCM of the tuning polarizer.

Close to one thousand detune sequences (with either 27 or 31 positions) have been taken since December 2005.

3.3. Wavelength Dependence Derived from Ground Calibration Tests: Tunable Part of HMI

This is arguably the most important part of the wavelength-dependence calibration, because the knowledge of the relative phases of the tunable elements is key to co-tuning HMI and obtaining high-quality data.

3.3.1. Phase and Contrast Maps

Tests with Laser Using the dye laser as the light source and taking a detune sequence allows for the production of phase and contrast maps of the three tunable elements (NB and WB Michelsons, and E1). Figures 5 and 6 show an example of results obtained in CALMODE and OBSMODE.

In OBSMODE, the phase gradient across the tunable elements is less than $\approx 25°$, which is a high degree of spatial uniformity: for the Michelsons, the relative phase ranges are 16° and 24°, while for E1 it is 16°. The contrasts of the Michelsons and E1 are high (respectively

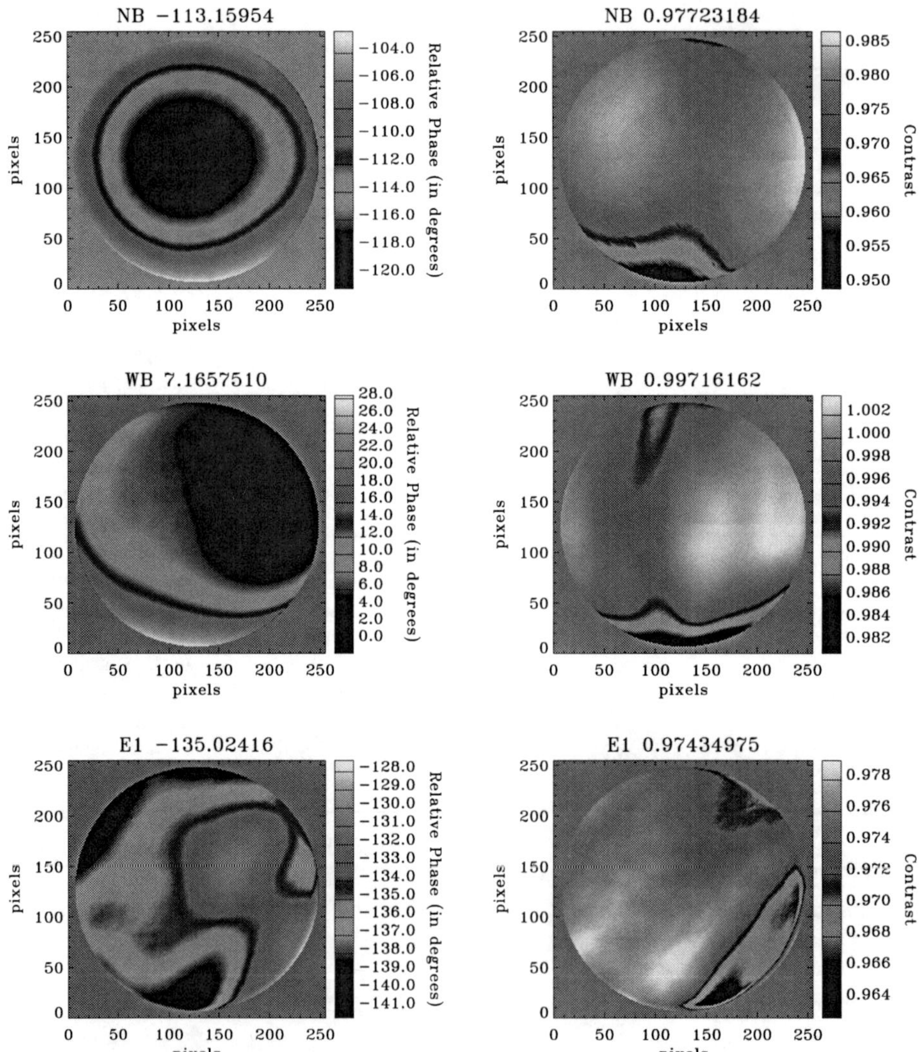

Figure 5 Relative phase (left column) and contrast (right column) maps of the tunable elements obtained in CALMODE with the dye laser as light source in October 2007. The spatially averaged phases (in degrees) and contrasts are written next to the element name on each panel.

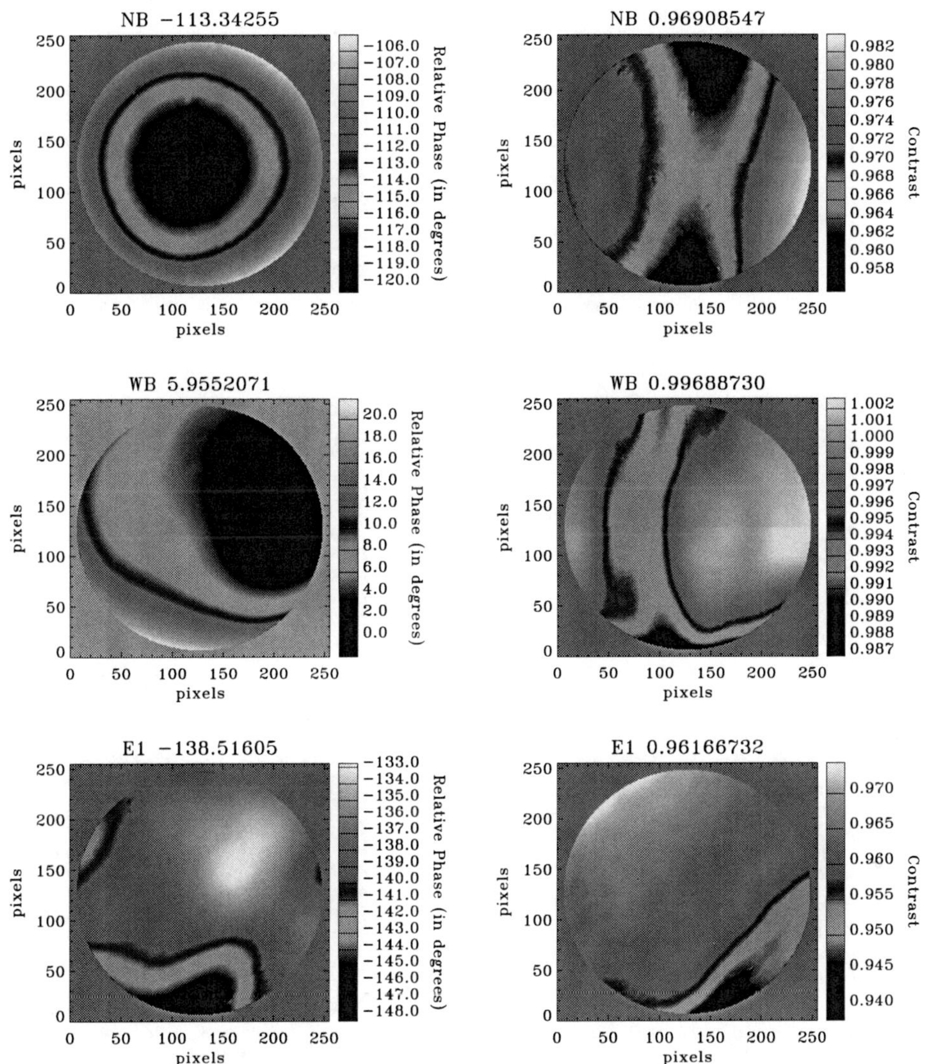

Figure 6 Relative phase (left column) and contrast (right column) maps of the tunable elements obtained in OBSMODE with the dye laser as light source in October 2007. The spatially averaged phases (in degrees) and contrasts are written next to the element name on each panel.

0.969, 0.996, and 0.962 on average). For comparison, the two Michelson interferometers M1 and M2 in the SOHO/MDI instrument had contrasts that were, on average, 0.92 and 0.98, respectively. Similarly, the range of relative phases for these MDI Michelsons were 17.9° and 45.4° (for sunlight results). Therefore, taken as a whole, the Michelson interferometers of HMI are more spatially uniform than those of MDI and have better contrasts.

Both OBSMODE and CALMODE sets of maps are very similar, even though the average values are not identical (see Section 3.3.3). A comparison of phase maps obtained from data taken by the front camera with data taken by the side camera shows no systematic difference for the WB Michelson and the Lyot element E1. However, there seems to be a small

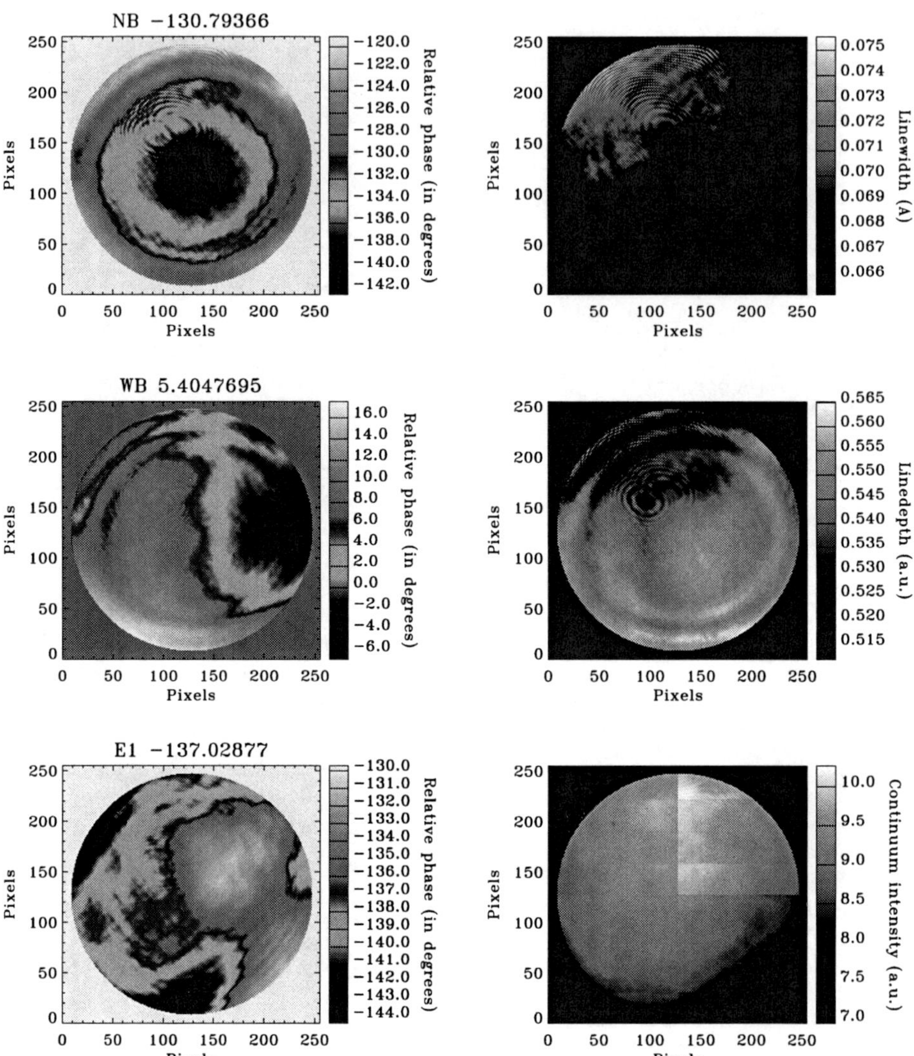

Figure 7 Left: relative phase. Right: line-width, solar-continuum intensity, and line-depth maps obtained with sunlight through a heliostat, in CALMODE, in September 2007. The spatially averaged phases (in degrees) are written next to the element name on the panels of the left column.

systematic difference for the NB Michelson (about 0.7° in the sense front camera minus side camera). Detune sequences will be taken during the commissioning phase (the first two months after SDO is launched) to check whether or not this difference is still present. In any case, such a difference is not an issue, as it does not hamper our ability to properly tune the instrument, and each camera is expected (at least most of the time) to be dedicated to a specific observable category: the one camera to the l.o.s. observables, the other camera to the full vector magnetic field.

Figure 7 shows phase maps obtained with sunlight data, in CALMODE. The fringe patterns produced by interferences in the front window and blocking filter are clearly visible on the line-width and line-depth maps. As previously mentioned, no contrast maps can be derived from sunlight data, but the solar continuum intensity, solar line width, and solar line depth are fitted at the same time as the relative phases. For the solar line, we need an analytical profile as simple as possible to minimize the number of parameters to fit, and to avoid degeneracies between these parameters. Hence the adoption of a basic Gaussian profile, neglecting any asymmetry present in the actual Fe I line. However, this creates a small systematic error of the phases of the tunable elements, which also translates into a systematic (zero-point) error on the Doppler-velocity determination. This systematic error depends on the velocity at the solar surface and on the position on the solar disk. At disk center, simulations based on a MDI-like algorithm with six sampling positions and averaging the Doppler velocities returned by the first and second Fourier coefficients and by the left and right circular polarizations (see Section 4.5 for details), provide an estimate of the average systematic error of about 80 $\mathrm{m\,s^{-1}}$, with a change of about 20 $\mathrm{m\,s^{-1}}$ across the velocity range $[-6.5, +6.5]$ $\mathrm{km\,s^{-1}}$ (the systematic error decreases with increasing velocities).

The non-linear least-squares fit algorithm we apply to fit the relative phases and solar-line parameters is a basic Gauss – Newton algorithm that was extensively tested on artificial data. The relative-phase maps obtained in sunlight and laser light are very similar. In sunlight we only work in CALMODE, because in this mode each location on the front window receives light from the entire – integrated – Sun: therefore, the solar rotation and oscillation velocities cancel out and the solar-line central wavelength is, to a first approximation, only shifted by the Sun – Earth velocity, unlike OBSMODE images for which the value of the solar wavelength is unknown. The solar-limb effect (a redshift of the solar line close to the limb; see *e.g.* Dravins, 1982) and the solar convective blueshift (a blueshift of the solar line, especially at disk center; see *e.g.* Dravins, 1982) both contribute to the l.o.s. Doppler velocity. Due to the dependence of these effects on the resolution of the instrument, it is difficult to precisely quantify what their impact is on HMI data, especially for ground-based data sensitive to the additional impact of atmospheric seeing. However, neglecting these two effects when calculating the spatially averaged velocity on the solar disk in CALMODE creates a systematic error (fixed in time) on the relative phases, and therefore a zero-point error on the Doppler-velocity maps.

The average phases shown on Figures 5 and 7 are not identical for sunlight and laser light because the laser wavelength is not known accurately (the wavemeter was often off, by as much as a few hundred mÅ), the Fe I solar line is approximated by a Gaussian profile only shifted by the Sun – Earth velocity, and the Michelson interferometer phases drifted in between the detune sequences shown here (taken more than a month apart) as is further explained in the next section. Finally, any residual differences in the phase maps between sunlight and laser light are likely due to the difference in illumination between these two light sources.

To illustrate the accuracy of our transmittance models of the optical-filter elements, we reconstructed a spatially averaged detune sequence taken in sunlight and CALMODE. Based on the relative phases, continuum intensity, solar line width, and line depth returned by the fitting code, we calculated at each position of the detune sequence the expected solar intensity (reconstructed intensity). The comparison between reconstructed and original sequences shows that our transmittance models (Equations (12) and (22)) manage to accurately reproduce the data: the maximum relative difference between reconstructed and actual intensities is less than 0.9% and there is no systematic trend in this difference over the sequence.

3.3.2. Drifts of the Phases of the Michelson Interferometers

The relative phases $\Phi(x, y)$ of the Michelson interferometers drift with time. A comparison of these phases obtained during calibration tests in September and October 2007 (both with HMI in the vacuum chamber, with the same oven temperature, and in Sunlight) shows a drift as large as 0.43° a day for the NB Michelson, and 0.15° a day for the WB Michelson. Other analyses based on dye-laser light returned somewhat smaller drifts (about 0.2° a day for the NB Michelson). In any case, these results warrant regular calibration runs once SDO is launched, to make sure that the cotune sequence we use is still optimal.

For comparison, since the launch of SOHO, the phases of the MDI Michelson interferometers have drifted by 218° (for M1) and 43° (for M2) (as of February 2009). The drift rates were much higher immediately after launch than they are now. The tentative explanation for these drifts is that the thickness of the glue holding the mirrors in the Michelsons varies with time, thus changing the optical-path difference.

3.3.3. Phase Difference between CALMODE *and* OBSMODE

We observed a systematic difference in the relative phases of the tunable elements obtained in OBSMODE and CALMODE, especially for E1 (the spatially averaged relative phases differing by about 3° between these two modes) and, to a lesser extent, for the Michelsons. Figure 8 shows these phase differences obtained between two detune sequences taken in laser light in October 2007, for the three tunable elements. The origin of this phase difference is not fully understood. A discrepancy in illumination of the imaged object (whether the front window in CALMODE or a field stop in OBSMODE) and, more significantly, the change in the angle-of-incidence distribution between OBSMODE and CALMODE play a role. A raytrace of HMI obtained with the Zemax software (developed by ZEMAX Development Corporation, www.zemax.com) shows that the rays of light reaching a given point on a HMI CCD are located in a cone during their propagation through the instrument. The diameter of this cone, and, therefore, the distribution of angles of incidence of rays, is different in CALMODE and OBSMODE. For instance, at the quarter-wave plate of E1, the Zemax model shows that the maximum angle of incidence of sunrays is 0.9737° in CALMODE, while it is 1.0270° in OBSMODE (with the first focus block). Since the angular dependence of the tunable elements is non-zero, this small difference between angles of incidence in OBSMODE and CALMODE impacts the relative phases (see Section 3.7).

Simulations show that the observed phase difference introduces an almost-constant (peak-to-peak variation of $5\ \mathrm{m\,s^{-1}}$) systematic error in the solar oscillation Doppler-velocity determination (zero-point error) because this Doppler velocity is computed based on phase maps obtained in CALMODE, while the observations are performed in OBSMODE. Indeed, by applying the MDI-like algorithm with six sampling positions and averaging the Doppler velocities returned by the first and second Fourier coefficients and by the left and right circular polarizations (see Section 4.5 for details), we estimate that this systematic error varies between ≈ 26 and $\approx 31\ \mathrm{m\,s^{-1}}$ in the velocity range $[-6.5, +6.5]\ \mathrm{km\,s^{-1}}$. The impact on the rms variation (produced by photon noise) of this Doppler velocity is hardly noticeable across the velocity range.

3.3.4. Free Spectral Range Calibration

The FSRs of the Lyot elements and Michelson interferometers were measured using spectrographs and collimated light, before HMI was assembled. It is possible to determine the FSRs

Figure 8 Relative phase difference CALMODE minus OBSMODE for two detune sequences taken in vacuum and with laser light in October 2007. Upper panel: NB Michelson; middle panel: WB Michelson; lower panel: Lyot element E1.

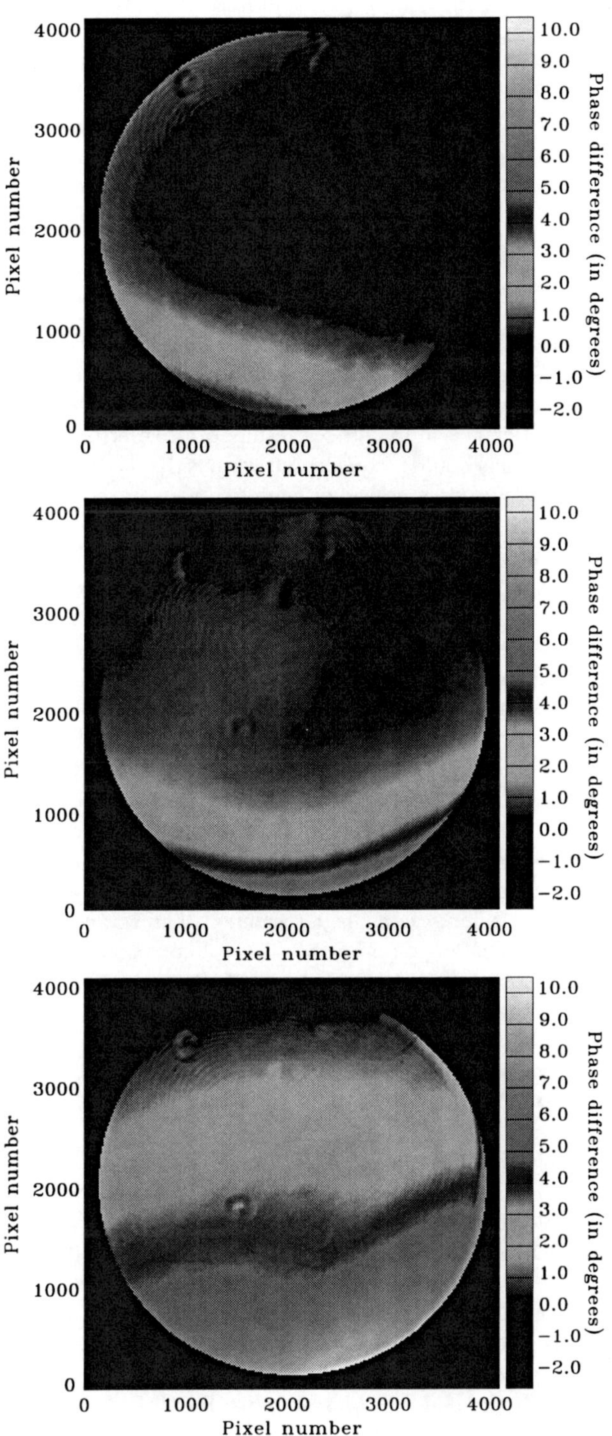

Table 4 Full Spectral Ranges of the Michelson interferometers and Lyot elements. FSR_n are the nominal FSRs, and FSR_m are the measured ones (measured by the manufacturer for the Michelson interferometers, and individually measured at LMSAL for the Lyot elements), in mÅ at 6173 Å.

Element	FSR_n	FSR_m
NB	172	172
WB	344	343.1
E1	690	693
E2	1380	1407
E3	2758	2779
E4	5516	5682
E5	11 032	11 354

of the tunable elements for the proper distribution of incidence angles with a greater precision, for instance by taking two detune sequences at different dye-laser wavelengths. From these two sequences, two sets of relative phases [Φ] of the tunable elements are derived. Any difference in the phases results from an error on the FSRs. We applied this method, taking many detune sequences at various dye-laser wavelengths and averaging the resulting FSRs. However, it proved difficult to obtain an accurate measurement of these FSRs on the ground, because, among other issues, the dye-laser wavelengths were not always stable. The laser intensity was fluctuating, and there was a generally high level of noise. We derived the following, possibly underestimated, estimates of the FSRs: 0.1709 Å for the NB Michelson (instead of the nominal 0.1720 Å), 0.3419 Å for the WB Michelson (instead of 0.3440 Å), and 0.6935 Å for E1 (instead of 0.6900 Å). These values are well within the specifications. Measurements of these FSRs at LMSAL while the Michelsons and E1 were outside of the instrument – and therefore for a different distribution of angles of incidence – returned: 0.1724 Å for the NB Michelson and 0.6930 Å for E1. The FSR of the WB Michelson was not measured at LMSAL, but reports from the manufacturer, LightMachinery, give an estimate of 0.3431 Å (and, for information, the same manufacturer reports 0.172 Å for the FSR of the NB Michelson). The FSRs of the non-tunable elements (E2 to E5) were measured at LMSAL while they were outside of the instrument, and are shown in Table 4.

After launch, we plan to use the orbital velocity of SDO to determine the FSRs of the tunable elements with a better accuracy than was obtained from the ground. For instance, a 1 $m\,s^{-1}$ accuracy on the Sun – SDO radial velocity translates into an accuracy of 0.02 mÅ on the solar line central wavelength, far better than the accuracy on the dye-laser wavelength (of the order of 10 mÅ).

3.4. Wavelength Dependence Derived from Ground Calibration Tests: Non-Tunable Part of HMI

Taking detune sequences at different dye-laser wavelengths spanning a wide enough wavelength range allows for a determination of the relative phase and contrast maps of the non-tunable elements. Examples of such maps are shown on Figures 9, 10, 11, and 12, in OBSMODE and CALMODE. They were obtained from detune sequences taken in vacuum at 30 different laser wavelengths in October 2007, ranging from 6167.5 to 6179.0 Å.

The determination of these phase and contrast maps is difficult because the intensity levels on the CCDs drop quickly for wavelengths far from the target one. These wavelengths are needed for an accurate computation of the characteristics of the elements with the largest FWHMs (E4 and E5). Moreover, the FWHM of E5 (about 5.7 Å) is relatively close to the one of the blocking filter (about 8 Å), which complicates the determination of its phases and contrasts. Thus the center of the blocking filter transmission profile needs to be known to

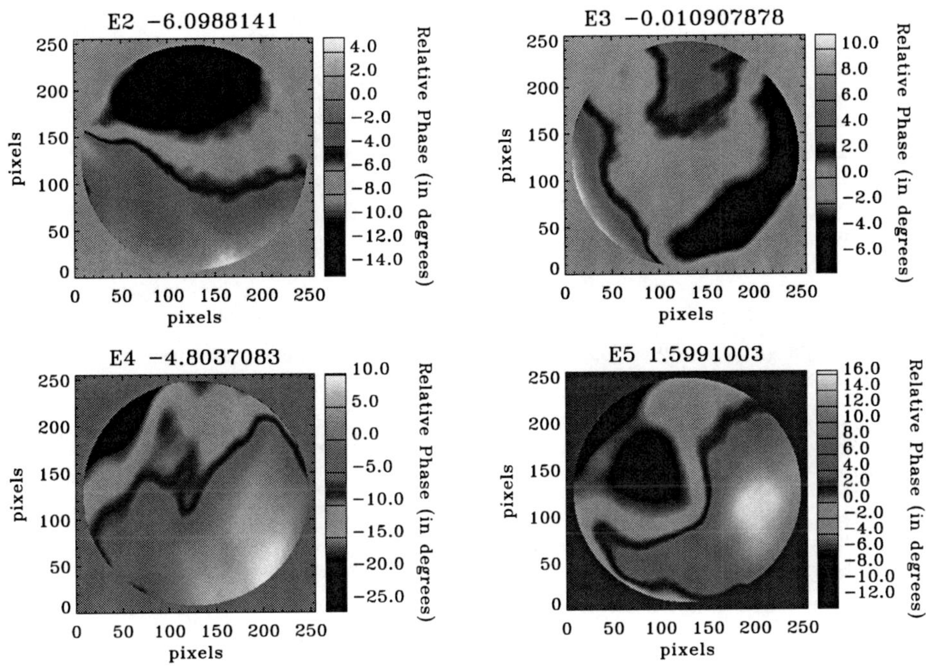

Figure 9 Relative phases of the non-tunable elements, measured of 15 October 2007: OBSMODE images. The blocking filter center is at ≈ -1 Å from target wavelength. The average phases (in degrees) are written next to the element name on each panel.

avoid a degeneracy with the peak transmittance of the profile of E5 (and, to a lesser extent, E4). To estimate the center location, we used three transmission profiles of the blocking filter provided by the Andover Corporation at three different temperatures and two angles of incidence. We assumed a quadratic dependence on temperature of the blocking filter wavelength shift [$\delta\lambda$], and an angular dependence following Equation (30). An uncertainty of a few hundred mÅ on the center of the blocking filter is not an issue as the fitting code computing the phases and contrasts compensates by shifting the average phase of E5 (and to a lesser extent E4) so that the goodness of fit and the non-tunable transmission profile remain essentially unchanged. For a proper fitting, the laser wavelengths and the fluctuations in laser intensity need to be accurately known. The fitting procedure, a Gauss–Newton algorithm, favors the high intensities obtained with wavelengths close to the target one. This results in a poor determination of the contrasts. To improve this determination, we added a weighting scheme to our code. The weights [W] are set to $W = 1/\sqrt{1+KI}$ where $K = 256$ and I is the intensity (I is the value in DNs from the filtergram, normalized by a factor 10 000 and corrected to account for changes in the laser intensity and exposure time between the different detune sequences): we multiply the data intensities by these factors. The weights were chosen to better represent the actual errors of the intensities, which have photon, gain, and constant noise. Several values of K were tested before selecting $K = 256$. This is the lowest value for which the spatially averaged contrasts on the non-tunable elements are lower than, or equal to, 1 (instead of average values as high as 1.2). However, locally some fitted contrasts are still unrealistically larger than one. The results of Figures 9, 10, 11, and 12, show that the non-tunable elements have a good degree of spatial uniformity in terms of phases $\Phi(x, y)$: the range is $\approx 19°$ for E2, $\approx 20°$ for E3, $\approx 36°$ for E4, and $\approx 30°$ for E5.

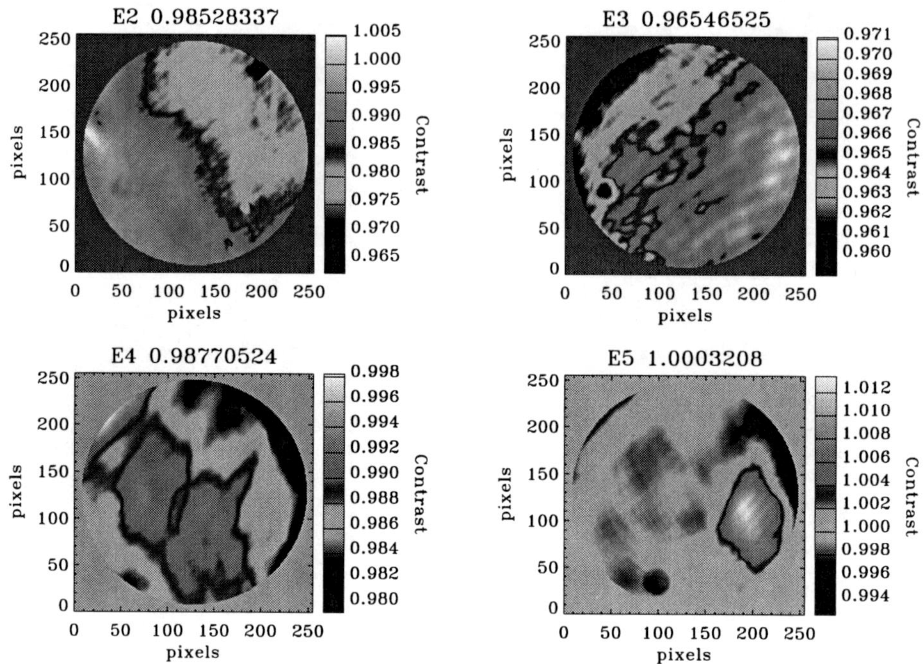

Figure 10 Contrasts of the non-tunable elements, measured 15 October 2007: OBSMODE images. The blocking filter center is at ≈ -1 Å from target wavelength. The average contrasts are written next to the element name on each panel.

These elements also have high contrasts (on average, larger than 0.96). The phase maps are very similar in OBSMODE and CALMODE. Other measurements of the phase maps of the non-tunable elements were performed at LMSAL in October 2005 while these elements were outside of the instruments. These data were obtained in a collimated light beam for which all of the rays are normal ($\theta = 0$). Despite this difference, the same features are visible on both sets of maps, and they display relatively similar phase gradients after we convolve the LMSAL maps with a mask to simulate the impact of the proper light distribution at an image point (except for the element E2 for which there are some noticeable differences).

The spatially averaged transmission profile of the non-tunable part of HMI is shown on Figure 13 (black line). This profile can be compared to measurements obtained in October 2006 with lamp light and a spectrograph (red line). During these measurements, performed at LMSAL, the blocking filter was absent, the lamp had a background infrared intensity that contaminated the transmission profiles, and the actual profile was convolved by the resolution window of the spectrograph. E1 was tuned at 36 different positions with a polarizer. Because the light from the lamp was also slightly polarized, the resulting non-tunable transmission profile (obtained by averaging the 36 profiles) is contaminated by the E1 profile. Despite all these differences, both transmission profiles look rather similar (note that we shifted the red curve in wavelength so that it overlaps the black one). The maximum transmittance is located at about $+16$ mÅ from the target wavelength, and the FWHM of the non-tunable profile is 624.5 mÅ (slightly larger than the nominal value of 612 mÅ).

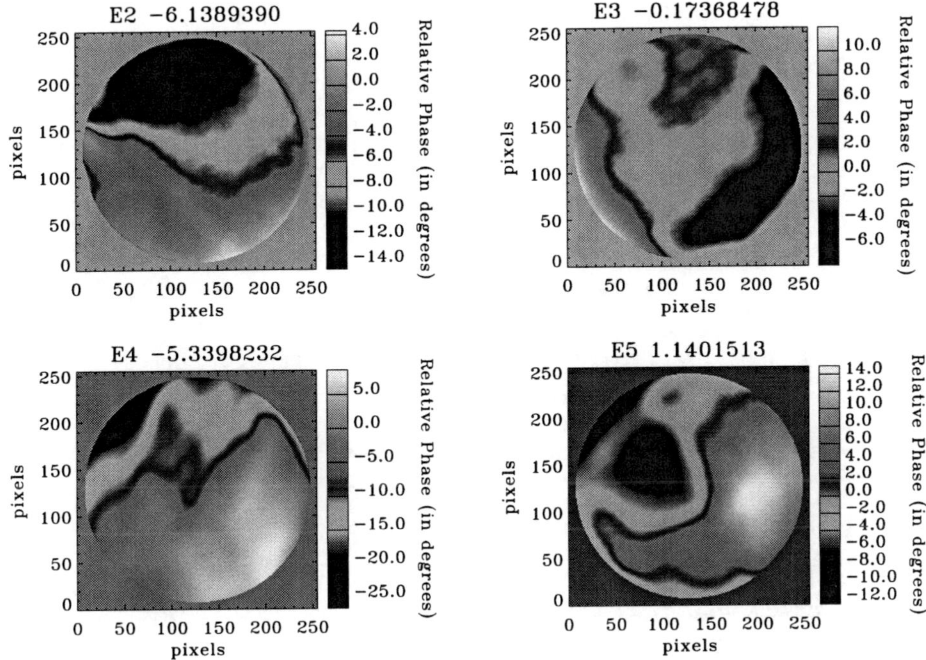

Figure 11 Relative phases of the non-tunable elements, measured 15 October 2007: CALMODE images. The blocking filter center is at ≈ -1 Å from target wavelength. The average phases (in degrees) are written next to the element name on each panel.

3.5. Interference Fringes Produced by the Front Window and Blocking Filter

The use of the lamp as light source while taking detune sequences allows for a characterization of the fringe patterns produced by the front window and blocking filter. In CALMODE, the front window is in focus, unlike the blocking filter. In OBSMODE, both the front window and blocking filter are out of focus. Therefore, a comparison between the results of CALMODE and OBSMODE detune sequences emphasizes the fringes produced by the front window only. Our fitting code expands the non-tunable transmission profile of HMI in the limited range $[-2\,\mathrm{FSR_{NB}}, +2\,\mathrm{FSR_{NB}}]$ (where $\mathrm{FSR_{NB}}$ is the FSR of the NB Michelson) into a Fourier series, and fits for the first seven Fourier coefficients, in addition to the mean value (zero frequency).

Indeed, at any position of a detune sequence, the intensity on the CCD is equal to the wavelength-independent lamp intensity times the integral (over the wavelength) of the transmittance of the non-tunable part of HMI multiplied by the three transmittances of the tunable elements. Here, these three transmittances are modeled following Equation (22). Applying trigonometric identities to separate the terms in $2\pi\lambda/\mathrm{FSR}$ from the terms in $\Phi + 4\phi$ (the tuning positions), and assuming that the FSR of the WB Michelson is twice the FSR of NB and the FSR of E1 is four times the FSR of NB (which is close to the actual values), we obtain an expression in which the total transmittance is expressed in terms of cosines and sines of $2\pi\lambda K/(4\,\mathrm{FSR_{NB}})$ where K ranges from 0 to 7, and in terms of tuning positions. Therefore, the transmittance of HMI at each detune position can be expressed as a combination of 14 Fourier coefficients (seven cosines and seven sines) of the non-tunable transmission

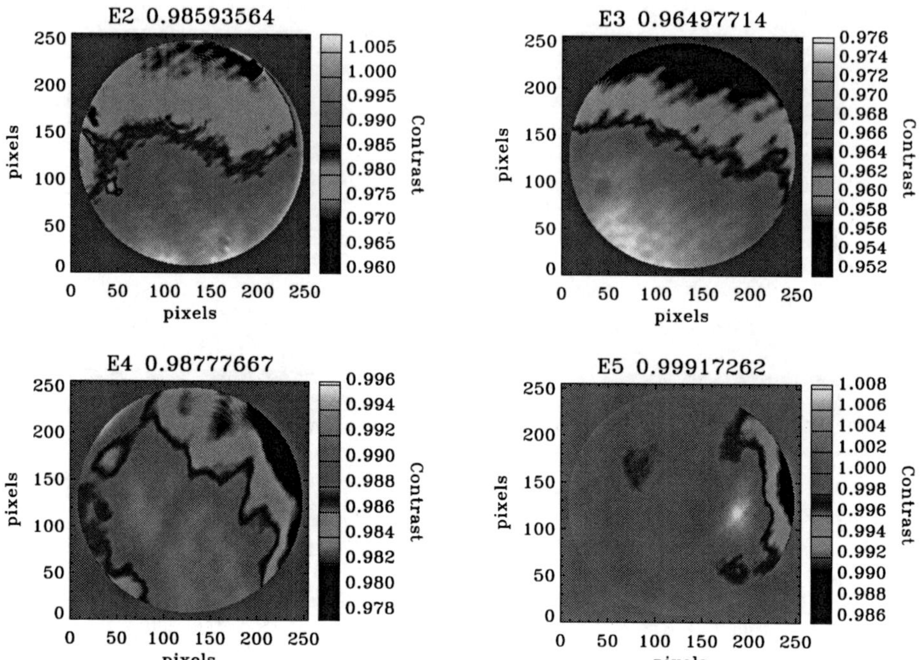

Figure 12 Contrasts of the non-tunable elements, measured 15 October 2007: CALMODE images. The blocking filter center is at ≈ -1 Å from target wavelength. The average contrasts are written next to the element name on each panel.

profile (in the range $[-2\,\mathrm{FSR_{NB}}, +2\,\mathrm{FSR_{NB}}]$), and of constants depending on the tuning positions. Since these constants can be calculated and the intensities on the CCD at each tuning position can be measured, it is theoretically possible to derive the seven Fourier coefficients of the non-tunable transmission profile, plus its mean value. The corresponding seven spatial frequencies f range from $1/(4\,\mathrm{FSR_{NB}}) \approx 1.462$ Å^{-1} to $7/(4\,\mathrm{FSR_{NB}}) \approx 10.234$ Å^{-1}. Figures 14 and 15 show the fringe patterns at some selected spatial frequencies expressed in terms of $\mathrm{FSR_{NB}}$, in OBSMODE and CALMODE. On these figures, the amplitude of the Fourier coefficients is very low for $f = 3/(4\,\mathrm{FSR_{NB}})$.

Because the interference-fringe patterns are connected to different glass-block thicknesses (see, *e.g.*, Equation (31)), we can theoretically deduce where the interferences occur in the front window and/or blocking filter. It is easier with the front window, composed of glass blocks of thicknesses 6, 3, and 6 mm. Partial reflections at the interfaces of these blocks can create fringes with the spatial frequencies $f = 1/\mathrm{FSR}$, with FSR expressed by Equation (31). The refractive index of the glass blocks is $n = 1.516$ at the target wavelength λ. The three thicknesses $[d]$ for which significant fringe patterns are observed (assuming normal rays) are 3, 6, and 9 mm (the respective spatial frequencies being $f = 2.39$, 4.77, and 7.16 Å^{-1}). Figure 16 shows the cosine and sine transform coefficients at these specific frequencies.

Both the front window and the blocking filter produce fringes. The main issue is that these fringes move with temperature, as was confirmed by thermal tests in which the front-window temperature, and (not at the same time) the oven temperature, were changed. The overall amplitude of the interference patterns does not seem to be affected by these temperature

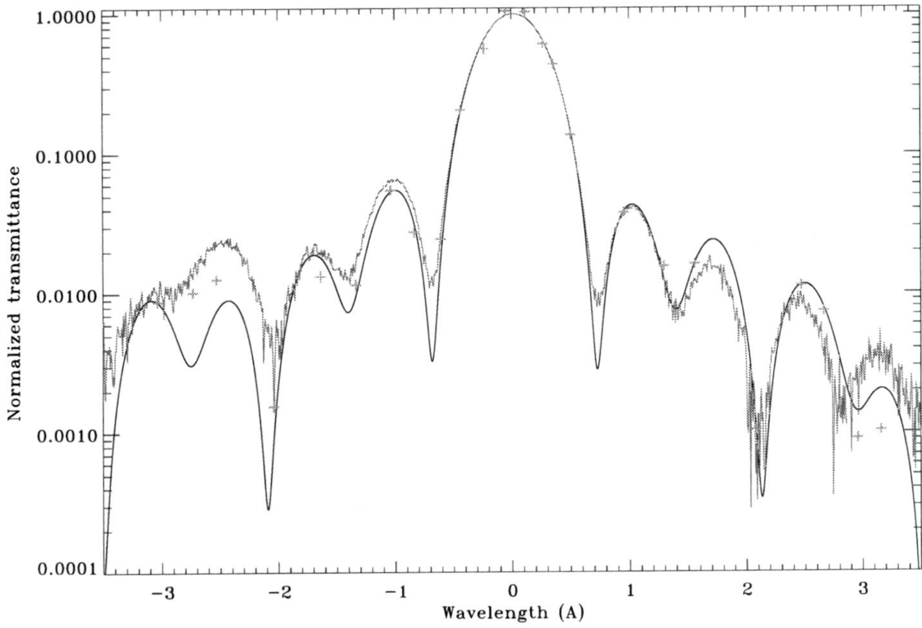

Figure 13 Spatially averaged transmission profiles of the non-tunable part of the HMI filter. Black line: model obtained by fitting for the phases and contrasts of the non-tunable elements, using detune sequences in laser light and CALMODE taken at different laser wavelengths in October 2007; green crosses: datapoints (each point is an average of the 27 intensities of the corresponding detune sequence); red line: measurements obtained with lamp light and a spectrograph while the Lyot filter was outside the instrument (the blocking filter was absent).

changes, but the phases of the patterns are. This is a problem after the regular eclipses affecting SDO, or any event changing the temperature gradient and/or temperature average on the front window.

3.6. I-ripple

As mentioned in Section 2.4, even with a perfect HMI optical filter the output intensity is expected to vary slightly from one filtergram sequence position to another. The resulting I-ripple is estimated at a few tenths of a percent, by calculating the theoretical intensities resulting from perfect transmittances of the different HMI elements (*e.g.*, phases equal to 0, contrasts equal to 1 for the Lyot elements and the Michelsons, and Gaussian profile centered at the target wavelength for the blocking filter), for all the 27 positions of a detune sequence. In July 2006, a significant I-ripple ($> 10\%$) was detected with detune sequences in white light. The root causes were misalignments in the Lyot element E1. This element was de-assembled, re-aligned, and then re-assembled. The overall I-ripple we derive from detune sequences, taken after re-assembly of the Lyot filter, in OBSMODE and CALMODE, is of the order of 1.9%, much lower than the previous I-ripple but still higher than what is expected for a perfect instrument. Figure 17 shows an example of I-ripple measurement obtained in vacuum in September 2007, on spatially averaged intensities of two detune sequences.

If we adopt Equation (33) as a model of the HMI individual I-ripples, the values of K_0, K_1, and K_2 can be fitted for each tunable element separately using a fine-tuning sequence taken in white light. A fine-tuning sequence is a sequence of filtergrams in which one tuning

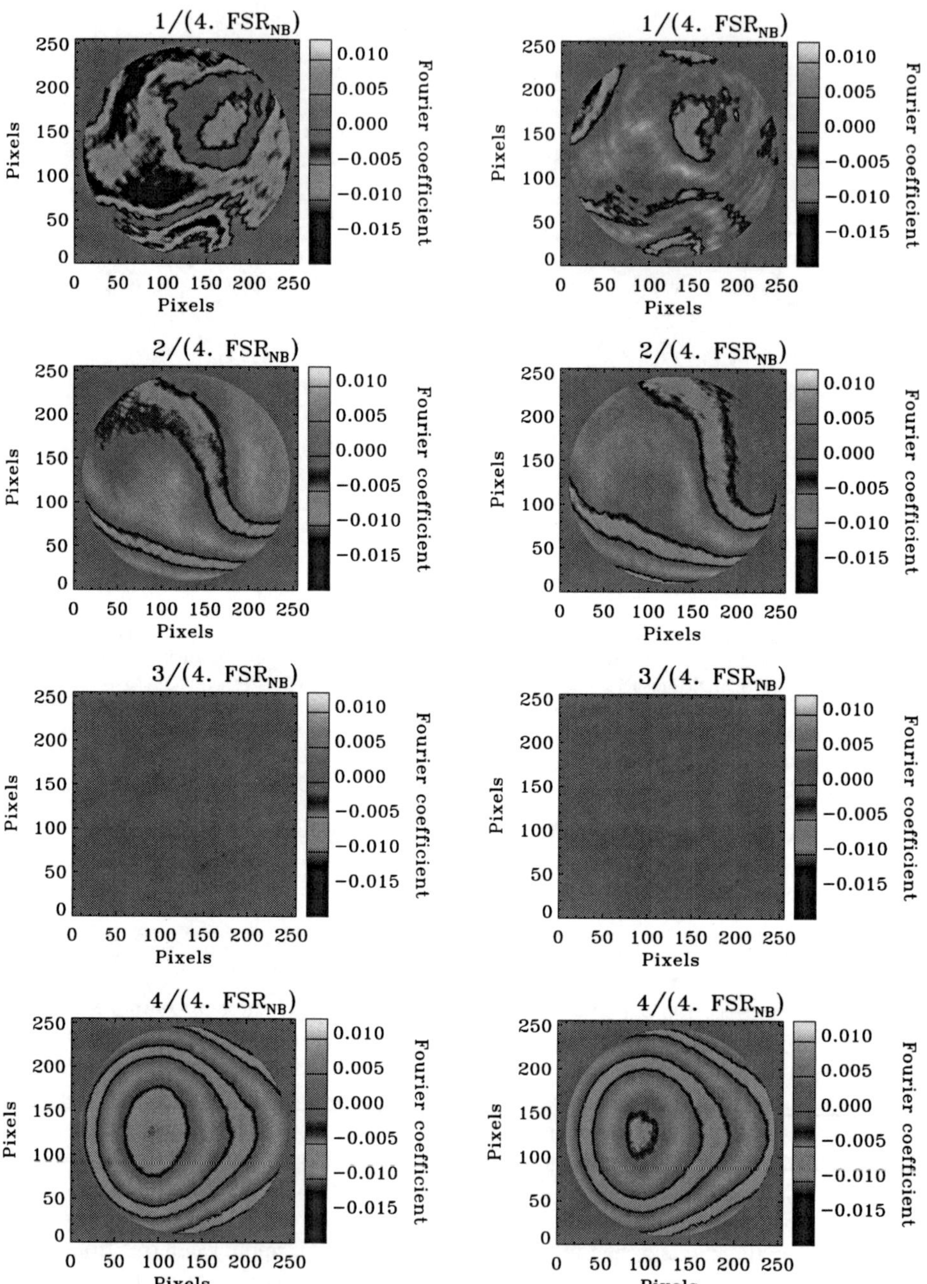

Figure 14 Fourier coefficients of the expansion of the transmission profile of the non-tunable part of HMI into a Fourier series, at a few selected spatial frequencies. Left panels: Fourier coefficients for the cosines; right panels: Fourier coefficients for the sines. The title of each panel is the spatial frequency (in terms of the FSR of the NB Michelson) corresponding to the coefficient. September 2007 data obtained in vacuum and OBSMODE.

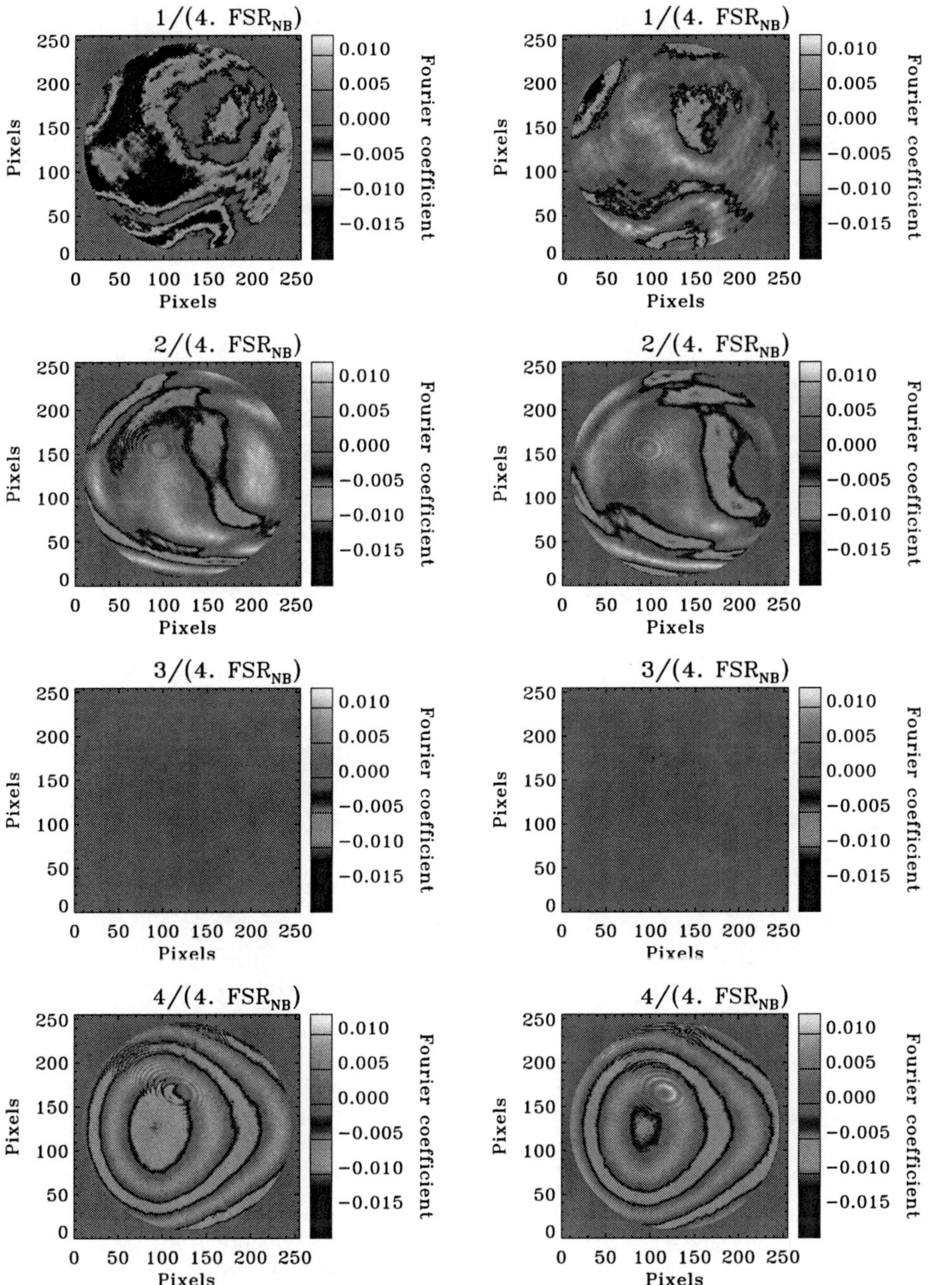

Figure 15 Fourier coefficients of the expansion of the transmission profile of the non-tunable part of HMI into a Fourier series, at a few selected spatial frequencies. Left panels: Fourier coefficients for the cosines; right panels: Fourier coefficients for the sines. The title of each panel is the spatial frequency (in terms of the FSR of the WB Michelson) corresponding to the coefficient. September 2007 data obtained in vacuum and CALMODE.

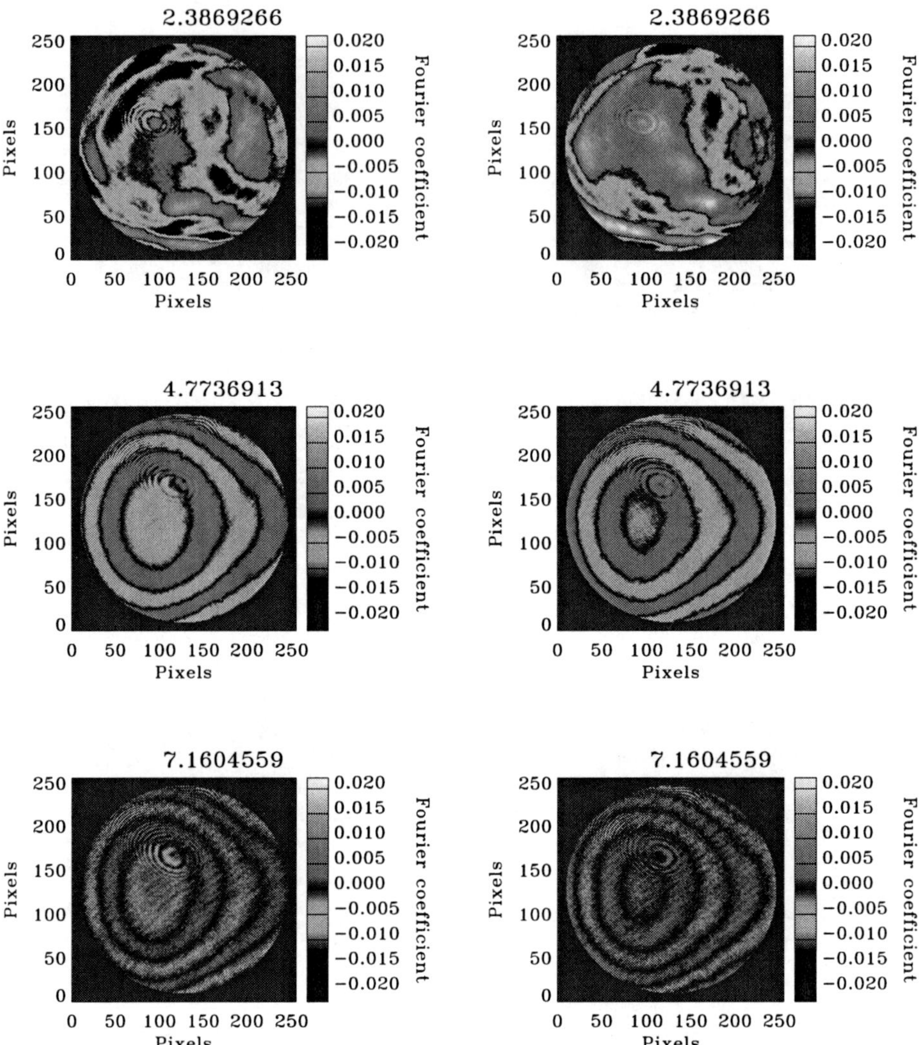

Figure 16 Cosine and sine coefficients of the transmission profile of the non-tunable part of HMI, at spatial frequencies corresponding to the front-window components. Left panels: cosine coefficients; right panels: sine coefficients. The title of each panel is the spatial frequency corresponding to the coefficient (in Å^{-1}). September 2007 data obtained in vacuum and CALMODE.

HCM is rotated all the way by a given number of steps while the other tuning HCMs are fixed. Such a sequence was taken in October 2006 after re-assembly of the Lyot filter, and returned the following estimates for the individual I-ripples: 2.03% for E1, 1.15% for the NB Michelson, and 0.99% for the WB Michelson. Figure 18 shows these individual I-ripples. In August 2009, other fine-tuning sequences were taken which returned somewhat smaller I-ripples: 0.98% for E1, 0.81% for NB, 0.78% for WB, and an overall I-ripple of 1.87%. A better (less flickering) light bulb could account for part of this difference.

With the simple models of transmittance for the Lyot elements and the Michelsons used in this article (Equations (12) and (22)), the impact of an I-ripple is neglected. Simula-

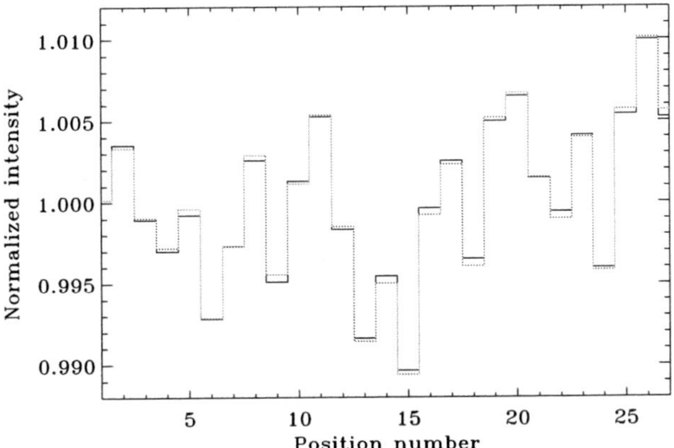

Figure 17 Variation of the spatially averaged intensity measured on the HMI CCDs during detune sequences in white light. Black curve: OBSMODE detune sequence; red curve: CALMODE detune sequence. Data of September 2007, taken in vacuum. No field stop was used for the OBSMODE sequence, while the 24.6 mm stop was used for the CALMODE sequence.

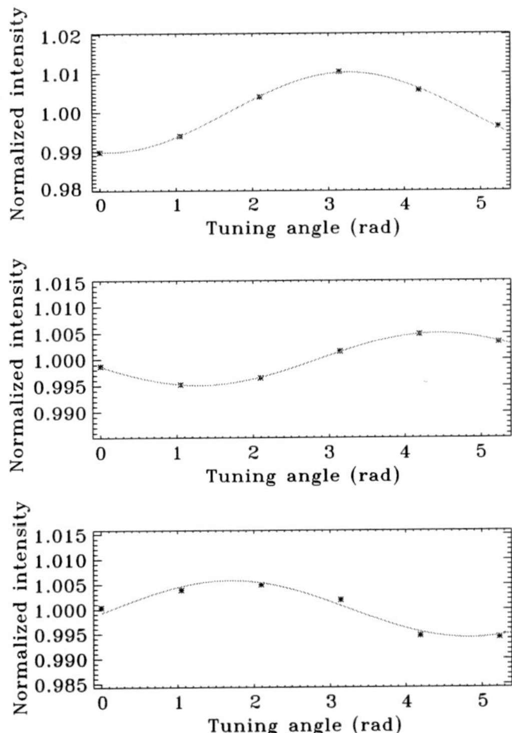

Figure 18 Variation of the spatially averaged intensity measured on the HMI CCDs during a fine-tuning sequence in white light in October 2006. Black stars: data points; red curves: fit by Equation (33). Upper panel: Lyot element E1; middle panel: WB Michelson; lower panel: NB Michelson.

tions show that, in the case of a misaligned half-wave plate in E1, this neglect produces a zero-point error (systematic error) on the determination of the Doppler velocity of, at most, several tens of $\mathrm{m\,s^{-1}}$. This systematic error varies almost linearly across the range $[-6.5, +6.5]$ $\mathrm{km\,s^{-1}}$. Therefore, it is theoretically possible to correct most of it by using the orbital velocity of SDO once the satellite is launched. Moreover, not taking into account the I-ripple in this case does not seem to have any impact on the rms variation of the solar Doppler velocity resulting from photon noise.

3.7. Angular Dependence

Angular-dependence tests were performed with the dye laser as light source. The setup is different in OBSMODE and CALMODE. In OBSMODE, an opaque mask with a circular hole (about 25 mm in diameter) mounted on the Polarization Calibration Unit (PCU) is placed in between the stimulus telescope and the front window (aperture stop), and the 24.6-mm field stop is used. Depending on the position of the hole (its radial distance from the center of the HMI aperture and its polar angle on the plane of this aperture), specific incidence and azimuthal angles are selected. In CALMODE, the 5-mm field stop is placed at the focal plane of the stimulus telescope. Detune sequences were taken at nine different hole/field-stop locations, and the corresponding phase maps of the tunable elements were derived.

The main issue with these setups is that a change in incidence and azimuthal angles is intertwined with a change in the path of the rays of light inside the instrument; therefore, spatial and angular dependences are entangled. Using ray traces of the instrument modeled with the Zemax software, we access the theoretical image sizes in OBSMODE and CALMODE at different locations in HMI, and also the diameters of the light cones (rays of an image point are distributed in a cone-shaped beam inside HMI). These data are needed to determine how to shift the resulting phase maps to cancel out the spatial-dependence effect. The value of the selected incidence angle at each tunable element can also be obtained from the Zemax software. Angular-dependence tests were performed in March 2006 (in air) and February 2007 (in vacuum). The results of these tests were challenging to analyze, and at times confusing, due to the spatial- and angular-dependence entanglement and to changes in the laser wavelength in between detune sequences. These results seem to show a weak angular and azimuthal dependence for the Michelsons. E1 seems to be the tunable element with the strongest angular dependence, followed by the NB Michelson, and the WB Michelson. This is hardly surprising, as the retardance of the Michelson interferometers is expected to vary mainly as θ^4 (for small θ) if the WF compensation is perfectly achieved, while the retardance of E1 is expected to vary as θ^2 (see Sections 2.1 and 2.2). Moreover, there seems to be a significant amount of azimuthal dependence with E1, which implies that some elements of the WF components are not perfectly aligned (either the split crystals or, more likely, the half-wave plate). Such a misalignment could explain the existence of the observed I-ripple (based on Equation (33), a tilt by 2° of the half-wave plate creates a $\approx 1.25\%$ I-ripple for E1).

Figure 19 shows the wavelength change [$\delta\lambda$] produced by a non-zero incidence angle [θ]. We fit the data points by a second-order polynomial [$\delta\lambda = a\theta^2 + c$] where a and c are the two polynomial coefficients (there is no term in θ because we assume symmetry around the axis $\theta = 0$). We present two cases for the tunable element E1: one for the azimuthal angle (among those measured) where the angular dependence is the strongest (corresponding, on the coordinate system of a HMI filtergram and counting the angles positively counterclockwise from the horizontal axis to the right, to the hole being centered at $-90°$), and one for which it is the weakest (corresponding to the hole centered at $+90°$). The maximum wavelength change measured is at $\theta \approx 0.9°$ (close to the edge of the field of view) and is $\delta\lambda \approx 51$ mÅ (the minimum change at $\theta \approx 0.9°$ appears to be only ≈ 5.5 mÅ). For the NB Michelson, the maximum wavelength change measured is much less, $\delta\lambda \approx 14$ mÅ, and for the WB Michelson it is only $\delta\lambda \approx 8.5$ mÅ.

Another way of observing the angular dependence in CALMODE at the center of the aperture of the tunable elements is to compare their phase maps obtained through a detune sequence taken with the solar-size field stop to phase maps obtained with the small field stop (centered). Raytraces in CALMODE show that the maximum angle of incidence in the

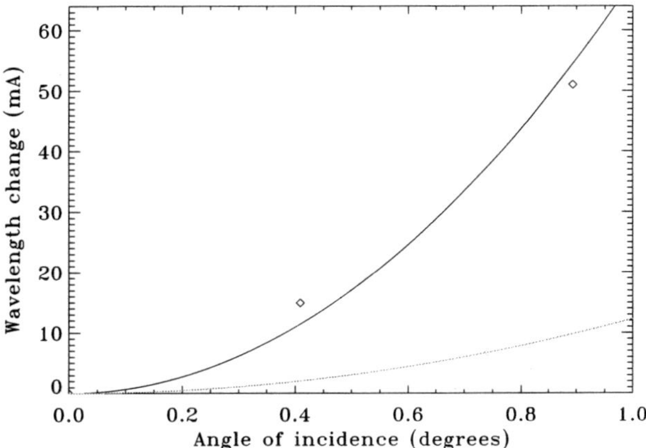

Figure 19 Angular dependence of the tunable element E1 in CALMODE. The diamonds show the data points, and the solid lines are the results of second-order polynomial fits (we assumed the data are symmetrical around $\theta = 0$). We show the results for the azimuthal angle, amongst those measured, corresponding to the maximum angular dependence (black line), and for the one corresponding to the minimum angular dependence (red line).

light beams inside the instrument varies, at the center of E1, from 0.17° (with the small field stop) to 0.96° (with the solar-size field stop). Therefore, any phase difference derived from these two detune sequences results mostly from the change in the distribution of the angles of incidence. This test confirms that E1 has a stronger angular dependence than the Michelsons. For instance, the results of two detune sequences taken three days apart with the different field stops and in similar conditions (same laser wavelength, same oven temperature, and in vacuum), in February 2007, show an average phase shift of 14.3° for E1, corresponding to a 27.5 mÅ wavelength change. This change, when extrapolated to $\theta = 0.9°$, is about 70 mÅ, larger than the maximum 51 mÅ change previously reported.

This relatively strong angular dependence partly explains the $\approx 3°$ average phase difference for E1 between CALMODE and OBSMODE. It is difficult to quantify the contribution of the angular dependence to this phase difference, but by using the angles of incidence in OBSMODE and CALMODE returned by the Zemax software for E1, and applying the strongest dependence displayed on Figure 19, it seems that, at most, $\approx 1.5°$ can be explained by the angular dependence. The Michelsons show a smaller phase difference between OBSMODE and CALMODE; this seems consistent with their weaker dependence on θ.

3.8. Temperature Dependence

During calibration runs in vacuum, the oven temperature was changed from 28 to 35°C. Taking detune sequences with the dye laser at three different, stable oven temperatures, allows the derivation to be made of the temperature dependence of E1 and the Michelsons. Figure 20 shows some results. We fit the three temperature points with a second-order polynomial because the temperature dependence is theoretically quadratic (see for instance Equation (15)). The slope is the lowest around 31 – 31.5°C. At 30°C, the nominal oven temperature, the slopes are -9.5 mÅ °C^{-1} for the NB Michelson, -9.7 mÅ °C^{-1} for the WB Michelson, and -12.4 mÅ °C^{-1} for the Lyot element E1. This is a relatively weak temperature dependence and is therefore fully satisfactory. In terms of relative phases, Φ increases with the oven temperature for temperatures below ≈ 31.5°C, and then decreases.

The temperature dependence of the Lyot elements was also investigated at the LMSAL facility in October 2006 while the Lyot filter was outside of the instrument, using sunlight and a spectrograph. Fits of the transmission profiles returned by this spectrograph at different temperatures show that the Lyot elements are thermally well compensated: the absolute value of the slope around 30°C is less than ≈ 10 mÅ °C^{-1} (see Table 2). In particular, for E1,

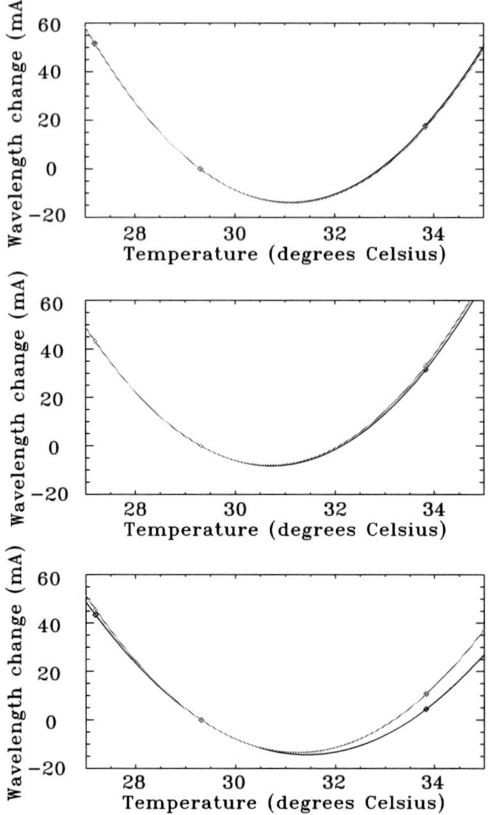

Figure 20 Temperature dependence of the tunable elements. Upper panel is for the NB Michelson, middle panel is for the WB Michelson, and lower panel is for the Lyot element E1. The diamonds show the data points, while the solid lines are second-order polynomial fits. The black lines are for OBSMODE data, the red lines are for CALMODE data. The zero wavelength change was arbitrarily assigned to the $T_e \approx 29.5°\mathrm{C}$ point.

the data from October 2006 show a decrease in the wavelength of peak transmittance from 25 to $\approx 31-32°\mathrm{C}$, and then an increase, in agreement with the results presented here.

Finally, we also checked the dependence of the I-ripple on the oven and front-window temperatures, using lamp-light detune sequences taken at different stable oven temperatures or stable front-window temperatures. This confirmed that the I-ripple is largely independent of any oven temperature change within the range 28 to 35°C, and is similarly largely independent of any front-window temperature change within the range 20 to 30°C.

4. Other Calibration Results

4.1. Thermal Stability of the Tunable Elements

The phases of the tunable elements exhibit a temperature dependence, as detailed in the previous section. However, this temperature dependence was obtained for stable oven temperatures: the temperatures were constant during a detune sequence, and *a priori* uniform across the tunable element apertures. A change in the oven temperature propagates in a finite time to the different tunable elements, and most likely creates a temperature gradient across their aperture. Therefore, here, we are interested in answering the questions: how do the tunable-element phase maps behave when the oven temperature changes and creates a

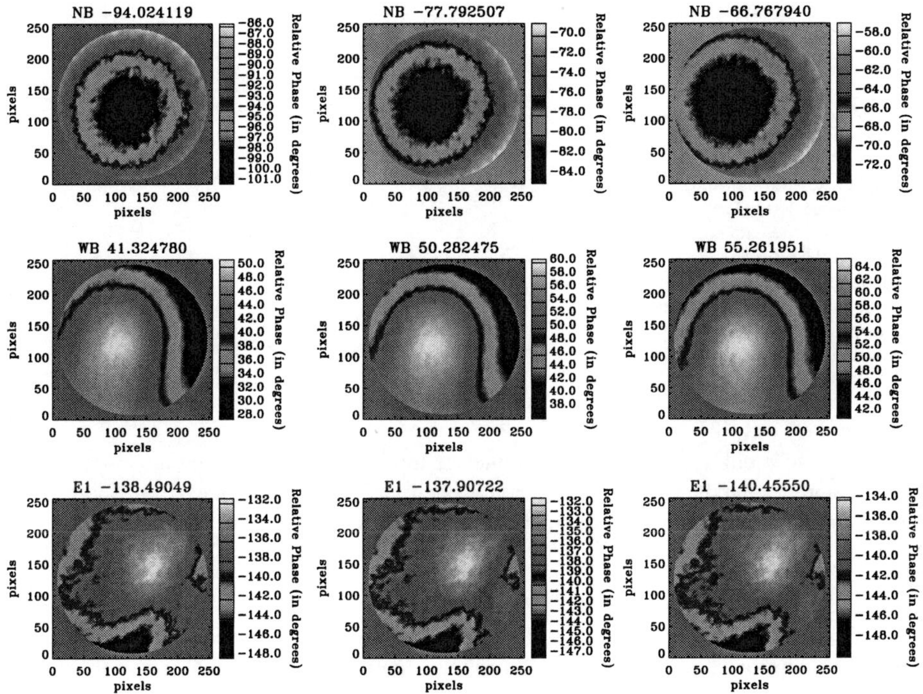

Figure 21 Change in the relative phases of the tunable elements during an oven-temperature transition from 30 to 35°C. The numbers in the panel titles are the average phases for the NB, WB, and E1 elements. The left column is for an average oven temperature of 29.9°C; the central column is for 31.3°C; and the right column is for 32.0°C. Detune sequences in OBSMODE taken in vacuum in February 2007, at a laser wavelength of 6173.3046 Å (estimate).

gradient across the aperture, and how long does it take for these phases to stabilize once the oven temperature stops changing (time it takes for the gradient to disappear).

In February 2007, during an oven-temperature transition from 30 to 35°C, about seven and a half hours of laser-light detune sequences were taken with HMI in the vacuum tank. These data show that the features on the phase maps of the tunable elements "drift" slowly across the element apertures before coming back to normal once the temperature is stabilized. Figure 21 shows three different sets of phase maps obtained for the tunable elements at three different unstable oven temperatures: 29.9°C, 31.3°C, and 32.0°C. In this case, it took less than the 7-hour run for the phase patterns to stabilize (probably around five hours, but sudden jumps in the dye-laser wavelength make it difficult to estimate more precisely the time of stabilization): this is more than the roughly 3.5 hours it took for the oven temperature to reach 35°C. Therefore, thermal stability of the phase maps (and therefore transmission profiles) of the tunable elements is only reached several hours after the oven temperature stops changing.

4.2. Instrument Throughput

The overall throughput, or ratio of the transmitted to input intensity, of the instrument needs to be assessed to determine the appropriate exposure times. Four detune sequences were taken at LMSAL in sunlight and CALMODE in the course of a same day, at different times

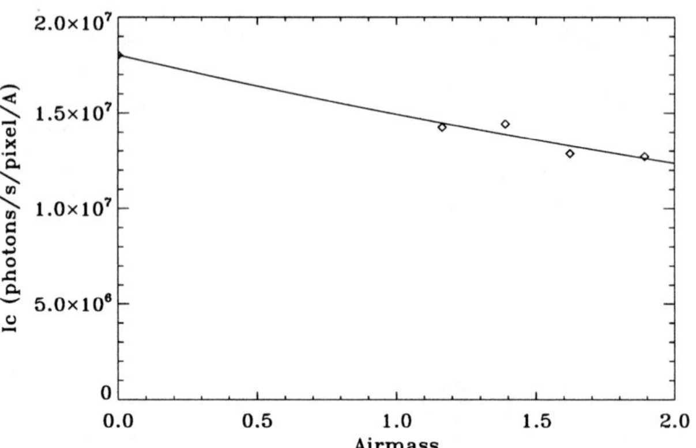

Figure 22 Measurement of the throughput of HMI: shown are the solar continuum intensities (in photons s^{-1} $pixel^{-1}$ $Å^{-1}$) derived as a function of the airmass from four detune sequences in CALMODE and sunlight taken during a single day in September 2007. The diamonds are the data points, except for the point at airmass = 0 which is extrapolated, while the solid line is the result of a fit using the Beer–Bougher–Lambert law of atmospheric absorption.

(to access different airmasses). To compute the throughput we calculate the attenuation of the incoming solar flux by the Earth atmosphere. We apply the Beer–Bougher–Lambert absorption law, assuming, among others, that the atmosphere is homogeneous:

$$\ln I_c = \ln I_{c0} - kX, \tag{35}$$

where I_c is the solar continuum intensity obtained from the detune-sequence data by the fitting code mentioned in Section 3.3.1, I_{c0} is the quantity that we need to derive (continuum intensity in the absence of atmospheric absorption/scattering, *i.e.* at airmass zero), k is the extinction coefficient at the solar line-center wavelength, and X is the airmass (returned by an ephemeris). The four fitted I_c values have to be corrected for the changes in the ratio inside-LMSAL to outside-LMSAL sunlight intensities. These ratios were measured with a power meter before each detune sequence, and their difference is due to the heliostat mirrors. Using the previous equation, I_{c0} is obtained by extrapolation to airmass zero (see Figure 22). With the four detune sequences analyzed, the lowest airmass reached is only 1.16, and therefore the extrapolation result has a large uncertainty. The number of photo-electrons detected on a CCD pixel per second in the HMI passband and at airmass zero is estimated from I_{c0}, the gain of the CCD (here taken as 16.36 for the side camera; see Wachter *et al.*, 2011), and the HMI sampling-position transmittances. The number of photons per second per pixel in the HMI passband, received at the entrance of the HMI instrument, is estimated from the black-body radiation law and the equivalent width of the HMI sampling-position transmittances. The ratio of these two quantities provides the throughput, here estimated at about 1.35%. This is on the low side of the expected range; a conservative radiometric analysis based on the specification numbers of the HMI components returns a low estimate of 1.15% for the throughput, and, considering that many components have efficiencies significantly exceeding their specifications, a tentative high estimate is about 2% (*e.g.* using 0.99 instead of 0.9 for all the mirror reflectivities). With a 1.35% throughput, the exposure time corresponding to an exposure level of 125 000 electrons (for a CCD full well of 150 000) is about 160 ms, much smaller than the maximum exposure time allowing the 45-second cadence required to produce the Doppler-velocity maps.

4.3. Artifact Check

It is necessary to check whether or not the fringe patterns produced by the front window and blocking filter, as well as the I-ripple value, vary with the focus blocks. This is the purpose

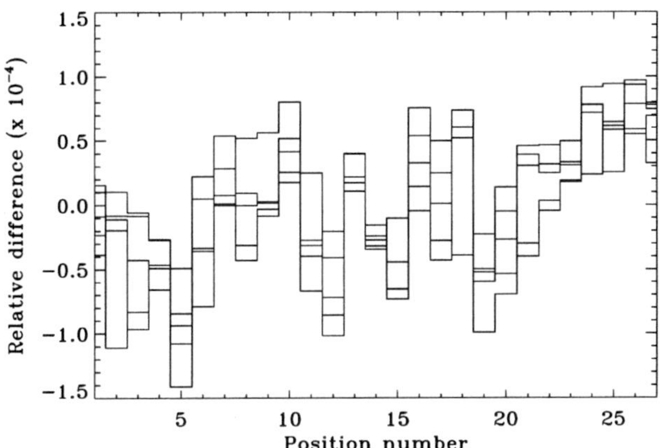

Figure 23 Artifact check for the I-ripple: relative differences in the spatially averaged intensities of seven detune sequences taken with different focus blocks (1, 2, 3, 4, 5, 9, and 13) in September 2007 with lamp light, in vacuum, and at an oven temperature of 29.5°C.

of the artifact-check test. To this end, detune sequences were taken in white light for focus positions 1 to 13 (OBSMODE). There was no significant difference in the fringe patterns and the value of the I-ripple. For instance, Figure 23 shows how stable this I-ripple is with respect to the focus blocks: the relative differences in intensities between detune sequences taken at seven different blocks are very small (maximum relative difference of 0.015%).

4.4. Tuning Polarizer Check

The two Michelson interferometers can be tuned by rotating a combination of a half-wave plate, a polarizer, and a second half-wave plate. Under normal conditions only the half-wave plates are used. The tuning polarizer is a redundant mechanism: were one of the two wave plates to fail, we could still tune HMI. To check whether the tuning polarizer is working properly, we use a detune sequence with four extra positions added to the usual 27-position sequence: the HCM of the tuning polarizer is set at the angles $\phi = 30$, 60, 90, and 120° (corresponding to tuning phases of 60, 120, 180, and 240°). This detune sequence is taken with a laser as light source and in the presence of the stimulus telescope. We fit the sequence filtergrams for the laser relative intensity, and the phases and contrasts of the tunable elements. We then reconstruct the detune sequence. The reconstruction shows whether or not the polarizer is working: a poor reconstruction is a sign of defect in the tuning polarizer. An example of such an analysis is shown on Figure 24. It confirms that the tuning polarizer is working as expected.

4.5. HMI Sampling-Position Profiles and the Computation of the Doppler Velocity and l.o.s. Magnetic-Field Strength

The results presented in this article allow a derivation of the tuning-position transmission profiles used in the observable sequence. The nominal plan is that a HMI observable sequence will consist of filtergrams taken at six different tuning positions, roughly spanning the range [− 172.5, + 172.5] mÅ around the target wavelength, and at different polarizations (Left Circular Polarization [LCP] and Right Circular Polarization [RCP] for the l.o.s. observables, or some combinations of the Stokes vector components I, Q, U, and V for the full vector magnetic field). The Doppler velocities and l.o.s. magnetic-field strengths will therefore be calculated using 12 filtergrams (six tuning positions, and two polarizations). At first, the algorithm converting the intensities on these filtergrams into l.o.s. observables will

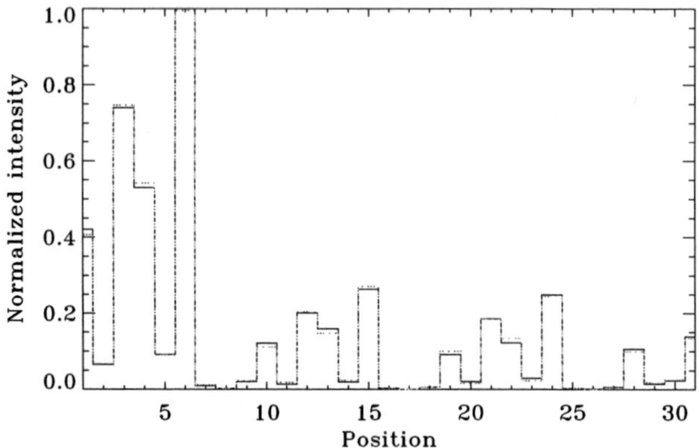

Figure 24 Reconstruction (in red and dashed line) of a spatially averaged 31-position detune sequence (in black solid line), taken in October 2007 in OBSMODE. This sequence includes four different positions for the tuning polarizer (the last four). The reconstruction is based on phases and contrasts of the tunable elements that were fitted on the actual detune sequence. The good agreement between actual and reconstructed sequences confirms that the tuning polarizer is working properly.

be based on the one used by MDI (see Scherrer *et al.*, 1995). Later, we plan to use a least-squares fit of an appropriate model of the Fe I line profile. The MDI-like algorithm is based on a discrete estimate of the first and second Fourier coefficients of the Fe I profile. The velocities corresponding to the phases of these Fourier coefficients, obtained by arctangent, are calculated. These velocities are then corrected using look-up tables. Correction is needed because the sampling-position transmission profiles are not δ functions. We only use six tuning positions to estimate the Fourier coefficients, and the formula applied to convert Fourier coefficients into Doppler velocities is only approximate (*e.g.*, it assumes that the solar line has a Gaussian profile). The look-up tables (two of them: one for the velocities returned by the first Fourier coefficient, and one for the velocities returned by the second Fourier coefficient) are computed separately in a program based on the HMI sampling-position profiles derived from the results presented here, and based on profiles of the solar Fe I line that were provided by R.K. Ulrich at three angular distances from disk center (profiles publicly available as text files on http://hmi.stanford.edu/data/calib/ironline/). Figure 25 shows an example of six HMI sampling-position profiles at CCD center and their location with respect to the solar Fe I line at rest and at disk center (the zero wavelength corresponds to the target wavelength 6173.3433 Å).

4.6. Preliminary Plans for the On-Orbit Calibration

It is theoretically possible to improve and further our knowledge of the HMI-filter transmission profiles by using on-orbit calibration procedures. SDO will orbit the Earth in a geosynchronous orbit. The Sun – SDO radial velocity at any given time will be known with high precision. This velocity should vary in the maximum range $\approx [-3.5, +3.5]$ km s^{-1} daily. The exact range changes during the year. We plan to use this velocity range to accurately determine the FSRs of the tunable elements, by taking series of detune sequences in CALMODE and OBSMODE at different times of a day. Rolls and offpoints may also be used.

Moreover, as previously mentioned, the l.o.s. observables will be determined from HMI filtergrams using a MDI-like algorithm based on six sampling positions. With this algorithm a set of look-up tables are computed, which are an estimate of the Doppler velocity

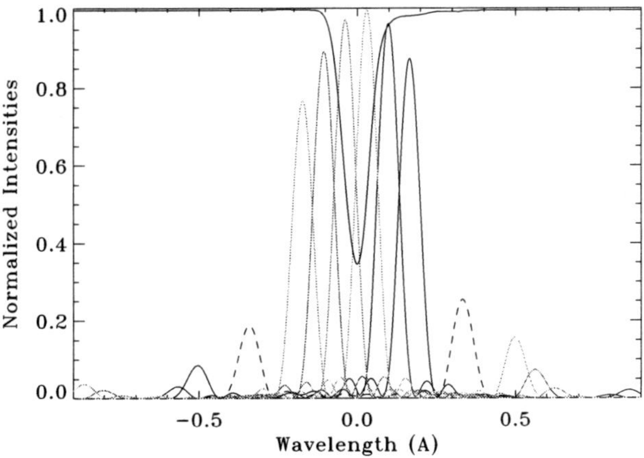

Figure 25 Example of HMI sampling-position profiles obtained from the wavelength-dependence calibration procedure (these profiles vary across the CCD and depend on the tuning angles used). Six tuning positions are shown here with respect to the Fe I solar line (black line) at disk center and at rest, on top of the continuum tuning position (dashed line). The continuum tuning position can be used to estimate the solar continuum intensity. The Fe I line profile was provided by R.K. Ulrich and obtained at the Mount Wilson Observatory.

returned by the algorithm as a function of the actual velocity, based on our knowledge of the sampling-function transmission profiles. Using the known Sun – SDO velocity and taking cotune sequences, it is possible to check the accuracy of, and partly correct, these look-up tables. Therefore, it should be possible to account for some effects not modeled in the transmittance formulae (*e.g.* the I-ripples, the CALMODE – OBSMODE phase differences) and the Fe I line profile used. Finally, it will also be possible to monitor the drift with time in the wavelengths of peak transmittance of the Michelson interferometers.

5. Conclusion

A thorough investigation of the wavelength dependence of the HMI instrument was conducted during the course of years of calibration tests performed at the LMSAL facility in California, and later at the NASA Goddard Space Flight Center in Maryland, and Astrotech Space Operations in Florida. The main results obtained on the assembled instrument are summarized in Table 5. Using various light sources (sunlight, laser light, lamp light) and a combination of detune and cotune sequences, it was possible to derive the transmittances of the elements of the HMI optical-filter system, their temperature dependence, and their angular dependence. These wavelength-calibration tests showed the very high quality of the HMI optical-filter system, and helped in detecting and correcting some defects. In particular, the early detection of a large I-ripple ($> 10\%$) was instrumental in highlighting a problem with the Lyot filter. This problem was successfully corrected. The knowledge of the transmission profiles of the filter elements is key to deriving accurate solar Doppler velocities and vector magnetic fields. The Michelson interferometers have a high degree of spatial uniformity, with a phase range of 16° across the aperture for the NB Michelson (equivalent to a wavelength range of 7.6 mÅ, or a velocity range of 369 m s^{-1}), and 24° for the WB (equivalent to a wavelength range of 22.9 mÅ, or a velocity range of 1113 m s^{-1}). Similarly the Lyot tunable element E1 has a phase range of 16° (30.7 mÅ, or 1492 m s^{-1}). The contrasts of these tunable elements are also very high (larger than 0.96 on average). This compares favorably

Table 5 Summary of selected ground calibration results for tests performed on the assembled instrument.

Test	OBSMODE	CALMODE
Phase gradient of NB	$\approx 16°$	$\approx 19°$
Phase gradient of WB	$\approx 24°$	$\approx 29°$
Phase gradient of E1	$\approx 16°$	$\approx 15°$
Phase gradient of E2	$\approx 20°$	$\approx 18.5°$
Phase gradient of E3	≈ 20	$\approx 20°$
Phase gradient of E4	$\approx 36.5°$	$\approx 36°$
Phase gradient of E5	$\approx 30°$	$\approx 28°$
Average contrast of NB	≈ 0.969	≈ 0.977
Average contrast of WB	≈ 0.997	≈ 0.997
Average contrast of E1	≈ 0.962	≈ 0.974
Average contrast of E2	≈ 0.985	≈ 0.986
Average contrast of E3	≈ 0.965	≈ 0.965
Average contrast of E4	≈ 0.988	≈ 0.988
Average contrast of E5	≈ 1.000	≈ 0.999
Phase drift of NB	$0.2° - 0.43°$ per day	$0.2° - 0.43°$ per day
Phase drift of WB	$\approx 0.15°$ per day	$\approx 0.15°$ per day
Phase drift of E1	0	0
FSR of NB	≈ 170.9 mÅ	≈ 170.9 mÅ
FSR of WB	≈ 341.9 mÅ	≈ 341.9 mÅ
FSR of E1	≈ 693.5 mÅ	≈ 693.5 mÅ
Total I-ripple	$\approx 1.9\%$	$\approx 1.9\%$
I-ripple of NB	$0.81 - 1.15\%$	$0.81 - 1.15\%$
I-ripple of WB	$0.78 - 0.99\%$	$0.78 - 0.99\%$
I-ripple of E1	$0.98 - 2.03\%$	$0.98 - 2.03\%$
Angular dependence of NB at $\theta = 0.9°$	≈ 14 mÅ	≈ 14 mÅ
Angular dependence of WB at $\theta = 0.9°$	≈ 8.5 mÅ	≈ 8.5 mÅ
Angular dependence of E1 at $\theta = 0.9°$	$\approx 51 - 70$ mÅ	$\approx 51 - 70$ mÅ
Temperature dependence of NB at 30°C	≈ -9.5 mÅ °C^{-1}	≈ -9.5 mÅ °C^{-1}
Temperature dependence of WB at 30°C	≈ -9.7 mÅ °C^{-1}	≈ -9.7 mÅ °C^{-1}
Temperature dependence of E1 at 30°C	≈ -12.4 mÅ °C^{-1}	≈ -12.4 mÅ °C^{-1}
Throughput	not measured	$\approx 1.35\%$

with the tunable elements of the MDI instrument onboard SOHO. The angular dependence of the tunable elements proved difficult to measure, and seems higher than expected for E1 (which also displays some azimuthal dependence). The maximum wavelength change of the transmission profile, compared to the profile for normal rays, was measured at $\theta = 0.9°$ for E1 and is larger than 50 mÅ. The angular dependence of the Michelsons appears to be much smaller. The temperature dependence of the tunable elements is relatively weak, less than -12.4 mÅ °C^{-1} at the nominal oven temperature. The Michelson transmission profiles drift in wavelength over time, but the drift rate is expected to slow down with time. Each tunable element has less than 1% of I-ripple (E1 being the worst), which is a further proof of a very good manufacturing and assembly. The overall I-ripple is of the order of 1.9%. The non-tunable transmission profile was also characterized and the Lyot elements E2 to E5 are of high quality in terms of phase gradient and contrast. Finally, the throughput of the instrument is about 1.35%, resulting in an exposure time well below the maximum value allowing the 45-second cadence required to produce the l.o.s. observables.

Regular calibration tests will be performed to ensure that HMI is constantly providing the best observables possible. Moreover, the SDO – Sun radial velocity, known with a precision better than 1 $\mathrm{m\,s^{-1}}$, will be used to, at least partly, correct for residual errors in the wavelength calibration.

Acknowledgements This work was supported by NASA Grant NAS5-02139 (HMI). We thank the HMI team members for their hard work, A. Title for providing us with notes regarding Jones calculus applied to the Lyot elements, and R.K. Ulrich for providing us with profiles of the solar Fe I line at 6173 Å. The HMI project is grateful to Karel Urbanek, Carsten Langrock, Joe Schaar and Robert Byer at Stanford University's Ginzton Laboratory for designing and building the tunable solid-state laser. We also thank Todd Hoeksema, Brett Allard, and countless others at LMSAL for their help with the taking of the calibration data. Finally, we acknowledge the comments of the anonymous referee for improving the quality of this article.

References

Dravins, D.: 1982, *Ann. Rev. Astron. Astrophys.* **20**, 61.

Dravins, D., Lindegren, L., Nordlund, Å.: 1981, *Astron. Astrophys.* **96**, 345.

Evans, J.W.: 1949, *J. Opt. Soc. Am.* **39**, 229.

Gault, W.A., Johnston, S.F., Kendall, D.J.W.: 1985, *Appl. Opt.* **24**, 1604.

Norton, A.A., Graham, J.P., Ulrich, R.K., Schou, J., Tomczyk, S., Liu, Y., Lites, B.W., Lopez Ariste, A., Bush, R.I., Socas-Navarro, H., Scherrer, P.H.: 2006, *Solar Phys.* **239**, 69.

Scherrer, P.H., Bogart, R.S., Bush, R.I., Hoeksema, J.T., Kosovichev, A.G., Schou, J., *et al.*: 1995, *Solar Phys.* **162**, 129.

Schou, J., Scherrer, P.H., Bush, R.I., Wachter, R., Couvidat, S., Rabello-Soares, M.C., *et al.*: 2011, *Solar Phys.*, in preparation.

Title, A.M.: 1974, *Solar Phys.* **39**, 505.

Title, A.M., Ramsey, H.E.: 1980, *Appl. Opt.* **19**, 2046.

Title, A.M., Rosenberg, W.J.: 1979, *Appl. Opt.* **18**(20), 3443.

Title, A.M., Pope, T.P., Ramsey, H.E., Schoolman, S.A.: 1976, In: Development of Birefringent Filters for Spaceflight, Lockheed Missiles and Space Co., Palo Alto, CA, publicly available on the NASA Technical Reports Server (document ID 19780011984).

Wachter, R., Schou, J., Rabello-Soares, M.C., Miles, J.W., Duvall, T.L. Jr., Bush, R.I.: 2011, *Solar Phys.* doi:10.1007/s11207-011-9709-6.

Solar Phys (2012) 275:327–355
DOI 10.1007/s11207-010-9639-8

Polarization Calibration of the *Helioseismic and Magnetic Imager* (HMI) onboard the *Solar Dynamics Observatory* (SDO)

J. Schou · J.M. Borrero · A.A. Norton · S. Tomczyk · D. Elmore · G.L. Card

Received: 29 May 2010 / Accepted: 8 September 2010 / Published online: 30 October 2010

Abstract As part of the overall ground-based calibration of the *Helioseismic and Magnetic Imager* (HMI) instrument an extensive set of polarimetric calibrations were performed. This paper describes the polarimetric design of the instrument, the test setup, the polarimetric model, the tests performed, and some results. It is demonstrated that HMI achieves an accuracy of 1% or better on the crosstalks between Q, U, and V and that our model can reproduce the intensities in our calibration sequences to about 0.4%. The amount of depolarization is negligible when the instrument is operated as intended which, combined with the flexibility of the polarimeter design, means that the polarimetric efficiency is excellent.

The Solar Dynamics Observatory
Guest Editors: W. Dean Pesnell, Phillip C. Chamberlin, and Barbara J. Thompson.

J. Schou (✉)
W.W. Hansen Experimental Physics Laboratory, Stanford University, Stanford, CA 94305-4085, USA
e-mail: schou@sun.stanford.edu

J.M. Borrero · S. Tomczyk · G.L. Card
High Altitude Observatory, National Center for Atmospheric Research, 3080 Center Green CG-1, Boulder, CO 80301, USA

S. Tomczyk
e-mail: tomczyk@ucar.edu

G.L. Card
e-mail: card@ucar.edu

Present address:
J.M. Borrero
Kiepenheuer-Institut für Sonnenphysik, Schöneckstr. 6, 79104 Freiburg, Germany
e-mail: borrero@kis.uni-freiburg.de

A.A. Norton
James Cook University, School of Engineering & Physical Sciences, Townsville, QLD 4810, Australia
e-mail: aimee.norton@jcu.edu.au

D. Elmore
National Solar Observatory/Sacramento Peak, 3010 Coronal Loop, Sunspot, NM 88349, USA
e-mail: elmore@nso.edu

Keywords Instrumentation and data management · Magnetic fields, photosphere · Solar Dynamics Observatory

1. Introduction

Precise polarimetry requires the calibration of the polarimeter and associated telescope. Polarimeter calibration is typically performed by measuring the response of the polarimeter to known polarization states created by calibration optics placed at the output of the telescope (Skumanich *et al.*, 1997; Beck *et al.*, 2005). Telescope polarization is measured by placing polarizers over the entrance aperture of the telescope (Beck *et al.*, 2005; Selbing, 2005) or by using the symmetry and anti-symmetry expected, or more precisely, not expected, in spectral Stokes profiles in Zeeman-sensitive lines (Collados, 2003; Kuhn *et al.*, 1994).

In-flight calibration optics were avoided on the *Helioseismic and Magnetic Imager* (HMI) instrument, as they were on *Hinode*, due to the risk of a mechanical failure leaving a calibration optic in the beam and also for complexity, mass, and power considerations. Absence of calibration optics limits the types of calibrations that can be performed after the instrument has been placed into operation and places particular importance on ensuring that ground-based calibrations are comprehensive and consider the range of possible environmental and instrumental drifts.

As with the *Hinode* ground-based calibration (Ichimoto *et al.*, 2008), the HMI polarimeter and telescope are calibrated together by placing calibration optics in front of the telescope and treating the entire optical path as the polarimeter.

The HMI instrument differs from most other instruments by having redundant polarization selectors (see Section 2.3). This was done to improve the reliability but has the consequence that it is necessary to derive a calibration valid for all settings of the polarization selectors rather than only for the expected combinations.

The design of observing sequences is discussed by Borrero *et al.* (2007).

As a matter of convention, the angles in the system are measured counterclockwise from horizontal as viewed from behind the instrument toward the Sun, with horizontal being the plane in which the light travels from the window through the filter system, *i.e.* the plane at which the cut in Figure 1 is made (see Sections 2 and 5.1.1). Horizontal polarization is $I+Q$, while $I+U$ corresponds to polarization at a 45 degree counterclockwise angle from horizontal, again as seen from the instrument.

The remainder of this section contains a brief description of the requirements of the instrument. Sections 2 and 3 contain descriptions of the instrument with an emphasis on the polarimetric aspects and the test setup, respectively. Section 4 contains a description of the polarimetric model used to describe the instrument and test equipment. Section 5 contains descriptions of the various tests and the analysis, while Section 6 describes the main results obtained. Section 7 describes how the results obtained will be used to calibrate the instrument, the accuracy achieved, and the plans for on-orbit calibrations. Section 8 presents the conclusion. In the appendices, details of the polarimetric models of different types of optical elements are given and the sequences run are described.

1.1. Requirements

HMI measures a linear combination L_{HMI} of the solar Stokes vector:

$$L_{\mathrm{HMI}} = \mathcal{O} I_{\mathrm{Sun}}, \tag{1}$$

where $\mathcal{O}$ is the modulation matrix. In order to demodulate and obtain the Stokes vector, the demodulation matrix $[\mathcal{D}]$ must be known and applied:

$$I_{\mathrm{HMI}} = \mathcal{D} L_{\mathrm{HMI}} = \mathcal{D}\mathcal{O} I_{\mathrm{Sun}}. \tag{2}$$

In reality there are systematic errors and the error matrix $\mathcal{E} = \mathcal{D}\mathcal{O}_{\mathrm{true}} - \mathcal{I}$, where $\mathcal{O}_{\mathrm{true}}$ is the actual modulation matrix and $\mathcal{I}$ the identity matrix, will not be identically zero.

To derive a requirement for maximum allowable elements in $\mathcal{E}$, it may be noted that the requirement on the noise in Q, U, and V is 0.3% of I in ten minutes and that systematic errors of less than 0.1% of I on Q, U, and V will thus only result in a marginal increase in the overall error. With typical values of Q/I, U/I, and V/I of around 0.1, it thus follows that the cross terms between Q, U, and V of less than 0.01 are desirable. From I to Q, U, and V values less than 0.001 are needed. The terms into I are less critical as they more or less correspond to a flat-field error and a total contribution of 0.01 should be adequate. This results in a requirement that the absolute values of $\mathcal{E}$ should be less than

$$\mathcal{E}_{\max} = \begin{pmatrix} 0.01 & 0.1 & 0.1 & 0.1 \\ 0.001 & 0.01 & 0.01 & 0.01 \\ 0.001 & 0.01 & 0.01 & 0.01 \\ 0.001 & 0.01 & 0.01 & 0.01 \end{pmatrix}. \tag{3}$$

To be able to construct left- and right-circular polarization [LCP and RCP] from a single camera and to keep a 50-second or better overall cadence for the Doppler and Line-of-Sight (LOS) field, a requirement was imposed to be able to derive V with less than 5% leakage from each of Q and U, across the field of view using only the motor positions available (see Sections 2.3 and 7.2.4).

Some of the measurements required to finally verify some of these requirements are deferred to on-orbit operations. This is discussed in Section 7.3.1.

How errors relate to magnetic-field properties is discussed by Norton *et al.* (2006).

2. Instrument Description

The overall layout of the instrument is described by Schou *et al.* (2010). The parts of the instrument relevant to the polarimetry are shown in Figure 1. In order, the light passes through a front window, the primary lens, the secondary lens, a focus block (or calibration lens), three polarization-selector waveplates (PS1, PS2, and PS3), an Image Stabilization System [ISS] fold mirror, another focus block (or calibration lens), and a polarizing beamsplitter, all of which are described below. Following these elements are a Lyot filter, two Michelson interferometers, and a beamsplitter dividing the light to a pair of shutters and cameras (front and side) for higher cadence and redundancy. It may be noted that since the light is essentially monochromatic, the instrument is not required to be achromatic.

2.1. Telescope

The telescope consists of the front window, primary, and secondary lenses. The front window and primary are very close to a pupil and are thus effectively indistinguishable. The secondary lens is not at the same equivalent position, but we have nonetheless chosen to treat the telescope as a unit.

The front window is a sandwich consisting of 6 mm of BK7G18 radiation-hard glass, 3 mm of GG495 green glass, and 6 mm of BK7G18 glass. There are two heaters on the mounting ring, allowing the temperature to be controlled. Given the poor thermal conductivity of glass this, of course, does not ensure that the rest of the window is at any particular temperature, although it will eventually reach a nearly steady state with a slowly varying radial temperature gradient.

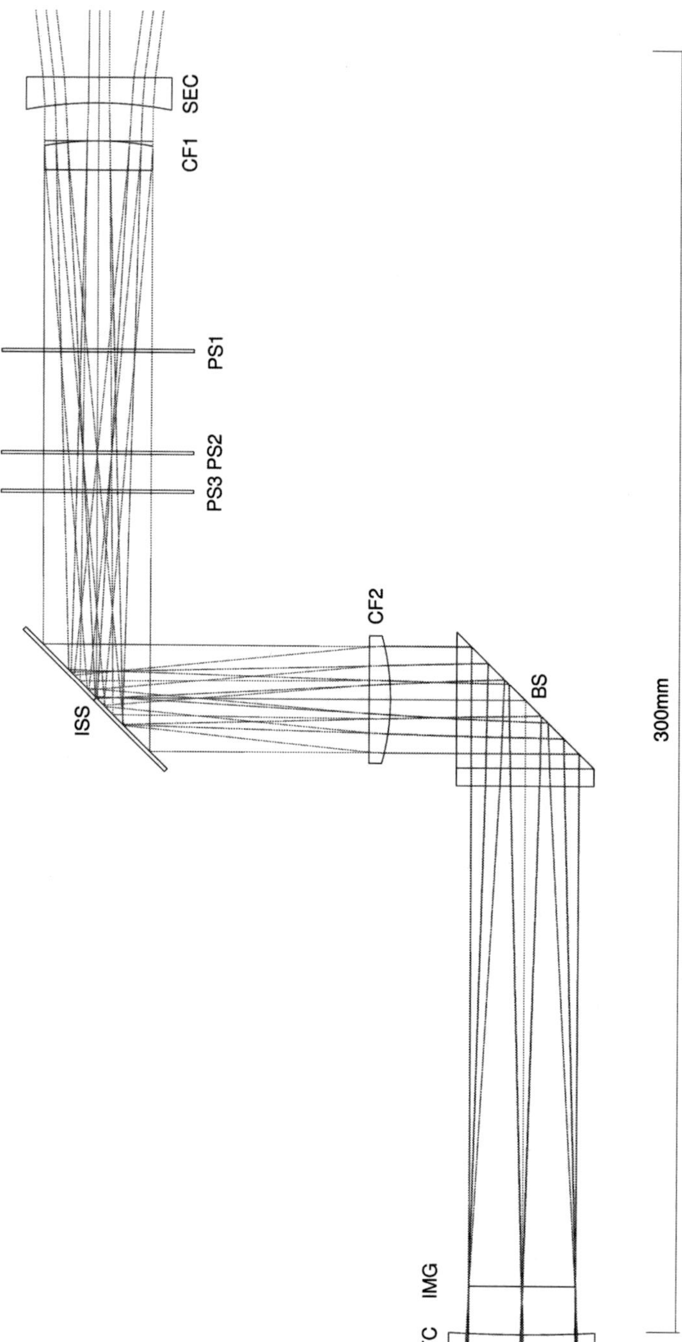

Figure 1 Raytrace with observing mode (obsmode) rays shown in blue and calibration mode (calmode) (see Section 2.2) rays shown in red. In the order that the light travels, the elements shown are: Secondary lens (SEC), first calibration lens (CF1), three polarization selectors (PS1, PS2, and PS3), ISS fold mirror (ISS), second calibration lens (CF2), polarizing beamsplitter (BS), a line showing the primary focus (IMG), and telecentric lens (TC).

The primary lens is coupled both conductively and radiatively to the front window and to the separately temperature-controlled telescope tube and is thus fairly well controlled, but will see some temperature gradients.

Some effects of the temperature gradients are discussed in Section 6.2.

2.2. Focus Blocks

The focus blocks are mounted on two filter wheels and different thicknesses of glass (including zero) can be inserted to adjust the overall focus. It is also possible to replace each focus block with a lens causing the entrance pupil to be imaged onto the detector. This mode is known as calibration mode (calmode), while the regular mode is referred to as observation mode (obsmode).

As shown by the ray trace in Figure 1, the introduction of the calibration lenses dramatically changes the angles of incidence through the polarization selectors, a subject that will be discussed in more detail in Section 4.4.

2.3. Polarization Selectors

The polarization selectors are high-order-crystal quartz waveplates mounted in hollow core motors (HCMs) capable of rotating each waveplate to 240 equally spaced positions. The choice of high-order waveplates was made to avoid mechanical problems associated with low-order waveplates, but it does have some implications, as discussed in Section 4.4.

The three waveplates have nominal retardances of 10.50, 10.25, and 10.50 waves, in the order encountered. Using three waveplates allows for some amount of redundancy. A study of possible combinations of waveplates revealed that several combinations achieve full three for two redundancy, meaning that any one of the three waveplates can be stuck at any position while still allowing for the measurement of an arbitrary polarization state (limited, of course, by the coarseness of the angular settings and any inaccuracies in the retardances). Of these combinations, two involve only simple retardances: $(1/2, 1/4, 1/2)$ and $(1/2, 1/4, 1/4)$. These result in equal motion of the HCMs (and thus presumably wear) through a standard observing sequence with the optimal choice of which waveplate (the middle) to keep fixed. The former causes slightly less wear in case one of the other waveplates cannot be rotated, so it was chosen.

For waveplates 1 and 2, increasing motor steps correspond to clockwise ($\mathrm{dir} = -1$) rotation of the waveplates as seen from the back of the instrument toward the Sun. For waveplate 3 the motion is counterclockwise ($\mathrm{dir} = +1$). In all cases one motor step is 1.5°. It thus follows that the angle of the fast axis of the waveplate (relative to horizontal) is given by $C_i - \theta_i$, where $C_i = 1.5° \times \mathrm{dir}_i \times \mathrm{N}_i$, N_i is the commanded motor position, and θ_i is the angle at which the fast axis is horizontal. The algorithm by which the motors are commanded is given in Schou *et al.* (2010).

During the assembly of the instrument, the motor positions at which the fast axis of each waveplate is horizontal were measured to be *roughly* 62, 89, and 9. It thus follows that *roughly* $\theta_1 = -93°$, $\theta_2 = -133.5°$, and $\theta_3 = 13.5°$. These values are unlikely to be accurate to better than one or two steps.

2.4. Polarizing Beamsplitter and ISS Mirror

The ISS mirror is used to stabilize the images and has essentially no effect from a polarimetric point of view (see Sections 4.5 and 7.1).

The polarizing beamsplitter selects a linear polarization state (roughly vertical) to measure from the light entering it and directs it to the Lyot filter. The orthogonally polarized light goes to the limb sensor (not shown in Figure 1). The choice of a polarizing beamsplitter (as opposed to a polarizer) causes a field dependence of the direction of this linear state, as discussed in Section 4.5. To avoid effects on the filter profiles of this and of any Lyot filter to beamsplitter rotational misalignment, a cleanup polarizer after the telecentric lens ensures that the direction of polarization delivered to the Lyot filter is unchanging and uniform.

3. Test Setup

The test setup consists of a stimulus telescope followed by a polarization calibration unit (PCU) described below, which allows for the generation of various states of polarized light.

The stimulus telescope, which is described in more detail by Schou *et al.* (2010) and Wachter *et al.* (2010), consists of a reversed HMI telescope (secondary plus primary lens), an optional target at the prime focus, and a light source. The light source consists of a fiber bundle, one end of which is fed by a stabilized lamp and the other end of which is imaged through the focus of the stimulus telescope onto the pupil. This setup results in a uniform illumination of both the pupil and the image with very little polarization (see Section 6.1.1).

Sunlight was also used, but the results are not discussed here as the weak but rotating polarization caused by the heliostat mirrors causes artifacts. The vacuum tests performed are not discussed due to artifacts introduced by the vacuum-chamber window.

3.1. Polarization Calibration Unit

The PCU, designed by the High Altitude Observatory, consists of a dichroic polymer linear polarizer followed by a roughly quarter-waveplate zero-order plastic retarder, both of which are mounted on fused-silica substrates. Each element can be inserted (or not) into the light beam and rotated (see Figure 2) to a commanded angle.

The commanded rotation angle at which the polarization passed by the polarizer is horizontal was precisely determined by an independent measurement to be -5.8° with an uncertainty of about 0.2°. The polarizer rotates clockwise (negative in angle) when commanded to rotate in the positive direction. The PCU-to-instrument alignment is discussed in Section 5.1.1

The retarder was determined to have the fast axis horizontal at a commanded angle of $\theta_{\mathrm{ret}} = 90.6^\circ$ (also with an uncertainty of about 0.2°) and rotates counterclockwise (positive in angle) when commanded to rotate in the positive direction. However, a correction to this angle is determined as part of the calibration, and so the accuracy of this value is of little consequence, except as a consistency check.

4. Polarimetric Model

The light emitted by the stimulus telescope [$I_{\mathrm{stim}} = (I, Q, U, V)$] goes through the PCU and HMI, being finally detected at the CCD as L_{cal}:

$$L_{\mathrm{cal}} = \mathcal{O} \times \mathcal{M}_{\mathrm{pcu}} \times I_{\mathrm{stim}} = (P_{\mathrm{det}}^T \times \mathcal{M}_{\mathrm{sel}} \times \mathcal{M}_{\mathrm{tel}}) \times \mathcal{M}_{\mathrm{pcu}} \times I_{\mathrm{stim}} \tag{4}$$

where $\mathcal{O}$ is the modulation matrix given by Equation (1) and the matrices $\mathcal{M}_{\mathrm{pcu}}$, $\mathcal{M}_{\mathrm{sel}}$, and $\mathcal{M}_{\mathrm{tel}}$ are the Mueller matrices for the PCU, polarization selectors, and telescope, respectively. P_{det} gives the Stokes vector passed by the polarizing beamsplitter, *i.e.* the first row of

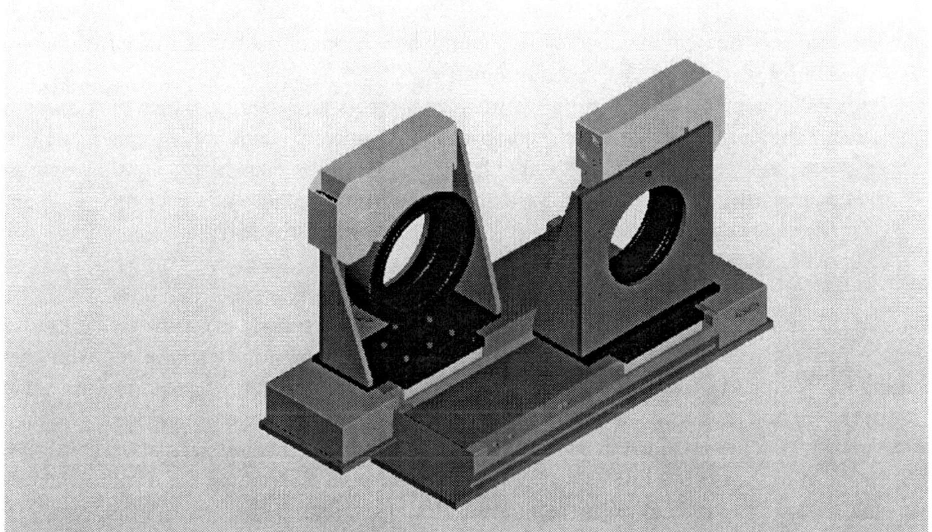

Figure 2 Schematic of the PCU. Each unit consists of a linear stage and a rotary stage with a piece of optics. The stage closest to the stimulus telescope contains a polarizer while the stage closest to the instrument contains a retarder. Otherwise, the two units are identical. The optics are removable and the polarizer unit can be fitted with black plates with aperture stops, as discussed in Section 5.1.1. In normal operation, the stages are enclosed in boxes for protection and to provide baffling of stray light.

the Mueller matrix of a polarizer. The dependency of the various components on the field position, instrument, and PCU settings has been suppressed for simplicity. Unless noted, the center ray hitting a given pixel is used and no integration over incidence angle is performed.

We present below details of the modeling of the components. In some cases, different models taking into account various physical effects and imperfections have been used. Further details of the models of various types of optical elements are given in Appendix A.

4.1. Stimulus Telescope

The stimulus-telescope polarization vector does not need any special modeling and simply has four components of a vector (I, Q, U, V).

4.2. PCU

The PCU is modeled as a polarizer followed by a retarder, but several effects need to be taken into account. First of all both elements have a transmission less than 100%.

$$\mathcal{M}_{\text{pcu}} = t_{\text{ret}} W(\mathrm{C}_{\text{ret}} - \theta_{\text{ret}}, r_{\text{ret}}) \times P(C_{\text{pol}}, t_{\text{pol}}), \tag{5}$$

where W and P are given in Appendices A.3 and A.1, respectively, t_{ret} is the retarder transmission, C_{ret} the commanded retarder rotation angle (corrected for the known offset discussed in Section 3.1), θ_{ret} is the retarder angle at which the fast axis is horizontal, r_{ret} the retardance, and C_{pol} is the polarizer rotation angle (again corrected for the known offset).

The two rotation angles [$C_{\rm ret}$ and $C_{\rm pol}$] are known from the PCU commanding (and telemetry) and thus vary from position to position in the test sequences. The rest of the parameters [$t_{\rm ret}$, $\theta_{\rm ret}$, $r_{\rm ret}$, and $t_{\rm pol}$] are fitted. There is no fitted offset for the polarizer as it is used as the absolute coordinate reference. When a component is out of the beam, the corresponding Mueller matrix is replaced by the identity matrix.

Both PCU elements have significant wavefront errors and thus introduce distortions of the stimulus-telescope image which, due to the non-uniform illumination, cause intensity changes unrelated to the polarization. We have found that the main errors are those caused by the image shifts. To model those we multiply the overall intensity by a factor

$$I_{\rm PCU} = (1 + \Delta x I_x)\times(1 + \Delta y I_y), \tag{6}$$

where Δx and Δy are the image offsets and I_x and I_y the (fitted) intensity gradients. The image offsets are separately determined by installing a field stop, tracking the edge, and fitting the results to a simple sinusoidal variation with retarder and polarizer rotation angle. Since the motions are repeatable, this is not done for every sequence run.

This effect is not included in calmode since the PCU elements are near a pupil.

Changes in the overall image size caused by wavefront errors (and the resulting change in intensity) are captured in the element transmissions.

4.3. Front Window and Telescope

Two basic models have been used for the front window. In the first the window is modeled as a general six-term Mueller matrix ($\mathcal{M}_{\rm tel} = M_6$ in Appendix A.3), allowing for both retardance and depolarization. The second uses a combination of a perfect retarder and a partial depolarizer: $\mathcal{M}_{\rm tel} = D_3 \times W_M$, where D_3 is the depolarizer described in Appendix A.2 and W_M is described by specifying M_{QV} and M_{UV} (see Appendix A.4).

4.4. Polarization Selectors

The polarization selectors would ideally be represented by $W(\theta, r)$ (see Equation (11) in Appendix A.3). However, they are high order and the incidence angles are significant, especially in calmode (see Figure 1), and thus a model with incidence angle $W_I(\theta, r, \alpha, \beta)$ (see Equation (12) in Appendix A.3) is used.

$$\mathcal{M}_{\rm sel} = W_I(C_3 - \theta_3, r_3, \alpha, \beta) \times W_I(C_2 - \theta_2, r_2, \alpha, \beta) \times W_I(C_1 - \theta_1, r_1, \alpha, \beta), \tag{7}$$

where the suffix is the selector number in the order encountered, C_i is the commanded motor position, θ_i is the position at which the fast axis is horizontal, r_i is the retardance, α is the incidence angle, and β is the angle from horizontal between the center of the field of view and the point of interest. Of these C_i, α, and β are known from commanding or optical ray traces, while θ_i and r_i are fitted for. Note that the r_i have to be the true retardances (*e.g.* 10.50, not 0.50) and that α is different in obsmode and calmode (see red and blue rays in Figure 1). Transmissions are assumed to be 1.0 since the waveplates are always in the beam.

As discussed in Section 5.1 the θ_i are near by degenerate and only two out of three can realistically be fitted for.

Issues related to the temperature dependence of the r_i are discussed in Section 6.2.

4.5. ISS Mirror, Polarizing Beamsplitter, and Cleanup Polarizer

The polarizing beamsplitter is treated as a perfect polarizer. Since the beam is not telecentric, the angle of incidence onto the beamsplitter coating, and thus the direction of polarization selected, depends on the vertical field position. At the center of the field, the direction is vertical and it changes by ± 0.0174 radians or $\pm 1.00^{\circ}$ at the edge of the image (CCD). In calmode the maximum angle is 0.059 radians or 3.39°. This variation is taken as given (an error will appear as an apparent vertical dependence of θ_3).

The polarization state selected from the light exiting the third waveplate is also affected by the ISS fold mirror since the direction of polarization of the light picked up by the beamsplitter is not perfectly parallel to the axis around which the mirror rotates the direction of the beam. However, this effect is very small and has not been included in the model.

Since the beamsplitter is assumed perfect, the cleanup polarizer should have no effect from a polarimetric point of view. However, it does cause a small amount ($< 0.1\%$) of light to be lost due to the varying direction of the output from the beamsplitter.

5. Test Description and Analysis Procedure

The main calibrations were done using a couple of standard sequences. These were run during dedicated calibration times as well as during the Comprehensive Performance Tests, which were run several times to verify that the instrument survived transport and various environmental tests unchanged. These tests are described in Section 5.1, which also describes the way in which the overall coordinate system is determined. Tests for focus-block birefringence are described in Section 5.2 and those used for establishing the value of the degenerate combination of angular zero points in Section 5.3.

5.1. Standard Tests

These test sequences were initially constructed by combining a number of shorter sequences, each designed to measure the properties of one element assuming that the properties of the others are known.

A Singular Value Decomposition of the sensitivity matrix (Hessian) was then performed to determine if there were any degeneracies and if additional observations could be added to break those. This analysis revealed two issues. The first is that regardless of which observations are added, one linear combination of parameters is degenerate: The zero points θ_1, θ_2, and θ_3 cannot be independently determined, as discussed in Sections 5.3 and 6.3.

The other issue is that some parameters are degenerate unless sequences moving more than one waveplate at a time are included.

Finally a number of redundant observations were removed.

In the end, the sequences used consist of these parts:

- Polarization sequences to determine the input polarization
- Sequences to determine properties of PCU polarizer
- Sequences to determine properties of PCU waveplate
- Sequences rotating each waveplate separately
- Sequences moving all waveplates together

In addition, observations were added between each block to track the stability of the light source. All observations also include darks to be able to account for temperature drifts in air with a high dark current (see Section 5.4.2).

Table 1 List of polarization sequences used. N indicates the number of good images from the front and side cameras. Type gives long (L) and short (S) sequence, as well as obsmode (O) and calmode (C). Roll is the PCU roll angle. It is left blank when unknown. Tests with FSN below 1 000 000 were performed at LMSAL. Tests with FSN between 1 000 000 and 1 941 258 were at GSFC. Tests after that at ASO. For definitions of the various sequences, etc., please see the following sections and Appendix B.

FSN	N	Type	Roll (°)	Box (C)	Window (C)	Leg (C)
440 560	262/262	LO		24.34	25.23	21.28
441 084	140/140	SO		24.35	25.02	21.29
441 418	262/262	LC		24.34	25.19	21.23
449 822	140/140	SC		20.83	20.25	20.39
450 102	140/140	SO		20.81	20.16	20.31
563 654	140/140	SO		25.99	26.01	20.93
563 934	262/262	LO		26.09	26.06	21.11
564 466	140/140	SC		26.20	26.09	21.26
878 140	140/140	SC	0.02	23.02	19.66	20.06
879 596	140/140	SO	0.02	22.82	19.42	19.83
947 806	140/140	SO	0.01	24.91	25.15	19.62
963 389	140/140	SC	0.01	19.09	35.01	18.77
963 689	140/140	SO	0.01	19.23	35.17	19.06
1 007 964	140/140	SO	90.17	25.25	21.45	21.25
1 010 487	140/140	SC	90.17	24.88	21.51	21.72
1 011 869	140/140	SC	90.17	25.78	30.95	21.66
1 012 311	140/140	SC	90.17	25.81	35.19	21.56
1 012 679	139/139	SC	90.17	25.80	38.95	21.51
1 019 605	140/140	SO	90.17	25.77	25.12	21.36
1 020 025	140/140	SC	90.17	25.62	24.91	21.33
1 230 144	140/140	SC	89.55	24.68	18.96	19.50
1 230 432	140/140	SO	89.55	24.53	19.13	19.67
1 765 703	140/140	SO	90.47	24.51	19.88	20.35
1 765 983	140/140	SC	90.47	25.10	20.54	20.76
1 955 567	140/140	SC	89.56	26.67	24.17	24.90
1 956 735	140/140	SO	89.56	26.94	24.46	25.32
1 988 033	140/140	SO	89.56	27.03	32.21	24.85
1 991 769	140/140	SO	89.56	26.96	24.05	24.88

Appendix B summarizes the standard sequences used, while Table 1 gives a list of runs of the various sequences including temperatures *etc.* for future reference.

5.1.1. PCU–HMI Roll Determination

In order to get the zero point for the (Q, U) coordinate system, it is necessary to determine the roll angle between the instrument and the PCU. This is done by moving an aperture stop in the PCU horizontally using the linear stage and taking images in calmode. The direction of motion of the image of this stop on the CCD gives the PCU–HMI roll. The rotation angles of the PCU polarizer and retarder are then corrected by this amount. A consequence of this is that operationally, horizontal is defined as the direction mapped into the row direction on the CCDs.

This test sequence was run near the time of the calibration sequences taken beginning with Filtergram Sequence Number (FSN) 678 221.

5.2. Focus Block Tests

Since the instrument is focused by inserting fixed-thickness focus blocks in the light path between the polarization selectors and the ISS beamsplitter, it is essential to determine if these blocks have any birefringence or cause other polarization artifacts. To this end the PCU is commanded to go through (approximately) $I + Q$, $I + U$, and $I + V$, with the instrument, at each, taking a polarization sequence with none and each of the six focus blocks individually in the beam.

5.3. Determination of Degenerate Angle

As mentioned in Section 5.1, θ_1, θ_2, and θ_3 are degenerate and it is only possible to determine two distinct linear combinations. There are several possible solutions to this problem. One would be to decide that since the numbers are degenerate anyway one can pick an arbitrary θ for one of the motors, such as the numbers from Section 2.3, but as described those are not very accurate. Since there may be high-order effects (such as the incidence-angle effect mentioned in Section 4.4) depending on the actual angles this option is unsatisfactory. It is also possible that some high-order effect or the effects mentioned in Section 4.4 could be exploited but this was not pursued.

In the end it was decided to perform a one-time, special test taking four sequences:

- A standard polarization sequence with all selectors present
- A sequence with all polarization selectors removed
- A sequence with only the first selector present
- Another standard sequence with all selectors present

Note that the polarization selectors were taken in and out by hand, which means that there may have been small alignment changes.

5.4. Data Analysis Procedure

The data analysis consists of several steps:

5.4.1. Bad Image Rejection

First any incomplete images caused by problems with the data collection are detected and rejected. As shown in Table 1, no more than one image per camera had to be rejected. Given the redundancy in the sequences this is not a significant problem.

5.4.2. Dark Subtraction

For observations in air, the CCDs are at room temperature and the dark current is thus large, at times exceeding the signal. Furthermore the temperature (and thus dark current) often changes during the time of the observations and images taken after a long pause (*e.g.* after moving PCU elements) often show slightly different dark levels. For these reasons two darks are taken at the beginning of each sequence of images of which the first is discarded and the second is used to make a model of the dark level as a function of time, which is then subtracted from the other images.

A rough flat field was also applied, although this has little effect on other parameters than those describing the light source.

5.4.3. Binning

Since individual pixels are noisy and since the spatial dependencies of the parameters are gradual, the images are binned to either 1×1 and 32×32 for the results presented here.

5.4.4. Fitting

Although it makes little physical sense (since many elements are rotating), each pixel is fitted separately. Given an initial guess for each of the parameters and the state of each mechanism, the model intensity and its derivatives with respect to each of the parameters are calculated. The best fit is then found using a non-linear minimization procedure.

6. Results

The number of results obtained from the numerous calibrations is very large, and thus only the main results are reported below. Except for determining temperature dependencies, results are typically only shown for one run of the sequences and only for the front camera. The other runs, including those for the side camera were analyzed and the instrumental results were found to be very similar.

6.1. Parameter Maps

Figures 3 through 8 show maps of the various fitted parameters near the planned operating temperature of 20°C. Unless otherwise noted obsmode results use the series at FSN = 450 102, while calmode results are for the series at FSN = 449 822.

6.1.1. Stimulus Telescope Parameters

Figure 3 shows variations of the order of 10% in the obsmode intensity, due to imperfections in the light feed, while the calmode intensity is more uniform. The patterns show significant variations, likely from differences in alignment during setup for the different runs.

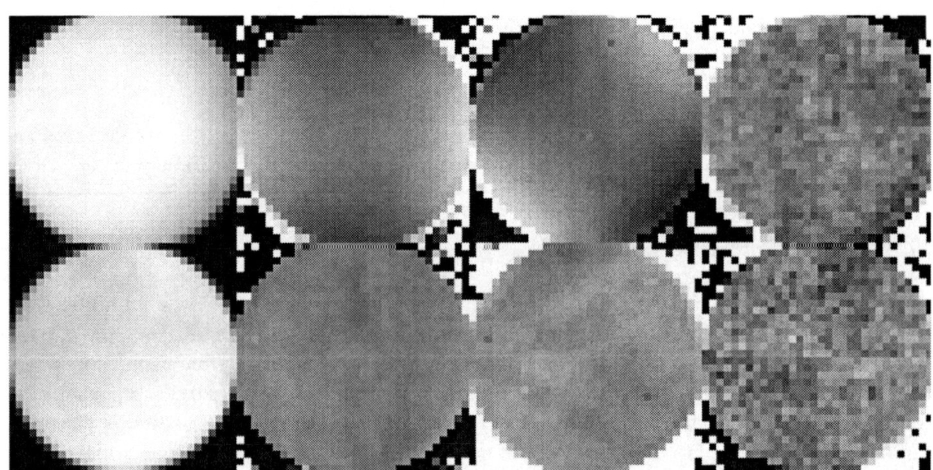

Figure 3 Parameters for the stimulus telescope light. Front left to right: Intensity (0 to maximum), Q/I (± 0.01), U/I (± 0.01) and V/I (± 0.001). The top shows obsmode, the bottom calmode.

The input polarization in obsmode shows an axisymmetric linear-polarization pattern with an amplitude around 0.5%, likely caused by the angle of incidence on some optics having a strong field dependence combined with an imperfect anti-reflective (AR) coating. The likely culprit is the condenser lens in the stimulus telescope, which is very close to focus and has a small radius of curvature. There is little input circular polarization. The polarization is very small in calmode, where the obsmode numbers have, effectively, been spatially averaged.

6.1.2. PCU Parameters

Figure 4 shows I_x and I_y from Equation (6), which describe the effect of the image offsets caused by the two PCU elements. As expected the patterns are roughly proportional to the corresponding derivatives of I. There are significant outliers in these plots, caused by dust specks and scratches on the condenser lens and, as expected, part of the pattern rotated by 90° as the instrument was rotated relative to the stimulus telescope from the Lockheed Martin Solar and Astrophysics Laboratory (LMSAL) to the Goddard Space Flight Center (GSFC)/Astrotech Science Operations (ASO) configuration (see Table 1). Unfortunately the linearization is not perfect and there are thus artifacts in some of the other parameters.

The remaining PCU parameters are shown in Figure 5. The polarizer transmits about 88.9% of the desired polarization, with a standard deviation of around 0.1%. For unknown

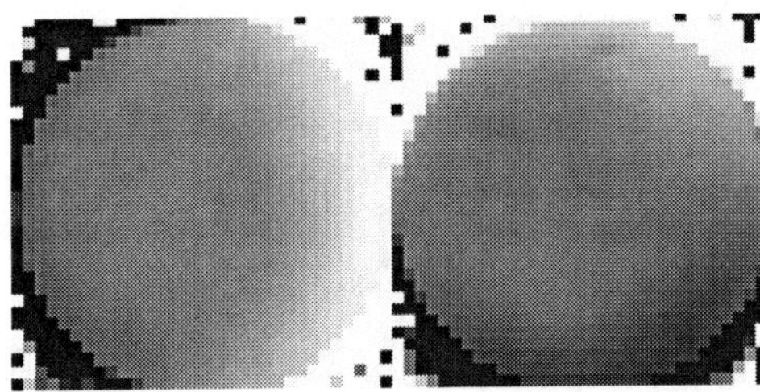

Figure 4 PCU offset parameters (see Equation (6)). Left panel shows I_x, right panel I_y. Scale is arbitrary.

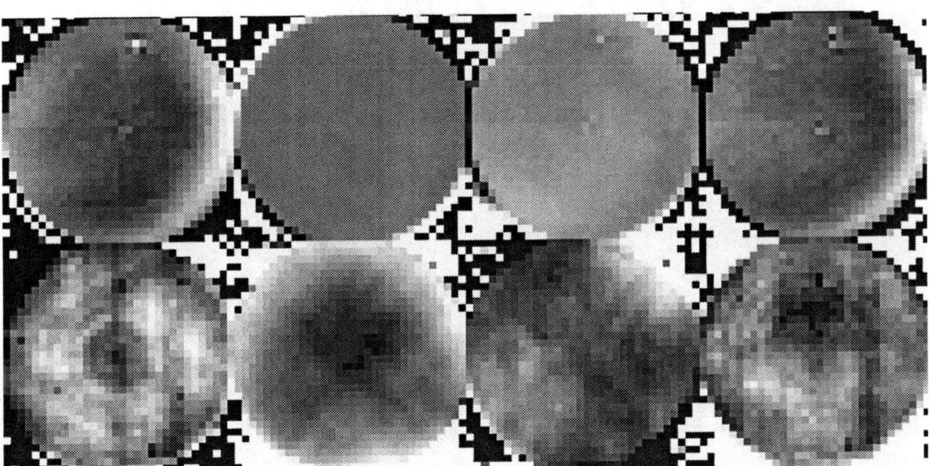

Figure 5 PCU element parameters. From left to right: Polarizer transmission (scale from 0.886 to 0.891), retarder retardance (0.209 to 0.214), retarder angle (−0.011 to −0.003 radians) and retarder transmission (0.990 to 0.997). Obsmode is in the top row, calmode in the bottom.

reasons the transmission was derived to be higher by about 0.5% for two of the runs. The parameters show little variation in obsmode (where any non uniformities are averaged out). In calmode (where the elements are near focus) there is a clear radial dependence for some of the parameters. The retarder angle shows some variations, but nothing systematic from run to run. The mean of -0.0055 radians or -0.3° is roughly consistent with the error estimate of 0.2°.

The PCU parameters appear to be consistent between obsmode and calmode when averaged spatially.

6.1.3. Window Parameters

Two types of parameters describe the HMI front window: one giving the amount of depolarization, the other the birefringence. As described in Section 4.3 a six parameter model is normally used, although a five-parameter model with explicit depolarization and birefringence parameters was also implemented. Figure 6 shows the result of such a fit (which at nominal operating temperature is as good as the six-parameter fit). The first thing to note is that the depolarization is very low. In obsmode the averages of D_Q, D_U, and D_V (see Equation (9) in Appendix A.2) are within 0.25% of 1, with D_Q slightly larger than 1, likely due to a small systematic error of unknown origin. The total birefringence is less than 0.1% of a wave.

The calmode results are more interesting. The birefringence, while small, shows a triangular pattern, likely due to a small amount of stress induced by the three-point mountings of the front window and primary lens. During the early part of the integration a test (FSN = 183462) was run without the window installed and the maps show that several of the features are still present, implicating the primary lens for at least some of the effects.

The effects of the window parameters are temperature dependent, as discussed in Sections 6.2.2 and 7.2.2.

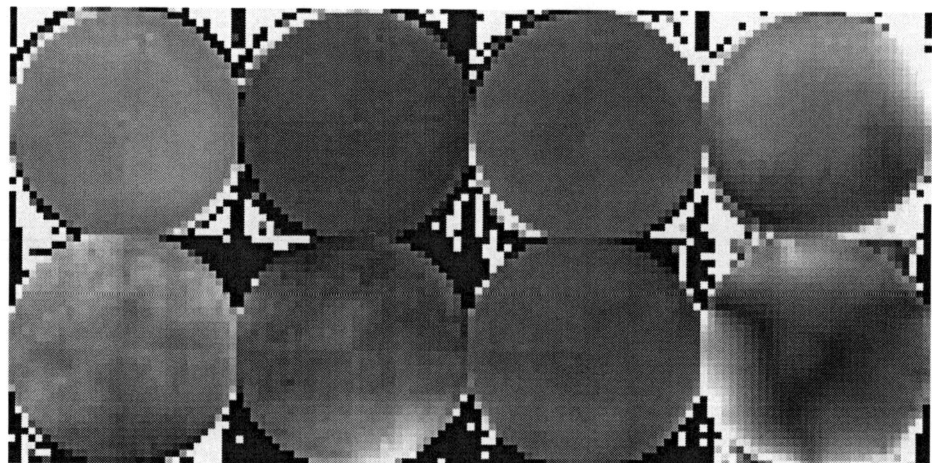

Figure 6 Front-window parameters in obsmode (top) and calmode (bottom). Three left columns show the depolarization terms on a scale from 0.99 to 1.01 (see text). The right-hand plots show the total birefringence $\sqrt{M_{QV}^2 + M_{UV}^2}$ on a scale from 0 to 0.01 (0 to 0.0016 waves) in obsmode and 0 to 0.15 (0 to 0.024 waves) in calmode.

6.1.4. Polarization Selector Parameters

The waveplate retardances $[r_i]$ and zero points $[\theta_i]$ are shown in Figures 7 and 8, respectively. The average for PS1 corresponds to a zero point in steps of 63 and the value for PS3 to 9, in good agreement with the estimates given in Section 2.3.

All parameters show some variations, most of which are consistent from test to test, but not between obsmode and calmode. Since the waveplates rotate between measurements, the spatial variations clearly cannot represent the true variations and at best represent averages. In any case the variations are small and not a cause of concern.

The value of θ_3 is degenerate with the angle selected by the polarizing beamsplitter. If this angle varies spatially due to unmodeled properties of the beamsplitter or ISS mirror, this will appear as an error in θ_3. Similarly, it is assumed that the illumination is independent of

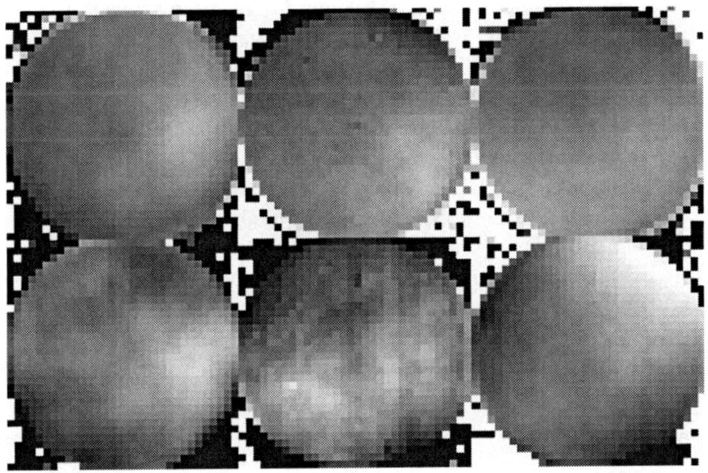

Figure 7 Retardances minus averages in obsmode (top) and calmode (bottom). From left to right are PS1 (at ±0.001 waves), PS2 (at ±0.001 waves), and PS3 (at ±0.0025 waves). Averages and standard deviations in obsmode are 10.5000±0.0002, 10.2520±0.0003, and 10.4981±0.0005. In calmode 10.4999±0.0003, 10.2517±0.0003, and 10.4980±0.0010.

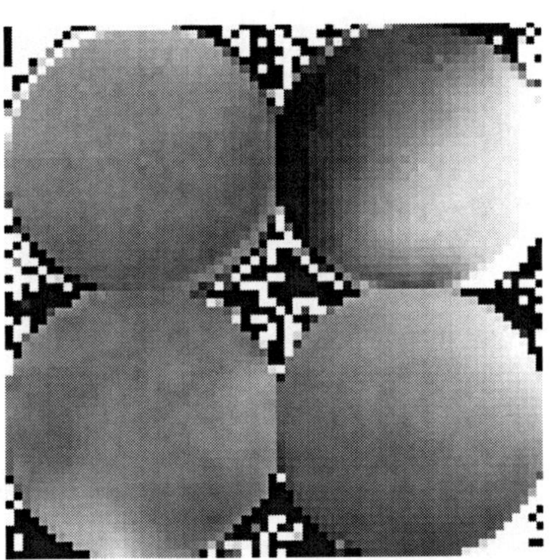

Figure 8 Retarder zero points θ_1 (left) and θ_3 (right) in obsmode (top at ±0.001 radians relative to mean) and calmode (bottom at ±0.005 radians relative to mean). Temperature for this measurement was 20.8°C. Averages and standard deviations are -1.6559 ± 0.0002, -2.3797 (see Section 6.3) and 0.2408 ± 0.0006 radians in obsmode and -1.6556 ± 0.0008, -2.3797 (see Section 6.3) and 0.2399 ± 0.0016 in calmode.

pupil position in obsmode and of the field position in calmode and that the average angle selected is thus, to first order, that of the central ray. However, the illumination is not totally uniform and varies from test to test, which may explain some of the temporal variations seen.

6.1.5. Quality of Fit

One of the ways in which one can assess the overall accuracy of the calibration is by examining the residuals. Figure 9 shows the quality of the fits for integrated light. The rms deviation is of the order 0.44% of the mean intensity.

A possible cause of residuals is the lamp intensity. To test for this, a sequence taking two darks, 50 identical images, and a dark on both cameras, all at the same cadence as used for the regular observations, was taken before each test. The resulting variations range from better than 100 ppm to somewhat over 1000 ppm. Larger values generally indicate that the bulb is about to fail or has not been adequately warmed up. In those cases the bulb is replaced and/or the lamp is warmed up and the test repeated. In any case the variations fail to explain a significant fraction of the residuals observed. Figure 9 shows a linear temporal trend in the residuals, likely due to the lamp changing intensity. However, this change is small, and correcting for it does not improve the fit significantly.

This is consistent with the residuals being correlated from test to test in the same configuration. The residuals between obsmode and calmode are significantly correlated in the same configuration. When the configuration changed from LMSAL to GSFC/ASO (causing the instrument to rotate 90° relative to the stimulus telescope) the residuals are almost uncorrelated.

Figure 10 shows the rms residuals as a function of position. In un-averaged light the rms over the images is 0.38% in obsmode and 0.69% in calmode, both with significant

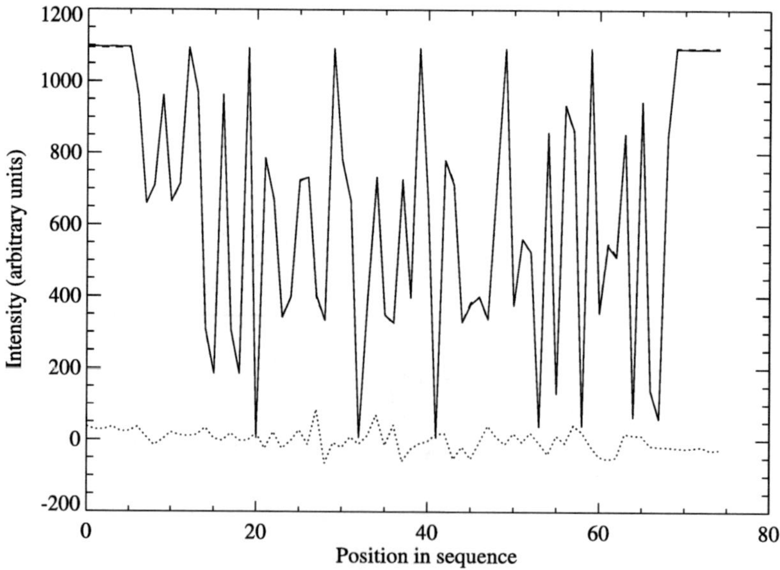

Figure 9 Spatially averaged intensity (solid line) and fit (dashed line) as a function of position in a standard (short) test sequence. Darks have been excluded. Note that the dashed line is barely visible indicating an excellent fit. This is further illustrated by the dotted line which shows ten times the residuals.

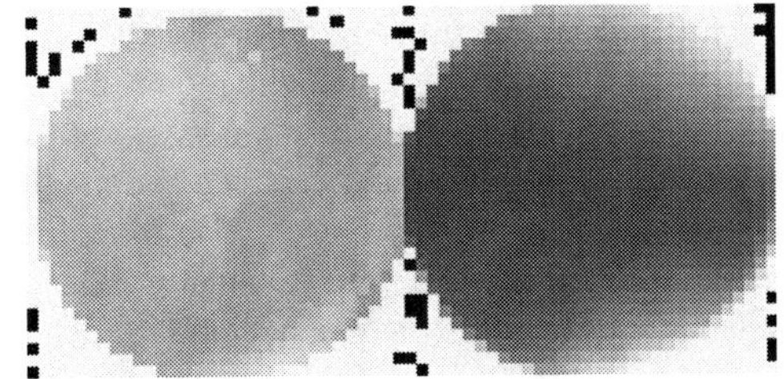

Figure 10 Rms residuals, divided, at each point, by the mean intensity at that point. Obsmode (left from 0 to 0.6%) and calmode (right from 0 to 1.6%).

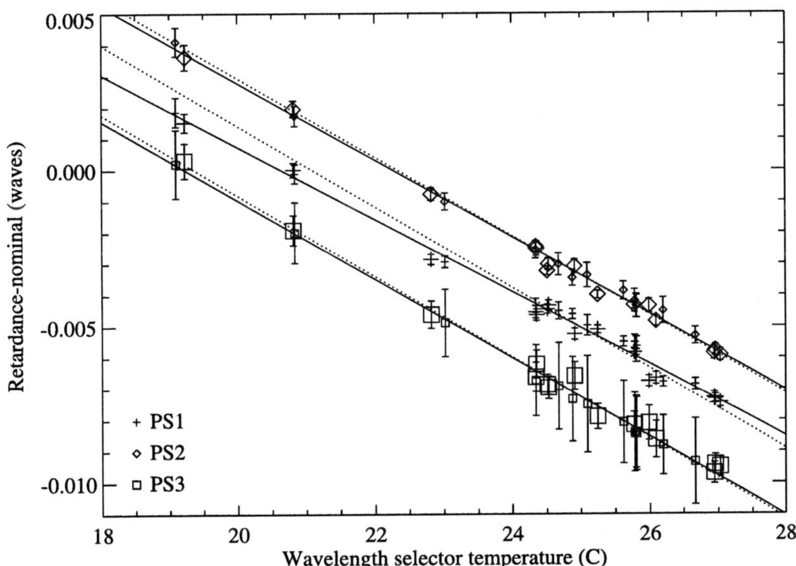

Figure 11 Waveplate retardances as a function of temperature. The values were averaged over $r \leq 0.95$ and the nominal values (10.5, 10.25, and 10.5 waves) subtracted. Error bars show the standard deviation over the averaging area. Large symbols are obsmode, small symbols calmode. Solid lines are fits. The slopes are -0.00116 waves C^{-1}, -0.00123 waves C^{-1}, and -0.00126 waves C^{-1} for PS1, PS2, and PS3, respectively. Dotted lines show the values assuming a fractional change of $-1.23 \times 10^{-4} C^{-1}$ relative to the numbers at 25°C.

variation. The obsmode residuals are fairly uniform while the calmode residuals have a very systematic pattern, likely due to some neglected physical effect, possible the same one causing the systematic residuals mentioned above.

6.2. Temperature Dependence

In principle, many components vary with temperature. However, we found only two significant temperature effects on the polarization properties: the waveplate retardances and front-window birefringence.

6.2.1. Waveplate Temperature Dependence

Figure 11 shows that the retardance variations are close to linear and to that expected from a fractional change of $-1.23 \times 10^{-4} C^{-1}$ (Etzel, Rose, and Wang, 2000). The rms deviations of the individual means from the fit are of the order of 0.0002 waves, meaning that it is straightforward to model them with the required accuracy.

The retardances are essentially independent of the configuration, in other words the values are the same in obsmode and calmode and in air and vacuum (not shown).

The residuals from the fit could be due to temperature gradients inside the instrument. The temperature used for Figure 11 is that of a sensor on the motor mount, and there could be a systematic difference between that and the actual optics.

When the front window is heated to create a large temperature gradient (see the next section), θ_1 shows large spatial variations in calmode. This is likely to be spurious due to the window model being inadequate in that case and is of little consequence since there are no plans to operate with a large gradient.

6.2.2. Front Window Birefringence

Maps of M_{QU}, M_{QV}, and M_{UV} (see Appendix A.4) at selected environment-to-window temperature differences are shown in Figure 12. Clearly there is significant stress birefringence, which could be acquired in at least two ways.

Bulk temperature change. As described in Section 2.1 the front window is a sandwich of different materials with different temperature-expansion coefficients that will induce stress in each other when the temperature is different from some reference value (presumably close to the assembly temperature).

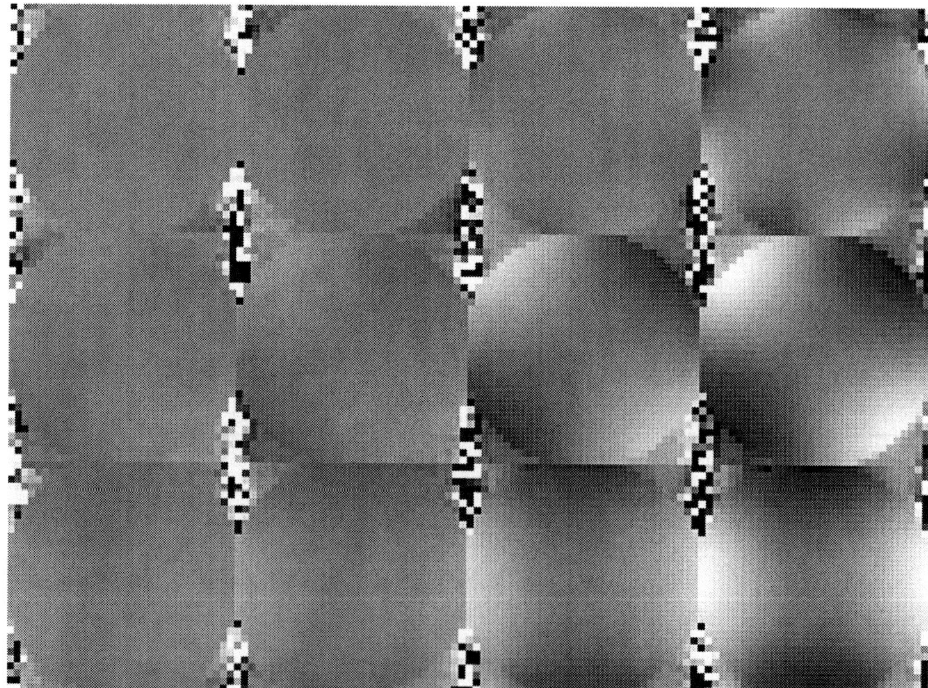

Figure 12 Window parameters in calmode: M_{QU} (top row), M_{QV} (middle row), and M_{UV} (bottom row) for four values of the leg-to-window temperature difference: −0.1°C, 4.8°C, 9.3°C, and 17.4°C. Scale is from −1 to 1.

Temperature gradients. Since the edge and center of the window are coupled differently to the environment, a radial temperature gradient can develop. In air one may reasonably assume that the dominant term determining the gradient is the difference between the air temperature and the temperature of the rim of the window, perhaps coupled with the amount of air flow. In vacuum, the gradient is presumably determined by a competition between the heat applied by the rim heater, on the one hand, and radiation to the environment and the interior, on the other. In either case the expected temperature distribution is, to lowest order, given by a cosh function of radius, which over some parameter range may be approximated by a quadratic.

This effect, in principle, also applies to other optical elements, but those are internal to the box, and see a well-regulated and homogeneous thermal environment. A likely exception to this is the primary lens, which is coupled closely to the window but which is in any case treated together with the window.

Determining which effect is dominant requires modeling or tests in different environments. Unfortunately it was not possible to test at a grid of air and rim temperature combinations (see Figure 13), and so it is not possible to do a clean analysis.

Assuming that the stress birefringence is small and linear in the temperature (or temperature difference), it is possible to correlate M_{QV} and M_{UV} against the relevant temperatures. The result is that M_{QV} and M_{UV} have correlation coefficients of 96.7% and 96.9% against the front-window temperature and 98.7% and 98.8% against the difference between the environment temperature and that of the front window. This strongly points to the temperature gradient being the culprit. M_{QU} is quadratic in the birefringence to lowest order and was therefore not used for the correlation, but it is given quite accurately by Expression (10) in Appendix A.3 from M_{QV} and M_{UV}.

Figure 14 shows that the birefringence temperature dependence is roughly quadratic with radius. Also shown is a fit of a cosh function which, as discussed earlier, is the expected

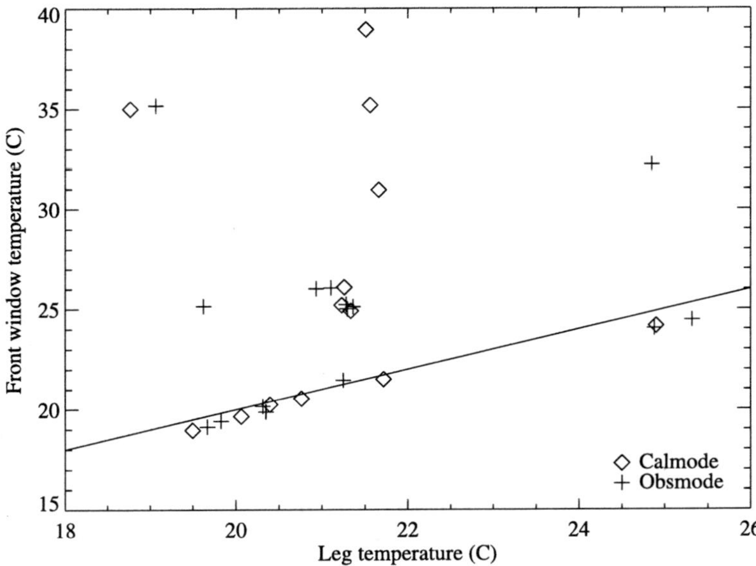

Figure 13 Available leg (environment) and front-window temperatures combinations. Straight line shows a one-to-one correspondence. Points below the line are likely due to the leg temperature not being a perfect representation of the environment and/or imperfect calibration of the temperature sensors.

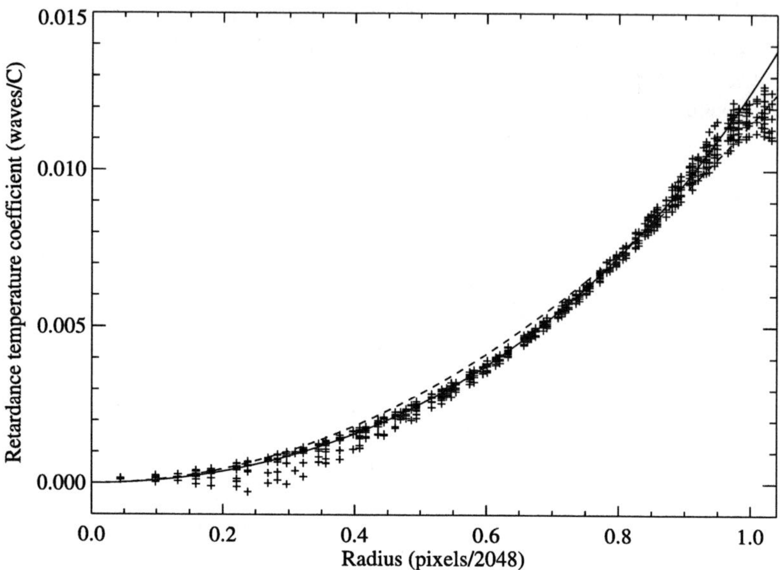

Figure 14 The temperature coefficient of the birefringence. The turnover near $r = 1.0$ is likely caused by the birefringence being close to 1/4 wave at the highest temperature causing the simple equation retardance $= \sin^{-1}(\sqrt{M_{QV}^2 + M_{UV}^2})$ to break down. Also shown is a quadratic $0.0115r^2$ (dashed) and a fit of a cosh function $0.0125(\cosh(1.85r) - 1)/(\cosh(1.85) - 1)$ (solid).

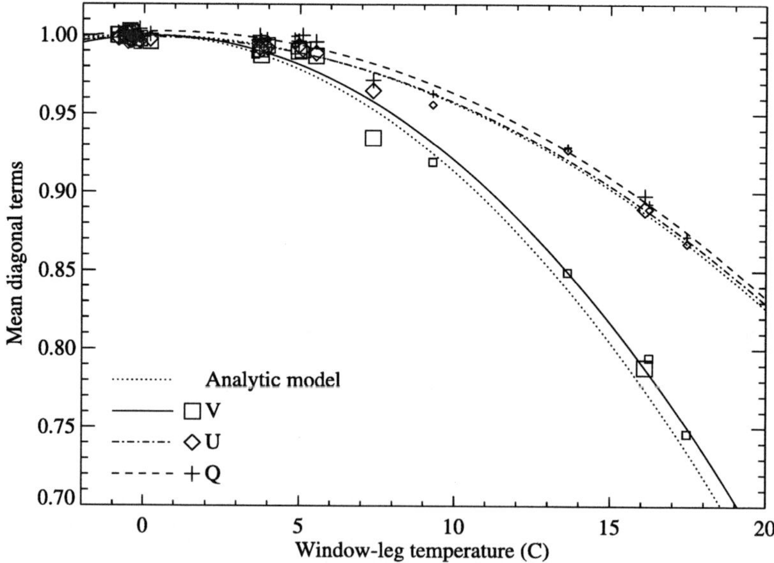

Figure 15 Temperature coefficient of the diagonal terms of the window Mueller matrix. Large symbols show obsmode and small symbols calmode. The dotted lines shows the values obtained from the fit in Figure 14 expanding the Mueller matrix of a waveplate to second order in the retardance and integrating the result over the aperture. Solid, dashed, and dash-dotted lines show quadratic fits.

functional form of the radial temperature dependence. The minimum birefringence (in an rms sense) is reached for a temperature difference of −1.5°C, that is when the window rim is colder than the environment. However, the proxy for the environment temperature (that of a mounting leg) is not necessarily an accurate representation, and so it is not clear that this is significantly different from zero.

Figure 15 shows the effect of a temperature gradient on the diagonal elements of the window Mueller matrix, which roughly gives the amount of depolarization. The numbers are reasonably well represented by the quadratic fits, as well as from the value expected from the quadratic relation shown in Figure 14. The maxima of the fitted curves are located between 0.2°C and 0.7°C with values between 0.999 and 1.0025. In other words the depolarization is very small when the window is at a uniform temperature. The deviations from the fit may be caused by the difference between the true environment and that measured by the leg temperature, and by the fact that the temperatures sometimes vary by 1°C or more during the observations.

It is clear that the window effects are significant and will require attention when constructing observables. A calibration to find the effective environmental temperature will have to be performed on orbit (see Section 7.3.2). A large amount of birefringence will not only result in depolarization (which may in principle be corrected for), it will also result in what effectively amounts to a polarization-dependent point-spread function.

Some implications of the temperature dependencies are discussed in Section 7.2.2.

6.3. Determination of Degenerate Waveplate Rotation Angle

The instrument was in an unusual configuration for this test with the Lyot filter removed, resulting in unusual light paths through parts of the instrument. For that reason only a single spatially averaged number (with unknown averaging) was determined.

The images were analyzed with the standard analysis code modified to allow for a fit of θ_2 and for missing elements. The result was $\theta_2 = -2.379742$ radians or −136.349°, within the uncertainty of the value of −133.5° determined before assembly (see Section 2.3).

6.4. PCU Roll Determination

As mentioned in Section 5.1.1, measurements of the PCU to HMI roll were taken in connection with the main polarization measurements, starting with FSN 678 221. Occasionally the automated algorithm used yields a bad value, which has to be rejected, but otherwise the offsets fall on a straight line to within better than one pixel. Since the two cameras are not perfectly aligned, the angle derived is different, typically by around 0.083° or 0.0015 radians, consistent with the value found by Wachter *et al.* (2010). For the calculations here, the mean angle is used for both cameras.

To test this procedure, a pair of tests (FSN 678 221 and 678 361, which are not otherwise used) were performed. One with a normal setup, the other with one side of the PCU propped up by 12 mm, which with a lever arm of about 800 mm should result in a tilt of 0.015 radians. The measured roll change was 0.0147 radians, well within the knowledge of the tip. After correcting the PCU angles by this number, θ_1 changed by 0.00081 radians, indicating a very accurate correction.

For the series taken before FSN 678 221, a constant was added to θ_1 to make the mean of each determination equal to the ones taken after FSN 678 221.

6.5. Focus Blocks

The images for each input polarization and focus block are demodulated using a simple polarization model. The case with no focus blocks is then subtracted from the cases with focus blocks in the beam, thereby canceling out other PCU and instrument defects to lowest order.

The results typically show spatially uniform differences from zero to 0.5% or so. However, there does not appear to be a persistent pattern to the changes and the effects seen are most likely due to the variable dark current mentioned in Section 5.4.2. Consistent with this, the spatial variations in polarization are significantly smaller, typically around 0.05% to 0.1%.

7. Discussion

In the previous sections the polarimetric model was described and the results presented. This section starts with a discussion in Section 7.1 of the overall accuracy of the estimated parameters, followed in Section 7.2 by a discussion of the features that are currently planned to be in the polarimetric model used for the calibration of the science data. Finally, Section 7.3 describes the plans for updates to the calibration after launch.

7.1. Accuracy of Derived Parameters

It is difficult to reliably assess the errors in the various measured quantities, but there are a number of hints about how large they may be. Among these are the rms residuals *versus* settings and image position, the (unphysical) spatial variations of the waveplate parameters, the variations between runs, the residuals from the fits *versus* temperature (Section 6.2), and how these errors propagate into the error matrix (Section 7.2.3).

The reconstruction errors were shown in Figure 9 and discussed in Section 6.1.5. These probably represent a lower bound on the errors in the combined PCU–HMI system. Unfortunately, it has not proven possible to determine their origin, and so it is unclear if they are related to the PCU or the instrument.

The up-down variation in calmode residuals in Figure 10 likely indicates the neglect of some physical effect. That it does not appear in obsmode could indicate that it is related to the steep angles through the area around the polarization selectors. The angles have been taken into account for the polarization selectors and the polarizing beamsplitter, but a possible candidate is the ISS mirror, which is likely not perfect (*e.g.* the reflectivity and the phase delay on reflection are different for s and p polarization). However, since the effect is negligible in obsmode, this issue has not been pursued.

The spatial variations of the waveplate retardances and zero points are, as discussed in Section 6.1.4, very small and are in some cases degenerate with other (unfitted) parameters and may thus, to some extent, reduce the errors in the calibrated results.

7.2. Resulting Polarimetric Model

7.2.1. High-Incidence-Angle Effects

Figure 16 shows the inferred retardances with and without including the effect of the finite incidence angles on the polarization selectors. The effect is quite significant in calmode,

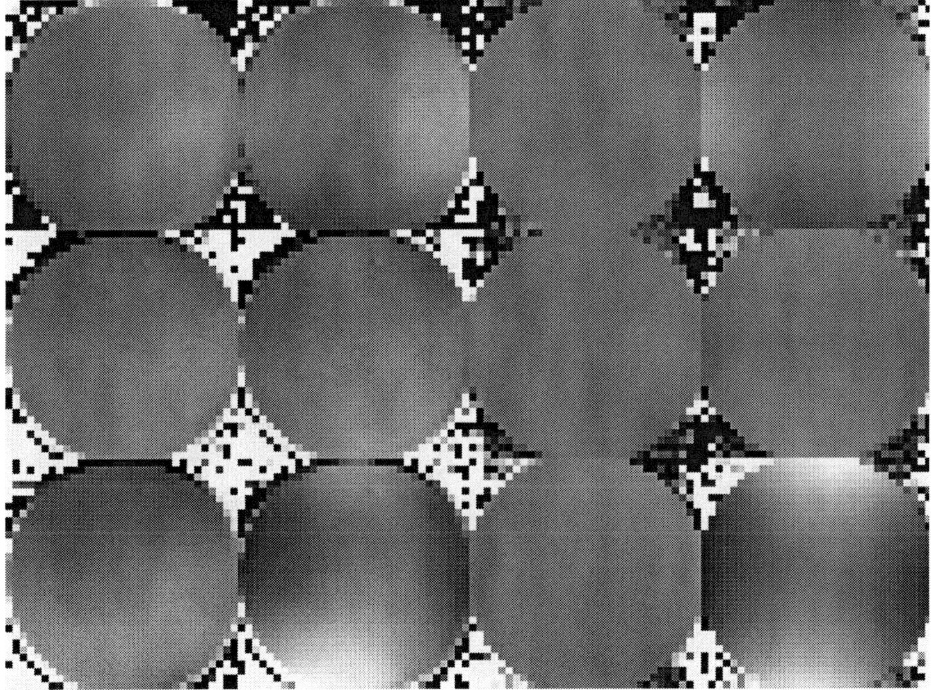

Figure 16 Waveplate retardances with and without correcting for the angle-of-incidence effect in the waveplates. Rows are the retardances of PS1 (top), PS2 (middle) and PS3 (bottom). Two left columns are in obsmode with and without the correction on a scale of ±0.002 waves. The two right columns the corresponding calmode results on a scale of ±0.02 waves.

especially for PS3, where including the effect makes the rms variation decrease from 0.0072 waves to 0.00097 waves. In obsmode the corresponding decrease is from 0.00072 to 0.00048 waves. The mean rms reconstruction error in calmode is reduced from 1.71% to 0.71%, again indicating a significant effect. The obsmode improvement from 0.48% to 0.45% is more modest.

Given the magnitude of the improvements, it is clear that this effect should be included, even if there is no requirement to calibrate calmode data.

7.2.2. *Models of Spatial and Temperature Dependence*

To perform the demodulation described in Section 1.1 it is necessary to have a model that will reliably predict the model parameters given instrument settings and environmental conditions, such as the window and polarization-selector temperatures. The prescription described below represents the current plans and will be revisited after on-orbit data have been taken.

To obtain the spatial dependence, it is necessary to smooth over the minor defects in the parameters due to dust specks *etc.* and to extrapolate the data between the edge of the valid area and the edge of the FOV used. This is done by fitting a low (fifth) order polynomial in x and y to the data out to a fractional radius of 0.95.

The window is modeled using the six-parameter model with the temperature effects for the (Q, U, V) part of the matrix given by a quadratic in window-environment temperature

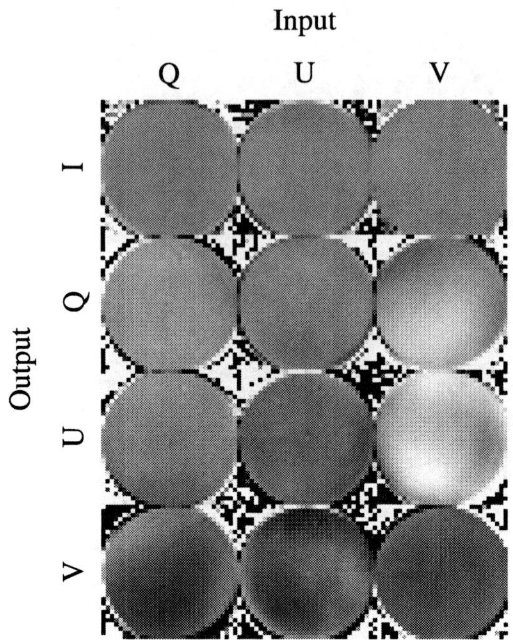

Figure 17 The demodulation errors resulting from modulating using a set of derived parameters (FSN 450 102) and demodulating using the model described in Section 7.2 for the corresponding temperatures. Columns have the different input polarizations, rows the outputs. Each sub image shows a spatial map of the corresponding element of the matrix $\mathcal{E} = \mathcal{D}\mathcal{O} - \mathcal{I}$ (see Section 1.1). Scale is $\pm 1\%$.

for the diagonal (depolarization) terms and by a constant for the off-diagonal terms. The determination of any window effects related to inherent polarization has been deferred to on-orbit, as has the determination of the effective on-orbit environment temperature.

The retardances of the polarization selectors will be modeled by the fits shown in Figure 11. The zero points will be assumed to be constant.

7.2.3. Demodulation Accuracy

Given the polarimetric model mentioned above, it is possible to calculate the expected parameters and the demodulation matrix $\mathcal{D}$ (obtained by a least-squares method from $\mathcal{O}$) for each calibration run, similar to what would be done for on-orbit data. For a given run and a given observing sequence, one can similarly calculate the modulation matrix $\mathcal{O}$ from the actually determined parameters and demodulate it using $\mathcal{D}$, thereby getting an estimate of the error matrix $\mathcal{E} = \mathcal{D}\mathcal{O}_{\text{true}} - \mathcal{I}$. Figure 17 shows a representative result.

Excluding the calibration sequences for which the window was excessively hot the rms deviation from 0 (over the sequences) of the error matrix is of the order 0.1% to 0.7% for the nine matrix elements mixing Q, U, and V, with a mean of 0.4%, thus meeting the requirement of 1% (Section 1.1). For the Q, U, and V to I elements the values are less than 0.2%, thus easily meeting the 10% requirement. As mentioned in Section 7.3.1, the determination of the instrument polarization is deferred to on-orbit and the model does not include it, resulting in the I to Q, U, and V errors being zero.

7.2.4. Selection of Optimal Settings for LCP and RCP

For a general polarization sequence that measures at least four non-degenerate states of polarization, the calibration ensures that the calibration errors are zero, assuming that the instrumental properties are known.

However, as mentioned in Section 1.1, there is a requirement to be able to determine LCP and RCP from only two observations with less that 5% crosstalk from Q and U across the FOV and with only the available motor positions. In this case it is not possible to calibrate out any linear-to-circular crosstalk and so the only way to meet the requirement is to rely on the uniformity of the optical properties across the FOV.

To determine if it is possible to satisfy these constraints, a search was performed. The result is that there are many such sequences, some of which keep not just the individual crosstalks but also the total linear-to-circular crosstalk below 3.8%.

7.3. Plans for On-Orbit Calibration

As mentioned earlier, some parts of the calibration have been deferred to on-orbit. Also, the performance of the instrument will have to be checked and monitored for any aging-related effects or other types of changes.

7.3.1. Items Deferred to On-Orbit

There are two main items deferred to on-orbit tests: the telescope polarization, and the effective environment temperature to use to correct for the window depolarization. The latter is discussed in the following section.

The former should be straightforward to determine. Spatially and/or temporally averaged sunlight is almost perfectly unpolarized, especially in the continuum. It is thus possible to determine the telescope polarization by demanding that the incoming sunlight is unpolarized on average. The averaging may either be done using calmode or long-term averages in obsmode (possibly excluding active regions).

7.3.2. Continuous Monitoring

The instrument will change after launch. Experience with other missions indicates that some change and degradation is bound to happen. In the case of HMI, a likely change from a polarization point of view is an increase in the front-window absorptivity, as seen on the *Michelson Doppler Imager* (MDI) onboard the *Solar and Heliospheric Observatory*. The effect of this is that the window will absorb more energy, heat up, and cause the center-to-edge gradient to change. As discussed in Section 6.2.2, this has significant consequences and it is desirable to keep the gradient low by adjusting the setpoint of the window heaters. There are various ways that this degradation can be tracked.

The amount of energy supplied to the window heaters in order to maintain a specified temperature is monitored. From this a thermal model may be able to provide an estimate of the gradient as a function of window temperature.

Another effect of a temperature gradient is that the focus of the instrument will change, as was seen with MDI. In the absence of any other changes, the window temperature that keeps the focus fixed in time is thus likely to be the same one that will cause the gradient to remain constant.

Finally it may be possible to measure the depolarization by observing a sunspot and varying the window temperature. The polarization at best focus should then vary quadratically as seen in Figure 15, with the maximum indicating the desired operating temperature. Assuming that a suitable spot is available, an attempt to do this will be made during commissioning, otherwise another time will have to be found. It is undesirable to perform this test frequently as it takes considerable time to execute.

8. Conclusion

The HMI instrument has been designed to measure full Stokes I, Q, U, and V for the full disk, for several wavelengths with $0.5''$ pixels.

We have demonstrated that the polarimetric efficiency of the instrument is almost perfect and allows for these measurements to be done with the required accuracy. We have also demonstrated that we have been able to build a polarimetric model that will allow us to calibrate the data independent of the details of the observing mode chosen and the environment.

Acknowledgements This work was supported by NASA Contract NAS5-02139 to Stanford University. The authors are grateful to Dick Shine, Bruce Lites, and Jack Harvey for critical advice and Tony Darnell for help with the design of the test equipment. The authors would also like to thank numerous engineers and other staff members at Stanford University and Lockheed Martin Solar and Astrophysics Laboratory including John Miles, Brett Allard, Dave Kirkpatrick, Glenn Gradwohl, Thomas Nichols, Noah Katz, Sebastien Couvidat, Richard Wachter, Rock Bush, Phil Scherrer, Todd Hoeksema, and Yang Liu, for designing the instrument, running tests, making the data system run smoothly, and helping with analysis. NCAR is supported by the National Science Foundation.

Appendix A: Models of Various Types of Optical Elements

The following subsections collect the information about how the various types of optical elements are modeled.

A.1. Polarizer Models

The matrix of a linear polarizer P with an angle θ and a transmission t is given by:

$$P(\theta,t)=\frac{t}{2}\begin{pmatrix} 1 & \cos 2\theta & \sin 2\theta & 0 \\ \cos 2\theta & \cos^2 2\theta & \cos 2\theta \sin 2\theta & 0 \\ \sin 2\theta & \cos 2\theta \sin 2\theta & \sin^2 2\theta & 0 \\ 0 & 0 & 0 & 0 \end{pmatrix}. \tag{8}$$

A.2. Depolarizer Model

For the results presented, only a three-term model has been used:

$$D_3(D_Q,D_U,D_V)=\mathrm{diag}(1,D_Q,D_U,D_V). \tag{9}$$

A.3. Retarder Models

The matrix of a retarder W with angle θ and retardance r (in waves) is given by:

$$W(\theta,r)=\begin{pmatrix} 1 & 0 & 0 & 0 \\ 0 & c^2+s^2\cos\delta & cs(1-\cos\delta) & -s\sin\delta \\ 0 & cs(1-\cos\delta) & s^2+c^2\cos\delta & c\sin\delta \\ 0 & s\sin\delta & -c\sin\delta & \cos\delta \end{pmatrix}, \tag{10}$$

where $c=\cos 2\theta$, $s=\sin 2\theta$, and $\delta=2\pi r$.

A problem with Equation (10) is that θ becomes poorly defined as r gets smaller, causing large spurious variations in θ in the presence of noise, as well as convergence problems. This can be avoided by using an alternate expression:

$$W_M(M_{QV}, M_{UV}) = W(\theta, r), \tag{11}$$

where, formally, $M_{QV} = -\sin 2\theta \sin r$ and $M_{UV} = \cos 2\theta \sin r$. In reality one avoids the calculation of θ and r and calculates the other matrix elements and derivatives directly.

Another problem with Equation (10) is that it is only valid for normal incidence. For non-zero incidence angle the retardance has to be modified, with the result that

$$W_I(\theta, r, \alpha, \beta) = W(\theta, r'), \tag{12}$$

where

$$r' = r\bigl(1 + \alpha^2 \cos(2(\beta - \theta))/2\bigr), \tag{13}$$

α is the incidence angle and β is the angle between the horizontal and the direction from the center of the field to the point of interest. In other words, $\beta - \theta$ is the angle between the fast axis of the waveplate and the direction from the center of the field to the point of interest. It is important that r is the true retardance (*e.g.* 10.25 waves not 0.25 waves). Also note that the sign of the correction term is different for other waveplate materials.

A.4. Combined Models

If an optical element has a varying fast-axis direction along the ray path, the resulting Mueller matrix can not, in general, be described as a simple retarder, as given above. If there is also an amount of depolarization due to, for example, variation of the retardance direction or magnitude over the beam, the Mueller matrix is further complicated. This Mueller matrix

$$M_6(M_{QQ}, M_{QU}, M_{QV}, M_{UU}, M_{UV}, M_{VV}) = \begin{pmatrix} 1 & 0 & 0 & 0 \\ 0 & M_{QQ} & M_{QU} & M_{QV} \\ 0 & M_{QU} & M_{UU} & M_{UV} \\ 0 & -M_{QV} & -M_{UV} & M_{VV} \end{pmatrix}, \tag{14}$$

which captures an average of an arbitrary set of retarders, is useful in that case, although it may be noted that it has more degrees of freedom than needed, in that $M_{VV} = M_{QQ} + M_{UU} - 1$ for an arbitrary combination of retarders.

Appendix B: Polarization Calibration Sequences

Two main sequences have been used for the calibration, the so-called short and long sequences. For each of a number of PCU configurations, one or more simple instrument sequences are taken. These are designed to measure simple polarization states with nominal waveplate properties. In other words they have been kept fixed over time and use initial guesses for the zero points *etc.* (rather than correcting them based on later information). Angles given below are in degrees and given relative to the numbers in Section 2.3. Also note that the PCU angles are nominal and refer to the LMSAL configuration where the bottom of the instrument and the bottom of the PCU are parallel. For the GSFC/ASO case 90° has to be added to the PCU angles.

Standard sequences:

- Single linear: Two darks and roughly $I + Q$ (waveplates at 0°, 357°, and 45°).
- Double linear: Two darks and roughly $I \pm Q$ (above and 0°, 357°, and 0°).
- Sequence A: Two darks, a four position sequence, a dark.

$PL1$	0	0	23	23
$PL2$	357	357	357	357
$PL3$	36	9	54	81

- Sequence C: Two darks, a six-position sequence, a dark.

$PL1$	0	0	23	23	0	0
$PL2$	357	357	357	357	357	357
$PL3$	45	0	45	0	23	68

- Sequence AC: Two darks, a four-position sequence, a six-position sequence, a dark.

$PL1$	0	0	23	23	0	0	23	23	0	0
$PL2$	357	357	357	357	357	357	357	357	357	357
$PL3$	36	9	54	81	45	0	45	0	23	68

- Short combo: Two darks, the following sequence of waveplate angles, and a dark.

$PL1$	0	120	240	0	0	0	0	120	240
$PL2$	357	357	357	117	237	357	357	117	237
$PL3$	0	0	0	0	0	120	240	120	240

- Long combo: Two darks, the following sequences of waveplate angles, and a dark.

$PL1$	0	60	120	180	240	300	0	0	0	0	0
$PL2$	357	357	357	357	357	357	57	117	177	237	297
$PL3$	0	0	0	0	0	0	0	0	0	0	0

$PL1$	0	0	0	0	0	60	120	180	240	300
$PL2$	357	357	357	357	357	57	117	177	237	297
$PL3$	60	120	180	240	300	60	120	180	240	300

Note that sequences A and C are both complete polarization sequences able to determine I, Q, U, and V, similar to those intended for normal use.

In addition, sequences to monitor the stability of the setup are used:

- Short clear: Single linear with PCU out.
- Long clear: Double linear with PCU out.

B.1. Short Sequence (75 images and 65 darks)

- Short clear
- Without PCU: Sequence A
- Short clear
- Polarizer at 0°, retarder at 90°, 150°, 210°, 270°, 330°, 30°: Single linear
- Short clear
- Polarizer only at 0°, 60°, 120°, 180°, 240°, 300°: Single linear
- Short clear

- Polarizer only at 0°, 120°, 240°: Short combo and short clear
- Polarizer at 0°, retarder at ∓45°: Short combo and short clear
- Without PCU: Sequence A
- Short clear

B.2. Long Sequence (192 images and 70 darks)

- Long clear
- Without PCU: Sequence AC
- Long clear
- Polarizer at 0°, retarder at 90°, 150°, 210°, 270°, 330°, 30°: Double linear
- Long clear
- Polarizer only at 0°, 60°, 120°, 180°, 240°, 300°: Double linear
- Long clear
- Polarizer only at 0°, 120°, 240°: Long combo and long clear
- Polarizer at 0°, retarder at ∓45°: Long combo and long clear
- Without PCU: Sequence AC
- Long clear

References

Beck, C., Schlichenmaier, R., Collados, M., Bellot Rubio, L., Kentischer, T.: 2005, A polarization model for the German vacuum tower telescope from in situ and laboratory measurements. *Astron. Astrophys.* **443**, 1047 – 1053. doi:10.1051/0004-6361:20052935.

Beck, C., Schmidt, W., Kentischer, T., Elmore, D.: 2005, Polarimetric littrow spectrograph – instrument calibration and first measurements. *Astron. Astrophys.* **437**, 1159 – 1167. doi:10.1051/0004-6361:20052662.

Borrero, J.M., Tomczyk, S., Norton, A., Darnell, T., Schou, J., Scherrer, P., Bush, R., Liu, Y.: 2007, Magnetic field vector retrieval with the helioseismic and magnetic imager. *Solar Phys.* **240**, 177 – 196. doi:10.1007/s11207-006-0219-x.

Collados, M.V.: 2003, Stokes polarimeters in the near-infrared. In: S. Fineschi (ed.) *SPIE Conference Series*, **4843**, 55 – 65. doi:10.1117/12.459370.

Etzel, S.M., Rose, A.H., Wang, C.M.: 2000, Dispersion of the temperature dependence of the retardance in SiO_2 and MgF_2. *Appl. Opt.* **39**, 5796 – 5800. doi:10.1364/AO.39.005796.

Ichimoto, K., Lites, B., Elmore, D., Suematsu, Y., Tsuneta, S., Katsukawa, Y., Shimizu, T., Shine, R., Tarbell, T., Title, A., Kiyohara, J., Shinoda, K., Card, G., Lecinski, A., Streander, K., Nakagiri, M., Miyashita, M., Noguchi, M., Hoffmann, C., Cruz, T.: 2008, Polarization calibration of the solar optical telescope onboard *Hinode*. *Solar Phys.* **249**, 233 – 261. doi:10.1007/s11207-008-9169-9.

Kuhn, J.R., Balasubramaniam, K.S., Kopp, G., Penn, M.J., Dombard, A.J., Lin, H.: 1994, Removing instrumental polarization from infrared solar polarimetric observations. *Solar Phys.* **153**, 143 – 155. doi:10.1007/BF00712497.

Norton, A.A., Graham, J.P., Ulrich, R.K., Schou, J., Tomczyk, S., Liu, Y., Lites, B.W., López Ariste, A., Bush, R.I., Socas-Navarro, H., Scherrer, P.H.: 2006, Spectral line selection for HMI: a comparison of Fe I 6173 Å and Ni I 6768 Å. *Solar Phys.* **239**, 69 – 91. doi:10.1007/s11207-006-0279-y.

Schou, J., Scherrer, P.H., Bush, R.I., Wachter, R., Couvidat, S., Rabello-Soares, M.C., *et al.*: 2010, The helioseismic and magnetic imager instrument design and calibration. *Solar Phys.*, in press.

Selbing, J.: 2005, SST polarization model and polarimeter calibration. Master's thesis, Stockholm University, Stockholm, Sweden.

Skumanich, A., Lites, B.W., Martinez Pillet, V., Seagraves, P.: 1997, The calibration of the advanced stokes polarimeter. *Astrophys. J. Suppl.* **110**, 357. doi:10.1086/313004.

Wachter, R., Schou, J., Rabello-Soares, M.C., Miles, J.W., Duvall, T.L. Jr., Bush, R.I.: 2010, Image quality of the helioseismic and magnetic imager. *Solar Phys.*, in press.

Solar Phys (2012) 275:357–374
DOI 10.1007/s11207-010-9652-y

Implementation and Comparison of Acoustic Travel-Time Measurement Procedures for the *Solar Dynamics Observatory/Helioseismic and Magnetic Imager* Time – Distance Helioseismology Pipeline

S. Couvidat · J. Zhao · A.C. Birch · A.G. Kosovichev · T.L. Duvall Jr. · K. Parchevsky · P.H. Scherrer

Received: 28 December 2009 / Accepted: 29 September 2010 / Published online: 4 November 2010

Abstract The *Helioseismic and Magnetic Imager* (HMI) instrument onboard the *Solar Dynamics Observatory* (SDO) satellite is designed to produce high-resolution Doppler-velocity maps of oscillations at the solar surface with high temporal cadence. To take advantage of these high-quality oscillation data, a time – distance helioseismology pipeline (Zhao *et al.*, *Solar Phys.* submitted, 2010) has been implemented at the Joint Science Operations Center (JSOC) at Stanford University. The aim of this pipeline is to generate maps of acoustic travel times from oscillations on the solar surface, and to infer subsurface 3D flow velocities and sound-speed perturbations. The wave travel times are measured from cross-covariances of the observed solar oscillation signals. For implementation into the pipeline we have investigated three different travel-time definitions developed in time – distance helioseismology: a Gabor-wavelet fitting (Kosovichev and Duvall, *SCORE'96: Solar Convection and Oscillations and Their Relationship*, ASSL, Dordrecht, 241, 1997), a minimization relative to a reference cross-covariance function (Gizon and Birch, *Astrophys. J.* **571**, 966, 2002), and a linearized version of the minimization method (Gizon and Birch, *Astrophys. J.* **614**, 472, 2004). Using Doppler-velocity data from the *Michelson Doppler Imager* (MDI) instrument onboard SOHO, we tested and compared these definitions for the mean and difference travel-time perturbations measured from reciprocal signals. Although all three procedures return similar travel times in a quiet-Sun region, the method of Gizon and Birch (*Astrophys. J.* **614**, 472, 2004) gives travel times that are significantly different from the others in a magnetic (active) region. Thus, for the pipeline implementation we chose the procedures of Koso-

The Solar Dynamics Observatory
Guest Editors: W. Dean Pesnell, Phillip C. Chamberlin, and Barbara J. Thompson.

S. Couvidat (✉) · J. Zhao · A.G. Kosovichev · K. Parchevsky · P.H. Scherrer
W.W. Hansen Experimental Physics Laboratory, Stanford University, 491 S Service Road, Stanford, CA 94305-4085, USA
e-mail: couvidat@stanford.edu

A.C. Birch
CoRA Division, NorthWest Research Associates, 3380 Mitchell Lane, Boulder, CO 80301, USA

T.L. Duvall Jr.
Solar Physics Laboratory, NASA/GSFC, Greenbelt, MD 20771, USA

vichev and Duvall (*SCORE'96: Solar Convection and Oscillations and Their Relationship*, ASSL, Dordrecht, 241, 1997) and Gizon and Birch (*Astrophys. J.* **571**, 966, 2002). We investigated the relationships among these three travel-time definitions, their sensitivities to fitting parameters, and estimated the random errors that they produce.

Keywords Sun: helioseismology · Sun: time – distance analysis · HMI

1. Introduction

Time – distance helioseismology (Duvall *et al.*, 1993) allows access to the subsurface physical properties of the quiet Sun (see *e.g.* Kosovichev and Duvall, 1997), including supergranulation (see *e.g.* Duvall and Gizon, 2000; Zhao and Kosovichev, 2003; Hirzberger *et al.*, 2007), and also the subsurface properties of active regions (see *e.g.* Kosovichev, Duvall, and Scherrer, 2000; Zhao, Kosovichev, and Duvall, 2001; Jensen *et al.*, 2001; Couvidat, Birch, and Kosovichev, 2006). So far the main goal has been to determine the velocity of material flows in both the horizontal and vertical directions, and the sound-speed profile as a function of depth. As such, time – distance helioseismology is a major tool to achieve the primary science goal of the *Helioseismic and Magnetic Imager* (HMI) instrument onboard the *Solar Dynamics Observatory* (SDO) satellite: to study the origin of solar variability, and to characterize and understand the Sun's interior and various components of magnetic activity. Therefore, a time – distance helioseismology data-analysis pipeline (described by Zhao *et al.*, 2010) has been implemented by the authors, at the SDO Joint Science Operations Center (JSOC). It will process the line-of-sight Doppler-velocity (and potentially continuum intensity) maps obtained by HMI to generate wave travel-time maps of the solar surface and invert these maps to generate estimates of the flow velocities and sound-speed perturbations beneath the surface.

The travel times of acoustic or surface-gravity wavepackets are measured from temporal cross-covariances of solar-oscillation signals measured at different positions on the solar disk. Over the years, various definitions of these travel times have been designed (Kosovichev and Duvall, 1997 [hereafter KD97]; Gizon and Birch, 2002 [hereafter GB02]; Gizon and Birch, 2004 [hereafter GB04]). They generally return different times for a given cross-covariance function. Upon their implementation in the HMI time – distance pipeline we compare the results of these definitions using Doppler-velocity data from the *Michelson Doppler Imager* (MDI) instrument (Scherrer *et al.*, 1995). We compare these three travel-time definitions for both quiet-Sun and active-region data. We only compute the mean and difference travel-time perturbations in the center-to-annulus geometric scheme, neglecting the North – South and East – West travel-time differences (see Couvidat and Birch, 2009, for a comparison of these travel-time differences in the presence of artificial uniform and steady flows, and for the three travel-time definitions). Moreover, we only consider acoustic waves (p modes). Jackiewicz *et al.* (2007) already studied the travel times of surface-gravity waves (f modes). Finally, Roth, Gizon, and Beck (2007) compared the travel-time differences returned by the KD97 and GB04 methods in the quiet Sun. In this paper we expand part of their analysis to include the GB02 approach, to include mean travel-time perturbations, and we also work on active-region data. Other possible ways to measure the travel-time shifts of wavepackets are not considered here. For instance, it is possible to compute the instantaneous phase of the cross-covariance using the Hilbert transform, and to define the travel time as the time of zero-crossing of this instantaneous phase (see *e.g.* Duvall *et al.*, 1996). Other related techniques are phase-correlation holography (Lindsey and Braun, 2005) and acoustic imaging (Chou and Duvall, 2000).

In Section 2 we review the time – distance formalism and discuss the travel-time definitions. In Section 3 we present the implementation of these travel-time measurement procedures in the HMI JSOC data-analysis pipeline. In Section 4 we compare the results of these different approaches for quiet-Sun and active-region data from SOHO/MDI. Finally, we present our conclusions in Section 5.

2. Time – Distance Formalism

2.1. Computation of the Cross-Covariances of Solar Oscillations

In this section we outline the cross-covariance calculations used for testing our travel-time measurement procedures and codes. We apply the measurement procedure detailed by Couvidat, Birch, and Kosovichev (2006). The measurements of the travel times are based on fitting the cross-covariances $[C(\mathbf{r}_1, \mathbf{r}_2, t)]$ between source $[\mathbf{r}_1]$ and receiver $[\mathbf{r}_2]$ calculated from a line-of-sight-velocity data cube $[\phi(\mathbf{r}, t)]$ where $\mathbf{r}$ is the horizontal position vector, and t is time:

$$C(\mathbf{r}_1, \mathbf{r}_2, t) = \frac{1}{T} \int_0^T \phi(\mathbf{r}_1, t')\phi(\mathbf{r}_2, t' + t)\, \mathrm{d}t' \tag{1}$$

Here T is the temporal duration of the observation. A high-pass filter is applied at 1.7 mHz to the Fourier transform of the data cube, to discard supergranulation. An f-mode filter is applied to discard the f-mode signal. To reduce the noise level of the cross-covariances, $\phi(\mathbf{r}, t)$ is filtered in the Fourier domain using a Gaussian phase-speed filter (Duvall *et al.*, 1997) for short travel distance $\Delta = |\mathbf{r}_2 - \mathbf{r}_1|$. Then, the point-to-point cross-covariances are averaged over annuli of radius Δ centered on the source. Such point-to-annulus cross-covariances are computed for 55 distances $[\Delta]$ ranging from 3.7 Mm to 66.7 Mm. Only 11 phase-speed filters are applied for these 55 distances (one filter for five distances). These distances are not those selected for the time – distance pipeline: they are used here only to test our implementation of the travel-time definitions. To further increase the signal-to-noise ratio for the point-to-annulus cross-covariances, the latter are shifted in time – so that they are centered on specific reference times – and averaged in groups of five distances (the five Δ computed with the same phase-speed filter), thus giving only 11 final Δ values for measuring the acoustic travel times. For instance, five cross-covariances computed at distances ranging from 3.7 to 8.7 Mm are shifted in time to be centered on the same reference time (here the group travel time of a wavepacket propagating in the quiet Sun over the distance $\Delta = 6.2$ Mm, as measured by the Gabor-wavelet fit presented in the next section) and are averaged to produce the final point-to-annulus cross-covariance at $\Delta = 6.2$ Mm. The values of Δ and properties of the phase-speed filters used in this paper are summarized in Table 1.

2.2. Travel-Time Definitions Implemented in the Pipeline

The travel times of the acoustic wavepackets are measured from the point-to-annulus cross-covariances. Historically, the cross-covariances were fitted by a Gabor wavelet defined as (Kosovichev and Duvall, 1997) [KD97]:

$$G(A, \omega_0, \delta\omega, \tau_\mathrm{p}, \tau_\mathrm{g}; t) = A \cos\big(\omega_0(t - \tau_\mathrm{p})\big) \exp\bigg(-\frac{\delta\omega^2}{4}(t - \tau_\mathrm{g})^2\bigg) \tag{2}$$

Table 1 Parameters used for the time – distance analysis: central source – receiver distances [Δ], range of Δ, properties of the Gaussian phase-speed filters applied to the Fourier transform of the line-of-sight-velocity data cube (v_0 is the central phase speed and FWHM is the full width at half maximum), and centers [t_0] of the temporal window applied to isolate the first-bounce skip on the cross-covariances.

Δ (Mm)	range (Mm)	v_0 (km s^{-1})	FWHM (km s^{-1})	t_0 (min)
6.20	3.70 – 8.70	12.77	6.18	19.0
8.70	6.20 – 11.2	14.87	6.18	23.3
11.60	8.70 – 14.5	17.49	6.18	24.4
16.95	14.5 – 19.4	24.82	9.09	28.7
24.35	19.4 – 29.3	35.46	12.36	33.5
30.55	26.0 – 35.1	39.71	7.18	36.3
36.75	31.8 – 41.7	43.29	7.42	38.7
42.95	38.4 – 47.5	47.67	8.41	40.8
49.15	44.2 – 54.1	52.26	10.5	42.8
55.35	50.8 – 59.9	57.16	8.90	44.7
61.65	56.6 – 66.7	61.13	8.03	45.0

where A is the amplitude of the wavelet, ω_0 is the central temporal frequency of the wavepacket, $\delta\omega$ is a measure of its frequency width, τ_p is the phase-travel time, and τ_g is the group travel time. Commonly, only phase-travel times are used for the time – distance inversions, because they can usually be determined more accurately than the group travel times. However, τ_p is not unique and is defined modulo $2\pi/\omega_0$. Therefore, the following rule has been used in this paper to select the phase-travel times: the value of τ_p that is selected is the closest to – and if two are equally distant, smaller than – the group travel time τ_g.

Subsequently, two definitions of the wavepacket travel times based on studies in geophysics were added. Gizon and Birch (2002) [GB02] define the travel time $\tau(\mathbf{r}, \Delta)$ as:

$$\tau_\pm(\mathbf{r}, \Delta) = \arg\min_t X_\pm(\mathbf{r}, \Delta, t) \tag{3}$$

where $X_\pm(\mathbf{r}, \Delta, t)$ are the functions:

$$X_\pm(\mathbf{r}, \Delta, t) = \sum_{t'} f(\pm t')\big[C(\mathbf{r}, \Delta, t') - C^{\mathrm{ref}}(\Delta, t' \mp t)\big]^2 \tag{4}$$

and $f(t)$ is a one-sided window function that selects the first-bounce skip of the cross-covariance (the skip corresponding to waves traveling directly from the center to the annulus) for the positive-time (*i.e.* outgoing waves) or negative-time (*i.e.* ingoing waves) branches of the cross-covariance. The same window function is used when fitting the cross-covariance by a Gabor wavelet even though it does not explicitly appear in Equation (2). The function $C^{\mathrm{ref}}(\Delta, t)$ is a reference cross-covariance. Here, $C^{\mathrm{ref}}(\Delta, t)$ is calculated as a spatial average of the measured cross-covariances over a region of the quiet Sun.

Lastly, Gizon and Birch (2004) [GB04] proposed another definition of the travel time that is linear with respect to the cross-covariance:

$$\tau_\pm(\mathbf{r}, \Delta) = \sum_t W_\pm(\Delta, t)\big[C(\mathbf{r}, \Delta, t) - C^{\mathrm{ref}}(\Delta, t)\big] \tag{5}$$

where $W_\pm$ is a weight function:

$$W_\pm(\Delta,t) = \frac{\mp f(\pm t)\dot{C}^{\text{ref}}(\Delta,t)}{\sum_{t'} f(\pm t')[\dot{C}^{\text{ref}}(\Delta,t')]^2} \tag{6}$$

where $\dot{C}^{\text{ref}}(\Delta,t)$ denotes the temporal derivative of the reference cross-covariance. Essentially, GB04 is a linearization of GB02 with respect to a parameter ϵ when $\epsilon \to 0$, where ϵ is used to define a smooth cross-covariance $C^\epsilon(\mathbf{r},\Delta,t)$:

$$C^\epsilon(\mathbf{r},\Delta,t) = \epsilon C(\mathbf{r},\Delta,t) + (1-\epsilon)C^{\text{ref}}(\Delta,t) \tag{7}$$

The definition GB02 applied to $C^\epsilon(\mathbf{r},\Delta,t)$ simplifies to GB04 in the limit $\epsilon \to 0$. For noiseless data, *i.e.* for an observation time $T = +\infty$, and when the amplitude of the cross-covariance does not vary, GB02 and GB04 are equivalent. GB04 can be applied only when the amplitude variations of $C(\mathbf{r},\Delta,t)$ are small. This is not the case in active regions. While the travel-time variations in sunspots do not exceed, roughly, 20% of the typical wave period, the cross-covariance amplitude changes by more than 100%. With GB04 there is a linear relationship between the travel times and $C(\mathbf{r},\Delta,t) - C^{\text{ref}}(\Delta,t)$.

As emphasized by the use of the $\pm$ symbol, for the three definitions of the travel times we measure τ separately for the outgoing (τ_+) and ingoing (τ_-) waves. We define the mean travel time, $\tau_{\text{mean}}(\mathbf{r},\Delta) = (\tau_+(\mathbf{r},\Delta) + \tau_-(\mathbf{r},\Delta))/2$, and the difference travel time, $\tau_{\text{diff}}(\mathbf{r},\Delta) = \tau_+(\mathbf{r},\Delta) - \tau_-(\mathbf{r},\Delta)$. The corresponding travel-time perturbations $\delta\tau$ are produced by subtracting the travel time obtained from the reference cross-covariance. For instance, $\delta\tau_{\text{mean}}(\mathbf{r},\Delta) = \tau_{\text{mean}}(\mathbf{r},\Delta) - \tau_{\text{mean;ref}}(\Delta)$ where $\tau_{\text{mean;ref}}(\Delta)$ is the mean travel time measured for the reference cross-covariance.

3. Implementation of the Travel-Time Definitions in the HMI Time – Distance Pipeline

The three travel-time definitions were implemented in the Fortran codes of the HMI time – distance helioseismology pipeline, based on the Data Resource Management System (DRMS) at Stanford University. They represent subroutines of the main time – distance program, which is also written in Fortran and encapsulated inside a C-wrapper to access the DRMS I/O functions. The input parameters of the travel-time subroutines include the cross-covariances, and their output are the travel-time maps in FITS format. The mean and difference travel-time maps are calculated, as well as the North – South and East – West difference travel-time maps in a center-to-quadrant geometry (used to measure horizontal flows). The Gabor-wavelet fitting procedure (KD97) is a least-squares fit using the Levenberg – Marquardt algorithm. The GB02 and GB04 procedures use a Whittaker – Shannon interpolation formula to interpolate the temporal cross-covariances from a grid with a 45-second sampling rate (nominal HMI cadence) or a 60-second sampling rate (MDI cadence) onto a grid with a five-second sampling rate. The interpolation subroutine computes interpolated points only in a narrow time range around the temporal window chosen to select the first-bounce skip (see, *e.g.*, the last column of Table 1 for some typical values). This speeds up the computations.

For the GB02 definition, the code uses the interpolated cross-covariances to find the value of $t = t_{\min}$ at which $X_\pm(\mathbf{r},\Delta,t)$ is minimum. Following Gizon and Birch (2002) we then fit $X_\pm(\mathbf{r},\Delta,t)$ with a parabola around $t_{\min}$ (using five measurement points). The code returns an estimate of the parameters c_0, c_1, and c_2 of the parabola $y = c_0 + c_1 t + c_2 t^2$, and the travel-time measurement is given by the location of the minimum of this parabola $\tau_\pm = c_1/(2c_2)$.

For the GB04 definition, the time derivative $\dot{C}^{\rm ref}(\Delta, t)$ is numerically computed for the interpolated reference cross-covariance using a five-point equation.

The three travel-time definitions return outliers (misfits in the case of KD97 and GB02). To correct for these outliers, each travel-time subroutine computes the mean $[\mu]$ and standard deviation $[\sigma]$ of the ingoing and outgoing travel times. When $|\tau_\pm(\mathbf{r}, \Delta) - \mu_\pm|$ is greater than X times $\sigma_\pm$ (where X is a parameter that can be set in the code; its value is typically larger than three), the cross-covariance at location $\mathbf{r}$ is averaged over the eight closest neighboring locations. Then the travel times at $\mathbf{r}$ are re-computed using the averaged cross-covariances.

4. Comparison of the Travel-Time Definitions

To test our implementation of the travel-time definitions, we apply them to MDI Doppler-velocity data. This allows a comparison of the travel-time definitions in the quiet Sun and in an active region.

4.1. Quiet-Sun Region

We work on a line-of-sight Doppler-velocity data cube obtained from MDI high-resolution (Hi-Res) data taken in May 2001 for a quiet-Sun region (Gizon and Birch, 2004). By quiet Sun we mean a region devoid of any significant magnetic activity (*e.g.*, no sunspot, no plage). The observation duration is 512 minutes (or 8.5 hours). We choose the Hi-Res data because the spatial sampling rate ($0.6''$ per pixel) is close to the HMI sampling rate. The data were remapped onto the heliographic coordinates using Postel's projection, and the region was tracked at the Carrington rate. After rebinning, the spatial resolution of our data cube is $\delta x = \delta y = 0.826$ Mm, and the temporal resolution is $\delta t = 1$ minute. The dimensions of this data cube are $256 \times 256 \times 512$ where the first two dimensions are the horizontal ones, and the last dimension is the temporal one. We apply the time – distance formalism on this data cube. By default, the temporal windows used to select the first-bounce skip on the cross-covariances are 20 minutes wide. Figures 1 and 2 show the mean and difference travel-time perturbation maps obtained for two distances $\Delta = 6.2$ and $\Delta = 30.55$ Mm using the three travel-time definitions. All of the maps look very similar: they exhibit the same features at the same locations. However, the amplitude of these features varies. At $\Delta = 6.2$ Mm, the amplitude of $\delta\tau_{\rm diff}(\mathbf{r}, \Delta)$ in the supergranules (black spots on the maps) is larger in absolute value for GB04 than for GB02 and KD97. Table 2 lists the values of the Pearson linear correlation coefficients between the different travel-time maps obtained for the quiet Sun. These coefficients are high, especially between GB02 and GB04.

Figure 3 shows histograms of the mean and difference travel-time perturbations for $\Delta = 6.2$ and $\Delta = 30.55$ Mm. The travel-time distributions for the three definitions look very similar (more or less close to a Gaussian distribution). The standard deviations are larger for GB02 and GB04 than for KD97 at $\Delta = 6.2$ Mm, but they are smaller at $\Delta = 30.55$ Mm. Therefore, while GB02 and GB04 seem to produce slightly "noisier" travel-time maps at short distances than KD97, this is the opposite for larger distances (what matters is actually not the standard deviation – an assessment of the noise level – but the signal-to-noise ratio). This is consistent with the results of Roth, Gizon, and Beck (2007) for very large

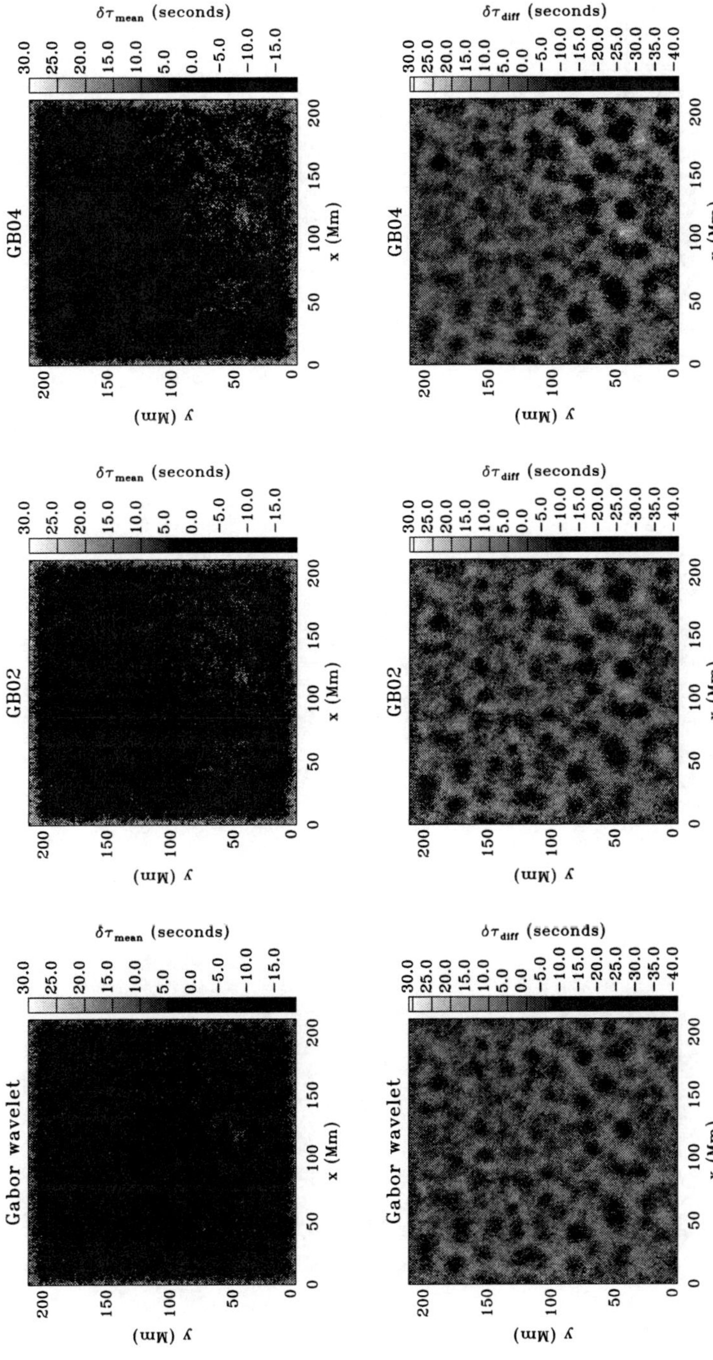

Figure 1 Mean (upper panels) and difference (lower panels) travel-time perturbations for the three travel-time definitions (left panels: KD97; central panels: GB02; right panels: GB04) and $\Delta = 6.2$ Mm.

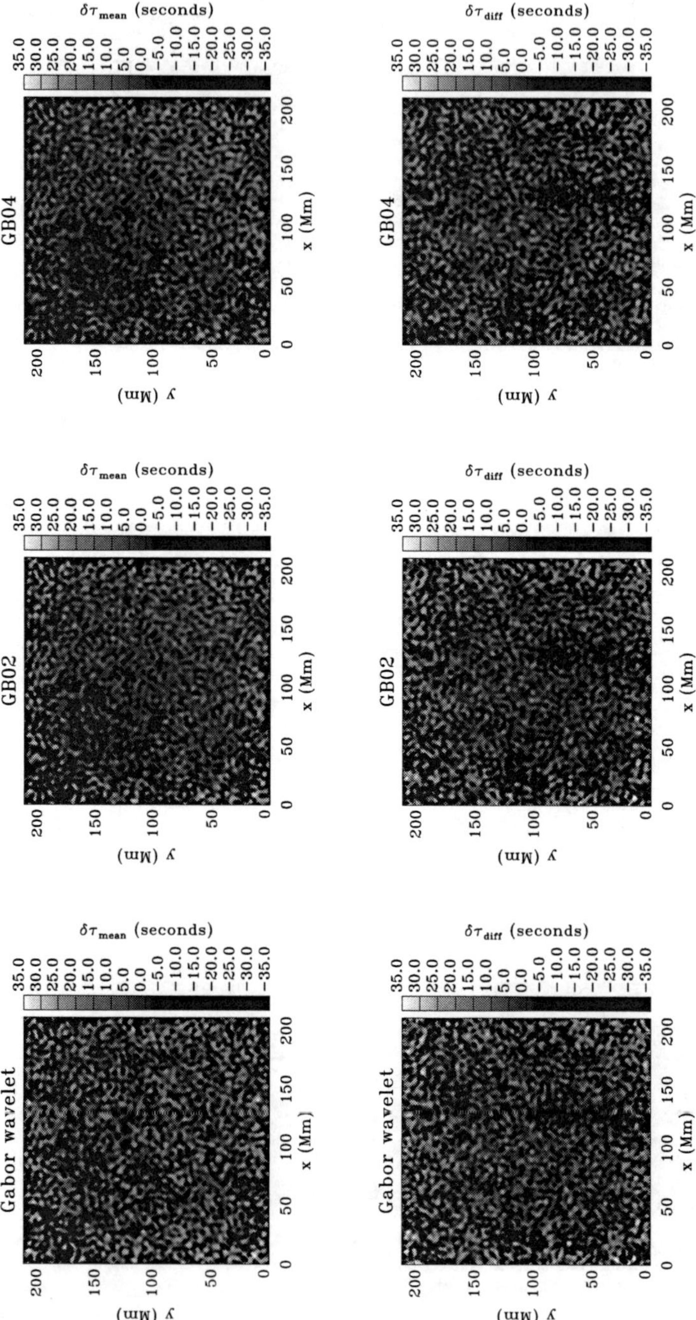

Figure 2 Mean (upper panels) and difference (lower panels) travel-time perturbations for the three travel-time definitions (left panels: KD97; central panels: GB02; right panels: GB04) and $\Delta = 30.55$ Mm. The scale has been truncated to $[-35, 35]$ seconds to highlight the features in the maps.

Table 2 Linear Pearson correlation coefficients between the different travel-time maps in the quiet Sun, for the entire 256 × 256 maps.

Map	KD97/GB02	KD97/GB04	GB02/GB04
$\delta\tau_{\text{mean}}$ $\Delta = 6.2$ Mm	0.781	0.748	0.956
$\delta\tau_{\text{mean}}$ $\Delta = 30.55$ Mm	0.783	0.726	0.940
$\delta\tau_{\text{diff}}$ $\Delta = 6.2$ Mm	0.876	0.839	0.945
$\delta\tau_{\text{diff}}$ $\Delta = 30.55$ Mm	0.849	0.797	0.913

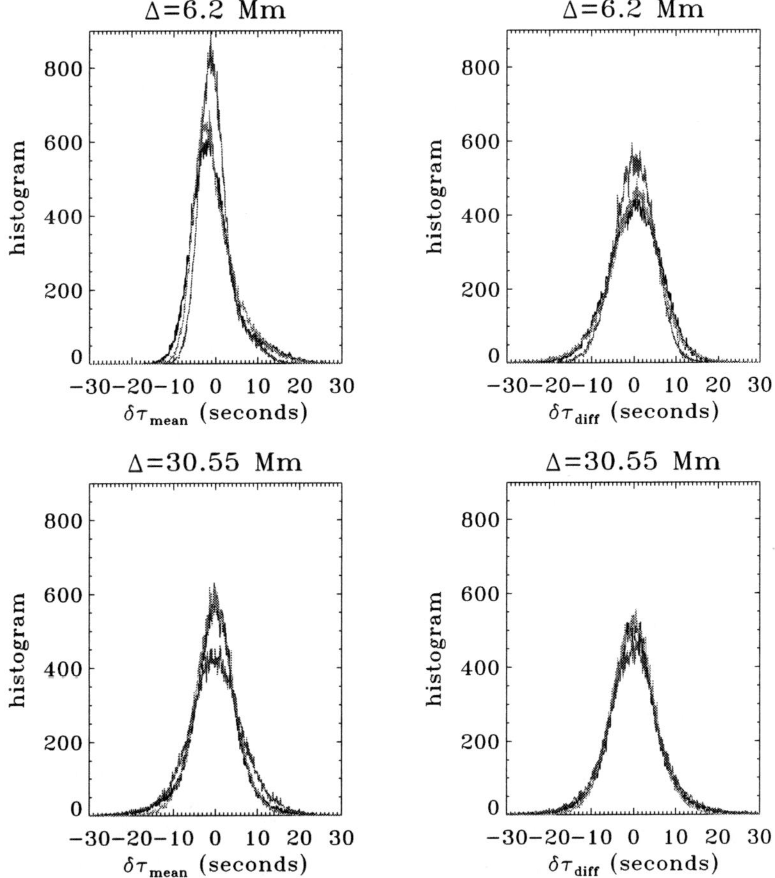

Figure 3 Histograms of the mean (left column) and difference (right column) travel-time perturbations for the two distances $\Delta = 6.2$ (upper panels) and $\Delta = 30.55$ (lower panels) Mm. Black lines are for GB02, green lines are for GB04, and red lines are for KD97.

distances (24 heliocentric degrees, or about 291 Mm). For short distance, here $\Delta = 6.2$ Mm, the distributions of the mean travel-time perturbations are noticeably different from a normal (Gaussian) distribution because of a tail at positive $\delta\tau_{\text{mean}}$. This is an edge effect: near the edges of the mean travel-time perturbation maps, $\delta\tau_{\text{mean}}$ is predominantly positive and large. Indeed, narrow white bands are clearly visible on these maps. These bands are produced by

Table 3 Linear Pearson correlation coefficients between the different travel-time maps in the quiet Sun, for truncated maps: instead of using the full 256 × 256 pixels, we restrict the calculation to an inner square of 215 × 215 pixels, thus avoiding the map edges.

Map	KD97/GB02	KD97/GB04	GB02/GB04
$\delta\tau_{\mathrm{mean}}$ $\Delta = 6.2$ Mm	0.550	0.556	0.958
$\delta\tau_{\mathrm{mean}}$ $\Delta = 30.55$ Mm	0.673	0.639	0.964
$\delta\tau_{\mathrm{diff}}$ $\Delta = 6.2$ Mm	0.887	0.852	0.953
$\delta\tau_{\mathrm{diff}}$ $\Delta = 30.55$ Mm	0.873	0.835	0.952

the algorithm we use to compute the cross-covariances: the center-to-annulus averaging is performed in the Fourier domain by applying the convolution theorem. However, because the Doppler-velocity data cube is not periodic in the horizontal direction, this creates a spurious behavior at the edges of the map. Ignoring these map edges produces distributions that are very close to Gaussian ones. It also somewhat reduces the correlation coefficients KD97/GB02 and KD97/GB04 for the mean travel times, but not for the difference travel times (see Table 3). The correlation coefficients for the difference travel times remain high. At $\Delta = 6.2$ Mm, this is probably because the difference maps, unlike the mean travel-time maps, include a large-scale regular signal (supergranular flows) and not just noise (caused by convection and random realization of solar oscillations). At $\Delta = 30.55$ Mm, there is no real signal, and what we see on the maps is realization noise (or, more exactly, filtered realization noise, since the data cube was phase-speed filtered). Therefore at this distance we are testing how the different methods are measuring the noise.

We also tested the impact of the location and width of the temporal window on the mean and difference travel-time values. Figures 4, 5, and 6 show scatter plots for KD97, GB02, and GB04. Figure 7 shows the standard deviation of the mean and difference travel-time perturbations as a function of the window center and width.

It appears that at short distance ($\Delta = 6.2$ Mm) the travel-time definitions are very sensitive to the temporal window used, especially for the KD97 definition. The standard deviation strongly depends on this temporal window: we can almost double the noise level on the travel-time maps by simply changing the width of the window. At 6.2 Mm, the first-bounce skip on the cross-covariances is entangled at short times $[t]$ with an artifact produced by the phase-speed filter (see *e.g.* Couvidat and Birch, 2009), and it is entangled at larger times $[t]$ with the second-bounce skip. The artifact produced by the filter is a group of peaks at small t, with the overall shape of a Gabor wavelet and that partly overlap with the first-bounce skip peaks, thus affecting their shape. Therefore by shifting the center of the temporal window or by changing its width, we more or less select the parts of the first-bounce skip contaminated by this filter artifact (and to a lesser extent by the second-bounce skip). A way of alleviating this problem is by using broader phase-speed filters: the filter artifact amplitude decreases with wider filters. For a given supergranule, the value of the difference travel-time perturbation obtained with GB02 and GB04 can be larger (in absolute value) or smaller than the value obtained with the Gabor-wavelet fit depending on how we center the temporal window.

The issue with the use of the phase-speed filter and its interaction with the temporal window may potentially be discarded with the HMI data. The phase-speed filter was introduced for analysis of MDI data because of the limited spatial resolution of the short distances on the cross-covariance diagram. It is expected that the HMI optics are significantly improved. Thus we may not need to do any phase-speed filtering at short distances (for an example

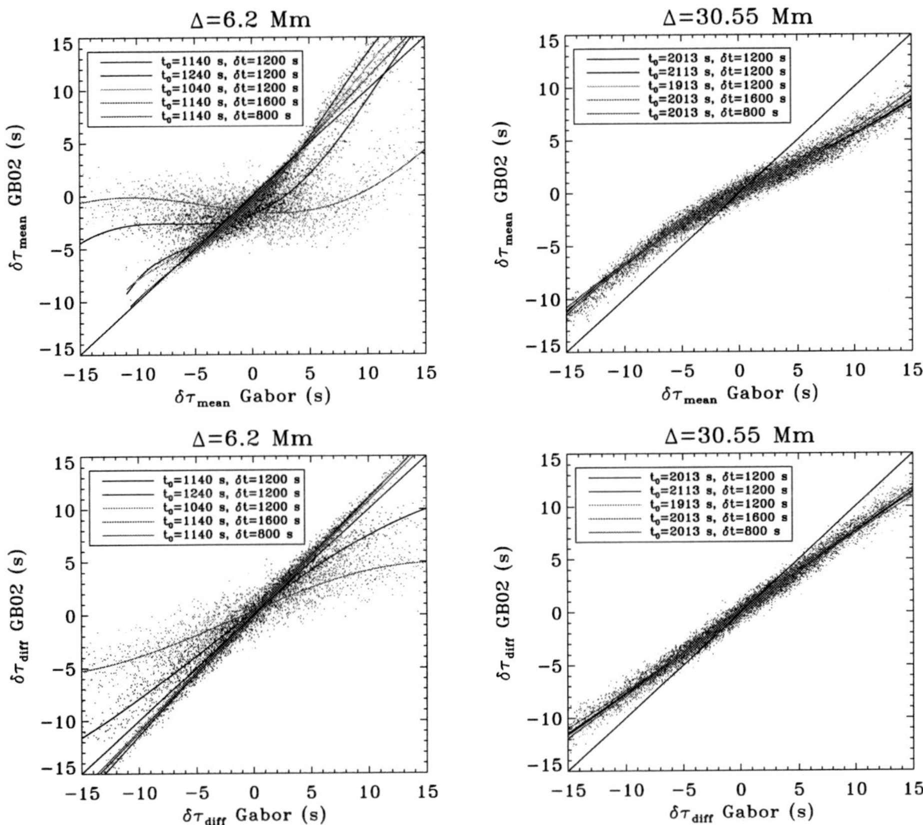

Figure 4 Scatter plots for the mean (upper panels) and difference (lower panels) travel-time perturbations obtained with the KD97 and GB02 travel-time definitions at two distances $\Delta = 6.2$ and $\Delta = 30.55$ Mm. The thick solid lines are fifth-order polynomials fitted to the data. The different colors are for different centers [t_0] and widths [δt] of the temporal window used to select the first-bounce skip during the travel-time computation.

of time – distance diagram obtained with no phase-speed filters, see the one derived from *Hinode* data by Sekii *et al.*, 2007). At $\Delta = 30.55$ Mm, the filter artifact, first-bounce and second-bounce skips, are clearly separated, explaining why the temporal window has less of an impact on the travel times.

The relationships between the KD97 travel times and GB02 or GB04 are not linear. The relationships between GB02 and GB04 are linear for small travel times ($\leq$ five seconds in absolute value), as expected since GB04 is a linearized version of GB02.

4.2. Active Region

Although GB02 and GB04 were designed for quiet-Sun data, it is possible to use them for active-region data, provided that we take into account the decrease in amplitude of the cross-covariances that occurs inside sunspots (due to a reduction of the acoustic power). A simple way of taking into account this decrease is to normalize the cross-covariances $C(\Delta, \mathbf{r})$ – for a given Δ – at each location $\mathbf{r}$ by dividing $C(\Delta, \mathbf{r})$ by its maximum amplitude (L. Gizon,

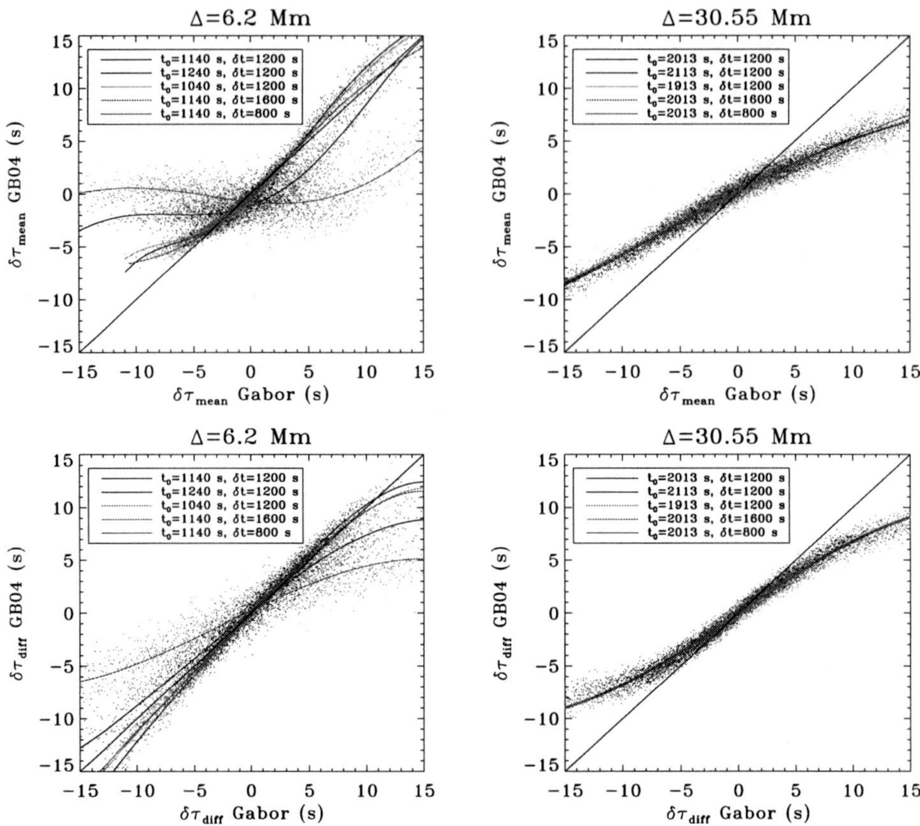

Figure 5 Scatter plots for the mean (upper panels) and difference (lower panels) travel-time perturbations obtained with the KD97 and GB04 travel-time definitions at two distances $\Delta = 6.2$ and $\Delta = 30.55$ Mm. The thick solid lines are fifth-order polynomials fitted to the data. The different colors are for different centers $[t_0]$ and widths $[\delta t]$ of the temporal window used to select the first-bounce skip during the travel-time computation.

private communication, 2009): $C_0(\Delta, \mathbf{r}) = C(\Delta, \mathbf{r})/\max(C(\Delta, \mathbf{r}))$. GB02 and GB04 are then applied to $C_0(\Delta, \mathbf{r})$ instead of $C(\Delta, \mathbf{r})$. The Gabor-wavelet definition is not affected by this normalization, because it fits for the amplitude $[A]$ of the wavelet at the same time as the travel times. Note that the combination of the decrease in the amplitude of oscillations and the use of phase-speed filters may cause systematic shifts of the travel times, particularly, the mean travel times (Rajaguru *et al.*, 2006; Parchevsky, Zhao, and Kosovichev, 2008; Nigam and Kosovichev, 2010).

For the active-region test, we use MDI Hi-Res data centered on active region NOAA 8243 of 18 June 1998. As for the quiet-Sun data cube, the line-of-sight Doppler-velocity data cube has the following dimensions: $256 \times 256 \times 512$, with a spatial resolution $\delta x = \delta y = 0.826$ Mm, and a temporal resolution $\delta t = 1$ minute. The observation duration is 512 minutes (or 8.5 hours). The reference cross-covariances $[C^{\mathrm{ref}}]$ are obtained by spatially averaging over the ranges $x = [20, 235]$ and $y = [20, 80]$ in pixel units. These ranges are selected because the corresponding MDI map of the line-of-sight magnetic field shows no significant magnetic activity there. The maximum value of the cross-covariances at each

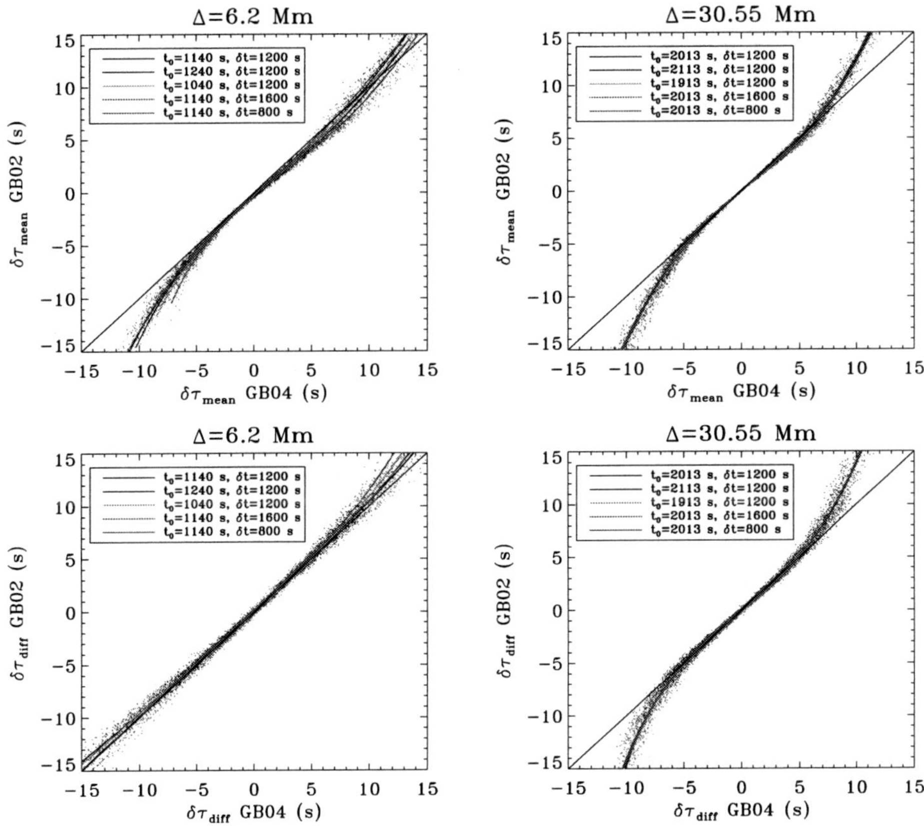

Figure 6 Scatter plots for the mean (upper panels) and difference (lower panels) travel-time perturbations obtained with the GB02 and GB04 travel-time definitions at two distances $\Delta = 6.2$ and $\Delta = 30.55$ Mm. The thick solid lines are fifth-order polynomials fitted to the data. The different colors are for different centers $[t_0]$ and widths $[\delta t]$ of the temporal window used to select the first-bounce skip during the travel-time computation.

location [**r**] is obtained by Fourier interpolating these cross-covariances on a fine temporal grid and then reading their maximum amplitude. Figures 8 and 9 show the mean and difference travel-time perturbations obtained with the three definitions. The travel-time maps are rather similar especially between KD97 and GB02, showing first a positive mean travel-time perturbation at short distances Δ and then a negative perturbation. Even though KD97 and GB02 agree well with each other overall, there can be some significant discrepancies locally, in some parts of the sunspot. These discrepancies may be caused by the realization noise of solar oscillations, which may be higher in sunspots due to the smaller number of excitation events in strong field regions, and, perhaps, by some systematic effects since these travel-time definitions are not identical. The amplitude of the travel times in the sunspot is underestimated by GB04 (in absolute value) compared to the other two definitions. It appears that for sunspots, GB02 and GB04 return rather different travel-time perturbations, which can be explained by the fact that GB04 strictly reduces to GB02 only when the cross-covariances are noiseless and do not have variations in amplitude (GB04 is a linearization of GB02 with respect to ϵ for a small ϵ), which is not the case in active regions. In addition to the strong non-linear amplitude variations of the cross-covariance in such regions,

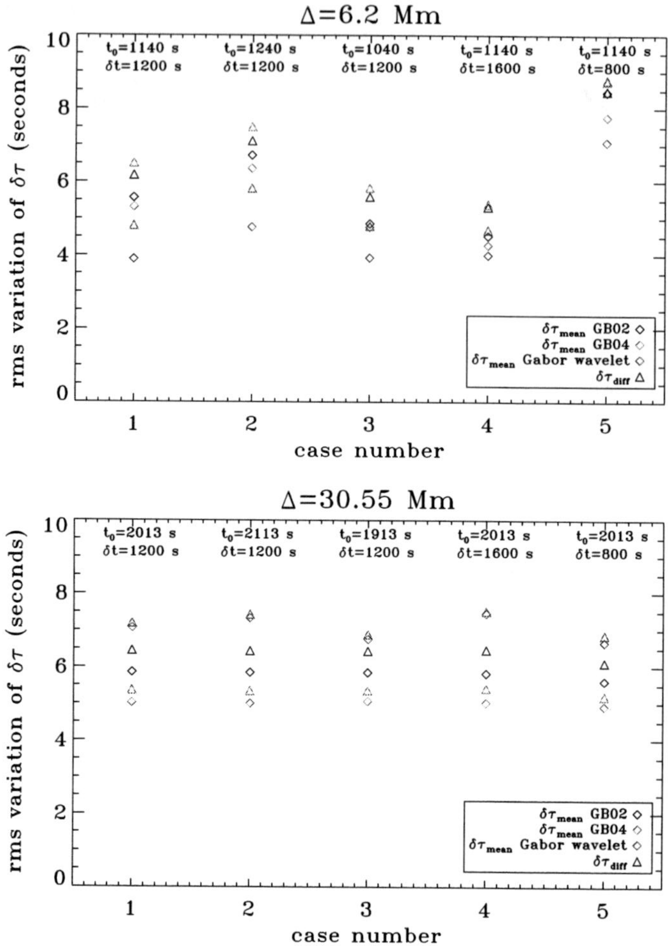

Figure 7 Standard deviations of the mean and difference travel-time perturbations at $\Delta = 6.2$ Mm (upper panel), and $\Delta = 30.55$ Mm (lower panel) for five different centers [t_0] and widths [δt] of the temporal window used to select the first-bounce ridge.

the noise level of $C(\Delta, \mathbf{r})$ is actually significantly higher than in the quiet Sun. Couvidat and Birch (2009) showed that for strong horizontal-flow velocities, the relationship between flow strength and resulting travel-time perturbation is not linear, and this non-linearity sets in much earlier (for slower flows) for GB04 than for GB02 and KD97. Therefore, we ultimately decided not to use GB04 in the pipeline. Of course, this does not preclude users of HMI data from using this or any other procedure in their work. For the linear inversions of the mean travel times in the HMI pipeline, we will use both the ray-path (Kosovichev and Duvall, 1997) and Born-approximation kernels (Birch, Kosovichev, and Duvall, 2004). Since the data-analysis pipeline will also provide the travel-time measurements described in this paper, the HMI data users will be able to develop their own inversion techniques.

Figure 10 shows the standard deviations of the mean and difference travel-time perturbations for the three travel-time definitions and for the 11 distances [Δ] studied here. These standard deviations are computed in a quiet-Sun region surrounding, but excluding, the ac-

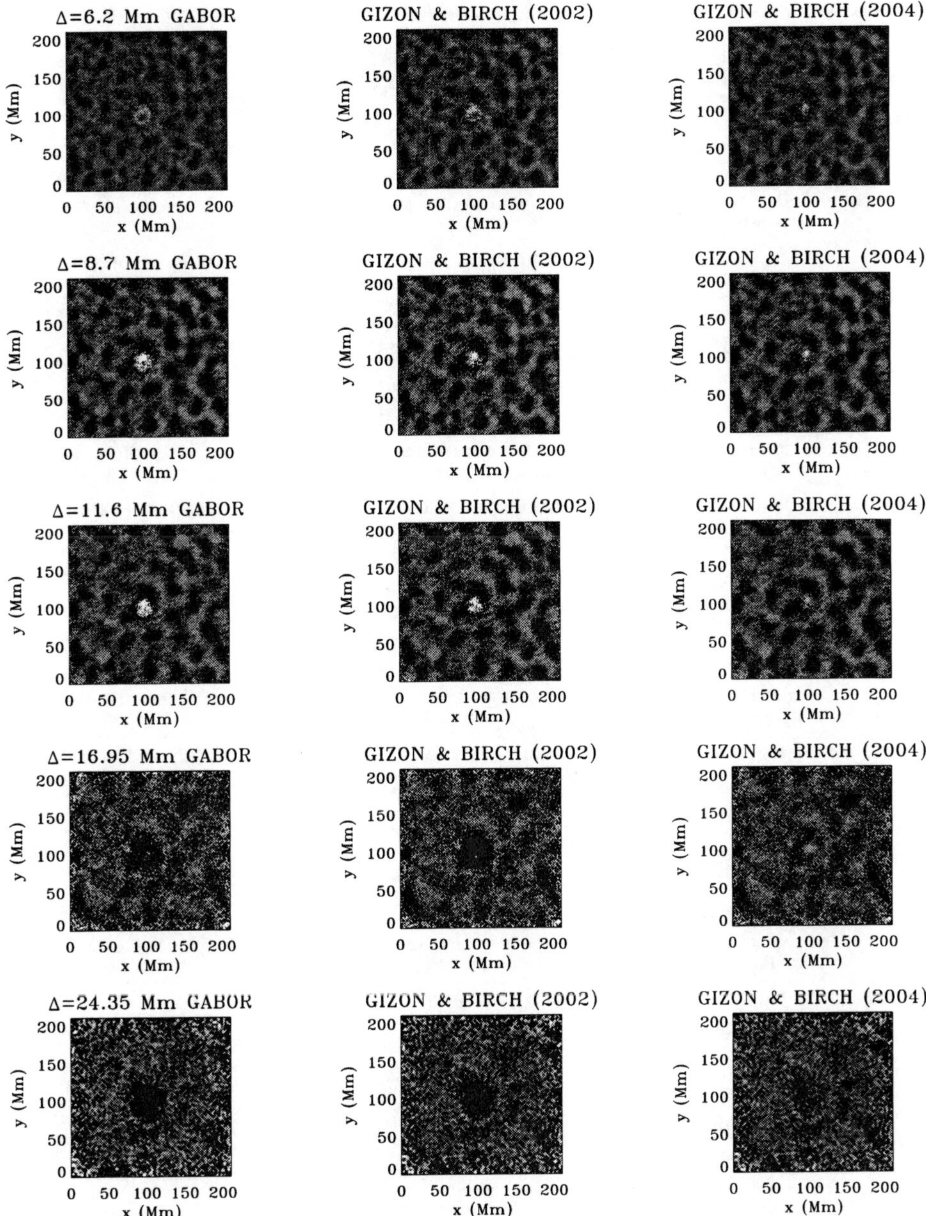

Figure 8 Difference travel-time perturbations for the active region NOAA 8243 obtained with KD97 (left column), GB02 (central column), and GB04 (right panel), and for five source – receiver distances [Δ]. The colorscale is the same for all the panels, and ranges from −40 seconds to +40 seconds.

tive region. They confirm the conclusion of the previous section in which we showed that at short distances [Δ] the travel-time maps produced by the Gabor-wavelet definition have a smaller standard deviation than those produced by GB02 and GB04, while the opposite is true at larger Δ.

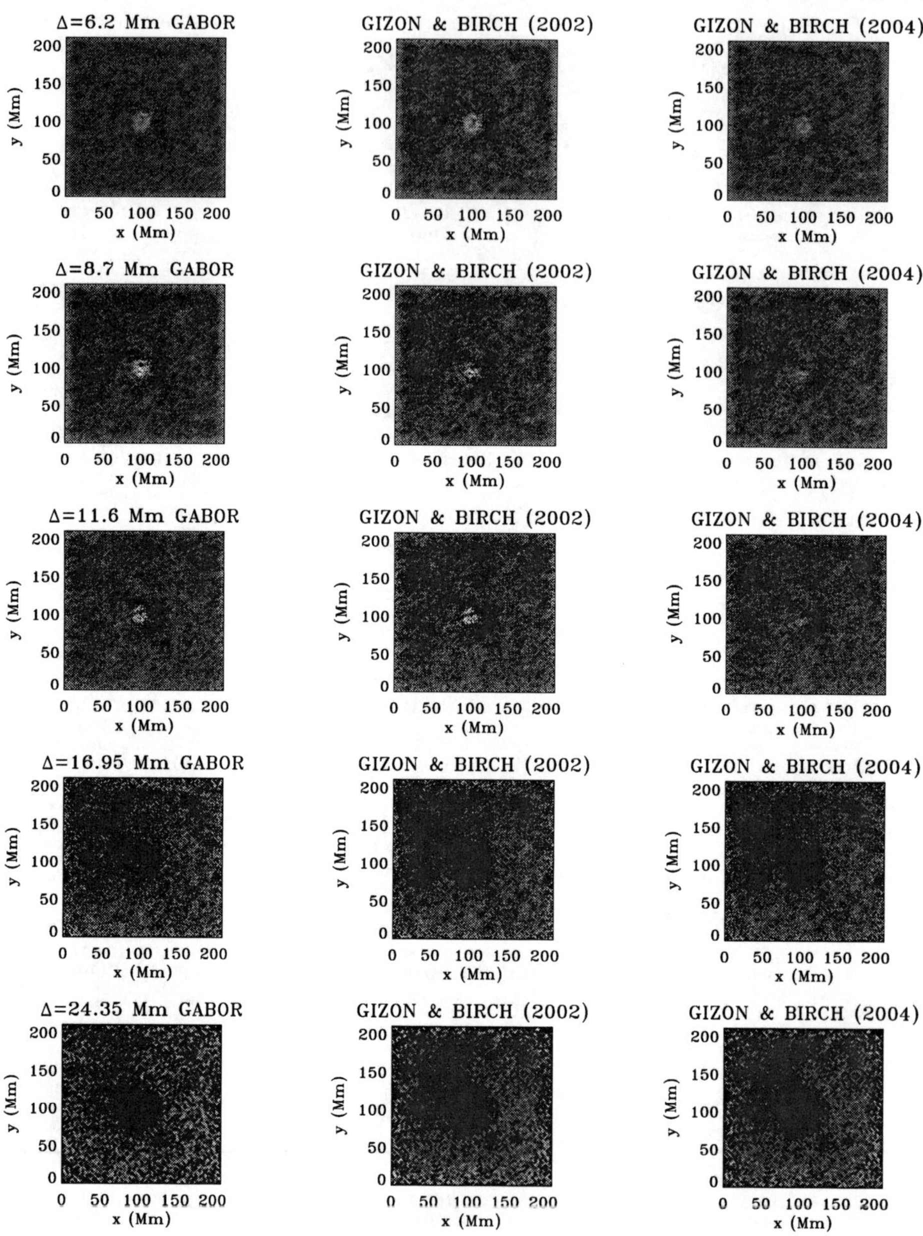

Figure 9 Mean travel-time perturbations for the active region NOAA 8243 obtained with KD97 (left column), GB02 (central column), and GB04 (right panel), and for five source–receiver distances [Δ]. The colorscale is the same for all the panels, and ranges from −40 seconds to +40 seconds.

5. Conclusion

We have implemented three travel-time measurement procedures (Kosovichev and Duvall, 1997; Gizon and Birch 2002, 2004). The subroutines computing these travel times of

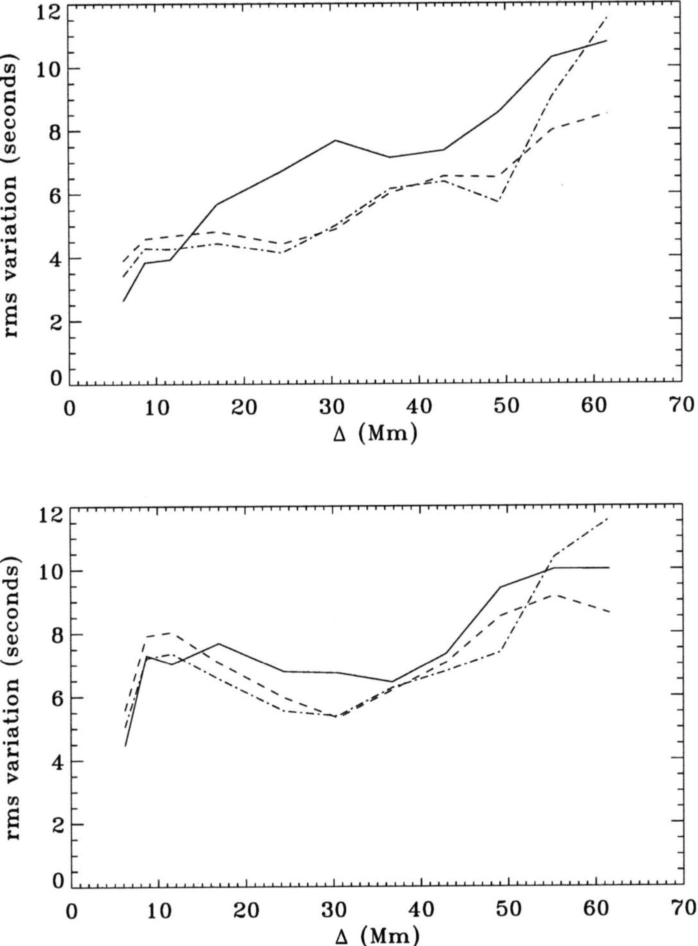

Figure 10 Upper panel: standard deviation of the mean travel-time perturbations for KD97 (solid line), GB02 (dashed line), and GB04 (dash-dotted line). Lower panel: same but for the difference travel-time perturbations.

wavepackets are part of the HMI time – distance helioseismology pipeline at the SDO Joint Science Operations Center at Stanford University. Using quiet-Sun and active-region data from the SOHO/MDI instrument we compared the mean and difference travel-time perturbations obtained by using these three definitions. They all generate similar travel times in the quiet Sun, provided that, at short distances [Δ], we use an appropriate temporal window to isolate the first-bounce skip on the cross-covariances. Indeed, at $\Delta = 6.2$ Mm, the travel times strongly depend on the location and width of this temporal window. This is mainly due to the phase-speed filter applied to the data cube. This phase-speed filter is used to isolate the wavefront signal from the artifact signals of the MDI instrument (Duvall *et al.*, 1997). Without such filter, the travel-time measurements at short distances are not possible from MDI data. However, this filter introduces issues of its own: the travel times measured at short distances depend on the parameters of these phase-speed filters. Using broader filters than those considered here reduces the magnitude of this problem. HMI data might not necessitate the application of phase-speed filtering, which should give us access to more robust travel-time estimates at short distances. The distributions of the mean and difference travel times obtained in the quiet Sun are close to a Gaussian distribution (especially when the map edges are discarded) and are very similar for the three travel-time definitions. The Gabor-wavelet technique (Kosovichev and Duvall, 1997) produces travel-time maps with a standard deviation smaller than the Gizon and Birch (2002, 2004) definitions for $\Delta < 15$ Mm, but with a larger standard deviation for larger Δ. The relationships between KD97 and GB02 or GB04 are not linear, while the relationships between GB02 and GB04 are linear for (absolute) times smaller than about five seconds. In the active region studied here, GB04 returned travel times that are significantly smaller (in absolute value) than those of GB02 and KD97. We ultimately decided not to include GB04 in the pipeline.

Acknowledgements This work was supported by NASA grant NAS5-02139 (HMI). The authors thank Laurent Gizon for suggesting a way of using the Gizon and Birch (2002, 2004) definitions in an active region, and for providing us with a MDI data cube of the quiet Sun.

References

Birch, A.C., Kosovichev, A.G., Duvall, T.L. Jr.: 2004, *Astrophys. J.* **608**, 580.
Chou, D.-Y., Duvall, T.L. Jr.: 2000, *Astrophys. J.* **533**, 568.
Couvidat, S., Birch, A.C.: 2009, *Solar Phys.* **257**, 217.
Couvidat, S., Birch, A.C., Kosovichev, A.G.: 2006, *Astrophys. J.* **640**, 516.
Duvall, T.L. Jr., Gizon, L.: 2000, *Solar Phys.* **192**, 177.
Duvall, T.L. Jr., Jefferies, S.M., Harvey, J.W., Pomerantz, M.A.: 1993, *Nature* **362**, 430.
Duvall, T.L. Jr., D'Silva, S., Jefferies, S.M., Harvey, J.W., Schou, J.: 1996, *Nature* **379**, 235.
Duvall, T.L. Jr., Kosovichev, A.G., Scherrer, P.H., Bogart, R.S., Bush, R.I., de Forest, C., Hoeksema, J.T., Schou, J., Saba, J.L.R., Tarbell, T.D., Title, A.M., Wolfson, C.J., Milford, P.N.: 1997, *Solar Phys.* **170**, 63.
Gizon, L., Birch, A.C.: 2002, *Astrophys. J.* **571**, 966.
Gizon, L., Birch, A.C.: 2004, *Astrophys. J.* **614**, 472.
Hirzberger, J., Gizon, L., Solanki, S.K., Duvall, T.L. Jr.: 2007, *Solar Phys.* **251**, 417.
Jackiewicz, J., Gizon, L., Birch, A.C., Duvall, T.L. Jr.: 2007, *Astrophys. J.* **671**, 1051.
Jensen, J.M., Duvall, T.L. Jr., Jacobsen, B.H., Christensen-Dalsgaard, J.: 2001, *Astrophys. J. Lett.* **553**, L193.
Kosovichev, A.G., Duvall, T.L. Jr.: 1997, In: Pijpers, F.P., Christensen-Dalsgaard, J., Rosenthal, C.S. (eds.) *SCORe'96: Solar Convection and Oscillations and Their Relationship*, *Astrophys. Space Science Lib.* **225**, Kluwer, Dordrecht, 241.
Kosovichev, A.G., Duvall, T.L. Jr., Scherrer, P.H.: 2000, *Solar Phys.* **192**, 159.
Lindsey, C., Braun, D.C.: 2005, *Astrophys. J. Lett.* **620**, L1107.
Nigam, R., Kosovichev, A.G.: 2010, *Astrophys. J.* **708**, 1475.
Parchevsky, K.V., Zhao, J., Kosovichev, A.G.: 2008, *Astrophys. J.* **678**, 1498.
Rajaguru, S.P., Birch, A.C., Duvall, T.L. Jr., Thompson, M.J., Zhao, J.: 2006, *Astrophys. J.* **646**, 543.
Roth, M., Gizon, L., Beck, J.G.: 2007, *Astron. Nachr.* **328**, 215.
Scherrer, P.H., Bogart, R.S., Bush, R.I., Hoeksema, J.T., Kosovichev, A.G., Schou, J., Rosenberg, W., Springer, L., Tarbell, T.D., Title, A., Wolfson, C.J., Zayer, I., MDI Engineering Team: 1995, *Solar Phys.* **162**, 129.
Sekii, T., Kosovichev, A.G., Zhao, J., Tsuneta, S., Shibahashi, H., Berger, T.E., Ichimoto, K., Katsukawa, Y., Lites, B., Nagata, S., Shimizu, T., Shine, R.A., Suematsu, Y., Tarbell, T.D., Title, A.M.: 2007, *Publ. Astron. Soc. Japan* **59**, S637.
Zhao, J., Kosovichev, A.G.: 2003, In: Sawaya-Lacoste, H. (ed.) *Soho 12/Gong+ 2002. Local and Global Helioseismology: The Present and Future* **SP-517**, ESA, Noordwijk, 417.
Zhao, J., Kosovichev, A.G., Duvall, T.L. Jr.: 2001, *Astrophys. J.* **557**, 384.
Zhao, J., Couvidat, S., Parchevsky, K.V., Bogart, R.S., Duvall, T.L. Jr., Beck, J.G., Birch, A.C., Kosovichev, A.G.: 2010, *Solar Phys.* submitted.

Solar Phys (2012) 275:375–390
DOI 10.1007/s11207-011-9757-y

THE SOLAR DYNAMICS OBSERVATORY

Time–Distance Helioseismology Data-Analysis Pipeline for *Helioseismic and Magnetic Imager* Onboard *Solar Dynamics Observatory* (SDO/HMI) and Its Initial Results

J. Zhao · S. Couvidat · R.S. Bogart · K.V. Parchevsky · A.C. Birch · T.L. Duvall Jr. · J.G. Beck · A.G. Kosovichev · P.H. Scherrer

Received: 16 September 2010 / Accepted: 23 March 2011 / Published online: 4 May 2011

Abstract The *Helioseismic and Magnetic Imager* onboard the *Solar Dynamics Observatory* (SDO/HMI) provides continuous full-disk observations of solar oscillations. We develop a data-analysis pipeline based on the time–distance helioseismology method to measure acoustic travel times using HMI Doppler-shift observations, and infer solar interior properties by inverting these measurements. The pipeline is used for routine production of near-real-time full-disk maps of subsurface wave-speed perturbations and horizontal flow velocities for depths ranging from 0 to 20 Mm, every eight hours. In addition, Carrington synoptic maps for the subsurface properties are made from these full-disk maps. The pipeline can also be used for selected target areas and time periods. We explain details of the pipeline organization and procedures, including processing of the HMI Doppler observations, measurements of the travel times, inversions, and constructions of the full-disk and synoptic maps. Some initial results from the pipeline, including full-disk flow maps, sunspot subsurface flow fields, and the interior rotation and meridional flow speeds, are presented.

Keywords Sun: helioseismology · Sun: oscillations · Sun: SDO

1. Introduction

The *Helioseismic and Magnetic Imager* onboard the *Solar Dynamics Observatory* (SDO/HMI: Schou *et al.*, 2011) observes the solar full-disk intensity, Doppler velocity, and vector

The Solar Dynamics Observatory
Guest Editors: W. Dean Pesnell, Phillip C. Chamberlin, and Barbara J. Thompson.

J. Zhao (✉) · S. Couvidat · R.S. Bogart · K.V. Parchevsky · J.G. Beck · A.G. Kosovichev · P.H. Scherrer
W.W. Hansen Experimental Physics Laboratory, Stanford University, Stanford, CA 94305-4085, USA
e-mail: junwei@sun.stanford.edu

A.C. Birch
NorthWest Research Associates, CoRA Division, 3380 Mitchell Lane, Boulder, CO 80301, USA

T.L. Duvall Jr.
Laboratory for Astronomy and Solar Physics, NASA Goddard Space Flight Center, Greenbelt, MD 20771, USA

magnetic field of the photosphere with high spatial resolution and high temporal cadence. Similar to the *Michelson Doppler Imager* (MDI: Scherrer *et al.*, 1995), an instrument onboard the *Solar and Heliospheric Observatory* (SOHO), the HMI Dopplergrams are primarily used for helioseismic analysis to investigate the interior structure and dynamics of the Sun. Helioseismology data-analysis pipelines are planned for near-real-time analyses of the observations in order to provide the analysis results to the helioseismology and solar physics communities. The time–distance analysis pipeline is one of the pipelines for local helioseismology studies, and other pipelines include ring-diagram analysis and far-side active region imaging. The time–distance pipeline is designed for the routine production of nearly full-disk subsurface wave-speed perturbations and horizontal flow fields every eight hours, as well as synoptic flow maps for every Carrington rotation. It can also be used to analyze specific target areas and time periods.

Time–distance helioseismology was first introduced by Duvall *et al.* (1993, 1996), and it has developed rapidly since then. Different inversion techniques were introduced and tested. The LSQR algorithm, introduced by Kosovichev (1996) and used later by Zhao, Kosovichev, and Duvall (2001), solves the inversion problem in the least-squares sense in the spatial domain by an iterative approach. The Multi-Channel Deconvolution (MCD) method, introduced by Jacobsen *et al.* (1999) and widely used in later studies (*e.g.* Couvidat *et al.*, 2004), solves the least-squares problems in the Fourier domain. Later, Couvidat *et al.* (2005) applied a horizontal-regularization procedure for this inversion technique. More recently, an optimally localized averaging (OLA) inversion scheme was introduced to study the solar subsurface flow fields (Jackiewicz, Gizon, and Birch, 2008).

Different types of sensitivity kernels, which describe the relationship between the travel times and interior properties, were also introduced and used in the time–distance inversion problems. Kosovichev (1996) first used ray-path approximation kernels, Jensen, Jacobsen, and Christensen-Dalsgaard (2000) introduced Fresnel-zone kernels; and Birch and Kosovichev (2000), Birch, Kosovichev, and Duvall (2004), and Birch and Gizon (2007) investigated Born-approximation kernels for both sound-speed structures and flow fields. Couvidat, Birch, and Kosovichev (2006) compared subsurface sound-speed perturbation structures inferred from these different types of kernels, and found that the inversion results obtained with the different kernels were basically consistent.

Important results on the solar interior properties have been obtained from the time–distance studies as well as from other local helioseismology techniques (*e.g.*, Komm *et al.*, 2004; Lindsey and Braun, 2000). The introduction that follows is limited to only time–distance results due to the scope of this paper. On global scales, poleward meridional flows were found below the photosphere (Giles *et al.*, 1997), and solar-cycle dependent meridional flow variations were also investigated and discussed (Chou and Dai, 2001; Beck, Gizon, and Duvall, 2002; Zhao and Kosovichev, 2004). On local scales, subsurface sound-speed perturbations and flow fields were derived for supergranulation (Kosovichev and Duvall, 1997; Duvall *et al.*, 1997; Duvall and Gizon, 2000; Sekii *et al.*, 2007; Jackiewicz, Gizon, and Birch, 2008) and for sunspots (Kosovichev, Duvall, and Scherrer, 2000; Gizon, Duvall, and Larsen, 2000; Zhao, Kosovichev, and Duvall, 2001; Couvidat, Birch, and Kosovichev, 2006; Zhao, Kosovichev, and Sekii, 2010). Additionally, time–distance helioseismology was used to detect the emergence of active regions before their appearances in the photosphere (Kosovichev, Duvall, and Scherrer, 2000; Jensen *et al.*, 2001; Zharkov and Thompson, 2008), to image large active regions on the far side of the Sun (Zhao, 2007; Ilonidis, Zhao, and Hartlep, 2009), and to measure sound-speed perturbations in the tachocline (Zhao *et al.*, 2009). These results are important for space-weather forecasting and understanding the mechanisms for the generation of solar magnetism. The time–distance helioseismology pipeline analyses, based on the high spatial-resolution and high temporal-cadence observations from HMI, will greatly advance our knowledge of the interior processes and their connections with solar activity above the photosphere.

However, one should keep in mind that the physics of solar oscillations in the turbulent magnetized plasma is very complicated, and that the helioseismic techniques are still in the process of being developed. Because of limited knowledge of the wave physics and complexity of the MHD turbulence, there may be systematic uncertainties in the local helioseismology inferences, particularly in strong magnetic-field regions of sunspots. For example, Lindsey and Braun (2005a, 2005b) argued that the outgoing and ingoing travel-time asymmetries observed in sunspot areas might be caused by a "shower-glass effect". Schunker *et al.* (2005) found that the inclined magnetic field in sunspot penumbrae might cause variations of measured acoustic travel times. Zhao and Kosovichev (2006) found that this inclined magnetic-field dependent effect does not exist in the measurements obtained from the MDI intensity observations. Later, Rajaguru *et al.* (2006) found that the non-uniform acoustic power distribution in the photosphere also contributed to measured travel-time shifts in active regions if a phase-speed filtering procedure was applied. This effect was then studied by Parchevsky, Zhao, and Kosovichev (2008) and Hanasoge *et al.* (2008) numerically, and also by Nigam and Kosovichev (2010) analytically. More recently, Gizon *et al.* (2009) derived a sunspot's subsurface flow fields after applying ridge filtering, and their inferred results did not agree with the previously inverted results with the use of phase-speed filtering. Using high spatial-resolution observations from *Hinode*, Zhao, Kosovichev, and Sekii (2010) showed that the principal results on sunspot structure did not depend on the use of phase-speed filtering. However, significant systematic uncertainties in sunspot seismology remain and need to be understood, and these are being actively studied by the use of numerical simulations. For a recent review of the sunspot seismology and uncertainties see Kosovichev (2010).

Despite the ongoing discussions of accuracy of time–distance measurements and interpretation of inversion results, it is useful to provide the measured travel times and the inversion results for investigations of structures and flows below the visible surface of the Sun. As the flow chart of the pipeline in Figure 1 shows, we apply phase-speed filtering to the HMI data, compute cross-covariances of the oscillations, and use two different travel-time fitting procedures to derive the acoustic travel times. We then perform inversions using two different sets of travel-time sensitivity kernels, based on the ray-path and Born approximations, to infer subsurface wave-speed perturbations and flow fields, using the MCD inversion method. We provide online access to the measured travel times, full-disk subsurface wave-speed perturbations and flow maps calculated every eight hours. In addition, we provide synoptic flow maps for each Carrington rotation. In this article, we describe details of the acoustic travel-time measurement procedure in Section 2, and the inversion procedure in Section 3. We describe the pipeline data products and present initial HMI results in Section 4, and summarize our work in Section 5.

2. Acoustic Travel-Time Measurement

2.1. Tracking and Remapping

SDO/HMI continuously observes the full-disk Sun, providing Doppler velocity, continuum intensity, line-depth, line-width, and magnetic-field maps with a 45-second cadence, and also vector magnetic-field measurements with a cadence of 12 minutes. Each full-disk image

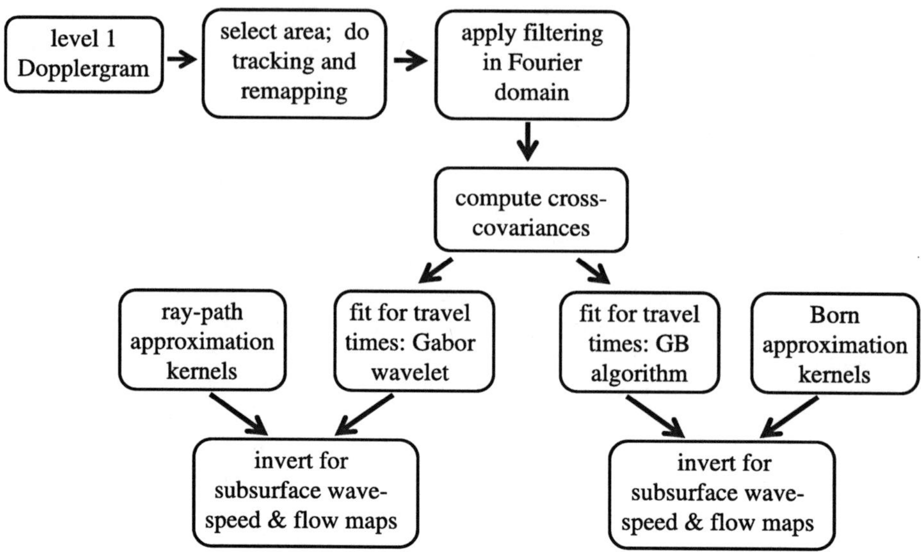

Figure 1 Flow chart for the HMI time–distance helioseismology data-analysis pipeline.

has 4096×4096 pixels with a spatial resolution of 0.504 arcsec pixel^{-1} (*i.e.*, approximately, 0.03 heliographic degree pixel^{-1} at the solar disk center). The Doppler observations are primarily used for helioseismic studies. The observing sequences, algorithms for deriving Doppler velocity and magnetic field, and other instrument calibration issues, are discussed by Schou *et al.* (2011).

As illustrated in Figure 1, the primary input for the pipeline is Dopplergrams, although in principle, the HMI intensitygrams and line-depth data can also be analyzed in the same manner. Users of the pipeline can select specific areas for analysis, preferably within 60° from the solar disk center. In practice, the users provide the Carrington longitude and latitude of the center of the selected area, and the mid-time of the selected time period, then the pipeline code selects an area of roughly $30° \times 30°$ centered at the given coordinate, and for a time interval of eight hours with the given time as the middle point. The data for this selected area and time period are then tracked to remove solar rotation, and remapped into the heliographic coordinates using Postel's projection (also known as azimuthal equidistant projection) relative to the given area center. Normally, the tracked area consists of 512×512 pixels with a spatial sampling of 0.06° pixel^{-1}; and the temporal sampling is the same as the observational cadence. Cubic interpolation is used for the pixels not located on the observational grid.

Figure 2 shows typical HMI k–ω and time–distance diagrams obtained from one tracked and remapped area. The selected area covers 30° in latitude and has an apparent differential rotation over this span. However, for fast computations, only one uniform tracking rate corresponding to the Snodgrass rotation rate at the center of this area is used. The Snodgrass rotation rate is $2.851 - 0.343\sin^2\phi - 0.474\sin^4\phi$ μrad s^{-1}, where ϕ is latitude (Snodgrass and Ulrich, 1990). The uniform tracking rate of the selected area results in a differential rotation velocity in the inverted horizontal velocity fields. This differential rotation velocity is removed from the full-disk flow maps after averaging over the whole Carrington rotation, and only the residual flow fields are given as the final results (see Section 4.1). But for the user-selected areas, the differential rotation is kept in the inversion results.

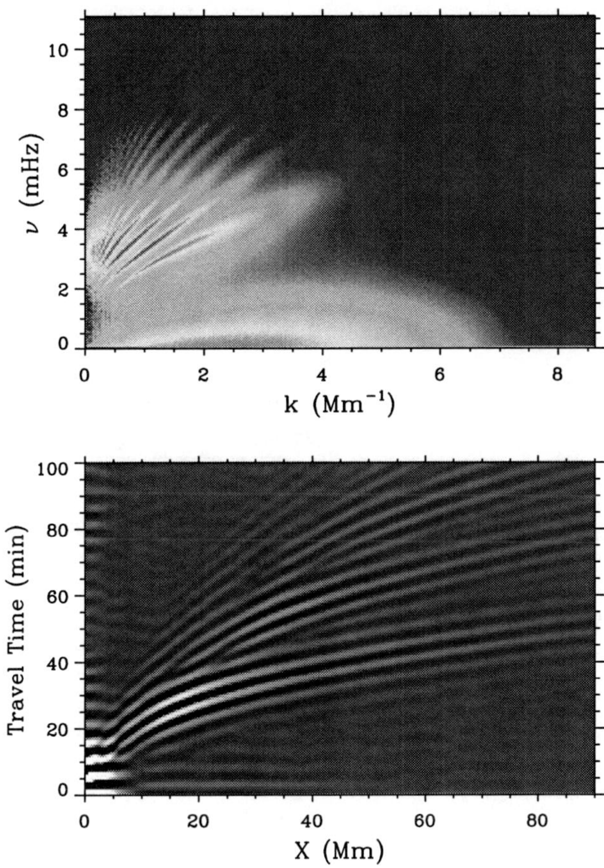

Figure 2 Typical power spectrum (k–ω) diagram (upper) and time–distance (cross-covariance) diagram (lower) made from eight hours of HMI Doppler observations.

2.2. Computing Cross-Correlations and Fitting for Travel Times

Each tracked and remapped Dopplergram datacube is filtered in the 3D Fourier domain. Solar convection and f-mode oscillation signals are removed first, and then phase-speed filtering is applied following the procedures prescribed by Couvidat *et al.* (2005). For the travel-time measurements, for each central point we select 11 annuli with various radii and widths chosen from our past experience with MDI analyses. All of the phase-speed filtering parameters, including the central phase speed, filter width, and the corresponding inner and outer annulus radii are presented in Table 1. The phase-speed filter is a Gaussian function of the wave's horizontal phase speed. It selects wave packets traveling between the central points and the annuli for the selected distances. This filter helps improve the signal-to-noise ratio, and also it removes leakage from low-degree oscillations. After the filtering, the data are transformed back to the space–time domain for cross-covariance computations. Two different fitting methods are used to derive the acoustic travel times from the cross-covariances: a Gabor-wavelet fitting (Kosovichev and Duvall, 1997), and a cross-correlation method based on seismology algorithms (Zhao and Jordan, 1998) adopted by Gizon and Birch (2002; GB algorithm hereafter). A detailed description of the filtering procedure, the cross-covariance computations, the two types of travel-time fittings, comparison of the travel times derived from the two fitting methods, and the measurement error estimates is given

Table 1 Phase-speed filtering parameters used for the selected travel distances (annulus ranges).

Annulus No.	Annulus range (heliographic degree)	Phase speed (μHz/ℓ)	FWHM (μHz/ℓ)
1	0.54 – 0.78	3.40	1.0
2	0.78 – 1.02	4.00	1.0
3	1.08 – 1.32	4.90	1.25
4	1.44 – 1.80	6.592	2.149
5	1.92 – 2.40	8.342	1.351
6	2.40 – 2.88	9.288	1.183
7	3.12 – 3.84	10.822	1.895
8	4.08 – 4.80	12.792	2.046
9	5.04 – 6.00	14.852	2.075
10	6.24 – 7.68	17.002	2.223
11	7.68 – 9.12	19.133	2.039

by Couvidat *et al.* (2010). In particular, it has been shown there that the two definitions of the acoustic travel times and the two fitting methods give generally consistent results in the quiet-Sun regions, but that they may give different results in active regions. The differences are currently not well understood, but in the pipeline we implement both travel-time definitions.

In the Postel-projected maps, the exact distance between two arbitrary points cannot be calculated using the Cartesian coordinates. Thus, when an annulus is selected around a given location, some additional computations are needed to determine the exact great-circle distance between the points. The formula to determine the great-circle distance is

$$\Delta = \arccos\bigl(\sin\theta_1 \sin\theta_2 + \cos\theta_1 \cos\theta_2 \cos(\phi_1 - \phi_2)\bigr), \tag{1}$$

where (θ_1, ϕ_1) and (θ_2, ϕ_2) are the heliospheric longitude and latitude coordinates for the two separate locations.

To facilitate the inversions for subsurface flow fields, each annulus is divided into four quadrants representing the North, South, East, and West directions (Kosovichev and Duvall, 1997). So, for each annulus and each travel-time fitting method, we obtain six measurements of acoustic travel times, corresponding to the outgoing and ingoing, East-, West-, South-, and North-going directions. We then combine these travel times to obtain one map for the mean travel time and three maps for the travel-time differences. These travel times are: τ_{mean} (average of outgoing and ingoing travel times), τ_{oi} (difference of outgoing and ingoing travel times), τ_{we} (difference of West- and East-going travel times), and τ_{ns} (difference of North- and South-going travel times). These four travel-time maps for each annulus are archived and available through the HMI Data Record Management System (DRMS).

3. Subsurface Wave-Speed Perturbation and Flow-Field Inversions

The acoustic travel times are derived by two different fitting methods: the Gabor-wavelet function and the GB algorithm. Then, as illustrated in Figure 1, to infer the subsurface wave-speed perturbations and flow velocities, the Gabor-wavelet travel times are inverted using the ray-path approximation sensitivity kernels, and the GB travel times are inverted using the Born-approximation sensitivity kernels. The Born-approximation kernels are calculated based with the filter and window parameters of the GB fits.

3.1. Inversions

Both the ray-path and Born-approximation kernels have been used in previous time–distance studies (see, *e.g.*, Zhao, Kosovichev, and Duvall, 2001; Couvidat, Birch, and Kosovichev, 2006). Details of the kernel calculations and their comparisons will be given in a separate article.

We employ the MCD inversion method (Jacobsen *et al.*, 1999) with a horizontal regularization (Couvidat *et al.*, 2005). For the wave-speed perturbation inversions, the linearized equation relating the measured mean travel times and the subsurface wave-speed perturbations are

$$\delta\tau_{\text{mean}}^{\lambda\mu\nu} = \sum_{ijk} A_{ijk}^{\lambda\mu\nu}\,\delta s_{ijk}, \tag{2}$$

where δs_{ijk} is the relative wave-speed perturbation $\delta c_{ijk}/c_{ijk}$ approximated by piece-wise constant functions on the inversion grid, and $A_{ijk}^{\lambda\mu\nu}$ is a matrix of the discretized sensitivity kernel. Here, λ and μ label the coordinates of the central points of the annuli in the observed areas, ν labels the different annuli, and i, j, and k are the indices for the discretized three-dimensional space for inversions. Usually, the horizontal coordinates of the inversion grid (i and j indices) are selected at the same locations as the central travel-time measurement points (λ and μ indices). In the first-order approximation, the sensitivity kernels are calculated for a spherically symmetric solar model and do not depend on the position on the solar surface. Therefore, in this case Equation (2) is actually equivalent to a convolution in the horizontal domain, which can be simplified as a direct multiplication in the Fourier domain:

$$\delta\tilde{\tau}^{\nu}(\kappa_\lambda, \kappa_\mu) = \sum_{k} \tilde{A}_{k}^{\nu}(\kappa_\lambda, \kappa_\mu)\,\delta\tilde{s}_k(\kappa_\lambda, \kappa_\mu), \tag{3}$$

where $\delta\tilde{\tau}$, $\tilde{A}$, and $\delta\tilde{s}$ are the 2D Fourier transforms of $\delta\tau$, A, and δs, respectively; k is the same as in Equation (2); κ_λ and κ_μ are the wavenumbers in the Fourier domain corresponding to λ and μ of the spatial domain. For each $(\kappa_\lambda, \kappa_\mu)$, the equation in the Fourier domain is a matrix multiplication:

$$d = Gm, \tag{4}$$

where

$$d = \{\delta\tilde{\tau}^{\nu}(\kappa_\lambda, \kappa_\mu)\}, \qquad G = \{\tilde{A}_{k}^{\nu}(\kappa_\lambda, \kappa_\mu)\}, \qquad m = \{\delta\tilde{s}_k(\kappa_\lambda, \kappa_\mu)\}.$$

Thus, we have a large number of linear equations describing the depth dependence of the Fourier components, and each linear equation can be solved in the least-squares sense. After all these equations are solved, and m is obtained for each $(\kappa_\lambda, \kappa_\mu)$, the values of δs_{ijk} are calculated by the inverse 2D Fourier transform.

Equation (4) is ill-posed, and regularization is required to obtain a smooth solution. The regularized least-squares algorithm is formulated as

$$\min\{\|(d - Gm)\|_2^2 + \lambda^2(\kappa)\|Lm\|_2^2\}, \tag{5}$$

where $\|\dots\|_2$ denotes the L2-norm, L is a regularization operator, and $\lambda(\kappa)$ is a regularization parameter. We choose L to be a diagonal matrix whose elements are the inverse of the square root of the spatial sampling Δz at each depth. Such weighting is necessary because the grid in the vertical direction is chosen to be approximately uniform in the acoustic depth, meaning that the spatial sampling of deep layers is larger than the sampling of shallower layers. The regularization parameter is taken in the form of $\lambda^2(\kappa) = \lambda_v^2 + \lambda_h^2(\kappa)$, where λ_v and λ_h are vertical and horizontal regularization parameters. The purpose of introducing λ_h is to damp the high-wavenumber components that may lead to noise amplification. Following the discussion of Couvidat *et al.* (2005), we choose $\lambda_h = \lambda_2\kappa^2$ with λ_2 as a constant.

Because the regularization is applied in the Fourier domain, it is quite difficult to use different regularization parameters for different horizontal locations of the same region. Sometimes different regularization parameters are needed, because in active regions the noise level may be different from the quiet-Sun regions, as we discuss in Section 3.3. Thus, it is necessary to implement into the analysis pipeline another inversion technique, the LSQR algorithm, which solves Equation (2) in the space domain by an iterative approach. This implementation is currently under development.

3.2. Inversion Depth and Validation of Inversions

For both the wave-speed and flow-field inversions, and for both the ray-path and Born-approximation inversions, we select a total of 11 inversion depths as follows: $0-1$, $1-3$, $3-5$, $5-7$, $7-10$, $10-13$, $13-17$, $17-21$, $21-26$, $26-30$, and $30-35$ Mm. There are a total of 11 depth intervals. The inversion results provide the wave-speed perturbations and flow velocities averaged in these layers. Due to the lack of acoustic-wave coverage in the deep interior, the reliability of inversion results decreases with the depth. Thus, only inversion results for the depths shallower than 20 Mm are included in the pipeline output. This may change in the future when more confidence is gained in the deeper interior inversion results.

In recent years, several studies have been carried out to validate the time–distance measurements and inversions. To validate the derived subsurface flow fields, Georgobiani *et al.* (2007) and Zhao *et al.* (2007) have analyzed realistic solar-convection simulations and found satisfactory inversion results for the shallow layers covered by the simulations. Validations of the wave-speed perturbation inversions based on numerical simulations with preset structures have also been performed. Meanwhile, numerical simulations for magnetic structures with flows are also under development (Rempel, Schüssler, and Knölker, 2009; Stein *et al.*, 2011; Kitiashvili *et al.*, 2011). Validations of the time–distance helioseismology techniques will be carried out as well using these simulations.

Cross-comparisons between different helioseismology techniques, *e.g.* comparing the mean rotation speed from our pipeline analysis with global helioseismology results, and comparing the subsurface flow fields with results from ring-diagram analyses, will also be important for the validation.

3.3. Error Estimate

There are two types of errors in the pipeline results: systematic errors due to our limited knowledge of the wave physics in the magnetized turbulent medium and simplified mathematical formulations, and statistical errors, which are mostly due to the stochastic nature of the solar oscillations ("realization noise"). Here, we only discuss the statistical errors.

To estimate the errors in the inversion results, we need first to estimate the uncertainties in the measured acoustic travel times. Here, we estimate the measurement uncertainties following the prescriptions of Jensen, Duvall, and Jacobsen (2003) and Couvidat, Birch, and Kosovichev (2006). We select 25 quiet-Sun regions, and use the rms variation of mean travel times for different measurement distances as an error estimate for the travel times. The estimated uncertainties for the Gabor-wavelet fitting are given in Table 2, and the uncertainties obtained for the GB algorithm are similar, but slightly larger for short distances and

Table 2 Measured mean travel times and uncertainties for both quiet-Sun regions and active regions.

Annulus No.	Mean travel time (min)	Uncertainty for quiet regions (min)	Uncertainty for active regions (min)
1	11.87	0.062	0.17
2	18.82	0.061	0.25
3	21.76	0.11	0.26
4	25.85	0.11	0.19
5	28.69	0.14	0.15
6	31.18	0.14	0.14
7	35.07	0.10	0.10
8	38.86	0.12	0.11
9	42.46	0.11	0.093
10	46.63	0.14	0.11
11	50.26	0.15	0.11

slightly smaller for long distances. Active regions have different measurement uncertainties. To estimate these, we selected a relatively stable sunspot, NOAA AR 11092, from 2 through 5 August 2010, and we assumed that the sunspot did not change during this period. Then we use the rms of the travel times measured inside the sunspot as an error estimate. Due to the evolution of the sunspot, this approach overestimates the measurement uncertainties, but can still give us an approximate estimate of measurement errors. These error estimates are presented in Table 2 as well.

Inversions are then performed for the same quiet-Sun regions and the active region to estimate the statistical errors in inversion results. Following the same approach, the rms of inverted wave-speed perturbations is assumed as the statistical error. The error estimates for both the quiet-Sun and active regions are shown in Table 3. Because supergranular flows are dominant in the flow fields, it is difficult to estimate errors of the inverted velocity for the quiet Sun by this approach. Instead, we estimate the velocity errors based on the rotational velocity profile, which has little change in a time scale of a few days. While the errors for the wave-speed perturbation inside active regions are roughly twice those for the quiet-Sun regions, the velocity errors inside active regions sometimes are seven times larger than the errors in the quiet Sun.

4. Data Products and Initial Results from HMI

The time–distance data-analysis pipeline is used for the routine production of nearly real-time full-disk (actually, nearly full-disk covering a $120° \times 120°$ area on the solar disk) wave-speed perturbation and flow field maps every eight hours. These maps are then used to construct the corresponding synoptic maps for each Carrington rotation. The pipeline can also be used for specific target areas, such as active regions. In this section, we introduce the data products from this pipeline and some initial results from it.

4.1. Routine Production: Full-Disk and Synoptic Maps

For each day of HMI observations, we select three eight-hour periods: 00:00 – 07:59 UT, 08:00 – 15:59 UT, and 16:00 – 23:59 UT. For each analysis period, we select 25 regions,

Table 3 Error estimates for the relative wave-speed perturbation and horizontal-velocity inferences in the quiet-Sun (QS) and active regions (AR).

Depth (Mm)	Wave speed for QS	Velocity for QS ($\mathrm{m\,s^{-1}}$)	Wave speed for AR	Velocity for AR ($\mathrm{m\,s^{-1}}$)
$0-1$	2.8×10^{-3}	7.8	6.3×10^{-3}	58.3
$1-3$	4.1×10^{-3}	7.5	10.9×10^{-3}	56.4
$3-5$	6.4×10^{-3}	8.9	8.7×10^{-3}	51.1
$5-7$	4.6×10^{-3}	9.4	9.7×10^{-3}	45.1
$7-10$	4.7×10^{-3}	13.1	6.7×10^{-3}	34.5
$10-13$	3.7×10^{-3}	12.9	3.1×10^{-3}	28.1

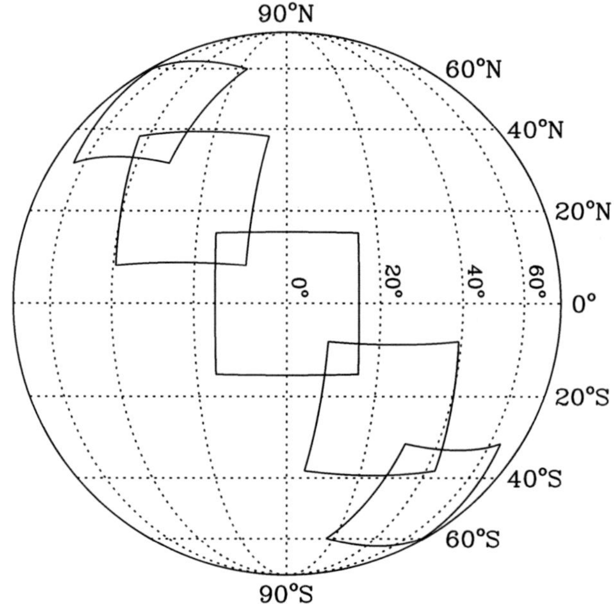

Figure 3 Schematic plot showing how areas are selected for a routine calculations of the full-disk wave-speed and flow maps. Not all of the 25 selected areas are shown.

with the central locations at 0°, ±24°, and ±48° in both longitude and latitude, where the longitude is relative to the central meridian at the mid-time of the selected period. Figure 3 shows locations of these areas on the solar disk. The total number of areas is 25: five rows and five columns. Due to the Postel's projection, the boundaries of these areas are often not parallel to the latitude or longitude lines. It is also evident that many areas overlap, some areas overlap twice, and some overlap four times. The travel times and inversion results are averaged in these overlapped areas. However, in the areas close to the solar limb, the foreshortening effect may become non-negligible, but the role of this is not yet systematically studied. Users of these maps are urged to be cautious when using the pipeline results in the areas close to the limb.

For each full-disk map and each synoptic map, the East–West velocity [v_x], the North–South velocity [v_y], and wave-speed perturbation [$\delta c/c$] in each depth layer are derived with a horizontal spatial sampling of 0.12° pixel^{-1}. For each of the 25 areas, the inversion results are first obtained in the Postel-projection coordinates, and then converted into the

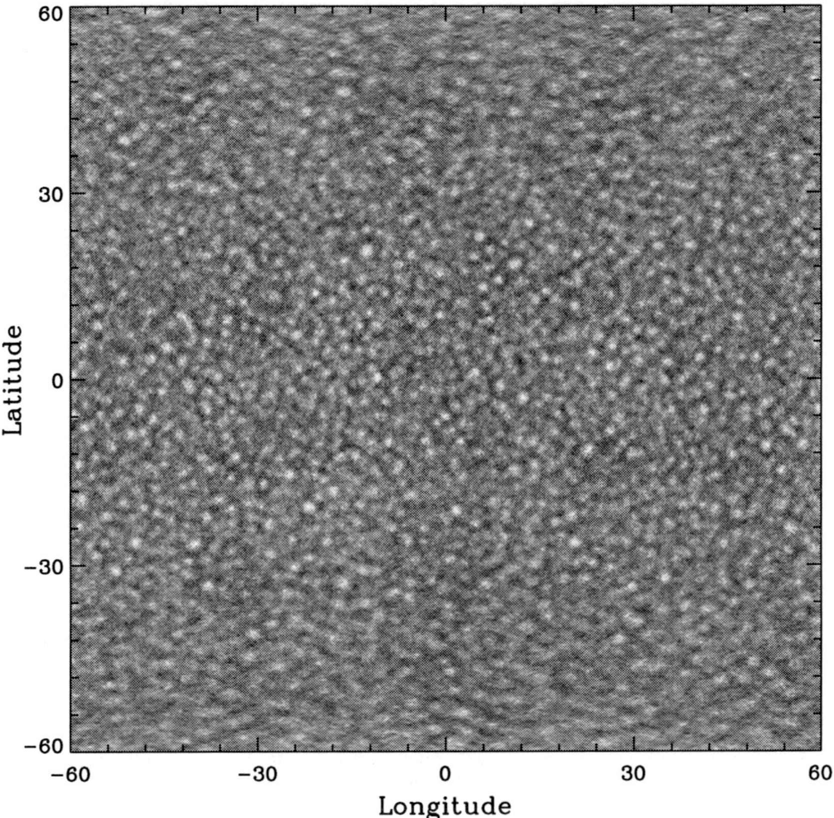

Figure 4 A map of horizontal flow divergence for a depth range of 1 – 3 Mm and a time period of 00:00 – 07:59 UT 19 May 2010. The display scale is from -6.2×10^{-4} to 6.2×10^{-4} s^{-1}. White areas with positive divergence correspond to supergranulation. Note that supergranules appear larger at high latitudes because of the rectangular longitude–latitude map projection.

longitude–latitude coordinates. This coordinate conversion is basically the inverse procedure of transforming the observed data into the Postel-projection coordinates for the travel-time measurements. Cubic spline interpolation is employed. The results in high-latitude regions are oversampled. After the coordinate transformation, the overlapped areas are averaged, and the final statistical errors are estimated for the whole procedure starting from the travel-time measurements. This includes all potential errors from the interpolation and remapping. The final full-disk results are saved on a uniform longitude–latitude coordinate grid, so each horizontal image of the subsurface layers has a total of 1000×1000 pixels covering 120° in both longitudinal and latitudinal directions. This coordinate choice is convenient for merging and analyzing results, but unavoidably distorts maps in high latitude areas.

Figure 4 shows an example of a full-disk map of the subsurface horizontal flow divergence calculated from v_x and v_y at the depth range of 1 – 3 Mm. The positive-divergence areas correspond to supergranulation. Such maps at various depths with continuous temporal coverage can be valuable for studying the structure and evolution of the supergranulation. Figure 5 displays the subsurface horizontal flow fields with full spatial resolution for a region located at the center of the map in Figure 4. Supergranular flows can be easily identified.

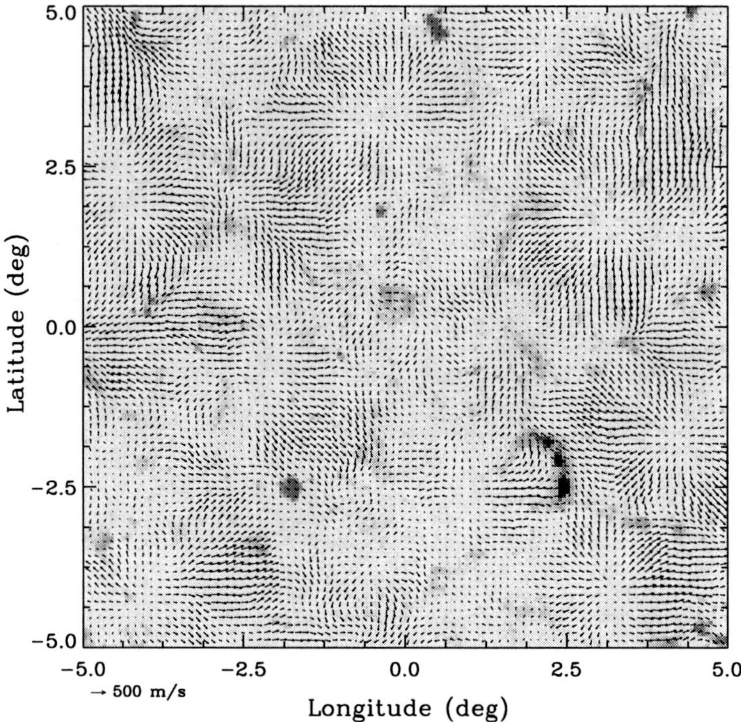

Figure 5 A sample of subsurface horizontal flow fields with full spatial resolution at the depth of 0 – 1 Mm. This area is sampled at the center of the map shown in Figure 4. The background image shows the line-of-sight magnetic field measured by HMI, with red as positive and blue as negative polarities. The range of the magnetic field is from −80 to 80 Gauss.

From the full-disk wave-speed perturbation and flow maps obtained every eight hours, we select an area 13.2° wide in longitude, *i.e.* −6.6° to 6.6° from the central meridian, to construct the synoptic wave-speed perturbation and flow maps. Since the Carrington rotation rate corresponds to a shift of approximately 4.4° every eight hours, each location in the constructed synoptic maps is averaged roughly three times (*i.e.* one whole day). The resultant synoptic map for each depth consists of 3000 × 1000 pixels. Since such a map is difficult to display, we show in Figure 6 a binned-down synoptic large-scale flow map obtained for the depth of 1 – 3 Mm for Carrington Rotation 2097 during the period from 20 May to 16 June, 2010. The vectors in the figure are obtained by averaging the flow fields in areas of 15° × 15° with a 3° sampling rate. The maps of this type are similar to the subsurface flow maps obtained from the ring-diagram analysis (Haber *et al.*, 2002). From Figure 6, it can be found that the large-scale flows often converge around magnetic regions.

Normally, the full-disk flow maps and the synoptic flow maps are provided as residual flow maps after subtracting the flows averaged over the entire Carrington rotation. The subtracted average maps contain the differential rotation, meridional flows, and possibly some systematic effects. Figure 7 shows examples of the subsurface differential rotation speed and meridional flow speed obtained by averaging the calculations for Carrington Rotation 2097. These results are also provided online together with the synoptic flow maps.

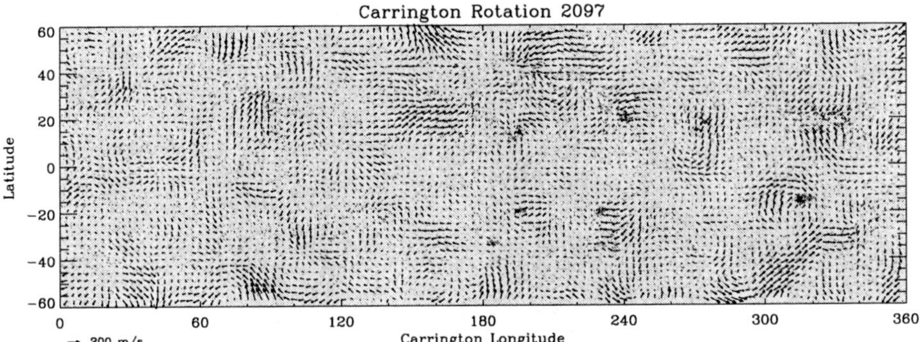

Figure 6 Synoptic map of large-scale horizontal flows at the depth of 1 – 3 Mm for CR 2097. The background image is the corresponding line-of-sight magnetic field, with red as positive and blue as negative polarities, in the range of −50 to 50 Gauss.

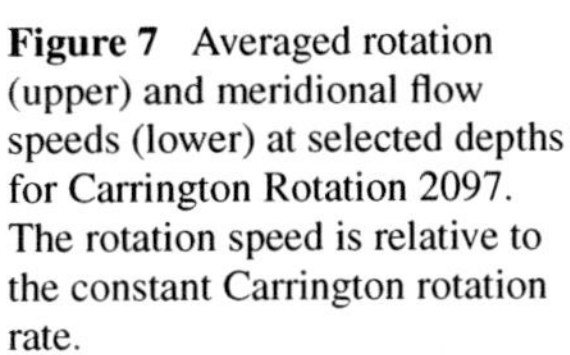

Figure 7 Averaged rotation (upper) and meridional flow speeds (lower) at selected depths for Carrington Rotation 2097. The rotation speed is relative to the constant Carrington rotation rate.

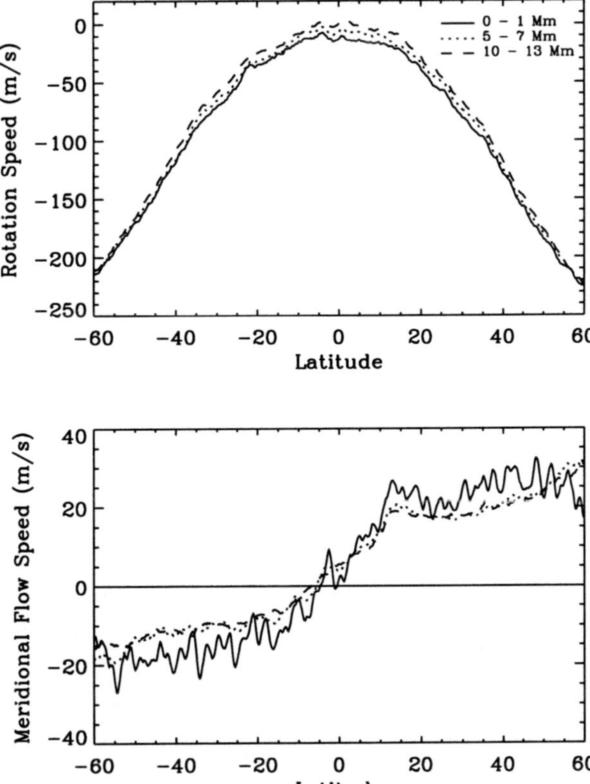

4.2. Target Areas

As already mentioned in Section 2, the pipeline can also be run for specific target areas and specific time intervals. The pipeline users are required to provide the Carrington coordinates of the center of the target area, and the mid-time of the time interval.

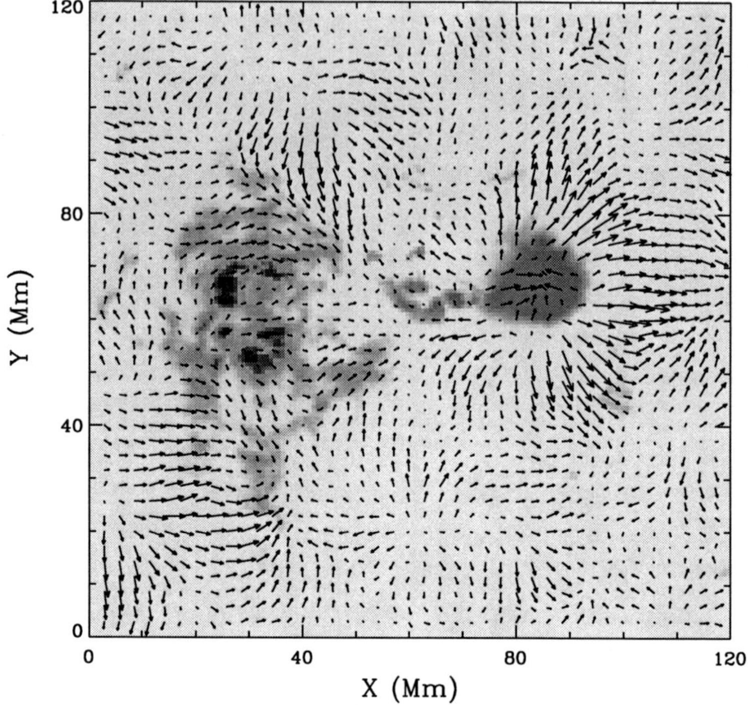

Figure 8 Subsurface flow fields of AR 11072 at the depth of 1 – 3 Mm, obtained from 16:00 – 23:59 UT 23 May 2010. The background image is line-of-sight magnetic field, with red positive and blue negative. The image is displayed with a scale from −1000 to 1000 Gs. The arrows are displayed after a 2 × 2 rebinning, and the longest arrow represents a speed of 300 $\mathrm{m\,s^{-1}}$.

Figure 8 shows an example of a small part (approximately 1/9) of a target area, with an active region, AR 11072, included in it. The subsurface flow field, at the depth of 1 – 3 Mm, is displayed after a 2 × 2 rebinning. The displayed time period, 16:00 – 23:59 UT on 23 May 2010, is approximately 2.5 days after the start of emergence of this active region that was still growing. Our results clearly show subsurface outflows around the leading sunspot, and some converging flows inside it. Comparing with the previous results of Kosovichev (2009) and Zhao, Kosovichev, and Sekii (2010), one may conclude that the subsurface flow fields of active regions evolve with the evolution of active regions. There may be prominent outflows around sunspots during their early growing phase and also their decaying phase, but there may also be strong converging flows during the stable period. Our continuous monitoring of the solar full-disk subsurface flows may give us an opportunity to statistically study the changes of the sunspot's subsurface flows with the sunspot evolution.

5. Summary

We have developed a time–distance helioseismology data-analysis pipeline for SDO/HMI Doppler observations. This pipeline performs acoustic travel-time measurements based on two different methods, and conducts inversions based on two different sensitivity kernels calculated in the ray-path and Born approximations. The pipeline gives nearly real-time routine products of full-disk wave-speed perturbations and flow-field maps in the range of depth 0 – 20 Mm every eight hours, and provides the corresponding synoptic wave-speed perturbation and flow-field maps for each Carrington rotation. The averaged rotation speed and meridional flow speed are also provided separately for each rotation. In addition to these routine production, the pipeline can also be used for analysis of specific target areas for specific time intervals. This data-analysis pipeline will provide important information about the subsurface structure and dynamics on both local and global scales, and its continuous coverage through years to come will be useful for understanding the connections between solar subsurface properties and magnetic activity in the chromosphere and corona. With the improvement of our understanding of acoustic-wave and magnetic-field interactions, and also the measurement and inversion techniques, the pipeline codes will be regularly revised.

References

Beck, J.G., Gizon, L., Duvall, T.L. Jr., 2002, *Astrophys. J. Lett.* **575**, L47.
Birch, A.C., Gizon, L.: 2007, *Astron. Nachr.* **328**, 228.
Birch, A.C., Kosovichev, A.G.: 2000, *Solar Phys.* **192**, 193.
Birch, A.C., Kosovichev, A.G., Duvall, T.L. Jr., 2004, *Astrophys. J.* **608**, 580.
Chou, D.-Y., Dai, D.-C.: 2001, *Astrophys. J. Lett.* **559**, L175.
Couvidat, S., Birch, A.C., Kosovichev, A.G.: 2006, *Astrophys. J.* **640**, 516.
Couvidat, S., Birch, A.C., Kosovichev, A.G., Zhao, J.: 2004, *Astrophys. J.* **607**, 554.
Couvidat, S., Gizon, L., Birch, A.C., Larsen, R.M., Kosovichev, A.G.: 2005, *Astrophys. J. Suppl.* **158**, 217.
Couvidat, S., Zhao, J., Birch, A.C., Kosovichev, A.G., Duvall, T.L. Jr., Parchevsky, K.V., Scherrer, P.H.: 2010, *Solar Phys.* doi:10.1007/s11207-010-9652-y.
Duvall, T.L. Jr., Gizon, L.: 2000, *Solar Phys.* **192**, 177.
Duvall, T.L. Jr., Jefferies, S.M., Harvey, J.W., Pomerantz, M.A.: 1993, *Nature* **362**, 430.
Duvall, T.L. Jr., D'Silva, S., Jefferies, S.M., Harvey, J.W., Schou, J.: 1996, *Nature* **379**, 235.
Duvall, T.L. Jr., Kosovichev, A.G., Scherrer, P.H., Bogart, R.S., Bush, R.I., DeForest, C., Hoeksema, J.T., Schou, J., Saba, J.L.R., Tarbell, T.D., Title, A.M., Wolfson, C.J., Milford, P.N.: 1997, *Solar Phys.* **170**, 63.
Georgobiani, D., Zhao, J., Kosovichev, A.G., Bensen, D., Stein, R.F., Nordlund, Å.: 2007, *Astrophys. J.* **657**, 1157.
Giles, P.M., Duvall, T.L. Jr., Scherrer, P.H., Bogart, R.S.: 1997, *Nature* **390**, 52.
Gizon, L., Birch, A.C.: 2002, *Astrophys. J.* **571**, 966.
Gizon, L., Duvall, T.L. Jr., Larsen, R.M.: 2000, *J. Astrophys. Astron.* **21**, 339.
Gizon, L., Schunker, H., Baldner, C.S., Basu, S., Birch, A.C., Bogart, R.S., Braun, D.C., Cameron, R., Duvall, T.L. Jr., Hanasoge, S.M., Jackiewicz, J., Roth, M., Stahn, T., Thompson, M.J., Zharkov, S.: 2009, *Space Sci. Rev.* **144**, 249.
Haber, D.A., Hindman, B.W., Toomre, J., Bogart, R.S., Larsen, R.M., Hill, F.: 2002, *Astrophys. J.* **570**, 885.
Hanasoge, S.M., Couvidat, S., Rajaguru, S.P., Birch, A.C.: 2008, *Mon. Not. Roy. Astron. Soc.* **391**, 1931.
Ilonidis, S., Zhao, J., Hartlep, T.: 2009, *Solar Phys.* **258**, 181.
Jacobsen, B.H., Møller, I., Jensen, J.M., Effersø, F.: 1999, *Phys. Chem. Earth, Part A, Solid Earth Geod.* **24**, 15.
Jackiewicz, J., Gizon, L., Birch, A.C.: 2008, *Solar Phys.* **251**, 381.
Jensen, J.M., Duvall, T.L. Jr., Jacobsen, B.H.: 2003, In: Sawaya-Lacoste, H. (ed.) *Proc. SOHO 12/ GONG+ 2002, Local and Global Helioseismology: the Present and Future* **SP-517**, ESA, Noordwijk, 315.
Jensen, J.M., Duvall, T.L. Jr., Jacobsen, B.H., Christensen-Dalsgaard, J.: 2001, *Astrophys. J. Lett.* **553**, L193.
Jensen, J.M., Jacobsen, B.H., Christensen-Dalsgaard, J.: 2000, *Solar Phys.* **192**, 231.
Kitiashvili, I.N., Kosovichev, A.G., Mansour, N.N., Wray, A.A.: 2011, *Solar Phys.* **268**, 283.
Komm, R., Corbard, T., Durney, B.R., Hernandez Gonzalez, I., Hill, F., Howe, R., Toner, C: 2004, *Astrophys. J.* **605**, 554.
Kosovichev, A.G.: 1996, *Astrophys. J. Lett.* **461**, L55.
Kosovichev, A.G.: 2009, *Space Sci. Rev.* **144**, 175.
Kosovichev, A.G.: 2010, *Solar Phys.* submitted. arXiv:1010.4927.
Kosovichev, A.G., Duvall, T.L. Jr.: 1997, In: Pijers, F.P., Christensen-Dalsgaard, J., Rosenthal, C.S. (eds.) *Proceedings of SCORe'96: Solar Convection and Oscillations and Their Relationship, Astrophysics and Astronomy Library* **225**, Kluwer, Dordrecht, 241.

Kosovichev, A.G., Duvall, T.L. Jr., Scherrer, P.H.: 2000, *Solar Phys.* **192**, 159.
Lindsey, C., Braun, D.C.: 2000, *Solar Phys.* **192**, 261.
Lindsey, C., Braun, D.C.: 2005a, *Astrophys. J.* **620**, 1107.
Lindsey, C., Braun, D.C.: 2005b, *Astrophys. J.* **620**, 1118.
Nigam, R., Kosovichev, A.G.: 2010, *Astrophys. J.* **708**, 1475.
Parchevsky, K.V., Zhao, J., Kosovichev, A.G.: 2008, *Astrophys. J.* **678**, 1498.
Rajaguru, S.P., Birch, A.C., Duvall, T.L. Jr., Thompson, M.J., Zhao, J.: 2006, *Astrophys. J.* **646**, 543.
Rempel, M., Schüssler, M., Knölker, M.: 2009, *Astrophys. J.* **691**, 640.
Sekii, T., Kosovichev, A.G., Zhao, J., Tsuneta, S., Shibahashi, H., Berger, T.E., Ichimoto, K., Katsukawa, Y., Lites, B., Nagata, S., Shimizu, T., Shine, R.A., Suematsu, Y., Tarbell, T.D., Title, A.M.: 2007, *Publ. Astron. Soc. Japan* **59**, S637.
Scherrer, P.H., Bogart, R.S., Bush, R.I., Hoeksema, J.T., Kosovichev, A.G., Schou, J., Rosenberg, W., Springer, L., Tarbell, T.D., Title, A., Wolfson, C.J., Zayer, I., MDI Engineering Team: 1995, *Solar Phys.* **162**, 129.
Schou, J., Scherrer, P.H., Watchter, R., Couvidat, S., Bush, R.I.: *et al.*: 2011, *Solar Phys.* in preparation.
Schunker, H., Braun, D.C., Cally, P.S., Lindsey, C.: 2005, *Astrophys. J. Lett.* **621**, L149.
Snodgrass, H.B., Ulrich, R.K.: 1990, *Astrophys. J.* **351**, 309.
Stein, R.F., Lagerfjärd, A., Nordlund, Å., Georgobiani, D.: 2011, *Solar Phys.* **268**, 271.
Zhao, J.: 2007, *Astrophys. J. Lett.* **664**, L139.
Zhao, L., Jordan, T.H.: 1998, *Geophys. J. Int.* **133**, 683.
Zhao, J., Kosovichev, A.G.: 2004, *Astrophys. J.* **603**, 776.
Zhao, J., Kosovichev, A.G.: 2006, *Astrophys. J.* **643**, 1317.
Zhao, J., Kosovichev, A.G., Duvall, T.L. Jr.: 2001, *Astrophys. J.* **557**, 384.
Zhao, J., Kosovichev, A.G., Sekii, T.: 2010, *Astrophys. J.* **708**, 304.
Zhao, J., Georgobiani, D., Kosovichev, A.G., Benson, D., Stein, R.F., Nordlund, Å.: 2007, *Astrophys. J.* **659**, 848.
Zhao, J., Hartlep, T., Kosovichev, A.G., Mansour, N.N.: 2009, *Astrophys. J.* **702**, 1150.
Zharkov, S., Thompson, M.J.: 2008, *Solar Phys.* **251**, 369.

Solar Phys (2012) 275:391–406
DOI 10.1007/s11207-011-9917-0

The *Solar Dynamics Observatory* (SDO) Education and Outreach (E/PO) Program: Changing Perceptions One Program at a Time

E. Drobnes · A. Littleton · W.D. Pesnell · K. Beck · S. Buhr · R. Durscher · S. Hill · M. McCaffrey · D.E. McKenzie · D. Myers · D. Scherrer · M. Wawro · A. Wolt

Received: 13 January 2010 / Accepted: 30 November 2011 / Published online: 21 December 2011

Abstract We outline the context and overall philosophy for the combined *Solar Dynamics Observatory* (SDO) Education and Public Outreach (E/PO) program, present a brief overview of all SDO E/PO programs along with more detailed highlights of a few key programs, followed by a review of our results to date, conclude a summary of the successes, failures, and lessons learned, which future missions can use as a guide, while incorporating their own content to enhance the public's knowledge and appreciation of science and technology as well as its benefit to society.

Keywords Solar dynamics observatory · Education and public outreach

The Solar Dynamics Observatory
Guest Editors: W. Dean Pesnell, Phillip C. Chamberlin, and Barbara J. Thompson

E. Drobnes (✉) · A. Littleton · D. Myers · M. Wawro · A. Wolt
ADNET Systems Inc., Greenbelt, MD 20771, USA
e-mail: emilie.drobnes@nasa.gov

E. Drobnes (✉)
e-mail: emilie.drobnes@gmail.com

W.D. Pesnell
NASA Goddard Space Flight Center, Greenbelt, MD 20771, USA

S. Buhr · M. McCaffrey
The Cooperative Institute for Research in Environmental Sciences (CIRES), Boulder, CO 80309, USA

K. Beck
Stanford University Haas Center for Public Service, Stanford, CA 94305, USA

R. Durscher · D. Scherrer
Stanford University, Stanford, CA 94305, USA

S. Hill
RS Information Systems, Lexington Park, MD 20653, USA

D.E. McKenzie
Montana State University, Bozeman, MT 59717, USA

1. Introduction

We present an overview of the education and public outreach (E/PO) program associated with the *Solar Dynamics Observatory* (SDO) mission. Reflecting on our efforts leading up to launch, we will provide insight into the programs developed, the lessons learned, and provide a model for future missions to build upon.

Many stakeholders (NASA, Department of Education, National Research Council, *etc.*) understand the inherent value of embedding E/PO programs into missions and scientific-research environments. The close connection to the science and data, coupled with the co-operation between scientists and education specialists, is imperative to successful engagement of the public and fostering their curiosity (NRC, 2011). NASA and the Heliophysics Science Division (HSD) have responded by making E/PO an integral part of every science mission.

In this section, we begin by presenting the state of science literacy and education in the United States and how NASA and the HSD at NASA are choosing to address needs for improving science education.

1.1. US Ranking in Science, Technology, Engineering, and Mathematics

For the past decade, US pre-college students have trailed their peers in other countries in their literacy levels and science and math scores. A recent Organization for Economic Co-operation and Development (OECD) report showed that the US ranked 31st in mathematics performance and 23rd in science in a major study of 65 countries (OECD, 2011). The resulting low numbers of US students pursuing science, technology, engineering, and mathematics (STEM) higher-education degrees means having to fill the US STEM-related workforce needs externally. In response, and to fill its own STEM workforce needs, NASA has identified E/PO as one of its priorities.

1.2. The NASA E/PO Umbrella

The NASA E/PO umbrella is continually evolving, often changing its approach and focus. However, the fundamentals tend to remain the same.

NASA has three main education goals: *i*) strengthen NASA and the nation's future workforce, *ii*) attract and retain students in science, technology, engineering, and math (STEM) disciplines, and *iii*) engage Americans in NASA's mission (NASA Office of Education, 2007).

NASA's Strategic Framework describes four main areas of E/PO (NASA, 2007):

i) Higher Education: targets undergraduate and graduate students inside and outside of the classroom.
ii) Formal Education: targets the K-12 system and involves standards-based (National Science Education Standards, 1996) curricula.
iii) Informal Education: involves science-education settings outside the classroom and is not tied to the National Science Education Standards.
iv) Public Outreach: includes activities aimed at reaching audiences in their own environments and providing unique opportunities for interacting with significantly larger audiences.

In an effort to support this strategic framework, NASA requires missions to spend a percentage of their total cost on E/PO programs closely tied to its mission. In the past, this

Table 1 SDO E/PO team and institutions

Institution	AIA	EVE	HMI	Mission
Laboratory for Atmospheric and Space Physics (LASP), Cooperative Institute for Environmental Science (CIRES)		Lead		
Goddard Space Flight Center (GSFC)				Lead
Harvard-Smithsonian Center for Astrophysics (CfA)	×			
Lockheed Martin Solar and Astrophysics Laboratory (LMSAL)	×			
Montana State University (MSU)	Lead			
Stanford University	×		Lead	

Consultant-authors were associated with GSFC.

amount has varied between 0.5% and 2%. The standard now is about 1% (SMD Science Management Council, 2010). This close link between missions and E/PO programs also helps ensure a tight connection between scientists, engineers, and the education community.

1.3. E/PO Within the Heliophysics Science Division

As part of the *Living With a Star* (LWS) program of the HSD at NASA, SDO must also align itself with the HSD E/PO goals and strategic communication objectives outlined within the Heliophysics Roadmap that builds upon the framework identified above. As is true with all science divisions, HSD sees E/PO as a key component of all flight and research programs (Christensen *et al.*, 2009). However, with HSD science in mind, it highly encourages a more integrated set of activities that reflect the system-wide approach to HSD science and how we study the Sun and its effects on the solar system. The four strategic communication objectives as outlined in the roadmap are:

i) Seek opportunities to increase and maintain public awareness of heliophysics science through activities, materials, and events.
ii) Engage students and sustain their interest in heliophysics-related STEM subjects.
iii) Collaborate with and engage educators to enhance their knowledge of heliophysics-related subjects and activities.
iv) Build awareness among students, educators, and the public of the diverse range of career opportunities related to heliophysics science and missions.

2. SDO Education and Public Outreach

Building upon the work and experience of previous solar missions (SOHO, STEREO, and *Yohkoh*) and in keeping with the NASA Framework, the strategic communications objectives, and the system-wide approach outlined in the heliophysics roadmap, the SDO E/PO team (Table 1) has developed an overall approach and philosophy by which it has built its portfolio of programs. In this section we will discuss our program strategies, provide a brief overview of our portfolio of programs, and conclude with a more in detail highlight of a select few of our programs.

2.1. SDO E/PO Overall Approach and Philosophy

Our guiding principles include *i*) using current educational research and best-practices approach regarding how people learn, *ii*) addressing misconceptions and changing the perceptions and behaviors of target audiences, *iii*) developing programs that can serve as backbones to other missions and partners, *iv*) relying on extensive evaluation, and *v*) thinking outside the box. We will now expand on each of these.

We develop our programs using current pedagogy and research into "How People Learn" (Bransford, Brown, and Cocking, 1999) and how best to integrate inquiry into every experience. Inquiry refers to participant activities that help develop the knowledge and understanding of scientific ideas, as well as an understanding of how scientists study the natural world (National Science Education Standards, 1996). Our research guided us to move away from teacher/moderator-centered resources (a model where teachers/moderators lecture to students/participants or run very scripted activities) and to rely instead on learner-centered resources, which are fundamental to effective learning experiences at all ages and levels.

Efforts within the SDO E/PO program aim to improve people's perceptions and behaviors toward science and scientists. We want to change the common misconceptions that all scientists are, "... white males ... who work alone in a laboratory ..." (Barman *et al.*, 1997), "eccentric" (McAdam, 1990), and that science is hard. We have found that the best approach to changing these perceptions is to help scientists connect their work and research with educators and the public. We do so by designing programs that make science more approachable and provide multiple opportunities for audiences to meet and interact with a variety of scientists and engineers. These experiences slowly break down the public's false perceptions of scientists and make science more accessible. In doing so, we also encourage a change in audience behaviors. Families are encouraged to participate in more science-related activities outside the classroom and our programs.

Programs are developed to serve as models or backbones. This approach provides two benefits: *i*) it allows for the development of big-picture programs and *ii*) allows other missions and institutions to easily plug into or adapt the program model to their own needs. This big-picture approach to program development aligns with the heliophysics roadmap desire to build E/PO programs that reflect the system-wide approach to heliophysics science and ensures the public is able to more easily connect to our story and science. The plug-in model helps ensure sustainability of programs long after missions or small funding sources have run out.

Although only a small percentage of resources was allocated to evaluate our programs, some form of evaluation was applied at all stages of the development cycle. Some programs have had a much more rigorous evaluation than others. An increased percentage of funds (20% – 30% of cost) have been allocated for evaluation of all post-launch efforts. This will provide us with the tools necessary to understand the effectiveness and impact of our current and future programs.

The SDO E/PO team strives to find new and innovative ways to reach out and share the excitement of the mission with a broad range of audiences. This includes exploring unexpected venues such as the "Did You Know" campaign that placed science factoids on the back of bathroom stalls and drink coasters as well as using social-media programs such as Facebook™ and Twitter™ to create a personality for the mission. Team members interact daily with followers to build a strong open and active dialogue with the public.

2.2. SDO E/PO Portfolio Overview – Higher Education

The HMI "Space Weather Monitor" program provides authentic space-science instruments and hands-on access to data to students in early university years (more below). A more sensitive research-quality monitor that tracks solar and lightning-induced changes in the ionosphere is available for upper level and graduate programs.

The AIA "Montana Space Public Outreach Team (SPOT)" program trains college students from a variety of disciplines in how to present the latest discoveries in space science and are then dispatched to schools and community organizations around the state of Montana.

The HMI and AIA "Science in Service" program aims to instill in college students the necessity for performing public service no matter what their future profession might be. Students from a variety of disciplines are trained in how to be effective science communicators and then connect them to youths in San Francisco Bay Area communities through science mentorship and after-school science clubs.

2.3. SDO E/PO Portfolio Overview – K-12

The GSFC "A Day at Goddard" field trip brings high-school (students aged 14 – 18) physics classes to Goddard for a day consisting of a facilities tour, a high-tech classroom laboratory experience, and a career-oriented meet-and-greet with scientists and engineers.

The GSFC "Think Scientifically" is a story-based science-literature series that integrates a classic storybook format with fundamental science to make an educational product that can be easily integrated into elementary-school (students aged 5 – 10) classrooms.

The GSFC "SDO Ambassador" program connects SDO scientists, engineers, and E/PO staff with teachers in the Washington, DC metropolitan area. SDO Ambassadors visit K-12 (students aged 5 – 18) classrooms for a day to present SDO, participate in a question and answer session, and conduct an SDO-related hands-on activity.

The GSFC, *Atmospheric Imaging Assembly* (AIA), *Helioseismic and Magnetic Imager* (HMI), and *EUV Variability Experiment* (EVE) "SDO Teacher's Guides" are a comprehensive set of solar- and space-weather-related activities developed for use in middle (students aged 11 – 13) and high-school (students aged 14 – 18) level classes.

The EVE "Solar Science Learning Kit" and associated Teacher Workshops were developed to support high-school (students aged 14 – 18) science courses. They include a collection of materials developed by the SDO E/PO team to bring solar and space-weather content into the classroom within the educational standards all teachers must follow.

The HMI "Sudden Ionospheric Disturbance (SID) Monitors" are part of the "Space Weather Monitor" program noted above and were designed to give students access to authentic research equipment, experiences, and data. The low-cost, high-school (students aged 14 – 18) version of the monitor allows students to track solar-induced changes in the ionosphere.

The AIA "Challenger Learning Center (CLC) Module Update" program integrated a space-weather scenario into the "Return to the Moon" and "Voyage to Mars" missions. During the simulated missions, students receive a coronal mass ejection alert from the SDO satellite. In determining if the event is headed towards them, students learn about the magnetosphere, space weather, and solar activity.

2.4. SDO E/PO Portfolio Overview – Informal Education

The GSFC "Family Science Night (FSN)" program is designed for middle-school children (children aged 11 – 13) and their families. Families explore space-science themes through a series of monthly two-hour events aimed at building their awareness of the importance of science in their daily lives.

2.5. SDO E/PO Portfolio Overview – Public Outreach

The GSFC "Sunday Experiment" is a program for families with young children designed to provide easy access to age-appropriate science content through easy hands-on activities and an opportunity to meet scientists and engineers.

The GSFC "AstroZone and Exploration Station" are four-hour, free-to-the-public open houses, held in conjunction with professional science meetings. The programs offer a variety of easy, family-friendly, hands-on activities and an opportunity to interact one-on-one with scientists, engineers, and education specialist.

The EVE "Science Writers Workshop" held at the Laboratory for Atmospheric and Space Physics (LASP, Boulder, CO, USA) prior to the launch of SDO brought 13 journalists from print and new media together with SDO/EVE scientists. Workshop themes included: "The Anatomy of the Sun"; "Impact of the Solar Minimum on Climate and the Space Environment"; "What will Solar Max look like?"; "Implications for Studying the Sun and for Space Weather"; and "Looking to the Future: The Great Observatory".

The EVE "Video Series" includes an overview of the EVE instrument, a virtual tour of LASP, and an overview of various space-science careers. Scientists attending the above-described Science Writers' Workshop were also interviewed about their solar-science research.

The GSFC and HMI "SDO Social Media" campaign is comprised of the SDO website, Twitter^{TM} feeds, Facebook^{TM} fan pages, YouTube^{TM} and Flickr^{TM} channels, and several topic-specific blogs. Each effort has been strategically designed and aims to interact with key target audiences in unique and efficient ways helping the mission engage the public in an informal conversation and create a sense of community around the SDO mission. For more information visit http://sdo.gsfc.nasa.gov/ and click on the social-media tab to the right of the screen.

The GSFC "Did You Know" campaign focuses on reaching non-traditional audience in unexpected locations by providing solar- and space-weather-related factoids on signs placed on tables, the back of bathroom stalls, and drink coasters.

2.6. SDO E/PO Portfolio Overview – Highlights

The "Space Weather Monitor" program: Studies confirm that the best way to learn science is to do science (AAAS, 1993; Boyer Commission, 1998). Based on this premise, this program was designed to provide authentic, space-science instruments to students in high school and early university years. The US National Science Foundation (NSF) developed Sudden Ionospheric Disturbance (SID) monitors to detect changes to the Earth's ionosphere caused by solar flares. While design of the low-cost instruments had been completed, no formal assessments had been made, and the instruments did not come with sufficient teacher-preparation or supporting educational materials. The "Space Weather Monitor" program addressed this by forming a team of E/PO specialists and HMI scientists to develop and implement a four-year test program.

The SID monitors function by tracking Very Low Frequency (VLF) radio transmissions as they bounce off the ionosphere. Solar activity dramatically affects the Total Electron Count (TEC) in the ionosphere and hence affects the refraction of the VLF waves. Students use the monitors to track VLF signal strength, directly related to TEC and the waves' bounce points. The monitors come preassembled, but students "buy in" by designing and building their own antenna. The systems then require a simple PC to record the data and an optional internet connection (where available) to share data. The package includes detailed teacher-training materials developed in partnership between the Stanford Solar Center and the Chabot Space and Science Center of Oakland, California, as well as an installation guide and research guide designed to train students to understand their data and encourage them to explore further research projects.

Year-1 β testing consisted of placing instruments in three local (San Francisco Bay area) schools (two high schools and one community college) with a high percentage of underserved students. Teachers were given direct access to HMI scientists for questions and clarifications. SDO scientists and other volunteers served as mentors to schools that requested them.

Year-2 was dedicated to creating a centralized data site where SID users could upload their data and to the distribution of over 150 instruments to high schools and universities throughout the US and at selected sites around the world.

In Year-3, the program was picked up by the United Nations for use during the International Heliophysical Year (IHY) where instruments were distributed to about 50 countries with a focus on developing nations (Scherrer *et al.*, 2008). The United Nations International Space Weather Initiative (IWSI) selected an enhanced version of the SID monitor as a participating program, and instruments continue to be distributed to universities worldwide. Data from the SID monitors have been tested for use in adjusting real-time models of the ionosphere.

In Year-4, the distribution and mentor mechanisms have been successfully handed off to the Society of Amateur Radio Astronomers (SARA) ensuring that the program continues in a self-sustaining manner.

The "Family Science Night (FSN)" program: Parents and families have the greatest influence on children's attitudes towards education and career choices. Research shows that when students have family members involved in their education, they show increased achievement: higher grades, and improved problem-solving and cognitive skills (Fan and Chen, 2001). Based on this result, we, in partnership with NASA's Astrophysics Science Division, developed the FSN program for middle school children and their families. Middle-school-aged children were chosen since they are at an age where interest in science and math can be nurtured and when they are still willing to participate in programs with their families. This program was designed as an informal education program that allows for more flexibility in its content and approach, and a deeper parent–child interaction as it was out of school and not tied to the science-education standards.

During the FSN program, families explore space-science themes through a series of monthly two-hour events. Our goal is to raise awareness in participating families of the importance of science in their daily lives and changing the way that families think about science, within the program and beyond. We use NASA science content with a focus on processes and infuse SDO mission science and data into sessions whenever appropriate. We also provide unique opportunities for participants to interact with volunteer scientists and engineers. We provide parents professional development to help parents become effective facilitators of inquiry with their own families and offer opportunities for the children to become comfortable teaching their parents and working with them as peers. Finally, we encourage family science-related activities outside of the program.

Years 1 – 4 were dedicated to the pilot testing of the program and refining of all facilitator guide materials. Our focus has shifted to the dissemination of the program to various locations across the USA in the year since launch.

The "SDO Social Media" campaign: Through the use of social media, the SDO E/PO team is able to engage a much wider audience, as well as engaging more deeply with the audience, than is possible through traditional outlets such as press releases (Mayfield,

2008). Elements of our new media plan include the use of online tools, such as Twitter™, Facebook™, YouTube™, blogs, and the SDO website. These services allow not only for dissemination of information, but also for two-way interaction and the building of conversations, crucial to extended engagement of the public and community building.

The NASA_SDO and NASA_SDO_EDU Twitter™ accounts have opened a constant dialogue between members of the public (*i.e.* followers) and members of the SDO E/PO, science, and engineering teams. These accounts post information on SDO, the Sun, NASA, and interesting developments in the STEM community. Followers can ask questions, comment on postings, and interact with members of the SDO team on a one-on-one basis. During the launch of SDO and first-light-related events, Twitter™ was used to include thousands of people from across the globe in the experience. Fifteen on-site "Twitter Correspondents" were trained to be effective science communicators and interact with those thousands of followers. Our audience included students from Romania, Mexico, and Alaska, home schooled families in Florida and Georgia, and 30 independently hosted events (events organized by members of the public not officially affiliated with SDO) including one in Second Life. For both launch and first light, the term "Solar Dynamics Observatory" was a "trending topic" on Twitter™ for over 24 hours. Twitter™ defines a trending topic as those topics that are occupying the most people's attention on Twitter™ at any given time. It is important to note that most trending topics only last between 5 – 11 minutes and are generally associated with a celebrity or noteworthy news topic.

The Facebook™ page has an intentionally more playful approach with updates made in the first person, from the satellite's perspective. "Little SDO" (now turned into a fan page) interacts daily with its friends (such as Camilla Corona) and followers around the world and shares a wide variety of solar and space-related information. This approach was chosen to give a personality to the spacecraft and allow followers to connect to it in the same way viewers connected with Disney's Wall-E character.

The YouTube™ channels stream SDO movies in a format that the public is intimately familiar with, and allows the viewer to easily share videos that they find interesting with members of their social circle, while the SDO website provides the background and details of the mission and serves as a resource to the public. The website also provides the public with easy access to the data (images and movies) through a variety of developed browse tools.

The SDO mission mascot, Camilla Corona SDO, is also on various social-media outlets and assists with sharing sun- and space-weather updates, but also focuses on getting girls and boys excited about careers in science, technology, and engineering fields. It follows the more playful approach to connect people to the SDO mission.

3. Results

This section presents the results obtained for those programs that performed formal and informal evaluations. Formative assessment is performed throughout a program's development and provides feedback to developers on how to improve a program as it is built. Summative assessment is a final evaluation at the end of a project life to indicate if it has met its stated goals. Programs that were evaluated are discussed below. Evaluation efforts for the "Think Scientifically" book series, the "SDO Teacher's Guides", and various classroom resources including the EVE video series have not yet begun. We are awaiting the final results from the "Science Writers' Workshop".

3.1. The Space Weather Monitor Program (SID Monitors)

Extensive formative assessment of this program and materials was done over the four-year study period, with significant involvement by SDO scientists. A California State University graduate student performed formal summative assessment of this program. Surveys were distributed to around 250 of the SID installations, with a 56% response rate. Among other findings, 78% of respondents rated the program "highly effective" or better for teaching about solar activity, and 70% rated the program as "highly effective" or better in conducting a scientific investigation. Summative assessment revealed that about 50% of sites did experience "some difficulty" with installation. The "SID" program is now successfully self-sustaining. The Space Weather Monitor program was one of only two projects highlighted for commendation in the National Research Council's (NRC) report on the NASA E/PO program (Quinn *et al.*, 2008)

3.2. The SPOT Program

A Ph.D. student from the Montana State University Education Department provides formal summative assessment of this program. Evaluation of this program focuses on the impact that presentations have on students and communities visited. The evaluator examines the teacher-submitted evaluations as well as the training and retention of student presenters via phone interviews. Although the K-12 students are not surveyed for their interest in the presentations, teachers are asked to describe the level of student interest, both during and after the "SPOT" presentations. The "SPOT" program has visited over 639 locations since it began to receive SDO funding in 2003. Almost three-quarters of the teachers surveyed commented that they could tell the students were interested because of the questions they were asking. This statement from a teacher is typical of these remarks: "Students asked many questions and were engaged during the program." Almost one quarter of the teachers indicated that their students continued to talk about the program after the presenters left.

3.3. The Science in Service Program

Formative assessment was carried out for each of the first four years of this program. Student participants and local community leaders were involved in evaluation discussions after each local visit, and also completed extensive questionnaires at the end of the term. The program was adjusted when weaknesses were identified. At the end of the fifth year, summative assessment was also carried out. The focus of the final evaluation was to determine whether the program was effective enough to warrant its funding and support by Stanford University. Extensive interviews were conducted, both with participants, involved HMI scientists, and local community leaders. A written report was generated, discussed, and evaluated. The final result of the assessment was that Stanford would assume responsibility for running the program and provide on-going funding to ensure its continuity. The successful program is now independent of SDO and self-sustaining.

3.4. A Day at Goddard

Magnolia Consulting is performing the ongoing formal evaluation of this program. Evaluation data collection activities for the "A Day At Goddard" field trips include student and teacher feedback surveys. Evaluators designed the student survey to capture students' perceptions of the field-trip components, and impacts on their interest and understanding of space-science and related careers. Evaluators designed the teacher survey to gather feedback from teachers on the event. Feedback included teachers' perceptions of impacts of the field trip on their students' awareness of careers, interest in space science, and understanding of the work of GSFC scientists and engineers. Initial results indicate that the events are effective at engaging student interest in space science and provided good exposure to the work of GSFC. Hands-on activities and interactions with scientists and engineers are particularly engaging for students. Four classes with a total of 52 students have participated in the program since evaluation began.

Student-survey results indicate that the program is successful in stimulating or maintaining interest in space-science and science careers for the majority of students who attended. 48% of students indicated that they were more interested in science careers as a result of the program while 21% indicated that interest was sustained. 54% of students also indicated that they were more interested in space science as a result of the program while 22% indicated that interest was sustained.

Teacher surveys found that 100% of teachers found that activities were effective in increasing students' understanding of the work at GSFC, that activities were effective in generating students' interest in space science, and that activities were effective in increasing students' awareness of STEM careers.

3.5. SDO Ambassador

Magnolia Consulting is performing the ongoing formal evaluation of this program. Data-collection activities for "SDO Ambassador" visits include teacher- and student-feedback surveys. Student-feedback surveys include Likert-scale (a scale that asks students to rate a question on a sliding scale usually $1-5$) items about their perceptions of the event and open-ended items about what they learned, what they liked best, and what they would change. Since evaluation of the program began, two schools with a total of five classrooms and 86 students have participated.

Of the respondents, 97.7% agreed or strongly agreed that they had fun learning about solar science and that they had learned new things about solar science from participating in the program. Ninety three percent agreed or strongly agreed that the event made solar science interesting while 87.2% indicated that they learned about the *Solar Dynamics Observatory* from participating.

3.6. The Solar Science Learning Kit and SDO ESL Space Science Course

The ATLAS Assessment and Research Center is conducting ongoing external evaluation of this program. The evaluator conducted interviews with five users of the "ESL Space Science Course." Interviews focused on perceived learning outcomes of students, fit with existing curriculum, ease of lesson implementation, organization, format and compatibility with state standards.

Results indicate that most lessons appear to be usable without major barriers for implementation and that content was viewed as meeting state standards. Concerns were expressed about the narrow range of content given the amount of time needed to complete lessons. Assessments were perceived as too long and lacking sufficient validity.

Lessons using the SID monitor were taught at three sites. Learning outcomes from these lessons (*e.g.* collection of real data, knowledge of content) were perceived as valuable and fitting with existing curriculum and standards. The logistical component of installing the SID monitor was perceived as somewhat burdensome and the amount of effort needed to install the system may not justify outcomes.

3.7. Family Science Night

An external evaluator at the Rochester Institute of Technology, conducted the evaluation for this program. A mixed method approach was used to evaluate the "FSN" program. Tools evaluated whether the program raised the awareness in middle-school children and their families of the importance of STEM in their daily lives, if the program enabled all members of a family to engage in active learning both within and beyond the program, and if the program promoted connections between parents and children. Since its inception, over 120 families and 500 individual participants have attended one or more "FSN" sessions.

Initial data showed that the "FSN" program does indeed motivate families to learn science together. Session observations reveal that 100% of the families engaged in learning science together for at least some part of every session. Participants were observed to be actively engaged (fully participating in activities with their families) in learning for approximately 75% of the total time. Almost all participants (90 – 100%), adults and children alike, were consistently able to state something new that they learned during each workshop as a result of the activities undertaken, and are able to relate what they learned to their everyday lives. "FSN" also promotes learning beyond the workshops. On average 90% of families report that they conducted STEM-related activities as a family between workshops as a result of the "FSN" program.

The "FSN" program received the 2008 Robert H. Goddard Award for Exceptional Achievement in Outreach.

3.8. The Sunday Experiment

"The Sunday Experiment" was handed off to the Goddard Visitor Center before formal evaluation could be conducted. However, it is important to note that one of its primary aims was to increase the public's awareness of Goddard in the local community by increasing the number of visitors that come to the Visitor Center. As a result of the program, the Goddard Visitor Center now sees between 250 and 1000 visitors during "The Sunday Experiment" program days as opposed to the 25 seen on other weekend days.

3.9. AstroZone/Exploration Station

Although there is no formal external evaluation of the "AstroZone" and "Exploration Station" events, participant cards have been collected along with demographic information and time spent at the event. These surveys help us to determine whether the stated objectives are being met. They measure levels of engagement, time spent at the event, as well as increased knowledge and interest in the available science topics and the potential for future interest in the STEM topics at the event.

More formal evaluation of the program started in the Fall of 2011. At the December 2010 Exploration Station, 119 participant cards and 68 participant evaluation forms were collected from an estimated 200 – 250 attendees. Feedback indicated that Exploration Station met or exceeded all of its stated objectives. The groups with children averaged 1.5 hours in the exhibit hall indicating a high level of engagement: over 75% of the respondents indicated that they had learned something new at the event and were able to express what it was. The majority of attendees also indicated there was a topic they were introduced to at the event that interested them to such an extent that they were "likely" to conduct further inquiry into that topic. The event also scored very high on overall enjoyment (4.5/5).

3.10. SDO Social-Media Campaign

We are currently working on a more integrated and effective social-media evaluation strategy. Without a formal evaluation methodology, concrete insights into efficacy and impact are difficult at this time. However, statistics and numbers are available that can serve as indications of levels of success.

Twitter®: As of March 2011 the @NASA_SDO Twitter™ feed has about 7000 followers with nearly 200 followers being added every month. When the account is put through many of the online Twitter™ analytics tools that look at the number of followers, interactions with followers, and overall popularity, it receives nearly perfect scores when being graded on use and shows up in the top 15% of Twitter™ accounts that have also been scored (twittergrader.com). The account has a consistent web influence Klout Score in the mid-50s out of 100, indicating that we "create content that is spread throughout [our] network and drives discussions". The Klout Score is the measurement of overall online influence. The scores range from 1 to 100 with higher scores representing a wider and stronger sphere of influence. Klout uses over 35 variables on Facebook™ and Twitter™ to measure True Reach, Amplification Probability, and Network Score (beta.klout.com).

Facebook®: SDO has over 14 000 followers between its Little SDO profile and Little SDO fan pages on Facebook™. Trends seem to indicate that the pages get upwards of 1 000 000 views a month and around 14 000 comments per month to the information posted. It is also interesting to note that there are almost equal numbers of male and female followers. On average 55% are male and 42% are female. It is assumed that the remaining missing percentage is due to the option in Facebook™ not to pick a gender.

YouTube®: SDO has multiple YouTube™ channels with over 15 000 subscribers and 3 500 000 total views between them.

4. Discussion

After significant evaluation, many programs have been shown to meet or exceed their stated desired outcomes. Those programs that have not been handed off to partner institutions will continue to operate and in most cases, useful insight was gained into how to improve each of the programs. Those that received more critical evaluation are described below and have been significantly overhauled accordingly and are awaiting the next round of evaluations. A small sub-set of programs is still scheduled for formal evaluations. However, informal evaluation and data collection of these programs still indicate that they are worth continuing and dedicating funds to formally evaluate each of them.

In response to the "Space Weather Program" (high-school SID monitors) evaluation where 50% of sites experienced "some difficulty" with installation, an even easier-to-install and less-expensive (US$50) monitor, SuperSID, was developed along with the improvement of all support materials, and the addition of an official mentorship component. This new monitor is now included in the "Solar Science Learning Kit." The program, which now includes over 500 sites, continues to grow and has now been handed off to the Society of Amateur Radio Astronomers. There has also been increased interest in these instruments as ways to increase student interest in science in developing nations, and the instruments continue to be distributed through the United Nations ISWI.

"SPOT program" evaluation indicated a desire for more content and longer-term connections to the presentations. Because of the geography of Montana and the program's statewide reach, it is often difficult to return to locations more than once, or provide on-going support to those interested. As a result, it was decided to develop pre- and post-presentation classroom supporting materials. Teachers and communities receive a packet of materials in advance of a presentation and are provided extra resources upon completion of the presentation. There are also discussions of starting a SPOT presentation teacher-training workshop series.

The "Solar Science Learning Kit" was developed as curriculum support material for the "SDO ESL Space Science Course." Formal evaluation provided the following recommendations: modify the curriculum to make it more modular (*e.g.*, make single lessons stand alone) and fit lessons so they can be used in after-school programs; standardize lesson format to more clearly emphasize objectives, materials, and information such as difficulty levels for each lesson, amount of time needed, and prior knowledge needed to complete lesson; and change assessments to shorter format and develop new questions. These changes have been made and the latest version of the Kit is being distributed through the Colorado Science Teachers Space Summit, the MESA program and *via* videoconferences through the NASA Digital Learning Network.

Evaluation results of the "FSN" program and four years of β testing and evaluations have allowed the development of a quite comprehensive and effective facilitator guide and program. Although it had not been thought about at the start of the program, a significant number of children participants with minor to severe mental disabilities participated in the program. The majority of these had ADD, ADHD, or Asperger Syndrome. The program inherently allowed the smooth and complete integration of these children who were able to participate fully in the program. In fact, moderators did not become aware of these students until a few sessions had passed and their parents commented that this was the first program that allowed their children to be equals and about the positive outcomes it had on their attitudes and self-esteem. Because this was not a goal of the program, there was no formal capturing of this finding other than through anecdotal evidence provided by parents. Along with funds for evaluation, a portion of the funds have also been reserved for research into how families learn. There is little research currently available on how families learn. What research can be found indicates how important families are to children's learning. As such, the "FSN" program hopes to contribute to this knowledge base. Formal evaluation of the program including dissemination sites will be completed in 2011. Final results should be available in 2012.

Both "AstroZone" and "Exploration Station" suffer from low turnout both in attendees and exhibitors. The programs were developed based on the American Meteorological Society's (AMS) WeatherFest that sees between 4000 and 5000 visitors come through its activity exhibit hall the day before its annual meeting held each year in January. WeatherFest has three advantages: *i*) access to televised meteorological staff that spread the word, *ii*) a local organizing committee, and *iii*) complete buy-in from the larger parent organizations (AMS and National Oceanic and Atmospheric Administration). These three areas are lacking in our own event. Along with improving the relationships with the American Astronomical Society (AAS) and the American Geophysical Union (AGU) the main areas of evaluation-based improvement noted by both the attendees and the exhibitors were increasing the number of attendees and increasing the number and variety of exhibitors. This does not come as a surprise and feeds into the increased efforts for our advertising campaigns for next year. The WeatherFest model confirms the draw such programs have and our informal survey results indicate those who do participate are quite engaged in what the program offers. However, attendance numbers must dramatically increase over the next two years for this program to be worth continuing. Hence, the majority of the recommendations for improvement are to assist in creating a more effective advertising campaign targeted to our audience and to potential exhibitors for future event.

Like most social-media efforts, the impact of the SDO social-media efforts is difficult to measure. This is compounded by the constantly changing use of individual methods and shifting sense of community. Funds were allocated to evaluate these efforts. The team is looking to work with partners to develop the best tools and solutions to more effectively use social media. It will probably take a team of computer scientists, psychologists, and evaluation specialists to truly understand the effectiveness and impact of our efforts. As some of our bigger programs ramp down, more efforts will be allocated to this issue.

5. Summary

The SDO E/PO team has designed a comprehensive and well-balanced portfolio of programs that serve target audiences in all four of NASA E/PO areas. All programs have also successfully aligned themselves with NASA and heliophysics goals and objectives while keeping within our guiding principles.

The success of our programs is due in part to a significant amount of research performed when developing each of them, looking into best practices, as well as what has been done before including what worked and did not work. Lessons learned are invaluable to any development process. As such, we have shared some of ours in this article and do so in more detail throughout many of the resources that we produce and any of the best-practices documents that we develop.

All programs are built to improve the perceptions and behaviors of target audiences. The SPOT and SIS programs teach science and non-science students to be better science communicators, changing not only their own perceptions and behaviors but also those of students who they reach. A Day at Goddard, SDO Ambassador, The Sunday Experiment, AstroZone, and Exploration Station are positively shifting participant interest in science through hands-on activities and one-on-one interactions with scientists and engineers, while FSN and the SDO ESL Space Science Course are significantly changing the perceptions and behaviors of participants through extensive sustained interaction.

Programs such as The Sunday Experiment, FSN, and A Day at Goddard were built to serve as models and backbone programs that others could plug into. The Sunday Experiment is a nine-month program that dedicates each monthly meeting to a different mission or engineering challenge. The program as a whole may cost up to US$50K a year in salaries or one half person facilities, and materials. However, a mission could choose to sponsor an event within an ongoing program for as little as US$1K. By having multiple missions participate, it also means that should one mission go away, the program is able to continue by finding new missions to participate. This is also the case for each of the other programs that allow for new content to be easily plugged into the format. Lunar and planetary sessions have already been developed for FSN, and three more astrophysics sessions are in the pipeline. Sustainability is also a big part of strategy, and multiple programs of ours are now self-sustaining beyond the SDO mission. The CLC Module Update, The Sunday Experiment, Science in Service, and the Space Weather Monitor programs have all been handed off to partner institutions and continue to be successfully run without financial support from the SDO mission.

Thinking outside the box and being able to imagine innovative and creative programs is extremely important to the SDO E/PO team. Through the SDO social-media campaign, the SDO E/PO team has played a key role in how NASA's use of social media has evolved, even hosting the first non-shuttle space-flight launch Tweet-up. Prior to SDO efforts, a few missions such as *Suzaku* and the *Mars Rovers* began blogging in the first person personifying the satellites and robots. Missions such as *Hubble* and the *Lunar Reconnaissance Orbiter* were also taking advantage of Facebook™ and Twitter™. However, there was no dialogue created, simply missions posting information for the public to read. As we were strategizing, there seemed to be clear signals from the public that their true desire was to be actively engaged in discussions. Through the SDO Facebook™ and Twitter™ accounts, the E/PO team is actively engaging the public and has built a strong and dedicated following that lets us know within seconds when one of our many online tools is no longer working, serving as our own core of science communicators, and even participating in our events at volunteers.

Now that SDO has launched and terabytes of data are being beamed back to Earth daily, the SDO E/PO team will adjust its focus to include student and public engagement with SDO data. We will look at different ways to get data into the hands of students via inquiry-based projects and into the public sphere through social media, the press, and "citizen-science" data-analysis projects.

Acknowledgements The SDO E/PO team would like to acknowledge and thank all those who have helped and supported our many efforts. Their contributions of writing and editing text, participating in question and answer sessions, volunteering at events, giving presentations, helping develop instruments and programs, as well as supporting teachers and students at all levels, have been invaluable to our efforts.

Special thanks to Elizabeth Citrin and Rob Lilly of the SDO Project Team for their support of E/PO during the development of the observatory. We would like to acknowledge members of the Science Investigation Teams for their contributions to this program. These include Philip Scherrer (HMI), Alan Title, Ed Deluca, and Bruce Ward (AIA), and Tom Woods (EVE).

We would like to say a special thank you to all of our volunteers, including Christopher Anderson, Caitlin Bacha, Monica Bobra, Phillip Chamberlain, Yaireska ColladoVega, Carmel Conaty, Priya Desai, László I. Etesi, Andrew Ingles, Leigh Jaynes, Rivers Lamb, Barbara Lambert, Haruko Makatani, Ryan O. Milligan, John W. Mitchell, Kim Ross, Andrew Singh, Hao Thai, Barbara Thompson, Jonathan Verville, Peter Williams, C. Alex Young, and the many teachers and students involved with our programs.

Finally, we would like to thank Phillip Chamberlin, Steven Christe, László I. Etesi, Todd Hoeksema, Jack Ireland, and Philip Scherrer for their contributions in preparing this article and providing insight into how to improve it.

This work was supported, in part, by the *Solar Dynamics Observatory* Project and NASA Grants NNH08CD27C-ROSES and NNH09ZDA001N-EPOESS.

References

American Association for the Advancement of Science: 1993, *Benchmarks for Science Literacy: Project 2061*, Oxford University Press, New York. http://www.project2061.org/publications/bsl/online/index.php?txtRef=&txtURIOld=%2Ftools%2Fbenchol%2Fbolframe.html.

Barman, C.R., Ostlund, K.L., Gatto, C.C., Halferty, M.: 1997, Fifth grade student's perceptions about scientists and how they study and use science, http://www.physics.ucsb.edu/~scipub/f2004/StudentPerceptions.pdf.

Bransford, J.D., Brown, A.L., Cocking, R.R.: 1999, *How People Learn: Brain, Mind, Experience, and School*, National Academy Press, Washington.

Boyer Commission on Educating Undergraduates in the Research University: 1998, *Reinventing Undergraduate Education: A Blueprint for America's Research Universities*, State University of New York, New York. http://naples.cc.sunysb.edu/Pres/boyer.nsf.

Christensen, A., Spann, J., St. Cyr, O.C., Cummings, A., Heelis, R., Hill, F., Immel, T., Kasper, J., Kistler, L., Kuhn, J., Reeves, G., Schwadron, S., Solomon, S., Strangeway, R., Tarbell, T.: 2009, Heliophysics: The solar and space physics of a new era, NP-2009-08-76-MSFC, Huntsville.

Fan, X., Chen, M.: 2001, Parental involvement and students' academic achievement: a meta-analysis. *Educ. Psychol. Rev.* **13**, 1 – 22.

Mayfield, A.: 2008, What is social media?, iCrossing, http://www.icrossing.co.uk/fileadmin/uploads/eBooks/What_is_Social_Media_iCrossing_ebook.pdf.

McAdam, J.E.: 1990, The persistent stereotype: children's images of scientists. *Phys. Educ.* **25**, 102.

NASA Office of Education: 2007, Strategic Coordination Framework: A Portfolio Approach, NP-2007-01-456-HQ, Washington, DC. http://www.nasa.gov/pdf/189101main_Education_Framework.pdf.

National Science Education Standards, National Research Council: 1996, http://www.nsta.org/publications/nses.aspx.

NRC: 2011, *Committee on the Planetary Science Decadal Survey, Space Studies Board, Division on Engineering and Physical Sciences: 2011, Vision and Voyages for Planetary Science in the Decade* 2013–2022, National Academy Press, Washington. http://www.nap.edu/catalog.php?record_id=13117#toc.

OECD: 2011, *Lessons from PISA for the United States, Strong Performers and Successful Reformers in Education*, OECD, Paris. http://dx.doi.org/10.1787/9789264096660-en.

Quinn, H.R., Schweingruber, H.A., Feder, M.A. (eds.), Committee for the Review and Evaluation of NASA's Precollege Education Program, National Research Council: 2008, *NASA's Elementary and Secondary Education Program*, National Academy Press, Washington. http://www.nap.edu/catalog.php?record_id=12081.

Scherrer, D., Cohen, M., Hoeksema, H., Inan, U., Mitchell, R., Scherrer, P.: 2008, Distributing space weather monitoring instruments and educational materials worldwide for the IHY 2007: The AWESOME and SID project. *Adv. Space Res.* **42**, 1777–1785.

SMD Science Management Council: 2010, Policy and Requirements for the Education and Public Outreach Programs of SMD Missions, Science Mission Directorate Policy, SMD Policy Document SPD-18, NASA, Washington, DC. http://science.nasa.gov/media/medialibrary/2011/04/13/SPD-18_MissionEPOPolicy_TAGGED.pdf.

CPSIA information can be obtained at www.ICGtesting.com
Printed in the USA
LVOW101907220512

282825LV00001B/3/P

9 781461 436720